AVIATION

WORLD AIRLINE FLEETS 19
The brand new edition of the
leading fleet yearbook -acknov
world wide as the numb
authority on the airlines of the world. Comprehensively covers the world's thousands of airlines, detailing, for each airline, registrations, type, construction number, engines, lease details and previous identity for each aircraft in the fleet with each airline featuring Head Office, telephone and telex numbers, brief write up, logo etc. Also includes photographs. An absolute must for anyone interested in airlines or airliners.

ISBN 0 946141 20 7 608 pages
£13.50 plus £1.35 postage Worldwide

WORLD AIRLINE COLOURS-at last we can announce the fabulous new colour book that you have been waiting for, World Airline Colours. Featuring superior quality photographs by some of the world's top photographers, of 148 of the world's leading airlines plus write-ups on each including brief details of operations, type of aircraft used, description and background of each livery. Absolutely essential for everyone involved or interested in airlines and to be joined by further books in the series.

ISBN 0 946141 16 9 152 pages
£9.95 plus £1.00 postage Worldwide

Name: ______ Address: ______ Total enclosed £ ______
______ (Inc. postage)

Credit Card (Delete as applicable) VISA/AMERICAN EXPRESS/EUROCARD/MASTERCHARGE/ACCESS

CARD No. ______ EXPIRY DATE ______

Credit Card orders can be telephoned through on 01 890 8933

How to pay: Cheques made payable to "Aviation Data Centre", Credit Card, Postal Order, International Money Order, Foreign Cheque payable on UK Bank, Girocheque on Direct Giro Account No. 344 3663. All payments must be in Pounds Sterling. Note Eurocheques accepted at no extra charge.

ORDER DIRECT FROM THE PUBLISHERS "IN PRINT MEANS IN STOCK"

USE OUR FAST EFFICIENT MAIL ORDER SERVICE

AVIATION DATA CENTRE
INNOVATIONS IN MAIL ORDER SERVICE

AVAILABLE FROM ALL GOOD BOOKSHOPS

SEND YOUR ORDER TO: AVIATION DATA CENTRE (MAIL ORDER CENTRE)
BROWCOM HOUSE, BROWELLS LANE, FELTHAM, MIDDLESEX, TW13 7EQ, ENGLAND. Telephone: 01-890 8933
(Use this form, a photocopy or write your order clearly on a separate sheet of paper)
ORDER FORM

BEWARE OF IMITATIONS!

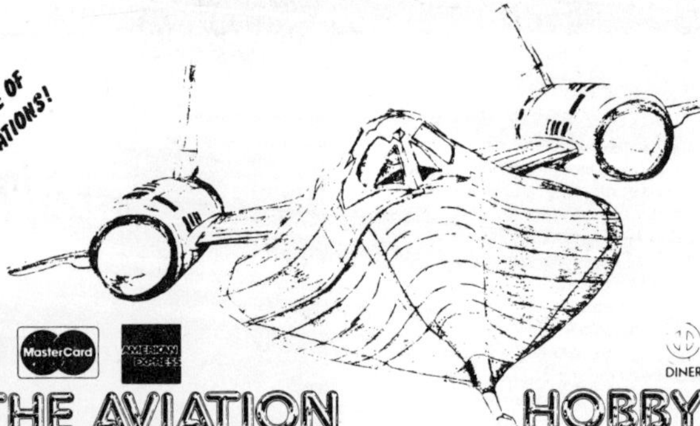

Access MasterCard AMERICAN EXPRESS DINERS CLUB VISA

THE AVIATION HOBBY SHOP

4, HORTON PARADE, HORTON ROAD,
WEST DRAYTON, MIDDLESEX, UB7-8EA.
W.D. 442123 (24 HOUR ANSWERING SERVICE)

YOU'VE TRIED THE REST
NOW TRY THE BEST!!
IF WE CAN'T HELP - WHO CAN ??

MODEL KITS MANY DELETED AIRFIX AND FROG ! ALSO; WELSH MODELS, RAREPLANES, REDSTAR, AEROCLUB AND MORE

PAINTS INCLUDING NEW Compucolor 2 RANGE. NEW LUFTWAFFE RAL F-4F (GREY MOUSE) COLOURS.

BOOKS AN EXTENSIVE RANGE OF ENTHUSIAST BOOKS FROM THE WRIGHT BROTHERS TO THE PRESENT DAY.

MAIL ORDER UNSURPASSED INTERNATIONAL MAIL ORDER, BOOKS AND KITS ***POST FREE*** IN THE U.K.

AIRBAND RADIOS NEW SIGNAL R532–528/537 A.O.R. ELECTRONICS AR 2001 UHF/VHF

DECALS MICROSCALE, FOWLER, A.T.P. ALWAYS THE MOST UP TO DATE RANGE FOR THE SERIOUS MODEL MAKER.

COLOUR SLIDES THOUSANDS OF TOP QUALITY SLIDES TO CHOOSE FROM AND THE RANGE IS ALWAYS EXPANDING!!

POSTCARDS MORE THAN 700 COLLECTORS CARDS ALL FULL COLOUR

PATCHES THE LARGEST SELECTION ANYWHERE IN THE U.K!!

AVIATION PRINTS TOP QUALITY!

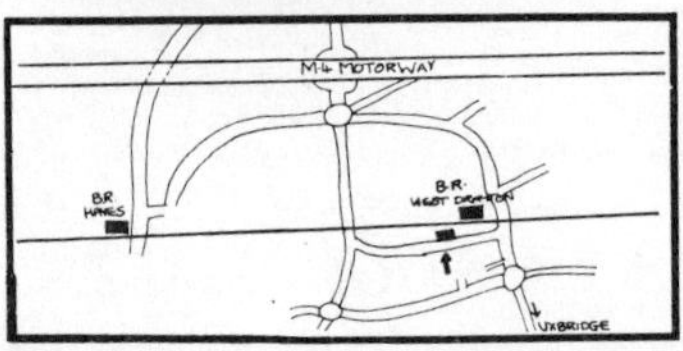

ALL OUR VAST RANGE OF GOODS ARE LISTED IN OUR 56 PAGE CATALOGUE, SENT FREE ON REQUEST

BUSES 222, 223, 724, 727, PASS AT THE TOP OF HORTON ROAD.

tahs

MODELS

AIRFIX
CROWN
ECSI
FROG
FUJIMI
HASAGAWA
HELLER
ITALERI
L & S
MATCHBOX
MONOGRAM
NOVO
REVELL
TAMIYA

BOOKS

AEROGRAPHS
AEROGUIDES
AIR BRITAIN
AIR FAN
AVIATION DATA CENTRE
CANADIAN PROFILES
DETAIL & SCALE
GLOBAL A.S.
IAN ALLAN
JANES
KOKO FAN
LAAS INTERNATIONAL
LUNDKVIST
MIDLAND COUNTIES
OSPREY
PUTNAMS
P.S.L.
SALAMANDER
SQUADRON SIGNAL
WARBIRDS

Alan J. Wright

Ian Allan abc

Civil Aircraft Markings 1986

LONDON

IAN ALLAN LTD

Contents

This thirty-seventh edition published 1986

ISBN 0 7110 1558 0

All rights reserved. No part of this book may be reproduced or transmitted in any form or by any means, electronic or mechanical, including photocopying, recording or by any information storage and retrieval system, without permission from the Publisher in writing.

©Ian Allan Ltd 1986

Published by Ian Allan Ltd, Shepperton, Surrey; and printed by Ian Allan Printing Ltd at their works at Coombelands in Runnymede, England.

Cover: *A scene at Heathrow taken in 1985 by award-winning BAA photographer, Arthur Kemsley.* BAA

Introduction

The 'G' prefixed four letter registration system was adopted in 1919 after a short-lived spell of about three months with serial numbers beginning at K-100. Until July 1928 the UK allocations were in the G-Exxx range, but as a result of further International agreements, this series was ended at G-EBZZ, the replacement being G-Axxx. From this point the registrations were issued in a reasonably orderly manner through to G-AZZZ, reached in July 1972. There were two exceptions. To avoid possible confusion with signal codes, the G-AQxx sequence was omitted, while G-AUxx was reserved for Australian use originally. In recent years however, an individual request for a mark in the latter range has been granted by the Authorities.

Although the next logical sequence was started at G-Bxxx, it was not long before the strictly applied rules relating to aircraft registration began to be relaxed. Permission was readily given for personalised marks to be issued incorporating virtually any four letter combination, while re-registration has also become a common feature, a practice almost unheard of in the past. In this book, where this has taken place at some time, the previous UK civil identity appears in parenthesis after the owner's/operator's name. An example of this is One-Eleven G-BBMG which originally carried G-AWEJ.

Some aircraft have also been allowed to wear military markings without displaying their civil identity. In this case the serial number actually carried is shown in parenthesis after the type's name. For example Gladiator G-AMRK flies as L8082 in RAF colours. As an aid to the identification of these machines, a military conversion list is provided.

Other factors caused a sudden acceleration in the number of registrations allocated by the Civil Aviation Authority in the early 1980s. The first surge came with the discovery that it was possible to register plastic bags and other items even less likely to fly, on payment of the standard fee. This erosion of the main register was checked in early 1982 by the issue of a special sequence for such devices commencing at G-FYAA. Powered hang-gliders provided the second glut of allocations as a result of the decision that these types should be officially registered. Although a few of the early examples penetrated the normal in-sequence register, the vast majority were given marks in other special ranges, this time G-MBxx, G-MJxx, G-MMxx and G-MNxx with G-MVxx, G-MWxx, G-MYxx and G-MZxx reserved for future use. At first it was common practice for microlights to ignore the requirement to carry their official identity. However the vast majority now display their registration somewhere on the structure, the size and position depending on the dimensions of the component to which it is applied.

Throughout the UK section of this book, there are many instances where the probable base of the aircraft has been included. This is positioned at the end of the owner/operator details preceded by an oblique stroke. It must of course be borne in mind that changes do take place and that no attempt has been made to record the residents at the many private strips. The base of airline equipment has been given as the company's headquarter's airport, although frequently aircraft are outstationed for long periods.

Non-airworthy preserved aircraft are shown with a star after the type.

Any new registrations issued by the CAA after this publication went to press will inevitably not be included until the next edition. To aid the recording of later marks logged, grids have been provided at the end of the book.

The two-letter codes used by airlines to prefix flight numbers in timetables, airport movements boards, etc are included for those carriers appearing in the book. Radio frequencies for the larger airfields/airports are also listed.

Acknowledgements

Once again thanks are extended to the Registration Department of the Civil Aviation Authority for their assistance and allowing access to their files. The comments and amendments flowing from Wal Gandy and Hans Kohne have as always proved of considerable value, while the help given by numerous airlines has been much appreciated. The work of A. S. Wright and C. P. Wright during the update of this edition must not go unrecorded, since without it, deadlines would probably become impossible.

AJW

International Civil Aircraft Markings

A2-	Botswana
A3-	Tonga
A5-	Bhutan
A6-	United Arab Emirates
A7-	Qatar
A9-	Bahrain
A40-	Oman
AP-	Pakistan
B-	China/Taiwan
C-F, C-G	Canada
C2-	Nauru
C3	Andora
C5-	Gambia
C6-	Bahamas
C9-	Mozambique
CC-	Chile
CCCP-*	Soviet Union
CN-	Morocco
CP-	Bolivia
CR-	Portuguese Overseas Provinces
CS-	Portugal
CU-	Cuba
CX-	Uruguay
D-	German Federal Republic (West)
D2-	Angola
D4	Cape Verde Islands
D6-	Comores Islands
DDR-	German Democratic Republic (East)
DQ-	Fiji
EC-	Spain
EI-, EJ-	Republic of Ireland
EL-	Liberia
EP-	Iran
ET-	Ethiopia
F-	France, Colonies and Protectorates
G-	United Kingdom
H4-	Solomon Islands
HA-	Hungarian People's Republic
HB-	Switzerland and Liechtenstein
HC-	Ecuador
HH-	Haiti
HI-	Dominican Republic
HK-	Colombia
HL-	Korea (South)
HP-	Panama
HR-	Honduras
HS-	Thailand
HZ-	Saudi Arabia
I-	Italy
J2-	Djibouti
J3-	Grenada
J5-	Guinea Bissau
J6-	St Lucia
JA-	Japan
JY-	Jordan
LN-	Norway
LQ-, LV-	Argentine Republic
LX-	Luxembourg
LZ-	Bulgaria
MI-	Marshall Islands
N-	United States of America
OB-	Peru
OD-	Lebanon
OE-	Austria
OH-	Finland
OK-	Czechoslovakia
OO-	Belgium
OY-	Denmark
P-	Korea (North)
P2-	Papua New Guinea
PH-	Netherlands
PJ-	Netherlands Antilles
PK-	Indonesia and West Irian
PP-, PT-	Brazil
PZ-	Surinam
RDPL-	Laos
RP-	Philippine Republic
S2-	Bangladesh
S7-	Seychelles
S9-	São Tomé
SE-	Sweden
SP-	Poland
ST-	Sudan
SU-	Egypt
SX-	Greece
T2	Tuvalu
T3-	Kiribati
TC-	Turkey
TF-	Iceland
TG-	Guatemala
TI-	Costa Rica
TJ-	United Republic of Cameroon
TL-	Central African Republic
TN-	Republic of Congo (Brazzaville)
TR-	Gabon
TS-	Tunisia
TT-	Chad
TU-	Ivory Coast
TY-	Benin
TZ-	Mali
V2-	Antigua
V3-	Belize
V8-	Brunei
VH-	Australia
VN-	Vietnam
VP-F	Falkland Islands
VP-LKA/LLZ	St Kitts-Nevis
VP-LMA/LUZ	Montserrat
VP-LVA/LZZ	Virgin Islands
VQ-T	Turks & Caicos Islands
VR-B	Bermuda
VR-C	Cayman Islands
VR-G	Gibraltar (not used: present Gibraltar Airways aircraft registered G-)
VR-H	Hong Kong
VT-	India
XA-, XB-, XC-,	Mexico
XT-	Upper Volta
XU-	Kampuchea
XY-, XZ-	Burma
YA-	Afghanistan
YI-	Iraq
YJ-	Vanuatu
YK-	Syria
YN-	Nicaragua
YR-	Romania
YS-	El Salvador
YU-	Yugoslavia
YV-	Venezuela
Z-	Zimbabwe

* Cyrillic letters for SSSR.

ZA-	Albania	5Y-	Kenya
ZK-	New Zealand	6O-	Somalia
ZP-	Paraguay	6V-, 6W-	Senegal
ZS-	South Africa	6Y-	Jamaica
3A-	Monaco	7O-	Democratic Yemen
3B-	Mauritius	7P-	Lesotho
3C-	Equatorial Guinea	7Q-	Malawi
3D-	Swaziland	7T-	Algeria
3X-	Guinea	8P-	Barbados
4R-	Sri Lanka	8Q-	Maldives
4W-	Yemen Arab Republic	8R-	Guyana
4X-	Israel	9G-	Ghana
4YB	Jordanian-Iraqi Co-op Treaty	9H-	Malta
5A-	Libya	9J-	Zambia
5B-	Cyprus	9K-	Kuwait
5H-	Tanzania	9L-	Sierra Leone
5N-	Nigeria	9M-	Malaysia
5R-	Malagasy Republic (Madagascar)	9N-	Nepal
5T-	Mauritania	9Q-	Zaïre
5U-	Niger	9U-	Burundi
5V-	Togo	9V-	Singapore
5W-	Western Somoa (Polynesia)	9XR-	Rwanda
5X-	Uganda	9Y-	Trinidad and Tobago

Aircraft Type Designations

(eg PA-28 Piper Type 28)

A.	Beagle, Auster	G.C.	Globe
AA-	American Aviation, Grumman American	GY	Gardan
AB	Agusta-Bell	H	Helio
AS	Aerospatiale	HM.	Henri Mignet
A.S.	Airspeed	HP.	Handley Page
A.W.	Armstrong Whitworth	HR.	Robin
B.	Blackburn, Bristol Boeing, Beagle	H.S.	Hawker Siddeley
BAC	British Aircraft Corporation	IL	Ilyushin
BAe	British Aerospace	J.	Auster
BN	Britten-Norman	L.	Lockheed
Bo	Bolkow	L.A.	Luton
Bu	Bucker	M.	Miles, Mooney
C.H.	Chrislea	MBB	Messerschmitt-Bölkow-Blohm
CLA	Comper	M.S.	Morane-Saulnier
CP.	Piel	P.	Hunting (formerly Percival), Piaggio
D.	Druine	PA-	Piper
DC-	Douglas Commercial	PC.	Pilatus
D.H.	de Havilland	R.	Rockwell
D.H.C.	de Havilland Canada	S.	Short, Sikorsky
DR.	Jodel (Robin-built)	SA., SE, SO.	Sud-Aviation, Aérospatiale, Scottish Aviation
EoN	Elliotts of Newbury	S.R.	Saunders-Roe, Stinson
EP	Edgar Percival	ST	SOCATA
F.	Fairchild, Fokker	T.	Tipsy
G.	Grumman	Tu	Tupolev
GA	Gulfstream American	UH.	United Helicopters (Hiller)
G.A.L.	General Aircraft	V.	Vickers-Armstrongs, BAC
		V.S.	Vickers-Supermarine
		W.S.	Westland
		Z.	Zlin

British Civil Aircraft Registrations

Notes	Reg.	Type	Owner or Operator
	G-EACN	BAT BK23 Bantam (K123) ★	Shuttleworth Trust/O. Warden
	G-EASO	Bristol Babe (replica) (BAPC87)★	Bomber County Museum
	G-EAVX	Sopwith Pup (B1807)	K. A. M. Baker
	G-EBHX	D.H.53 Humming Bird	Shuttleworth Trust/O. Warden
	G-EBIA	S.E.5A (F904)	Shuttleworth Trust/O. Warden
	G-EBIB	S.E.5A (F939) ★	Science Museum
	G-EBIC	S.E.5A (F938) ★	RAF Museum
	G-EBIR	D.H.51	Shuttleworth Trust/O. Warden
	G-EBJE	Avro 504K (E449)★	RAF Museum
	G-EBJG	Parnall Pixie III ★	Midland Aircraft Preservation Soc
	G-EBJO	ANEC II ★	Shuttleworth Trust/O. Warden
	G-EBKY	Sopwith Pup (N5180)	Shuttleworth Trust/O. Warden
	G-EBLV	D.H.60 Cirrus Moth	British Aerospace/Hatfield
	G-EBMB	Hawker Cygnet I ★	RAF Museum
	G-EBNV	English Electric Wren	Shuttleworth Trust/O. Warden
	G-EBQP	D.H.53 Humming Bird ★	Russavia Collection/Duxford
	G-EBWD	D.H.60X Hermes Moth	Shuttleworth Trust/O. Warden
	G-EBYY	Cierva C.8L ★	Musée de l'Air, Paris
	G-EBZM	Avro 594 Avian IIIA ★	Manchester Air & Space Museum
	G-AAAH	D.H.60G Gipsy Moth (replica) ★	Hilton Hotel/Gatwick
	G-AAAH	D.H.60G Gipsy Moth ★	Science Museum
	G-AACN	H.P.39 Gugnunc ★	Science Museum/Wroughton
	G-AAHY	D.H.60M Moth	I. M. White
	G-AAIN	Parnall Elf II	Shuttleworth Trust/O. Warden
	G-AAMY	D.H.60M Moth	C. C. & Mrs. J. M. Lovell
	G-AANG	Blériot Monoplane	Shuttleworth Trust/O. Warden
	G-AANH	Deperdussin Monoplane	Shuttleworth Trust/O. Warden
	G-AANI	Blackburn Monoplane	Shuttleworth Trust/O. Warden
	G-AANJ	L.V.G.-C VI (7198/18)	Shuttleworth Trust/O. Warden
	G-AANV	D.H.60G Moth	D. H. Ellis
	G-AAOK	Curtiss Wright Travel Air 12Q	Shipping & Airlines Ltd/Biggin Hill
	G-AAOR	D.H.60G Moth	J. A. Pothecary/Shoreham
	G-AAPZ	Desoutter I (mod.) ★	Shuttleworth Trust/O. Warden
	G-AAUP	Klemm L.25-IA	J. I. Cooper
	G-AAVJ	D.H.60GMW Moth	R. W. Livett/Sywell
	G-AAWO	D.H.60G Gipsy Moth	J. F. W. Reid
	G-AAXK	Klemm L.25-1A	C. C. Russell-Vick
	G-AAYX	Southern Martlet	Shuttleworth Trust/O. Warden
	G-AAZP	D.H.80A Puss Moth	R. P. Williams
	G-ABAA	Avro 504K (H2311) ★	RAF Museum/Henlow
	G-ABAG	D.H.60G Moth	Shuttleworth Trust/O. Warden
	G-ABDW	D.H.80A Puss Moth (VH-UQB)	Museum of Flight/E. Fortune
	G-ABDX	D.H.60G Moth	M. D. Souch
	G-ABEE	Avro 594 Avian IVM (Sports) ★	Aeroplane Collection Ltd
	G-ABEV	D.H.60G Moth	Wessex Aviation & Transport Ltd
	G-ABLM	Cierva C.24 ★	Mosquito Aircraft Museum
	G-ABLS	D.H.80A Puss Moth	R. C. F. Bailey
	G-ABMR	Hart 2 (J9941)★	RAF Museum
	G-ABNT	Civilian C.A.C.1 Coupe	Shipping & Airlines Ltd/Biggin Hill
	G-ABNX	Redwing 2	J. Pothecary
	G-ABOI	Wheeler Slymph ★	Midland Air Museum
	G-ABOX	Sopwith Pup (N5195)	Museum of Army Flying/Middle Wallop
	G-ABTC	CLA.7 Swift	P. Channon/Biggin Hill
	G-ABUS	CLA.7 Swift	R. C. F. Bailey
	G-ABUU	CLA.7 Swift	H. B. Fox/Booker
	G-ABVE	Arrow Active 2	J. D. Penrose
	G-ABWP	Spartan Arrow	R. E. Blain/Barton
	G-ABXL	Granger Archaeopteryx ★	Shuttleworth Trust/O. Warden
	G-ABYA	D.H.60G Gipsy Moth	Dr I. D. C. Hay & J. F. Moore
	G-ABZB	D.H.60G-III Moth Major	R. E. & B. A. Ogden
	G-ACCB	D.H.83 Fox Moth ★	Midland Aircraft Preservation Soc
	G-ACDC	D.H.82A Tiger Moth	Tiger Club Ltd/Redhill
	G-ACDJ	D.H.82A Tiger Moth	F. J. Terry
	G-ACGT	Avro 594 Avian IIIA ★	M. Rockliffe

Reg.	Type	Owner or Operator	Notes
G-ACIT	D.H.84 Dragon ★	Science Museum/Wroughton	
G-ACLL	D.H.85 Leopard Moth	D. C. M. & V. M. Stiles	
G-ACMA	D.H.85 Leopard Moth	S. J. Filhol/Sherburn	
G-ACMN	D.H.85 Leopard Moth	H. D. Labouchere	
G-ACOL	D.H.85 Leopard Moth	M. J. Abbott	
G-ACSP	D.H.88 Comet ★	Veteran & Vintage Aircraft (Engineering) Ltd/Chirk	
G-ACSS	D.H.88 Comet	Shuttleworth Trust/Farnborough	
G-ACTF	CLA.7 Swift	A. J. Chalkley/Booker	
G-ACUS	D.H.85 Leopard Moth	C. W. Annis	
G-ACUU	Cierva C.30A ★	G. S. Baker/Duxford	
G-ACUX	S.16 Scion ★	Ulster Folk & Transport Museum	
G-ACVA	Kay Gyroplane★	Glasgow Museum of Transport	
G-ACWP	Cierva C.30A (AP507) ★	Science Museum	
G-ACXE	B.K.L-25C Swallow	D. G. Ellis	
G-ACZE	D.H.89A Dragon Rapide	Wessex Aviation & Transport Ltd (G-AJGS)	
G-ADAH	D.H.89A Dragon Rapide ★	Museum of Flight/E. Fortune	
G-ADCG	D.H.82A Tiger Moth	D. A. Lowe	
G-ADEL	Spartan Cruiser III ★	Museum of Flight/E. Fortune	
G-ADEV	Avro 504K (H5199)	Shuttleworth Trust (G-ACNB)/ O. Warden	
G-ADFO	Blackburn B-2 ★	R. Cole	
G-ADFV	Blackburn B-2 ★	Humberside Aircraft Preservation Soc	
G-ADGP	M.2L Hawk Speed Six	R. H. Reeves/O. Warden	
G-ADGT	D.H.82A Tiger Moth	D. R. & Mrs M. Wood	
G-ADHA	D.H.83 Fox Moth	Wessex Aviation & Transport Ltd	
G-ADIA	D.H.82A Tiger Moth	March Helicopters Ltd/Sywell	
G-ADJJ	D.H.82A Tiger Moth	J. M. Preston	
G-ADKC	D.H.87B Hornet Moth	M. L. Roe/Halfpenny Green	
G-ADKK	D.H.87B Hornet Moth	C. P. B. Horsley & R. G. Anniss	
G-ADKL	D.H.87B Hornet Moth	L. J. Rice	
G-ADKM	D.H.87B Hornet Moth	F. R. E. Hayter	
G-ADLY	D.H.87B Hornet Moth	P. & A. Wood	
G-ADMT	D.H.87B Hornet Moth ★	Scottish Aircraft Collection/Perth	
G-ADMW	M.2H Hawk Major (DG590) ★	RAF Museum/Henlow	
G-ADND	D.H.87B Hornet Moth	Shuttleworth Trust/O. Warden	
G-ADNE	D.H.87B Hornet Moth	Shipping & Airlines Ltd/Biggin Hill	
G-ADNZ	D.H.82A Tiger Moth	R. W. & Mrs. S. Pullan	
G-ADOT	D.H.87B Hornet Moth ★	Mosquito Aircraft Museum	
G-ADPJ	B.A.C. Drone	P. G. Dunnington	
G-ADPR	P.3 Gull ★	Shuttleworth Trust *Jean*/O. Warden	
G-ADPS	B.A. Swallow 2	Wessex Aviation & Transport Ltd	
G-ADRA	Pietenpol Aircamper	A. J. Mason & R. J. Barrett	
G-ADRC	K. & S. Jungster J-1	J. J. Penney & L. R. Williams	
G-ADRH	D.H.87B Hornet Moth	I. M. Callier	
G-ADRY†	Pou-du-Ciel (Replica) (BAPC29) ★	M. L. Beach	
G-ADUR	D.H.87B Hornet Moth	R. C. Lenton	
G-ADWO	D.H.82A Tiger Moth (BB807)	Wessex Aviation Soc	
G-ADXS	Pou-du-Ciel ★	Rebel Air Museum/Andrewsfield	
G-ADXT	D.H.82A Tiger Moth	J. & J. M. Pothecary/Shoreham	
G-ADYS	Aeronca C.3	B. Cooper	
G-AEBB	Pou-du-Ciel★	Shuttleworth Trust/O. Warden	
G-AEBJ	Blackburn B-2	British Aerospace/Brough	
G-AEDB	B.A.C. Drone 2	M. C. Russell/Duxford	
G-AEEG	M.3A Falcon	Vintage Aircraft Magazine Ltd/Denham	
G-AEEH	Pou-du-Ciel ★	RAF/St Athan	
G-AEFG	HM.14 Pou-du-Ciel ★	N. Ponsford	
G-AEFT	Aeronca C-3	C. E. Humphreys & ptnrs/Henstridge	
G-AEGV	HM.14 Pou-du-Ciel ★	Midland Aircraft Preservation Soc	
G-AEHM	Pou-du-Ciel ★	Science Museum/Wroughton	
G-AEJZ	Mignet HM.14 Pou-du-Ciel (BAPC120) ★	Bomber County Museum/Cleethorpes	
G-AEKR	Flying Flea (replica) (BAPC 121) ★	S. Yorks Aviation Soc	
G-AEKV	Kronfield Drone	M. L. Beach/Booker	
G-AELO	D.H.87B Hornet Moth	S. N. Bostock	
G-AEML	D.H.89 Dragon Rapide	I. Jones	
G-AENP	Hawker Hind (K5414)	Shuttleworth Trust/O. Warden	
G-AEOA	D.H.80A Puss Moth	P. & A. Wood/O. Warden	
G-AEOF†	HM.14 Pou-du-Ciel (BAPC22) ★	Aviodome/Schiphol, Holland	
G-AEOF	Rearwin 8500	Shipping & Airlines Ltd/Biggin Hill	

† False registration

Notes	Reg.	Type	Owner or Operator
	G-AEOH	HM.14 Pou-du-Ciel ★	Midland Air Museum
	G-AEPH	Bristol F.2B (D8096)	Shuttleworth Trust/O. Warden
	G-AERD	P.3 Gull Six	N. C. Jensen/Redhill
	G-AERV	M.11A Whitney Straight ★	Ulster Folk & Transport Museum
	G-AESE	D.H.87B Hornet Moth	J. G. Green/Redhill
	G-AESZ	Chilton D.W.1	R. E. Nerou
	G-AETA	Caudron G.3 (3066) ★	RAF Museum
	G-AEUJ	M.11A Whitney Straight	R. E. Mitchell
	G-AEVS	Aeronca 100	A. J. E. Smith
	G-AEVZ	B.A. Swallow 2	E. H. S. Warner
	G-AEXD	Aeronca 100	Mrs M. A. & R. W. Mills
	G-AEXF	P.6 Mew Gull	T. M. Storey/Redhill
	G-AEXT	Dart Kitten II	C. A. Stubbings
	G-AEXZ	Piper J-2 Cub	Mrs M. & J. R. Dowson/Leicester
	G-AEYY	Martin Monoplane ★	Martin Monoplane Syndicate/Hatfield
	G-AFAP	C.A.S.A. C.352L ★	Aerospace Museum/Cosford
	G-AFBS	M14A Hawk Trainer ★	G. D. Durbridge-Freeman (G-AKKU)/ Duxford
	G-AFCL	B. A. Swallow 2	A. M. Dowson/O. Warden
	G-AFDX	Hanriot HD.1 (75) ★	RAF Museum
	G-AFEL	Monocoupe 90A	Fortune Holdings Ltd/Barton
	G-AFFD	Percival Q-6 ★	B. D. Greenwood
	G-AFFI	Pou-du-Ciel (BAPC76) ★	Bomber County Museum
	G-AFGC	B. A. Swallow 2	H. Plain
	G-AFGD	B. A. Swallow 2	A. T. Williams & ptnrs/Shobdon
	G-AFGE	B. A. Swallow 2	Donald G. Ellis/Sandown
	G-AFGH	Chilton D.W.1.	M. L. & G. L. Joseph
	G-AFGI	Chilton D.W.1. ★	J. E. McDonald
	G-AFGM	Piper J-4A Cub Coupé	A. J. P. Marshall/Carlisle
	G-AFHA	Mosscraft M.A.1. ★	C. V. Butler
	G-AFIN	Chrislea Airguard ★	Aeroplane Collection Ltd
	G-AFIU	Parker C.A.4 Parasol (LA-3 Minor)	S. P. Connatty
	G-AFJA	Watkinson Dingbat ★	K. Woolley
	G-AFJB	Foster-Wickner G.M.1. Wicko (DR613) ★	K. Woolley
	G-AFJR	Tipsy Trainer 1	M. E. Vaisey
	G-AFJV	Mosscraft MA.2	C. V. Butler
	G-AFLW	M.17 Monarch	N. I. Dalziel/Biggin Hill
	G-AFNG	D.H.94 Moth Minor	R. W. Livett/Sywell
	G-AFNI	D.H.94 Moth Minor	B. M. Welford
	G-AFOB	D.H.94 Moth Minor	R. E. Ogden
	G-AFOJ	D.H.94 Moth Minor	R. M. Long
	G-AFPN	D.H.94 Moth Minor	J. Black/Barton
	G-AFPR	D.H.94 Moth Minor	J. A. Livett
	G-AFRZ	M.17 Monarch	R. E. Mitchell (G-AIDE)
	G-AFSC	Tipsy Trainer 1	R. V. & M. H. Smith
	G-AFSV	Chilton D.W.1A	R. Nerou
	G-AFTA	Hawker Tomtit (K1786)	Shuttleworth Trust/O. Warden
	G-AFTN	Taylorcraft Plus C2 ★	Leicestershire County Council Museums
	G-AFVE	D.H.82 Tiger Moth	P. E. Swinstead
	G-AFVN	Tipsy Trainer 1	W. Callow & ptnrs
	G-AFWH	Piper J-4A Cub Coupé	G. R. G. Smith
	G-AFWI	D.H.82A Tiger Moth	N. E. Rankin & ptnrs
	G-AFWT	Tipsy Trainer 1	J. S. Barker/Redhill
	G-AFYD	Luscombe 8E Silvaire	J. D. Iliffe
	G-AFYO	Stinson H.W.75	S. R. Clarke
	G-AFZA	Piper J-4A Cub Coupé	E. H. S. Warner
	G-AFZE	Heath Parasol	K. C. D. St Cyrien
	G-AFZL	Porterfield CP.50	P. G. Lucas/White Waltham
	G-AFZN	Luscombe 8A	A. G. Palmer
	G-AGBN	G.A.L.42 Cygnet 2	Museum of Flight/E. Fortune
	G-AGEG	D.H.82A Tiger Moth	T. P. A. Norman
	G-AGFT	Avia FL.3	I. B. Willis/Panshanger
	G-AGIV	Piper J-3C-65 Cub	K. M. & H. Bowen
	G-AGJG	D.H.89A Dragon Rapide	Russavia Ltd/Duxford
	G-AGLK	Auster 5D	W. C. E. Tazewell
	G-AGNV	Avro 685 York 1 (MW100) ★	Aerospace Museum/Cosford
	G-AGOH	J/I Autocrat	Leicestershire County Council Museums
	G-AGOS	R.S.3 Desford I (VZ728)	Scottish Aircraft Collection/Perth
	G-AGOY	M.48 Messenger 3 (U-0247)	T. Clark

Reg.	Type	Owner or Operator	Notes
G-AGRU	V.498 Viking 1A ★	Aerospace Museum/Cosford	
G-AGSH	D.H.89A Dragon Rapide 6	Specialist Flying Training Ltd/Carlisle	
G-AGTM	D.H.89A Dragon Rapide 6 (NF875)	Russavia Ltd/Duxford	
G-AGTO	J/I Autocrat	M. J. Barnett & D. J. T. Miller/Duxford	
G-AGTT	J/I Autocrat	C. Wilson	
G-AGVN	J/I Autocrat	P. J. Elliott/Leicester	
G-AGVV	Piper L-4H Cub	A. R. W. Taylor & D. Lofts/Sleap	
G-AGWE	Avro 19 Srs 2 ★	Loughborough & Leicester Air Museum/Bruntingthorpe	
G-AGXN	J/IN Alpha	J. G. Theobald/Cranwell	
G-AGXT	J/IN Alpha ★	Nene Valley Aircraft Museum/Sibson	
G-AGXU	J/IN Alpha	Mrs J. Lewis	
G-AGXV	J/I Autocrat	F. Mumford	
G-AGYD	J/IN Alpha	P. Herring/Dishforth	
G-AGYH	J/IN Alpha	G. E. Twyman & P. J. Rae	
G-AGYK	J/I Autocrat	D. A. Smith	
G-AGYL	J/IN Autocrat ★	WWII Aircraft Preservation Society/ Lasham	
G-AGYT	J/IN Alpha	Lee Air Ltd/Lee-on-Solent	
G-AGYU	DH.82A Tiger Moth (DE208)	P. & A. Wood	
G-AGYY	Ryan ST.3KR	D. S. & I. M. Morgan	
G-AGZZ	D.H.82A Tiger Moth	G. P. LaT. Shea-Simonds/Netheravon	
G-AHAL	J/IN Alpha	Skegness Air Taxi Services Ltd/ Ingoldmells	
G-AHAM	J/I Autocrat	D. W. Philp/Goodwood	
G-AHAN	D.H.82A Tiger Moth	S. E. Bates	
G-AHAU	J/I Autocrat	B. J. W. Foley/Panshanger	
G-AHAV	J/I Autocrat	C. J. Freeman/Headcorn	
G-AHBL	D.H.87B Hornet Moth	Dr Ursula H. Hamilton	
G-AHBM	D.H.87B Hornet Moth	P. A. & E. P. Gliddon	
G-AHCK	J/IN Alpha	P. A. Woodman/Shoreham	
G-AHCN	J/IN Alpha	FTS Flying Group	
G-AHCR	Gould-Taylorcraft Plus D Special	D. E. H. Balmford & D. R. Shepherd/ Yeovil	
G-AHED	D.H.89A Dragon Rapide (RL962) ★	RAF Museum (Cardington)	
G-AHGD	D.H.89A Dragon Rapide (Z7258)	M. R. L. Astor/Booker	
G-AHGW	Taylorcraft Plus D (LB375)	C. V. Butler/Coventry	
G-AHGZ	Taylorcraft Plus D	S. J. Ball/Leicester	
G-AHHH	J/I Autocrat	H. A. Jones/Norwich	
G-AHHK	J/I Autocrat	W. J. Ogle (*stored*)/Newtownards	
G-AHHN	J/I Autocrat	KK Aviation	
G-AHHP	J/1N Alpha	D. J. Hutcheson (G-SIME)	
G-AHHT	J/IN Alpha	R.A.E. Aero Club/Farnborough	
G-AHIP	Piper J-3C-65 Cub	R. E. Nerou/Coventry	
G-AHIZ	D.H.82A Tiger Moth	C.F.G. Flying Ltd/Cambridge	
G-AHKX	Avro 19 Srs 2	British Aerospace PLC/Woodford	
G-AHKY	Miles M.18 Series 2	Scottish Aircraft Collection/Perth	
G-AHLI	Auster 3	G. A. Leathers	
G-AHLK	Auster 3	E. T. Brackenbury/Leicester	
G-AHLT	D.H.82A Tiger Moth	R. C. F. Bailey	
G-AHMJ	Cierva C.30A (K4235) ★	Shuttleworth Trust/O. Warden	
G-AHMN	D.H.82A Tiger Moth (N6985)	George House (Holdings) Ltd	
G-AHOO	D.H.82A Tiger Moth	G. W. Bisshopp	
G-AHRI	D.H.104 Dove 1 ★	Lincolnshire Aviation Museum	
G-AHRO	Cessna 140	R. H. Screen/Kidlington	
G-AHSA	Avro 621 Tutor (K3215)	Shuttleworth Trust/O. Warden	
G-AHSD	Taylorcraft Plus D	A. Tucker	
G-AHSO	J/IN Alpha	Skegness Air Taxi Services Ltd/ Ingoldmells	
G-AHSP	J/I Autocrat	D. S. Johnstone & ptnrs	
G-AHSS	J/IN Alpha	Parker Airways Ltd/Denham	
G-AHST	J/IN Alpha	P. H. Lewis/Henstridge	
G-AHSW	J/I Autocrat	K. W. Brown/Coventry	
G-AHTE	P.44 Proctor V	J. G. H. Hassell	
G-AHTW	A.S.40 Oxford (V3388) ★	Skyfame Collection/Duxford	
G-AHUF	D.H.82A Tiger Moth	D. S. & I. M. Morgan	
G-AHUI	M.38 Messenger 2A ★	A. Topen	
G-AHUJ	M.14A Hawk Trainer 3 (R1914)	Vintage Aircraft Team	
G-AHUV	D.H.82A Tiger Moth	W. G. Gordon	
G-AHVU	D.H.82A Tiger Moth (T6313)	Parker Airways Ltd/Denham	
G-AHVV	D.H.82A Tiger Moth	R. Jones	

Notes	Reg.	Type	Owner or Operator
	G-AHWJ	Taylorcraft Plus D	A. Tucker
	G-AHXE	Taylorcraft Plus D (LB312)	Museum of Army Flying/Middle Wallop
	G-AIBE	Fulmar II (N1854) ★	F.A.A. Museum/Yeovilton
	G-AIBH	J/IN Alpha	Subtec Aviation
	G-AIBM	J/I Autocrat	J. L. Goodley
	G-AIBW	J/IN Alpha	W. E. Bateson/Blackpool
	G-AIBX	J/I Autocrat	Wasp Flying Group/Panshanger
	G-AIBY	J/I Autocrat	D. Morris/Sherburn
	G-AIDL	D.H.89A Dragon Rapide 6	Southern Joyrides Ltd/Biggin Hill
	G-AIDN	V.S.502 Spitfire T.8 (MT818)	G. F. Miller/Coventry
	G-AIDS	D.H.82A Tiger Moth	K. D. Pogmore & T. Dann
	G-AIEK	M.38 Messenger 2A (RG333)	J. Buckingham
	G-AIFZ	J/IN Alpha	C. P. Humphries
	G-AIGD	J/I Autocrat	A. G. Batchelor/Finmere
	G-AIGF	J/IN Alpha	A. R. C. Mathie
	G-AIGM	J/IN Alpha	Wickenby Flying Club Ltd
	G-AIGR	J/IN Alpha	Merrett Aviation Ltd
	G-AIGT	J/IN Alpha	B. D. Waller
	G-AIGU	J/IN Alpha	T. Pate
	G-AIIH	Piper J-3C-65 Cub	J. A. de Salis
	G-AIIZ	D.H.82A Tiger Moth (T6645)	D. E. & J. A. Baker
	G-AIJI	J/IN Alpha ★	Humberside Aircraft Preservation Soc
	G-AIJM	Auster J/4	R. H. A. Thorne/Booker
	G-AIJR	Auster J/4	B. A. Harris/Halfpenny Green
	G-AIJT	Auster J/4 srs 100	Aberdeen Auster Flying Group
	G-AILL	M.38 Messenger 2A	H. Best-Devereux
	G-AIPR	Auster J/4	MPM Flying Group/Booker
	G-AIPV	J/I Autocrat	J. Linegar
	G-AIPW	J/I Autocrat	B. Hillman
	G-AIRC	J/I Autocrat	A. G. Martlew/Barton
	G-AIRI	D.H.82A Tiger Moth	E. R. Goodwin
	G-AIRK	D.H.82A Tiger Moth	R. C. Teverson & ptnrs
	G-AISA	Tipsy B Srs 1	B. T. Morgan & A. Liddiard
	G-AISB	Tipsy B Srs 1	D. M. Fenton
	G-AISC	Tipsy B Srs 1	Wagtail Flying Group
	G-AISD	M.65 Gemini 1A	J. E. Homewood
	G-AISS	Piper J-3C-65 Cub	G. R. Lovell
	G-AIST	V.S.300 Spitfire IA (AR213)	The Hon P. Lindsay/Booker
	G-AISX	Piper J-3C-65 Cub	R. I. Souch
	G-AITB	A.S.10 Oxford (MP425) ★	RAF Museum
	G-AITP	Piper J-3C-65 Cub	G. L. Owens
	G-AIUA	M.14A Hawk Trainer 3 ★	A. Topen
	G-AIUL	D.H.89A Dragon Rapide 6	I. Jones
	G-AIWA	P.28B Proctor 1 (R7524)	B. Wilmot/Biggin Hill
	G-AIXA	Taylorcraft Plus D	P. Stevens/Cranwell
	G-AIXD	D.H.82A Tiger Moth	Sark International Airways/Guernsey
	G-AIXH	D.H.82A Tiger Moth	L. M. Haran
	G-AIXN	Benes-Mraz M.1c Sokol	J. F. Evetts & D. Patel
	G-AIYR	D.H.89A Dragon Rapide	C. D. Cyster
	G-AIYS	D.H.85 Leopard Moth	Wessex Aviation & Transport Ltd/ Hamble
	G-AIZE	F.24W Argus 2 ★	RAF Museum/Henlow
	G-AIZG	V.S. Walrus (L2301) ★	F.A.A. Museum/Yeovilton
	G-AIZU	J/I Autocrat	A. H. R. Stansfield/Middle Wallop
	G-AIZY	J/I Autocrat	B. J. Richards
	G-AIZZ	J/I Autocrat	S. E. Bond/Biggin Hill
	G-AJAB	J/IN Alpha	Air Farm Ltd
	G-AJAC	J/IN Alpha	R. C. Hibberd
	G-AJAD	Piper J-3C-65 Cub	A. J. E. Ditheridge
	G-AJAE	J/IN Alpha	M. G. Stops
	G-AJAJ	J/IN Alpha	R. B. Lawrence
	G-AJAM	J/2 Arrow	D. A. Porter/Sturgate
	G-AJAO	Piper J-3C Cub	C. C. & J. M. Lovell
	G-AJAS	J/IN Alpha	C. J. Baker/Sturgate
	G-AJCP	D.31 Turbulent	H. J. Shaw
	G-AJDW	J/I Autocrat	D. R. Hunt
	G-AJEB	J/IN Alpha ★	Aeroplane Collection Ltd/Warmingham
	G-AJEE	J/I Autocrat	A. R. C. De Albanoz/Bournemouth
	G-AJEH	J/IN Alpha	D. Nieman & G. Haim/Booker
	G-AJEI	J/IN Alpha	Skegness Air Taxi Services Ltd/ Ingoldmells
	G-AJEM	J/I Autocrat	S. F. Griggs

Reg.	Type	Owner or Operator	Notes
G-AJES	Piper J-3C-65 Cub	P. Crawford & R. Bowes	
G-AJGJ	Auster 5	P. A. Lee/Shoreham	
G-AJHO	D.H.89A Dragon Rapide ★	East Anglian Aviation Soc Ltd	
G-AJHS	D.H.82A Tiger Moth	Machine Music Ltd/Redhill	
G-AJHU	D.H.82A Tiger Moth	F. P. Le Coyte	
G-AJID	J/I Autocrat	D. J. Ronayne	
G-AJIH	J/I Autocrat	D. F. Campbell & ptnrs	
G-AJIS	J/IN Alpha	A. Tucker	
G-AJIT	J/I Kingsland Autocrat	Kingsland Aviation Ltd	
G-AJIU	J/I Autocrat	A. Murfin/Netherthorpe	
G-AJIW	J/IN Alpha	N. A. Roberts	
G-AJJP	Jet Gyrodyne (XJ389) ★	Aerospace Museum/Cosford	
G-AJKK	M.38 Messenger 2A ★	A. M. Lambourne & T. C. Eaves	
G-AJOA	D.H.82A Tiger Moth (T5424)	F. P. Le Coyte	
G-AJOC	M.38 Messenger 2A ★	Ulster Folk & Transport Museum	
G-AJOE	M.38 Messenger 2A (RH378)	J. Eagles & P. C. Kirby/Staverton	
G-AJON	Aeronca 7AC Champion	C. C. Lovell	
G-AJOV	Sikorsky S-51 ★	Aerospace Museum/Cosford	
G-AJOZ	F.24W Argus 2 ★	Lincolnshire Aviation Museum	
G-AJPI	F.24R-41a Argus 3	J. F. Read/White Waltham	
G-AJPZ	J/I Autocrat	K. Pyle	
G-AJRB	J/I Autocrat	S. C. Luck/Sywell	
G-AJRC	J/I Autocrat	S. W. Watkins & ptnrs	
G-AJRE	J/I Autocrat (Lycoming)	R. Gammage/Headcorn	
G-AJRH	J/IN Alpha	Leicestershire County Council Museums	
G-AJRS	M.14A Hawk Trainer 3 (P6382)	Shuttleworth Trust/O. Warden	
G-AJTW	D.H.82A Tiger Moth	J. A. Barker	
G-AJUD	J/I Autocrat	C. L. Sawyer	
G-AJUE	J/I Autocrat	M. A. G. Westman	
G-AJUL	J/IN Alpha	M. J. Crees	
G-AJVE	D.H.82A Tiger Moth	M. J. Abbot & I. J. Jones	
G-AJVH	Swordfish (LS326)	F.A.A. Museum/Yeovilton	
G-AJVT	Auster 5	I. N. M. Cameron	
G-AJXC	Auster 5	J. E. Graves	
G-AJXV	Auster 4 (NJ695)	P. C. J. Farries/Tollerton	
G-AJXY	Auster 4	G. B. Morris	
G-AKAA	Piper L-4H Cub	P. Raggett	
G-AKAT	M.14A Magister (T9738) ★	Newark Air Museum	
G-AKAZ	Piper J-3C-65 Cub	J. B. Coxon	
G-AKBM	M.38 Messenger 2A ★	Bristol Plane Preservation Unit	
G-AKBO	M.38 Messenger 2A	R. E. Coates	
G-AKDN	D.H.C. IA Chipmunk 10	K. R. Nunn/Seething	
G-AKEL	M.65 Gemini 1A	J. M. Bisco	
G-AKER	M.65 Gemini 1A ★	Vintage Aircraft Team	
G-AKEZ	M.38 Messenger 2A (RG333) ★	Wales Aircraft Museum	
G-AKGE	M.65 Gemini 3C	*Stored*	
G-AKHP	M.65 Gemini 1A	P. G. Lee	
G-AKHW	M.65 Gemini 1A	Vintage Aircraft Magazine Ltd/Denham	
G-AKHZ	M.65 Gemini 7 ★	Vintage Aircraft Team	
G-AKIB	Piper J-3C-65 Cub	B. R. Rossiter/White Waltham	
G-AKIF	D.H.89A Dragon Rapide	Airborne Taxi Services Ltd/Booker	
G-AKIN	M.38 Messenger 2A	A. J. Spiller/Sywell	
G-AKIU	Proctor V ★	N. Weald Aircraft Restoration Flight	
G-AKJU	J/IN Alpha	R. C. Lewis	
G-AKKB	M.65 Gemini 1A	S.A.C. Bristol Ltd	
G-AKKH	M.65 Gemini 1A	M. C. Russell/Duxford	
G-AKKR	M.14A Magister (T9707) ★	Manchester Air & Space Museum	
G-AKOE	D.H.89A Dragon Rapide 4	J. E. Pierce	
G-AKOW	Auster 5 (TJ569)	Museum of Army Flying/Middle Wallop	
G-AKPF	M.14A Hawk Trainer (N3788) ★	L. N. D. Taylor	
G-AKPI	Auster 5 (NJ703)	B. H. Hargrave/Sherburn	
G-AKRA	Piper J-3C-65 Cub	W. R. Savin	
G-AKSZ	Auster 5	A. R. C. Mathie/Upavon	
G-AKUW	C.H.3 Super Ace	C. V. Butler	
G-AKVF	C.H.3 Super Ace	P. V. B. Longthorp/Bodmin	
G-AKVZ	M.38 Messenger 4B	Shipping & Airlines Ltd/Biggin Hill	
G-AKWS	Auster 5	J. E. Homewood	
G-AKWT	Auster 5 (MT360) ★	Humberside Aircraft Preservation Soc	
G-AKXP	Auster 5	F. E. Telling	
G-AKXS	D.H.82A Tiger Moth	P. A. Colman	
G-AKZN	P.30 Proctor 2E (Z7197) ★	RAF Museum	
G-ALAH	M.38 Messenger 4A (RH377) ★	Aeroplane Collection Ltd/Henlow	

Notes	Reg.	Type	Owner or Operator
	G-ALAX	D.H.89A Dragon Rapide ★	Durney Aeronautical Collection
	G-ALBJ	Auster 5	R. H. Elkington
	G-ALBK	Auster 5	S. J. Wright & Co (Farmers) Ltd
	G-ALBN	Bristol 173 (XF785) ★	RAF Museum/Henlow
	G-ALCK	P.34A Proctor 3 (LZ766) ★	Skyfame Collection/Duxford
	G-ALCS	M.65 Gemini 3C	*Stored*
	G-ALCU	D.H.104 Dove 2 ★	Midland Air Museum
	G-ALDG	HP.81 Hermes 4 ★	Duxford Aviation Soc (*Fuselage only*)
	G-ALEH	PA-17 Vagabond	A. D. Pearce/Redhill
	G-ALFA	Auster 5	Alpha Flying Group/Leicester
	G-ALFM	D.H.104 Devon C.2	N. J. Taaffe & A. D. Hemley
	G-ALFT	D.H.104 Dove 6 ★	Torbay Aircraft Museum
	G-ALFU	D.H.104 Dove 6 ★	Imperial War Museum/Duxford
	G-ALGT	V.S.379 Spitfire 14 (RM689)	Rolls-Royce Ltd
	G-ALIW	D.H.82A Tiger Moth	D. I. M. Geddes & F. Curry/Booker
	G-ALJF	P.34A Proctor 3	J. F. Moore/Biggin Hill
	G-ALJL	D.H.82A Tiger Moth	C. G. Clarke
	G-ALNA	D.H.82A Tiger Moth	A. W. Kennedy
	G-ALND	D.H.82A Tiger Moth (N9191)	Arrow Air Services (Engineering) Ltd/ Shipdham
	G-ALOD	Cessna 140	J. R. Stainer
	G-ALRH	EoN Type 8 Baby	P. D. Moran
	G-ALRI	D.H.82A Tiger Moth	Wessex Aviation & Transport Ltd
	G-ALSP	Bristol 171 (WV783) Sycamore ★	RAF Museum/Henlow
	G-ALSS	Bristol 171 (WA576) Sycamore ★	E. Fortune
	G-ALST	Bristol 171 (WA577) Sycamore ★	N.E. Aircraft Museum
	G-ALSW	Bristol 171 (WT933) Sycamore ★	Newark Air Museum
	G-ALSX	Bristol 171 (G-48-1) Sycamore ★	Rotorcraft Museum/Duxford
	G-ALTO	Cessna 140	J. E. Cummings/Popham
	G-ALTW	D.H.82A Tiger Moth	C. J. Musk/(*stored*)
	G-ALUC	D.H.82A Tiger Moth	D. R. & Mollie Wood
	G-ALWB	D.H.C.1 Chipmunk 22A	D. G. Mathes
	G-ALWC	Dakota 4	Visionair (International Aviation) Ltd
	G-ALWF	V.701 Viscount ★	Viscount Preservation Trust/Duxford
	G-ALWW	D.H.82A Tiger Moth	F. W. Fay & ptnrs/Long Marston
	G-ALXT	D.H.89A Dragon Rapide ★	Science Museum/Wroughton
	G-ALXZ	Auster 5-150	B. J. W. Thomas & R. A. E. Witheridge
	G-ALYB	Auster 5 ★	S. Yorks Aviation Soc
	G-ALYG	Auster 5D	A. L. Young
	G-ALZE	BN-1F ★	Aerospace Museum/Cosford
	G-ALZO	A.S.57 Ambassador★	Dan-Air Preservation Group/Lasham
	G-AMAW	Luton L.A.4 Minor	J. R. Coates
	G-AMBB	D.H.82A Tiger Moth	J. Eagles
	G-AMCA	Dakota 3	Air Atlantique Ltd/Coventry
	G-AMDA	Avro 652A Anson 1 (N4877)★	Skyfame Collection/Duxford
	G-AMEN	PA-19 Super Cub 95	A. Lovejoy & ptnrs/Lasham
	G-AMHF	D.H.82A Tiger Moth	L. Mayhead
	G-AMHJ	Dakota 6	Topflight Ltd
	G-AMIU	D.H.82A Tiger Moth	R. & Mrs J. L. Jones
	G-AMKU	J/IB Aiglet	Southdown Flying Group/Slinfold
	G-AMLZ	P.50 Prince 6E ★	J. F. Coggins/Coventry
	G-AMMS	J/5F Aiglet Trainer	D. Collyer
	G-AMOG	V.701 Viscount ★	Aerospace Museum/Cosford
	G-AMPF	PA-18 Super Cub 95	R. I. Souch
	G-AMPG	PA-12 Super Cruiser	R. I. Souch
	G-AMPI	SNCAN Stampe SV-4C	J. Hewett
	G-AMPO	Dakota 4	Topflight Ltd
	G-AMPP	Dakota 3 (G-AMSU) ★	Dan-Air Preservation Group/Lasham
	G-AMPY	Dakota 4	Air Atlantique Ltd/Coventry
	G-AMPZ	Dakota 4	Harvest Air Ltd/Southend
	G-AMRA	Dakota 6	Air Atlantique Ltd/Coventry
	G-AMRF	J/5F Aiglet Trainer	A. I. Topps/E. Midlands
	G-AMRK	G.37 Gladiator (L8032)	Shuttleworth Trust/O. Warden
	G-AMSG	SIPA 903	S. W. Markham
	G-AMSV	Dakota 4	Air Atlantique Ltd/Coventry
	G-AMSZ	Auster 5	G. & D. Knight & D. G. Pridham
	G-AMTA	J/5F Aiglet Trainer	H. J. Jauncey/Rochester
	G-AMTD	J/5F Aiglet Trainer	C. I. Fray
	G-AMTM	J/I Autocrat	R. Stobo & D. Clewley
	G-AMUF	D.H.C.I Chipmunk 21	Redhill Tailwheel Flying Club Ltd
	G-AMUH	D.H.C.I Chipmunk 21	W. A. Fernie/Cranwell
	G-AMVD	Auster 5	C. G. Clarke
	G-AMVP	Tipsy Junior	A. R. Wershat/Blackbushe

Reg.	Type	Owner or Operator	Notes
G-AMVS	D.H.82A Tiger Moth	F. D. Bichener	
G-AMXA	D.H.106 Comet C2 (XK655) ★	Strathallan Aircraft Collection	
G-AMXT	D.H.104 Sea Devon C.20 (XJ347)	Scoteroy Ltd/Southampton	
G-AMYD	J/5L Aiglet Trainer	G. H. Maskell	
G-AMYJ	Dakota 6	Harvest Air Ltd/Southend	
G-AMZI	J/5F Aiglet Trainer	J. F. Moore/Biggin Hill	
G-AMZT	J/5F Aiglet Trainer	D. Hyde & J. W. Saull/Cranfield	
G-AMZU	J/5F Aiglet Trainer	R. N. Goode & ptnrs/White Waltham	
G-ANAF	Dakota 4	Topflight Ltd	
G-ANAP	D.H.104 Dove 6 ★	Brunel Technical College/Lulsgate	
G-ANCF	B.175 Britannia 308F	(*Stored*)/Manston	
G-ANCS	D.H.82A Tiger Moth	M. F. Newman	
G-ANCX	D.H.82A Tiger Moth	D. R. Wood/Biggin Hill	
G-ANDE	D.H.82A Tiger Moth	Newbury Aeroplane Ltd	
G-ANDM	D.H.82A Tiger Moth	J. Green	
G-ANDP	D.H.82A Tiger Moth	A. H. Diver	
G-ANEF	D.H.82A Tiger Moth (T5493)	RAF College Flying Club Co Ltd/ Cranwell	
G-ANEL	D.H.82A Tiger Moth (N9238)	W. P. Maynall	
G-ANEM	D.H.82A Tiger Moth	P. J. Benest	
G-ANEW	D.H.82A Tiger Moth	A. L. Young	
G-ANEZ	D.H.82A Tiger Moth	D. G. Ellis & C. D. J. Bland/ Sandown	
G-ANFC	D.H.82A Tiger Moth (DE363) ★	Mosquito Aircraft Museum	
G-ANFH	Westland S.55 ★	British Rotorcraft Museum	
G-ANFI	D.H.82A Tiger Moth (DE623)	D. H. R. Jenkins	
G-ANFL	D.H.82A Tiger Moth	R. P. Whitby & ptnrs	
G-ANFM	D.H.82A Tiger Moth	S. A. Brook & ptnrs/Booker	
G-ANFP	D.H.82A Tiger Moth ★	Mosquito Aircraft Museum	
G-ANFV	D.H.82A Tiger Moth (DF155)	R. A. L. Falconer/Inverness	
G-ANFW	D.H.82A Tiger Moth	G. M. Fraser/White Waltham	
G-ANHK	D.H.82A Tiger Moth	J. D. Iliffe	
G-ANHR	Auster 5	N. A. Forde	
G-ANHS	Auster 4	G. A. Griffin	
G-ANHX	Auster 5D	D. J. Baker/(*stored*)	
G-ANHZ	Auster 5	J. H. D. Newman/Headcorn	
G-ANIE	Auster 5	S. A. Stibbard	
G-ANIJ	Auster 5D	Museum of Army Flying/Middle Wallop	
G-ANIS	Auster 5	J. Clarke-Cockburn	
G-ANJA	D.H.82A Tiger Moth (N9389)	J. J. Young	
G-ANJD	D.H.82A Tiger Moth	H. J. Jauncey/(*stored*)	
G-ANJK	D.H.82A Tiger Moth	Montgomery Ultra Light Flying Club	
G-ANJV	Westland S.55 Srs 3 ★	British Rotorcraft Museum	
G-ANKK	D.H.82A Tiger Moth (T5854)	P. W. Crispe/Halfpenny Green	
G-ANKT	D.H.82A Tiger Moth (T6818)	Shuttleworth Trust/O. Warden	
G-ANKZ	D.H.82A Tiger Moth (N6466)	Cillam Holdings Ltd/Barton	
G-ANLD	D.H.82A Tiger Moth	D. P. Parks	
G-ANLH	D.H.82A Tiger Moth	Antique Aeroplane Co Ltd	
G-ANLS	D.H.82A Tiger Moth	P. A. Gliddon/E. Fortune	
G-ANLW	W.B.1. Widgeon (MD497) ★	Helicopter Hire Ltd/Southend	
G-ANMV	D.H.82A Tiger Moth (T7404)	George House (Holdings) Ltd	
G-ANNK	D.H.82A Tiger Moth	Mrs P. J. Wilcox/Sywell	
G-ANOD	D.H.82A Tiger Moth	D. R. & Mrs M. Wood	
G-ANOH	D.H.82A Tiger Moth	D. H. Parkhouse & ptnrs/White Waltham	
G-ANOK	S.91 Safir ★	Strathallan Aircraft Collection	
G-ANOM	D.H.82A Tiger Moth	P. A. Colman	
G-ANON	D.H.82A Tiger Moth (T7909)	A. C. Mercer/Sherburn	
G-ANOO	D.H.82A Tiger Moth	T. J. Hartwell & ptnrs	
G-ANOR	D.H.82A Tiger Moth	G. A. Black	
G-ANOV	D.H.104 Dove 6 ★	Museum of Flight/E. Fortune	
G-ANPK	D.H.82A Tiger Moth	The D. & P. Group	
G-ANPP	P.34A Proctor 3	C. P. A. & Mrs J. Jeffery/Duxford	
G-ANRF	D.H.82A Tiger Moth	C. D. Cyster	
G-ANRN	D.H.82A Tiger Moth	J. J. V. Elwes	
G-ANRP	Auster 5 (TW439) ★	Warnham War Museum	
G-ANRX	D.H.82A Tiger Moth ★	Mosquito Aircraft Museum	
G-ANSM	D.H.82A Tiger Moth	J. W. Davy	
G-ANTE	D.H.82A Tiger Moth	T. I. Sutton & B. J. Champion/ Chester	
G-ANTK	Avro 685 York ★	Dan Air Preservation Group/Lasham	
G-ANTS	D.H.82A Tiger Moth (N6532) ★	Strathallan Aircraft Collection	
G-ANUO	D.H.114 Heron 2D	Topflight Aviation Ltd/Fairoaks	
G-ANUW	D.H.104 Dove 6 ★	Civil Aviation Authority/Stansted	

Notes	Reg.	Type	Owner or Operator
	G-ANVU	D.H.104 Dove 1B	Dove Air Europa/Cranfield
	G-ANWB	D.H.C.I Chipmunk 21	G. Briggs/Blackpool
	G-ANWX	J/5L Aiglet Trainer	Applied Fastenings & Components
	G-ANXB	D.H.114 Heron 1B ★	Newark Air Museum
	G-ANXC	J/5R Alpine	R. H. Malcolm-Green
	G-ANXR	P.31C Proctor 4 (RM221)	L. H. Oakins/Biggin Hill
	G-ANYP	P.31C Proctor 4 (NP184) ★	Torbay Aircraft Museum
	G-ANZJ	P.31C Proctor 4 (NP303) ★	A. Hillyard
	G-ANZR	D.H.82A Tiger Moth	D. R. & Mrs. M. Wood
	G-ANZU	D.H.82A Tiger Moth	P. A. Jackson/Sibson
	G-ANZZ	D.H.82A Tiger Moth	IBC Transport Containers Ltd
	G-AOAA	D.H.82A Tiger Moth	Tiger Club Ltd/Redhill
	G-AOAR	P.31C Proctor 4 (NP181) ★	Historic Aircraft Preservation Soc
	G-AOBH	D.H.82A Tiger Moth (T7997)	C. H. A. Bott
	G-AOBO	D.H.82A Tiger Moth	T. J. Bolt & J. N. Moore
	G-AOBU	P.84 Jet Provost ★	Shuttleworth Trust/O. Warden
	G-AOBV	J/5P Autocar	P. E. Champney
	G-AOBX	D.H.82A Tiger Moth	M. Gibbs/Redhill
	G-AOCR	Auster 5D	J. M. Edis
	G-AOCU	Auster 5	S. J. Ball/Leicester
	G-AODA	Westland S.55 Srs 3	Bristow Helicopters Ltd
	G-AODT	D.H.82A Tiger Moth	N. A. Brett & A. H. Warminger
	G-AOEH	Aeronca 7AC Champion	M. Weeks & ptnrs
	G-AOEI	D.H.82A Tiger Moth	C.F.G. Flying Ltd/Cambridge
	G-AOEL	D.H.82A Tiger Moth (N9510) ★	Museum of Flight/E. Fortune
	G-AOES	D.H.82A Tiger Moth	A. Twemlow & G. A. Cordery/Redhill
	G-AOET	D.H.82A Tiger Moth	Glylynn Ltd
	G-AOEX	Thruxton Jackeroo	A. T. Christian
	G-AOFE	D.H.C.1 Chipmunk 22A	B. Webster/Shobdon
	G-AOFJ	Auster 5	Miss M. R. Innocent/Perth
	G-AOFM	J/5P Autocar	C. M. Barnes
	G-AOFS	J/5L Aiglet Trainer	G. W. Howard/Stapleford
	G-AOGA	M.75 Aries	Irish Aviation Museum *(stored)*
	G-AOGE	P.34A Proctor 3	N. I. Dalziel/Biggin Hill
	G-AOGI	D.H.82A Tiger Moth	W. J. Taylor
	G-AOGR	D.H.82A Tiger Moth	H. C. Adkins & E. Shipley/N. Denes
	G-AOGV	J/5R Alpine	ABH Aviation
	G-AOHK	V.802 Viscount ★	Hotel de France/St Helier, Jersey
	G-AOHL	V.802 Viscount ★	British Air Ferries (Cabin Trainer)/ Southend
	G-AOHM	V.802 Viscount	British Air Ferries *Viscount de L'Isle*/ Southend
	G-AOHT	V.802 Viscount	British Air Ferries/Southend
	G-AOHZ	J/5P Autocar	M. R. Gibbons & G. W. Brown
	G-AOIL	D.H.82A Tiger Moth	Shuttleworth Trust/(*stored*)
	G-AOIM	D.H.82A Tiger Moth	R. M. Wade & F. J. Terry
	G-AOIR	Thruxton Jackaroo	Stevenage Flying Club/O. Warden
	G-AOIS	D.H.82A Tiger Moth	V. B. & R. G. Wheele/Shoreham
	G-AOIY	J/5G Autocar	P. E. Scott
	G-AOJC	V.802 Viscount ★	Wales Aircraft Museum
	G-AOJH	D.H.83C Fox Moth	J. S. Lewery/Bournemouth
	G-AOJJ	D.H.82A Tiger Moth (DF128)	E. Lay
	G-AOKH	P.40 Prentice 1	J. F. Moore/Biggin Hill
	G-AOKL	P.40 Prentice 1 (VS610)	J. R. Batt/Southend
	G-AOKO	P.40 Prentice 1 ★	J. F. Coggins/Coventry
	G-AOKZ	P.40 Prentice 1 (VS623) ★	Midland Air Museum
	G-AOLK	P.40 Prentice 1	Hilton Aviation Ltd/Southend
	G-AOLU	P.40 Prentice 1 (VS356) ★	Scottish Aircraft Collection/Perth
	G-AORB	Cessna 170B	G. C. Masterson
	G-AORL	D.H.C.1 Chipmunk 22	D. Gardner
	G-AORW	D.H.C.1 Chipmunk 22A	D. C. Budd/Netherthorpe
	G-AOSK	D.H.C.1 Chipmunk 22	J. G. Cullen
	G-AOSO	D.H.C.1 Chipmunk 22	Aviation Advisory Services Ltd/ Stapleford
	G-AOSU	D.H.C.1 Chipmunk 22 (Lycoming)	RAFGSA/Bicester
	G-AOSY	D.H.C.1 Chipmunk 22	J. A. W. Clowes/Barton
	G-AOSZ	D.H.C.1 Chipmunk 22A	D. C. Flavell/Denham
	G-AOTD	D.H.C.1 Chipmunk 22 (WB588)	Shuttleworth Trust/(*stored*)
	G-AOTF	D.H.C.1 Chipmunk 23 (Lycoming)	RAFGSA/Bicester
	G-AOTI	D.H.114 Heron 2D	Topflight Aviation Ltd/Fairoaks
	G-AOTK	D.53 Turbi	The T. K. Flying Group/Hatfield
	G-AOTR	D.H.C.1 Chipmunk 22	P. W. B. & M. A. R. Spearing

Reg.	Type	Owner or Operator	Notes
G-AOTY	D.H.C.1 Chipmunk 22A	West London Aero Services Ltd/ White Waltham	
G-AOUJ	Fairey Ultra-Light ★	British Rotorcraft Museum	
G-AOUO	D.H.C.1 Chipmunk 22 (Lycoming)	RAFGSA/Bicester	
G-AOUP	D.H.C.1 Chipmunk 22	Wessex Flying Group	
G-AOVF	B.175 Britannia 312F★	Aerospace Museum/Cosford	
G-AOVT	B.175 Britannia 312F ★	Duxford Aviation Soc	
G-AOVW	Auster 5	B. Marriott/Cranwell	
G-AOXG	D.H.82A Tiger Moth (XL717)	F.A.A. Museum/Yeovilton	
G-AOXN	D.H.82A Tiger Moth	S. L. G. Darch	
G-AOYG	V.806 Viscount	Guernsey Airlines *Viscount Guernsey*	
G-AOYL	V.806 Viscount	British Air Ferries *Viscount Churchill*/ Southend	
G-AOYN	V.806 Viscount	British Air Ferries *Viscount Rotterdam*/ Southend	
G-AOYP	V.806 Viscount	British Air Ferries *Viscount Corbiere*/ Southend	
G-AOYR	V.806 Viscount	British Air Ferries *Viscount Aberdeen*/ Southend	
G-AOZB	D.H.82A Tiger Moth	Structure Flex Ltd/Redhill	
G-AOZH	D.H.82A Tiger Moth (K2572)	V. B. & R. G. Wheele/Shoreham	
G-AOZL	J/5Q Alpine	E. A. Taylor/Southend	
G-AOZP	D.H.C.1 Chipmunk 22	M. E. Darlington	
G-APAF	Auster 5	Globalpost Ltd/Goodwood	
G-APAH	Auster 5	Executive Flying Services Ltd	
G-APAL	D.H.82A Tiger Moth (N6847)	Stevenage Flying Club	
G-APAM	Thruxton Jackaroo	R. P. Williams	
G-APAO	D.H.82A Tiger Moth	C. K. Irvine	
G-APAP	D.H.82A Tiger Moth	R. A. Slade	
G-APAS	D.H.106 Comet 1XB ★	Aerospace Museum/Cosford	
G-APBD	PA-23 Apache 160	E. A. Clack & T. Pritchard	
G-APBE	Auster 5	G. W. Clark/O. Warden	
G-APBI	D.H.82A Tiger Moth (EM903)	R. Devaney & ptnrs/Audley End	
G-APBO	D.53 Turbi	H. C. Cox	
G-APBW	Auster 5	J. R. Batt/Southend	
G-APCB	J/5Q Alpine	M. J. Wilson/Redhill	
G-APCC	D.H.82A Tiger Moth	L. J. Rice/Henstridge	
G-APCU	D.H.82A Tiger Moth (N9508)	K. C. K. Virtue	
G-APDB	D.H.106 Comet 4 ★	Duxford Aviation Soc	
G-APDT	D.H.106 Comet 4 ★	Fire School/Heathrow	
G-APEG	V.953C Merchantman	Air Bridge Carriers Ltd/E. Midlands	
G-APEJ	V.953C Merchantman	Air Bridge Carriers Ltd/E. Midlands	
G-APEK	V.953C Merchantman	Air Bridge Carriers Ltd *Dreadnought*/ E. Midlands	
G-APEP	V.953C Merchantman	Air Bridge Carriers Ltd/E. Midlands	
G-APES	V.953C Merchantman	Air Bridge Carriers Ltd *Swiftsure*/ E. Midlands	
G-APET	V.953C Merchantman	Air Bridge Carriers Ltd *Temeraire*/ E. Midlands	
G-APEY	V.806 Viscount	British Air Ferries *Viscount Shetland*/ Southend	
G-APFA	D.54 Turbi	A. Eastelow & F. J. Keitch/Dunkeswell	
G-APFG	Boeing 707-436 ★	*Instructional airframe*/Stansted	
G-APFJ	Boeing 707-436 ★	Aerospace Museum/Cosford	
G-APFU	D.H.82A Tiger Moth	M. R . Coward & D. M. White/Bristol	
G-APGL	D.H.82A Tiger Moth (NM140) ★	Strathallan Aircraft Collection	
G-APGM	D.H.82A Tiger Moth	H. A. N. Orde-Pavett	
G-APHV	Avro 19 Srs 2 (VM360) ★	Museum of Flight/E. Fortune	
G-APIE	Tipsy Belfair B	C. O'Brien	
G-APIH	D.H.82A Tiger Moth (R5086)	A. J. Detheridge	
G-APIK	J/1N Alpha	T. D. Howe	
G-APIM	V.806 Viscount	British Air Ferries *Viscount Stephen Piercey*/Southend	
G-APIY	P.40 Prentice 1 (VR249) ★	Newark Air Museum	
G-APJB	P.40 Prentice 1 ★	City Airways/Coventry	
G-APJJ	Fairey Ultra-light ★	Midland Aircraft Preservation Soc	
G-APJN	Hiller UH-12B	Bristow Helicopters Ltd	
G-APJO	D.H.82A Tiger Moth	D. R. & Mrs M. Wood	
G-APJZ	J/1N Alpha	L. Goddard & E. Amey	
G-APKH	D.H.85 Leopard Moth	P. Franklin (G-ACGS)/White Waltham	
G-APKN	J/1N Alpha	Felthorpe Auster Group	
G-APKY	Hiller UH-12B	D. A. George	

Notes	Reg.	Type	Owner or Operator
	G-APLG	J/5L Aiglet Trainer	B. Russel/Thruxton
	G-APLO	D.H.C.1 Chipmunk 22A (WD379)	Channel Islands Aero Holdings Ltd
	G-APMH	J/1U Workmaster	R. E. Neal & S. R. Stevens
	G-APML	Dakota 6	Air Atlantique Ltd
	G-APMM	D.H.82A Tiger Moth	R. K. J. Hadlow
	G-APMX	D.H.82A Tiger Moth	K. B. Palmer/Headcorn
	G-APMY	PA-23 Apache 160 ★	Kelsterton College (*instructional airframe*)/Deeside
	G-APNJ	Cessna 310 ★	Chelsea College/Shoreham
	G-APNS	Garland-Bianchi Linnet	Paul Penn-Sayers Model Services Ltd
	G-APNT	Currie Wot	L. W. Richardson & ptnrs
	G-APNZ	D.31 Turbulent	Tiger Club Ltd/Redhill
	G-APOA	J/1N Alpha	Bristow Helicopters Ltd
	G-APOD	Tipsy Belfair	R. K. A. Menage
	G-APOI	Saro Skeeter Srs 8	F. F. Chamberlain/Inverness
	G-APOL	D.36 Turbulent	T. A. S. Rayner
	G-APPL	P.40 Prentice 1	Miss S. J. Saggers/Biggin Hill
	G-APPM	D.H.C.1 Chipmunk 22	P. D. Evans & Co Ltd/Shoreham
	G-APRF	Auster 5	Sladburys Flying Group/Clacton
	G-APRJ	Avro 694 Lincoln B.2 (G-29-I) ★	D. W. Arnold
	G-APRL	AW.650 Argosy 101	Elan International Ltd/E. Midlands
	G-APRR	Super Aero 45	M. C. Searle/Elstree
	G-APRU	M.S.760 Paris	Cranfield Institute of Technology
	G-APSO	D.H.104 Dove 5	*Stored*/Shobdon
	G-APSR	J/1U Workmaster	D. & K. Aero Services Ltd/Shobdon
	G-APSZ	Cessna 172	M. J. Butler & ptnrs/Barton
	G-APTH	Agusta-Bell 47J	P. Lancaster & R. White/Shoreham
	G-APTP	PA-22 Tri-Pacer 150	J. R. Williams/Blackpool
	G-APTR	J/1N Alpha	C. J. & D. J. Baker
	G-APTS	D.H.C.1 Chipmunk 22A	B. R. Pickard/Biggin Hill
	G-APTU	Auster 5	D. A. Thackery-Tyers
	G-APTW	W.B.1 Widgeon ★	Cornwall Aero Park/Helston
	G-APTY	Beech G.35 Bonanza	G. E. Brennand & J. M. Fish
	G-APTZ	D.31 Turbulent	H. W. Raith/Wick
	G-APUD	Bensen B.7M (modified) ★	Manchester Air & Space Museum
	G-APUE	L-40 Meta Sokol	S. E. & M. J. Aherne
	G-APUK	J/1 Autocrat	P. L. Morley
	G-APUP	Sopwith Pup (N5182) (replica)★	RAF Museum
	G-APUR	PA-22 Tri-Pacer 160	G. A. Allen & ptnrs
	G-APUW	Auster J-5V-160	Anglia Auster Syndicate
	G-APUY	D.31 Turbulent	C. Jones & ptnrs/Barton
	G-APUZ	PA-24 Comanche 250	H. McCutcheon Clarke
	G-APVF	Putzer Elster B	R. I. Souch
	G-APVG	J/5L Aiglet Trainer	Cranfield Institute of Technology
	G-APVN	D.31 Turbulent	R. Sherwin/Shoreham
	G-APVS	Cessna I70B	P. E. L. Lamyman
	G-APVU	L-40 Meta-Sokol	D. Kirk
	G-APVV	Mooney M-20A	Telcon Associates/Barton
	G-APVY	PA-25 Pawnee 150	KK Aviation
	G-APVZ	D.31 Turbulent	A. F. Bullock/Staverton
	G-APWJ	HPR-7 Herald 201 ★	Duxford Aviation Soc
	G-APWR	PA-22 Tri-Pacer 160	Airads/Shoreham
	G-APWY	Piaggio P.166 ★	Science Museum/Wroughton
	G-APWZ	EP.9 Prospector ★	Museum of Army Flying/Middle Wallop
	G-APXJ	PA-24 Comanche 250	Trolmay Ltd/Wellesbourne
	G-APXR	PA-22 Tri-Pacer 160	A. Troughton
	G-APXT	PA-22 Tri-Pacer 150	K. A. Goodchild/Southend
	G-APXU	PA-22 Tri-Pacer 125	C. G. Stone/Biggin Hill
	G-APXW	EP.9 Prospector ★	Museum of Army Flying/Middle Wallop
	G-APXX	D.H.A.3 Drover 2 (VH-FDT) ★	World War II Aircraft Pres Soc
	G-APXY	Cessna 150	Merlin Flying Club Ltd/Hucknall
	G-APYB	T.66 Nipper 2	B. O. Smith
	G-APYD	D.H.106 Comet 4B ★	Science Museum/Wroughton
	G-APYG	D.H.C.1 Chipmunk 22	E. J. I. Musty & P. A. Colman
	G-APYI	PA-22 Tri-Pacer 135	T. Edwards
	G-APYN	PA-22 Tri-Pacer 160	W. D. Stephens
	G-APYT	7FC Tri-Traveller	C. H. Morris & R. W. Brown
	G-APYU	7FC Tri-Traveller	K. Collins
	G-APYW	PA-22 Tri-Pacer 150	D. R. & D. F. Smith
	G-APZE	PA-23 Apache 160	J. P. Dodd/Biggin Hill
	G-APZG	PA-24 Comanche 250	Steve Stephens Ltd
	G-APZJ	PA-18 Super Cub 150	Southern Sailplanes
	G-APZK	PA-18 Super Cub 95	W. T. Knapton

Reg.	Type	Owner or Operator	Notes
G-APZL	PA-22 Tri-Pacer 160	M. R. Coward & R. T. Evans	
G-APZS	Cessna 175A	A. J. House	
G-APZU	D.H.104 Dove 6	Acraman Holdings Ltd	
G-APZX	PA-22 Tri-Pacer 150	M. G. Montgomerie	
G-ARAB	Cessna 150	A. H. Nicholas/Elstree	
G-ARAI	PA-22 Tri-Pacer 160	J. E. Fox	
G-ARAJ	PA-22 Tri-Pacer 160	Wearside Flying Group	
G-ARAM	PA-18 Super Cub 150	E. Sussex Gliding Club Ltd	
G-ARAN	PA-18 Super Cub 150	D. E. S. Sarney/Booker	
G-ARAO	PA-18 Super Cub 95	G. Ashmore & ptnrs/Tollerton	
G-ARAP	7EC Traveller	P. J. Heron	
G-ARAS	7FC Tri-Traveller	A. Bruniges	
G-ARAT	Cessna 180C	R. E. Styles & ptnrs	
G-ARAU	Cessna 150	S. Lynn/Sibson	
G-ARAW	Cessna 182C Skylane	G. Greenall	
G-ARAX	PA-22 Tri-Pacer 150	Megacirc Ltd/Sywell	
G-ARAY	H.S.748 Srs 2	Dan-Air Services Ltd/Gatwick	
G-ARAZ	D.H.82A Tiger Moth (R4959)	M. V. Gauntlett/Goodwood	
G-ARBE	D.H.104 Dove 8	British Aerospace/Brough	
G-ARBG	T.66 Nipper 2	Felthorpe Tipsy Group	
G-ARBL	D.31 Turbulent	G. H. Matthews	
G-ARBN	PA.23 Apache 160	H. Norden & H. J. Liggins	
G-ARBO	PA-24 Comanche 250	D. M. Harbottle/Blackpool	
G-ARBP	T.66 Nipper 2	A. Cambridge & D. B. Winstanley	
G-ARBS	PA-22 Tri-Pacer 160	Garb Enterprises/Southend	
G-ARBV	PA-22 Tri-Pacer 150	C. R. Turner/Biggin Hill	
G-ARBZ	D.31 Turbulent	D. G. H. Hilliard	
G-ARCC	PA-22 Tri-Pacer 150	Fainville Ltd	
G-ARCF	PA-22 Tri-Pacer 150	A. L. Scadding	
G-ARCI	Cessna 310D	Sandtoft Air Services Ltd	
G-ARCS	Auster D6/180	E. A. Matty/Shobdon	
G-ARCT	PA-18 Super Cub 95	M. Kirk	
G-ARCV	Cessna 175A	W. F. H. Gough	
G-ARCW	PA-23 Apache 160	E. M. Brain & R. Chew/Wellesbourne	
G-ARCX	AW Meteor 14 ★	Museum of Flight/E. Fortune	
G-ARDB	PA-24 Comanche 250	R. A. Sareen	
G-ARDD	CP.301C1 Emeraude	A. Mackintosh/Shobdon	
G-ARDE	D.H.104 Dove 6	R. J. H. Small/Cranfield	
G-ARDG	EP.9 Prospector ★	Museum of Army Flying/Middle Wallop	
G-ARDJ	Auster D.6/180	J. D. H. Radford/Tollerton	
G-ARDO	Jodel D.112	P. J. H. McCraig	
G-ARDP	PA-22 Tri-Pacer 150	G. M. Jones	
G-ARDS	PA-22 Caribbean 150	D. V. Asher	
G-ARDT	PA-22 Tri-Pacer 160	A. A. Whiter	
G-ARDV	PA-22 Tri-Pacer 160	B. Hillman	
G-ARDY	T.66 Nipper 2	M. T. Groves	
G-ARDZ	Jodel D.140A	W. R. Dryden	
G-AREA	D.H.104 Dove 8	British Aerospace/Hatfield	
G-AREB	Cessna 175B Skylark	R. J. Postlethwaite & ptnrs/ Wellesbourne	
G-AREE	PA-23 Aztec 250	W. C. C. Meyer/Biggin Hill	
G-AREF	PA-23 Aztec 250	Express Aviation Services Ltd/ Biggin Hill	
G-AREH	D.H.82A Tiger Moth	T. Pate	
G-AREI	Auster 3 (MT438)	R. Alliker & ptnrs/Bodmin	
G-AREJ	Beech 95 Travel Air	D. Huggett/Stapleford	
G-AREL	PA-22 Caribbean 150	H. H. Cousins/Fenland	
G-AREO	PA-18 Super Cub 150	Lasham Gliding Soc Ltd	
G-ARET	PA-22 Tri-Pacer 160	P. & V. Slatterey	
G-AREV	PA-22 Tri-Pacer 160	Echo Victor Group/Barton	
G-AREX	Aeronca 15AC Sedan	R. J. Middleton-Turnbull & P. Lowndes	
G-AREZ	D.31 Turbulent	J. St. Clair-Quentin/Shobdon	
G-ARFB	PA-22 Caribbean 150	C. T. Woodward & ptnrs	
G-ARFD	PA-22 Tri-Pacer 160	C. Fergusson & ptnrs	
G-ARFG	Cessna 175A Skylark	C. S. & Mrs B. A. Frost/Panshanger	
G-ARFH	PA-24 Comanche 250	L. M. Walton	
G-ARFL	Cessna 175B Skylark	P. M. & R. G. Rees	
G-ARFM	Cessna 175B Skylark	Foyle Aviation Ltd	
G-ARFO	Cessna 150A	Moray Flying Club Ltd	
G-ARFT	Jodel D.R. 1050	R. Shaw	
G-ARFV	T.66 Nipper 2	C. G. Stone/Biggin Hill	
G-ARGB	Auster 6A	A. M. Witt	
G-ARGG	D.H.C.1 Chipmunk 22	B. Hook	

Notes	Reg.	Type	Owner or Operator
	G-ARGO	PA-22 Colt 108	B. E. Goodman/Liverpool
	G-ARGV	PA-18 Super Cub 150	Deeside Gliding Club (Aberdeenshire) Ltd/Aboyne
	G-ARGY	PA-22 Tri-Pacer 160	D. H. Tanner & I. J. Enoch/Coventry
	G-ARGZ	D.31 Turbulent	N. J. Horton
	G-ARHC	Forney F-1A Aircoupe	A. P. Gardner/Elstree
	G-ARHI	PA-24 Comanche 180	W. H. Entress/Swansea
	G-ARHL	PA-23 Aztec 250	J. J. Freeman & Co Ltd/Headcorn
	G-ARHM	Auster 6A	D. Hollowell & ptnrs/Finmere
	G-ARHN	PA-22 Caribbean 150	D. B. Furniss & A. Munro/Doncaster
	G-ARHP	PA-22 Tri-Pacer 160	W. Wardle
	G-ARHR	PA-22 Caribbean 150	J. A. Hargraves/Fairoaks
	G-ARHT	PA-22 Caribbean 150 ★	Moston Technical College
	G-ARHU	PA-22 Tri-Pacer 160	N. Kirk
	G-ARHW	D.H.104 Dove 8	British Aerospace/Woodford
	G-ARHZ	D.62 Condor	D. H. Wilson-Spratt
	G-ARIA	Bell 47G	C. P. Horsley/Shoreham
	G-ARID	Cessna 172B	G. R. Porter
	G-ARIE	PA-24 Comanche 250	W. Radwanski/Coventry
	G-ARIH	Auster 6A	A. L. Tuttle & R. Larder
	G-ARIK	PA-22 Caribbean 150	C. J. Berry
	G-ARIL	PA-22 Caribbean 150	G. N. Richardson Motors/Shobdon
	G-ARIN	PA-24 Comanche 250	S. J. Savage & ptnrs
	G-ARIV	Cessna 172B	C. Taylor & ptnrs
	G-ARIW	CP.301B Emeraude	CJM Flying Group/Wellesbourne
	G-ARJE	PA-22 Colt 108	J. Souch
	G-ARJF	PA-22 Colt 108	M. J. Collins
	G-ARJH	PA-22 Colt 108	A. Walmsley
	G-ARJR	PA-23 Apache 160 ★	*Instructional airframe*/Kidlington
	G-ARJS	PA-23 Apache 160	Bencray Ltd/Blackpool
	G-ARJT	PA-23 Apache 160	R. D. Dickson/Sywell
	G-ARJU	PA-23 Apache 160	Chantaco Ltd/Biggin Hill
	G-ARJV	PA-23 Apache 160	Gordon King (Aviation) Ltd/Biggin Hill
	G-ARJW	PA-23 Apache 160	Gordon King (Aviation) Ltd/Biggin Hill
	G-ARJZ	D.31 Turbulent	N. H. Jones
	G-ARKG	J/5G Autocar	M. M. James & C. M. G. Ellis
	G-ARKJ	Beech N35 Bonanza	P. S. Bubbear/Luton
	G-ARKK	PA-22 Colt 108	A. W. Baxter/Tollerton
	G-ARKM	PA-22 Colt 108	L. E. Usher
	G-ARKN	PA-22 Colt 108	J. H. Underwood & A. J. F. Tabenor
	G-ARKP	PA-22 Colt 108	C. J. & J. Freeman/Headcorn
	G-ARKR	PA-22 Colt 108	J. C. Rimell
	G-ARKS	PA-22 Colt 108	J. Dickinson
	G-ARLD	H-395 Super Courier	P. H. Hall
	G-ARLG	Auster D.4/108	Auster D4 Group
	G-ARLK	PA-24 Comanche 250	M. Walker & C. Robinson
	G-ARLP	A.61 Terrier	T. L. Gray & ptnrs
	G-ARLR	A.61 Terrier	A. Kennedy & D. Delaney
	G-ARLT	Cessna 172B Skyhawk	A. R. German & Sons
	G-ARLU	Cessna 172B Skyhawk ★	*Instructional airframe*/Irish AC
	G-ARLV	Cessna 172B Skyhawk	R. T. Hayward
	G-ARLW	Cessna 172B Skyhawk	S. Lancashire Flyers Ltd
	G-ARLX	Jodel D.140B	Meridian Drilling Co Ltd/Biggin Hill
	G-ARLY	J/5P Autocar	P. J. Elliott & G. Green/Leicester
	G-ARLZ	D.31A Turbulent	M. K. Crofts
	G-ARMA	PA-23 Apache 160 ★	*Instructional airframe*/Kidlington
	G-ARMB	D.H.C.1 Chipmunk 22A (WB660)	P. A. Layzell/Goodwood
	G-ARMC	D.H.C.1 Chipmunk 22A	W. London Aero Services Ltd
	G-ARMG	D.H.C.1 Chipmunk 22A	Chipmunk Preservation Group Ltd/ Wellesbourne
	G-ARMI	PA-23 Apache 160 ★	Stapleford Flying Club Ltd
	G-ARMJ	Cessna 185 Skywagon	British Skysports
	G-ARML	Cessna 175B Skylark	R. C. Convine
	G-ARMN	Cessna 175B Skylark ★	Southall College of Technology
	G-ARMO	Cessna 172B Skyhawk	Sangria Designs Ltd & BRM Plastics Ltd/Booker
	G-ARMP	Cessna 172B	Southport & Merseyside Aero Club (1979) Ltd
	G-ARMR	Cessna 172B Skyhawk	G. Burns & ptnrs
	G-ARMZ	D.31 Turbulent	G. L. Owens
	G-ARNA	Mooney M.20B	R. Travers/Blackpool
	G-ARNB	J/5G Autocar	M. T. Jeffrey
	G-ARND	PA-22 Colt 108	Richard Rimington Ltd
	G-ARNE	PA-22 Colt 108	T. D. L. Bowden/Shipdham

Reg.	Type	Owner or Operator	Notes
G-ARNI	PA-22 Colt 108	J. D. Crymble	
G-ARNJ	PA-22 Colt 108	MKM Flying Group/Leavesden	
G-ARNK	PA-22 Colt 108	A. R. Cameron	
G-ARNL	PA-22 Colt 108	J. A. & J. A. Dodsworth/White Waltham	
G-ARNN	GC-1B Swift	K. E. Sword	
G-ARNO	A.61 Terrier	M. B. Hill	
G-ARNP	A.109 Airedale	P. A. Gunn	
G-ARNY	Jodel D.117	Inverness Flying Services Ltd	
G-ARNZ	D.31 Turbulent	P. L. Cox & ptnrs	
G-AROA	Cessna 172B Skyhawk	D. E. Partridge/Andrewsfield	
G-AROD	Cessna 175B	Medical Co Hospital Supplies Ltd	
G-AROE	Aero 145	G. S. & Mrs P. Galt/White Waltham	
G-AROF	L.40 Meta-Sokol	B. G. Barber	
G-AROJ	A.109 Airedale	D. J. Shaw	
G-ARON	PA-22 Colt 108	R. W. Curtis	
G-AROO	Forney F-1A Aircoupe	W. J. McMeekan	
G-AROW	Jodel D.140B	Kent Gliding Club Ltd	
G-AROY	Stearman A.75N.1	W. A. Jordan	
G-ARPD	H.S.121 Trident 1C ★	CAA Fire School, Tees-side	
G-ARPH	H.S.121 Trident 1C ★	Aerospace Museum, Cosford	
G-ARPK	H.S.121 Trident 1C ★	Manchester Airport Authority	
G-ARPL	H.S.121 Trident 1C ★	British Airports Authority/Edinburgh	
G-ARPN	H.S.121 Trident 1C ★	British Airports Authority/Aberdeen	
G-ARPO	H.S.121 Trident 1C ★	CAA Fire School/Tees-side	
G-ARPP	H.S.121 Trident 1C ★	British Airports Authority/Glasgow	
G-ARPR	H.S.121 Trident 1C ★	CAA Fire School/Tees-side	
G-ARPW	H.S.121 Trident 1C ★	CAA Fire School/Tees-side	
G-ARPX	H.S.121 Trident 1C ★	Airwork Services Ltd/Perth	
G-ARPZ	H.S.121 Trident 1C ★	RFD Ltd/Dunsfold	
G-ARRD	Jodel DR.1050	N. L. E. Dupee/Dunkeswell	
G-ARRE	Jodel DR.1050	S. J. Pugh & M. Ridsdale	
G-ARRF	Cessna 150A	D. O. & P. A. Thirkell	
G-ARRI	Cessna 175B Skylark	C. L. Thomas	
G-ARRL	J/1N Alpha	G. N. Smith & C. Webb	
G-ARRM	Beagle B.206-X ★	Brighton Transport Museum	
G-ARRP	PA.28 Cherokee 160	M. J. Flynn/Cardiff	
G-ARRS	CP-301A Emeraude	J. Y. Paxton/Sibson	
G-ARRT	Wallis WA-116-1	K. H. Wallis	
G-ARRU	D.31 Turbulent	J. R. Edwards & D. D. Smith	
G-ARRW	H.S.748 Srs 1	Dan-Air Services Ltd/Gatwick	
G-ARRY	Jodel D.140B	K. G. Andrews/Southend	
G-ARRZ	D.31 Turbulent	R. J. Grimstead/Redhill	
G-ARSB	Cessna 150A	B. T. White/Andrewsfield	
G-ARSG	Avro Triplane (replica)	Shuttleworth Trust/O. Warden	
G-ARSJ	CP.301-C2 Emeraude	J. R. Ware	
G-ARSL	A.61 Terrier	R. A. Hutchinson & P. T. M. Hardy	
G-ARSP	L.40 Meta-Sokol	Classic Aerodrome Ltd/Staverton	
G-ARSU	PA-22 Colt 108	P. E. Palmer	
G-ARSW	PA-22 Colt 108	J. P. Smith/Shipdham	
G-ARSX	PA-22 Tri-Pacer 160	P. C. Riggs	
G-ARTD	PA-23 Apache 160	Dr. D. A. Jones	
G-ARTF	D.31 Turbulent	J. R. D. Bygraves/O. Warden	
G-ARTG	Hiller UH-12C ★	White Hart Inn/Stockbridge	
G-ARTH	PA-12 Super Cruiser	R. Hornby/Popham	
G-ARTJ	Bensen B.8 ★	Museum of Flight/E. Fortune	
G-ARTL	D.H.82A Tiger Moth (T7281)	P. A. Jackson	
G-ARTT	M.S.880B Rallye Club	D. C. S. Gunning & R. W. C. Simpson/ Panshanger	
G-ARUG	J/5G Autocar	N. P. Biggs	
G-ARUH	Jodel DR.1050	PFA Group/Denham	
G-ARUI	A.61 Terrier	Crossford Hotels (Dunfermline) Ltd	
G-ARUL	Cosmic Wind	P. G. Kynsey/Redhill	
G-ARUO	PA-24 Comanche 180	Uniform Oscar Group/Elstree	
G-ARUR	PA-28 Cherokee 160	The G-ARUR Group/Redhill	
G-ARUV	CP.301A Emeraude	J. Tanswell	
G-ARUY	J/1N Alpha	P. W. Metcalfe/Tollerton	
G-ARUZ	Cessna 175C Skylark	J. E. Sansome & J. T. Gout/Luton	
G-ARVF	V.1101 VC10 ★	Hermeskeil Museum (nr Trier)/ W. Germany	
G-ARVM	V.1101 VC10 ★	Aerospace Museum/Cosford	
G-ARVO	PA-18 Super Cub 95	J. O. Souch	
G-ARVS	PA-28 Cherokee 160	Stapleford Flying Club Ltd	
G-ARVT	PA-28 Cherokee 160	C. R. Knapton	
G-ARVU	PA-28 Cherokee 160	D. Hyland	

Notes	Reg.	Type	Owner or Operator
	G-ARVV	PA-28 Cherokee 160	Computer Services Ltd
	G-ARVW	PA-28 Cherokee 160	Bencray Ltd/Blackpool
	G-ARVZ	D62B Condor	C. Watson & W. H. Cole/Redhill
	G-ARWB	D.H.C.1 Chipmunk 200	Aero-Bonner Co Ltd/Shoreham
	G-ARWC	Cessna 150B	Worldwide Wheels Ltd/Exeter
	G-ARWH	Cessna 172C Skyhawk	P. E. Nunn/Fenland
	G-ARWM	Cessna 175C	Agricopters Ltd/Chilbolton
	G-ARWO	Cessna 172C Skyhawk	T. A. Cox & R. C. Jackman/Bodmin
	G-ARWR	Cessna 172C Skyhawk	The Devanha Flying Group Ltd
	G-ARWS	Cessna 175C Skylark	D. Q. Read & E. N. Skinner
	G-ARWW	Bensen B.8M	B. McIntyre
	G-ARWY	Mooney M.20A	B. P. Irish/Bodmin
	G-ARXD	A.109 Airedale	D. Howden
	G-ARXF	PA-23 Aztec 250B	Weendy Aviation (UK)
	G-ARXG	PA-24 Comanche 250	B. R. Gaunt & T. W. G. Frodsham/ Blackpool
	G-ARXH	Bell 47G	A. B. Searle
	G-ARXN	Tipsy Nipper 2	Griffon Flying Group
	G-ARXP	Luton LA-4A Minor	W. C. Hymas/Stapleford
	G-ARXT	Jodel DR.1050	G. D. Bowd
	G-ARXU	Auster 6A	Bath & Wilts Gliding Club Ltd
	G-ARXW	M.S.885 Super Rallye	M. A. Jones
	G-ARXX	M.S.880B Rallye Club	M. S. Bird
	G-ARYB	H.S.125 Srs 1★	British Aerospace PLC/Hatfield
	G-ARYC	H.S.125 Srs 1 ★	The Mosquito Aircraft Museum
	G-ARYD	Auster AOP.6 (WJ358)	Museum of Army Flying/Middle Wallop
	G-ARYF	PA.23 Aztec 250	I. J. T. Branson/Biggin Hill
	G-ARYH	PA-22 Tri-Pacer 160	Filtration (Water Treatment Engineers) Ltd
	G-ARYI	Cessna 172C	M. E. Mortlock & L. K. Heaton
	G-ARYK	Cessna 172C	Transavia Ltd/Luton
	G-ARYR	PA-28 Cherokee 180	Glenochill Engineering/Edinburgh
	G-ARYS	Cessna 172C Skyhawk	P. H. Preston & R. W. J. Andrews/ Halfpenny Green
	G-ARYV	PA-24 Comanche 250	P. Meeson
	G-ARYZ	A.109 Airedale	J. D. Reid
	G-ARZA	Wallis WA.116 Srs 1	N. D. Z. de Ferranti
	G-ARZB	Wallis WA.116 Srs 1	K. H. Wallis
	G-ARZE	Cessna 172C ★	*Parachute jump trainer*/Cockerham
	G-ARZF	Cessna 150B	M. M. James/Leicester
	G-ARZM	D.31 Turbulent	Tiger Club Ltd/Redhill
	G-ARZN	Beech N35 Bonanza	Beech Aircraft Ltd/Elstree
	G-ARZP	A.109 Airedale	G. B. O'Neill/(*stored*)/Booker
	G-ARZW	Currie Wot	D. F. Faulkner-Bryant/Redhill
	G-ARZX	Cessna 150B	Needteam Ltd
	G-ASAA	Luton LA-4A Minor	D. F. Lingard
	G-ASAI	A.109 Airedale	A. C. Watt
	G-ASAJ	A.61 Terrier 2 (WE569)	R. Skingley
	G-ASAK	A.61 Terrier 2	Rochford Hundred Flying Group/ Southend
	G-ASAL	SAL Bulldog 120	British Aerospace/Prestwick
	G-ASAM	D.31 Turbulent	Tiger Club Ltd/Redhill
	G-ASAN	A.61 Terrier 2	Truman Aviation Ltd/Tollerton
	G-ASAT	M.S.880B Rallye Club	R. J. Chinn & E. J. Kemp/Bristol
	G-ASAU	M.S.880B Rallye Club	W. J. Armstrong
	G-ASAV	M.S.880B Rallye Club	A. L. Averill & ptnrs
	G-ASAX	A.61 Terrier 2	G. Strathdee
	G-ASBA	Currie Wot	M. A. Kaye
	G-ASBB	Beech 23 Musketeer	D. Silver/Southend
	G-ASBH	A.109 Airedale	D. T. Smollett
	G-ASBS	C.P.301A Emeraude	D. M. Upfield
	G-ASBY	A.109 Airedale	M. J. Barnett & ptnrs
	G-ASCC	Beagle E.3 AOP Mk 11	M. D. N. & A. C. Fisher
	G-ASCJ	PA-24 Comanche 250	Telspec Ltd/Rochester
	G-ASCM	Isaacs Fury II (K2050)	D. Calabritto
	G-ASCU	PA-18A-150 Super Cub	Farm Aviation Services Ltd
	G-ASCZ	CP.310A Emeraude	Hylton Flying Group
	G-ASDA	Beech 65-80 Queen Air	Parker & Heard Ltd/Biggin Hill
	G-ASDF	Edwards Gyrocopter ★	B. King
	G-ASDK	A.61 Terrier 2	Applied Fastenings & Components
	G-ASDL	A.61 Terrier 2	T. J. Rilley & C. E. Mason
	G-ASDO	Beech 95-A55 Baron	Executive Aviation Ltd/Jersey
	G-ASDY	Wallis WA-116/F	K. H. Wallis

Reg.	Type	Owner or Operator	Notes
G-ASEA	Luton LA-4A Minor	C. W. N. Huke	
G-ASEB	Luton LA-4A Minor	R. K. Lynn	
G-ASEG	A.61 Terrier	J. C. Wilson/Liverpool	
G-ASEO	PA-24 Comanche 250	M.G.F. Co/Colchester	
G-ASEP	PA-23 Apache 235	Horizon Holdings Ltd/Guernsey	
G-ASEU	D.62A Condor	W. Grant & D. McNicholl	
G-ASEV	PA-23 Aztec 250	Selexpress Ltd	
G-ASFA	Cessna 172D	Clare's Garage Ltd	
G-ASFD	L-200A Morava	N. Price/Goodwood	
G-ASFK	J/5G Autocar	Orman (Carrolls Farm) Ltd	
G-ASFL	PA.28 Cherokee 180	P. S. Cottrell & M. G. Fountain	
G-ASFR	Bo.208A1 Junior	S. T. Dauncey	
G-ASFX	D.31 Turbulent	E. F. Clapham & W. B. S. Dobie	
G-ASGC	V.1151 Super VC10 ★	Imperial War Museum/Duxford	
G-ASHA	Cessna F.172D	R. Soar & ptnrs	
G-ASHB	Cessna 182F	RN & R Marines Sport Parachute Association/Dunkeswell	
G-ASHD	Brantly B-2A ★	British Rotorcraft Museum/Weston	
G-ASHH	PA-23 Aztec 250	Leicestershire Thread & Trimming Manufacturers Ltd	
G-ASHR	Beech B35-33 Debonair	G. D. Clark	
G-ASHS	Stampe SV.4B	L. W. Gruber & ptnrs/Goodwood	
G-ASHT	D.31 Turbulent	J. T. Taylor/Booker	
G-ASHU	PA-15 Vagabond	G. J. Romanes	
G-ASHV	PA-E23 Aztec 250	R. J. Ashley & G. O'Gorman	
G-ASHX	PA-28 Cherokee 180	D. Morris	
G-ASIB	Cessna F.172D	A. T. Jay	
G-ASII	PA-28 Cherokee 180	Worldwide Wheels Ltd & ptnrs/Lulsgate	
G-ASIJ	PA-28 Cherokee 180	J. G. Hinley/Coventry	
G-ASIL	PA-28 Cherokee 180	N. M. Barker & ptnrs/Leicester	
G-ASIS	Jodel D.112 Club	E. F. Hazell	
G-ASIT	Cessna 180	A. & P. A. Wood	
G-ASIY	PA-25 Pawnee 235	RAFGSA/Bicester	
G-ASJL	Beech H.35 Bonanza	P. M. Coulton	
G-ASJM	PA-30 Twin Comanche 160	Air & General Services Ltd/Biggin Hill	
G-ASJO	Beech B.23 Musketeer	A. P. Curtis & S. H. Kirkby/Thruxton	
G-ASJV	V.S.361 Spitfire IX (MH434)	Nalfire Aviation Ltd/Booker	
G-ASJY	GY-80 Horizon 160	A. D. Hemley	
G-ASJZ	Jodel D.117A	V. K. Hardy & R. B. Spink/Barton	
G-ASKC	D.H.98 Mosquito 35 (TA719) ★	Skyfame Collection/Duxford	
G-ASKH	D.H.98 Mosquito T.3 (RR299)	British Aerospace/Chester	
G-ASKK	HPR-7 Herald 211 ★	Norwich Aviation Museum	
G-ASKL	Jodel D.150A	J. M. Graty	
G-ASKP	D.H.82A Tiger Moth	Tiger Club Ltd/Redhill	
G-ASKS	Cessna 336 Skymaster	M. J. Godwin	
G-ASKT	PA-28 Cherokee 180	Capel & Co (Printers) Ltd/Biggin Hill	
G-ASKV	PA-25 Pawnee 235	Southdown Gliding Club Ltd	
G-ASLF	Bensen B.7	S. R. Hughes	
G-ASLH	Cessna 182F	Celahurst Ltd/Southend	
G-ASLK	PA-25 Pawnee 235	Crop Aviation Ltd/Wyberton	
G-ASLR	Agusta-Bell 47J-2	D. Jack	
G-ASLV	PA-28 Cherokee 235	C.S.E. (Aircraft Services) Ltd/Kidlington	
G-ASLX	CP.301A Emeraude	K. C. Green/Panshanger	
G-ASMA	PA-30 Twin Comanche 160	B. D. Glynn/Biggin Hill	
G-ASMC	P.56 Provost T.1.	W. Walker/Kidlington	
G-ASME	Bensen B.8M	R. K. Williamson	
G-ASMF	Beech D.95A Travel Air	Hawk Aviation Ltd	
G-ASMG	D.H.104 Dove 8	British Aerospace/Dunsfold	
G-ASMJ	Cessna F.172E	J. B. Stocks & ptnrs	
G-ASML	Luton LA-4A Minor	R. L. E. Horrell	
G-ASMM	D.31 Tubulent	Kenneth Browne	
G-ASMN	PA-23 Apache 160	W. London Aero Services Ltd/ White Waltham	
G-ASMO	PA-23 Apache 160	Aviation Enterprises/Fairoaks	
G-ASMS	Cessna 150A	K. R. & T. W. Davies	
G-ASMT	Fairtravel Linnet 2	A. F. Cashin	
G-ASMU	Cessna 150D	JRB Aviation Ltd/Southend	
G-ASMV	CP1310-C3 Super Emeraude	P. F. D. Waltham/Leicester	
G-ASMW	Cessna 150D	Yorkshire Light Aircraft Ltd/Leeds	
G-ASMY	PA-23 Apache 160	Suffolk Aero Club Ltd/Ipswich	
G-ASMZ	A.61 Terrier 2	Museum of Army Flying/Middle Wallop	
G-ASNA	PA-23 Aztec 250	Margate Motors Plant & Aircraft Hire Ltd/Headcorn	
G-ASNB	Auster 6A	M. Pocock & ptnrs	

Notes	Reg.	Type	Owner or Operator
	G-ASNC	Beagle D.5/180 Husky	Peterborough & Spalding Gliding Club/ Crowland
	G-ASND	PA-23 Aztec 250	Commercial Air (Woking) Ltd/ Fairoaks
	G-ASNE	PA-28 Cherokee 180	J. L. Dexter
	G-ASNH	PA-23 Aztec 250	Derek Crouch (Contractors) Ltd
	G-ASNI	CP1310-C3 Super Emeraude	D. Chapman
	G-ASNK	Cessna 205	Woodvale Parachute Centre
	G-ASNU	H.S.125 Srs. 1	Flintgrange Ltd
	G-ASNW	Cessna F.172E	J. A. Gibbs
	G-ASNY	Bensen B.8M	D. L. Wallis
	G-ASNZ	Bensen B.7M	W. H. Turner
	G-ASOC	Auster 6A	Aquila Gliding Club
	G-ASOH	Beech B.55A Baron	J. S. Goodsir & J. Mason/Biggin Hill
	G-ASOI	A.61 Terrier 2	R. H. Jowett
	G-ASOK	Cessna F.172E	Okay Flying Group/Denham
	G-ASON	PA-30 Twin Comanche 160	Avtec Enterprises Ltd
	G-ASOO	PA-30 Twin Comanche 160	Cold Storage (Jersey) Ltd
	G-ASOX	Cessna 205A	Dorglen Ltd/Halfpenny Green
	G-ASPF	Jodel D.120	W. S. Howell
	G-ASPI	Cessna F.172E	Icarus Flying Group/Rochester
	G-ASPK	PA-28 Cherokee 140	Hampshire Aeroplane Co Ltd/St Just
	G-ASPP	Bristol Boxkite replica	Shuttleworth Trust/O. Warden
	G-ASPS	Piper J-3C-65 Cub	A. J. Chalkley
	G-ASPU	D.31 Turbulent	I. Maclennan/Redhill
	G-ASPV	D.H.82A Tiger Moth	B. S. Charters/Shipdham
	G-ASPX	Bensen B-8S	L. D. Goldsmith/St Athan
	G-ASRB	D.62B Condor	Tiger Club Ltd (*stored*)
	G-ASRC	D.62B Condor	P. S. Milner/Redhill
	G-ASRF	Jenny Wren	G. W. Gowland
	G-ASRH	PA-30 Twin Comanche 160	Island Aviation & Travel Ltd
	G-ASRI	PA-23 Aztec 250	Graham Collins Associates Ltd
	G-ASRK	A.109 Airedale	M. J. Barnett & R. Skingley
	G-ASRO	PA-30 Twin Comanche 160	A. G. Perkins/Halfpenny Green
	G-ASRP	Jodel DR.1050	D. J. M. Edmondston
	G-ASRR	Cessna 182G	G. J. Richardson/Bourn
	G-ASRT	Jodel D.150	H. M. Kendall
	G-ASRW	PA-28 Cherokee 180	K. R. Deering/Shoreham
	G-ASRX	Beech 65 A80 Queen Air	Aero Charter (Midlands) Ltd/Coventry
	G-ASSB	PA-30 Twin Comanche 160	Villotel Ltd/Elstree
	G-ASSE	PA-22 Colt 108	J. B. King/Goodwood
	G-ASSF	Cessna 182G Skylane	Burbage Farms Ltd
	G-ASSP	PA-30 Twin Comanche 160	The Mastermix Engineering Co Ltd/ Coventry
	G-ASSR	PA-30 Twin Comanche 160	Direct Air Ltd/Leavesden
	G-ASSS	Cessna 172E	D. H. N. Squires/Bristol
	G-ASST	Cessna 150D	F. R. H. Parker
	G-ASSU	CP.301A Emeraude	R. W. Millward/Redhill
	G-ASSW	PA-28 Cherokee 140	C. J. Plummer/Biggin Hill
	G-ASSY	D.31 Turbulent	G-ASSY Group/Redhill
	G-ASTA	D.31 Turbulent	J. Gillespie & D. C. Writer
	G-ASTG	Nord 1002	L. M. Walton
	G-ASTI	Auster 6A	M. Pocock
	G-ASTL	Fairey Firefly 1 (Z2033) ★	Skyfame Collection/Duxford
	G-ASTP	Hiller UH-12C	L. Goddard
	G-ASUB	Mooney M.20E Super 21	T. J. Pigott/Doncaster
	G-ASUD	PA-28 Cherokee 180	H. J. W. Ellison/Biggin Hill
	G-ASUE	Cessna 150D	D. Huckle/Panshanger
	G-ASUG	Beech E18S ★	Museum of Flight/E. Fortune
	G-ASUH	Cessna F.172E	G. H. Willson & E. Shipley/Felthorpe
	G-ASUI	A.61 Terrier 2	D. Collyer
	G-ASUL	Cessna 182G Skylane	Halfpenny Green Parachute Centre Ltd
	G-ASUP	Cessna F.172E	GASUP Air/Swansea
	G-ASUR	Dornier Do 28A-1	Sheffair Ltd
	G-ASUS	Jurca MJ.2B Tempete	D. G. Jones/Coventry
	G-ASVG	CP.301B Emeraude	K. R. Jackson
	G-ASVM	Cessna F.172E	R. W. Tydeman & P. W. Clark/Cambridge
	G-ASVN	Cessna U.206 Super Skywagon	British Skysports
	G-ASVO	HPR-7 Herald 214	British Air Ferries *Herald Tribune*/ Southend
	G-ASVP	PA-25 Pawnee 235	Aquila Gliding Club Ltd
	G-ASVZ	PA-28 Cherokee 140	J. Yourell/Luton
	G-ASWB	A.109 Airedale	C. Gene & G. Taylor/Tees-side
	G-ASWF	A.109 Airedale	C. Baker/Goodwood

Reg.	Type	Owner or Operator	Notes
G-ASWG	PA-25 Pawnee 235	A.D.S. (Aerial) Ltd/Southend	
G-ASWH	Luton LA-5A Major	D. G. J. Chisholm/Coventry	
G-ASWJ	Beagle 206 Srs 1 (8449M) ★	RAF Halton	
G-ASWL	Cessna F.172F	C. Wilson	
G-ASWN	Bensen B.8M	D. R. Shepherd	
G-ASWP	Beech A.23 Musketeer	P. J. Annard/Aberdeen	
G-ASWW	PA-30 Twin Comanche 160	Bristol & Wessex Flying Club Ltd/ Bristol	
G-ASWX	PA-28 Cherokee 180	K. Hassell/Liverpool	
G-ASXB	D.H.82A Tiger Moth	G. W. Bisshopp	
G-ASXC	SIPA 901	B. L. Proctor	
G-ASXD	Brantly B2B	Brantly Enterprises	
G-ASXF	Brantly 305	Express Aviation Services Ltd/ Biggin Hill	
G-ASXI	T.66 Nipper 2	I. D. Daniels	
G-ASXJ	Luton LA-4A Minor	J. S. Allison	
G-ASXR	Cessna 210	S. G. Brady/Aberdeen	
G-ASXS	Jodel DR.1050	C. J. J. Blyth/Redhill	
G-ASXU	Jodel D.120A	R. W. & Mrs J. Thompsett/Fenland	
G-ASXV	Beech 65-A80 Queen Air	G. J. Ward	
G-ASXX	Avro 683 Lancaster 7 (NX611) ★	RAF Scampton Gate Guard	
G-ASXY	Jodel D.117A	P. A. Davies & ptnrs/Cardiff	
G-ASXZ	Cessna 182G Skylane	P. M. Robertson/Perth	
G-ASYD	BAC One-Eleven 670	British Aerospace/Bournemouth	
G-ASYJ	Beech D.95A Travel Air	Crosby Aviation Ltd	
G-ASYK	PA-30 Twin Comanche 160	E. W. Pinchbeck & D. F. Ranger	
G-ASYL	Cessna 150E	D. Mallinson	
G-ASYP	Cessna 150E	M. R. Spilman & A. Cowdery	
G-ASYV	Cessna 310G	Design & Production Props Ltd	
G-ASYW	Bell 47G-2	Bristow Helicopters Ltd	
G-ASYZ	Victa Airtourer 100	R. Fletcher/Halfpenny Green	
G-ASZB	Cessna 150E	H. J. Cox	
G-ASZD	Bo 208A2 Junior	D. R. Elphick/O. Warden	
G-ASZE	A.61 Terrier 2	P. J. Moore	
G-ASZJ	S.C.7 Skyvan 3A-100	Short Bros Ltd/Sydenham	
G-ASZR	Fairtravel Linnet	H. C. D. & F. J. Garner	
G-ASZS	GY.80 Horizon 160	T. B. W. Jeremiah & ptnrs/Stapleford	
G-ASZU	Cessna 150E	T. H. Milburn	
G-ASZV	T.66 Nipper 2	R. L. Mitcham/Elstree	
G-ASZX	A.61 Terrier	D. V. D. Reid	
G-ATAA	PA-28 Cherokee 180	Brendair/Elstree	
G-ATAD	Mooney M.20C	H. W. Walker/Swansea	
G-ATAF	Cesna F.172F	G. Bush	
G-ATAG	Jodel DR. 1050	T. J. N. H. Palmer & G. W. Oliver	
G-ATAI	D.H.104 Dove 8	Centrax Ltd	
G-ATAS	PA-28 Cherokee 180	D. R. Wood/Biggin Hill	
G-ATAT	Cessna 150E	The Derek Pointon Group/Coventry	
G-ATAU	D.62B Condor	M. A. Peare/Redhill	
G-ATAV	D.62C Condor	The Condor Syndicate	
G-ATBF	F-86E Sabre 4 (XB733) ★	T. Bracewell	
G-ATBG	Nord 1002	L. M. Walton	
G-ATBH	Aero 145	Colpak Aviation Ltd/Elstree	
G-ATBI	Beech A.23 Musketeer	R. F. G. Dent/Staverton	
G-ATBJ	Sikorsky S-61N	British Airways Helicopters Ltd/ China	
G-ATBK	Cessna F.172F	Viewmoor Ltd/Manchester	
G-ATBL	D.H.60G Moth	M. E. Vaisey	
G-ATBN	PA-28 Cherokee 140	M. R. McGregor/Stapleford	
G-ATBP	Fournier RF-3	M. J. Aherne & ptnrs	
G-ATBS	D.31 Turbulent	R. N. Crosland/Redhill	
G-ATBU	A.61 Terrier 2	P. R. Anderson	
G-ATBW	T.66 Nipper 2	N. J. Newbold & ptnrs/Fenland	
G-ATBX	PA-20 Pacer 135	M. R. Smith/Staverton	
G-ATBZ	W.S-58 Wessex 60	Westland Helicopters Ltd/Yeovil	
G-ATCC	A.109 Airedale	F. J. Lingham/Biggin Hill	
G-ATCD	D.5/180 Husky	Oxford Flying & Gliding Group	
G-ATCE	Cessna U.206	J. Fletcher & D. Hickling/Langar	
G-ATCI	Victa Airtourer 100	B. & C. Building Materials (Canvey Island) Ltd	
G-ATCJ	Luton LA-4A Minor	R. M. Sharphouse	
G-ATCL	Victa Airtourer 100	D. Alexander/Leicester	
G-ATCN	Luton LA-4A Minor	J. C. Gates & C. Neilson	

Notes	Reg.	Type	Owner or Operator
	G-ATCR	Cessna 310 ★	Holly Hill Service Station/ Swanton Novers
	G-ATCU	Cessna 337	University of Cambridge
	G-ATCX	Cessna 182H Skylane	K. J. Fisher
	G-ATDA	PA-28 Cherokee 160	D. E. Siviter (Motors) Ltd/Coventry
	G-ATDB	Nord 1101 Noralpha	J. B. Jackson
	G-ATDN	A.61 Terrier 2 (TW641)	Express Aviation Services Ltd/ Biggin Hill
	G-ATDO	Bo 208C Junior	D. L. Elite
	G-ATDS	HPR.7 Herald 209	Channel Express/Bournemouth
	G-ATED	Hiller UH-12E	North Scottish Helicopters Ltd
	G-ATEF	Cessna 150E	M. Smith & ptnrs/Popham
	G-ATEG	Cessna 150E	Knack Air Ltd
	G-ATEM	PA-28 Cherokee 180	G. Wyles & W. Adams
	G-ATEP	EAA Biplane	E. L. Martin/Guernsey
	G-ATES	PA-32 Cherokee Six 260 ★	*Parachute jump trainer*/Ipswich
	G-ATEV	Jodel DR. 1050	B. A. Mills & G. W. Payne
	G-ATEW	PA-30 Twin Comanche 160	Air Northumbria Group/Newcastle
	G-ATEX	Victa Airtourer 100	D. C. Giles & ptnrs/Southend
	G-ATEZ	PA-28 Cherokee 140	J. A. Burton/E. Midlands
	G-ATFA	Bensen B-8	J. Butler (*stored*)
	G-ATFD	Jodel DR. 1050	H. Fawcett & ptnrs/Netherthorpe
	G-ATFF	PA-23 Aztec 250	Tempus Aviation (Southern) Ltd
	G-ATFG	Brantly B2B	R. J. Chapman Ltd
	G-ATFK	PA-30 Twin Comanche 160	L. J. Martin/Redhill
	G-ATFL	Cessna F.172F	R. L. Beverley/Bournemouth
	G-ATFM	Sikorsky S-61N	British Airways Helicopters Ltd/ Aberdeen
	G-ATFR	PA-25 Pawnee 150	Miller Aerial Spraying Ltd/Wickenby
	G-ATFU	D.H.85 Leopard Moth	A. H. Carrington & C. D. Duthy-James
	G-ATFV	Agusta/Bell 47J-2A	Alexander Warren & Co
	G-ATFW	Luton LA-4A Minor	C. Kirk
	G-ATFX	Cessna F.172G	M. J. J. Fenwick
	G-ATFY	Cessna F.172G	E. Cure Ltd/Woodvale
	G-ATGE	Jodel DR.1050	Ambassador Aviation Ltd
	G-ATGF	M.S.892A Rallye Commodore 150	E. G. Bostock & R. A. Punter
	G-ATGG	M.S.885 Super Rallye	B&C Plant Hire Ltd/Southend
	G-ATGH	Brantly B2B	R. Crook/Kidlington
	G-ATGO	Cessna F.172G	W. H. Reeves
	G-ATGP	Jodel DR.1050	W. M. Haley/Tees-side
	G-ATGY	GY.80 Horizon	P. W. Gibberson/Birmingham
	G-ATGZ	GH-4 Gyroplane	G. Griffiths
	G-ATHA	PA-23 Apache 235	Avon Aviation Services Ltd
	G-ATHD	D.H.C.1 Chipmunk 22	Spartan Flying Group Ltd/Denham
	G-ATHF	Cessna 150F ★	Lincolnshire Aircraft Museum
	G-ATHG	Cessna 150F	G. T. Williams
	G-ATHK	Aeronca 7AC Champion	A. Corran
	G-ATHL	Wallis WA-116/F	W. Vinten Ltd
	G-ATHM	Wallis WA-116 Srs 1	Wallis Autogyros Ltd
	G-ATHN	Nord 1101 Noralpha	E. L. Martin/Guernsey
	G-ATHR	PA-28 Cherokee 180	Britannia Airways Ltd/Luton
	G-ATHT	Victa Airtourer 115	W. G. Hunt/Southend
	G-ATHU	A.61 Terrier 1	J. A. L. Irwin
	G-ATHV	Cessna 150F	Vectaphone Manufacturing Ltd
	G-ATHX	Jodel DR. 100A	T. S. Wilkins
	G-ATHZ	Cessna 150F	Merrett Aviation Ltd
	G-ATIA	PA-24 Comanche 260	India Alpha Partnership/Sywell
	G-ATIC	Jodel DR.1050	R. J. Hurstone & G. D. Kinnie
	G-ATID	Cessna 337	M. R. Tarrant/Stansted
	G-ATIE	Cessna 150F ★	*Parachute jump trainer*/Chetwynd
	G-ATIG	HPR-7 Herald 214	Janus Airways Ltd/Lydd
	G-ATIN	Jodel D.117	D. R. Upton & J. G. Kay/Barton
	G-ATIR	Stampe SV.4C	Mitchell Aviation
	G-ATIS	PA-28 Cherokee 160	Oxford Educational Resources Ltd Kidlington
	G-ATIZ	Jodel D.117	N. Chandler
	G-ATJA	Jodel DR.1050	S. J. Kew/Booker
	G-ATJC	Victa Airtourer 100	D. G. Palmer & D. C. Giles
	G-ATJL	PA-24 Comanche 260	M. J. Berry/White Waltham
	G-ATJM	Fokker DR.1 replica (152/17)	R. Lamplough/Duxford
	G-ATJN	Jodel D.119	J. K. S. Wills/Biggin Hill
	G-ATJP	PA-23 Apache 160	T. Hood & A. Mattacks/Biggin Hill
	G-ATJR	PA-E23 Aztec 250	W. A. G. Willbond

Reg.	Type	Owner or Operator	Notes
G-ATJT	GY.80 Horizon 160	P. J. Stephenson	
G-ATJV	PA-32 Cherokee Six 260	Doncaster Parachute Centre	
G-ATJX	Bu 131 Jungmann (5036)	J. G. Johnston	
G-ATKC	Stampe S.V.4B	Tiger Club Ltd/Redhill	
G-ATKF	Cessna 150F	Rogers Aviation Ltd/Netherthorpe	
G-ATKH	Luton LA-4A Minor	L. Hepper/Rochester	
G-ATKI	Piper J-3C-65 Cub	A. C. Netting	
G-ATKS	Cessna F.172G	Kevstart Ltd	
G-ATKT	Cessna F.172G	N. Y. Souster	
G-ATKU	Cessna F.172G	S. E. Ward & Sons (Engineers) Ltd/ Doncaster	
G-ATKX	Jodel D.140C	Tiger Club Ltd/Redhill	
G-ATKY	Cessna 150F	R. L. Beverley	
G-ATKZ	T.66-2 Nipper	M. W. Knights/Felthorpe	
G-ATLA	Cessna 182J Skylane	Shefford Transport Engineers Ltd/ Luton	
G-ATLB	Jodel DR.1050-M1	Tiger Club Ltd/Redhill	
G-ATLC	PA-23 Aztec 250	Alderney Air Charter Ltd/Bournemouth	
G-ATLD	Cessna E-310K	Pearlcat Ltd/Biggin Hill	
G-ATLG	Hiller UH-12B	Bristow Helicopters Ltd	
G-ATLM	Cessna F.172G	Yorkshire Flying Services Ltd/Leeds	
G-ATLN	Cessna F.172G	Routair Aviation Services Ltd/Southend	
G-ATLP	Bensen B.8M	C. D. Julian	
G-ATLR	Cessna F.172G	A. Wood & R. F. Patmore/Andrewsfield	
G-ATLT	Cessna U-206A	Lincoln Parachute Centre Ltd/Sturgate	
G-ATLV	Jodel D.120	G. Dawes/Headcorn	
G-ATLW	PA-28 Cherokee 180	P. T. Unden & ptnrs/Shoreham	
G-ATMB	Cessna F.150F	Mickey Bravo Group 84/Barton	
G-ATMC	Cessna F.150F	H. E. Peacock	
G-ATMG	M.S.893 Rallye Commodore 180	F. W. Fay & ptnrs/Wellesbourne	
G-ATMH	Beagle D.5/180 Husky	Devon & Somerset Gliding Club Ltd	
G-ATMI	H.S.748 Srs 2A	Dan-Air Services Ltd/Gatwick	
G-ATMJ	H.S.748 Srs 2A	Dan-Air Services Ltd/Gatwick	
G-ATML	Cessna F.150F	N. P. Newberry/Shobdon	
G-ATMT	PA-30 Twin Comanche 160	D. H. T. Bain/Newcastle	
G-ATMU	PA-23 Apache 160	Southend Flying Club	
G-ATMW	PA-28 Cherokee 140	Bencray Ltd/Blackpool	
G-ATMX	Cessna F.150F	H. M. Synge	
G-ATMY	Cessna 150F	D. C. Hyde/Little Snoring	
G-ATNB	PA-28 Cherokee 180	Van Diemen International Racing Service Ltd	
G-ATNE	Cessna F.150F	R. Gray/Leicester	
G-ATNI	Cessna F.150F	S. E. Marples	
G-ATNK	Cessna F.150F	Pegasus Aviation Ltd	
G-ATNL	Cessna F.150F	P. Crayfourd	
G-ATNU	Cessna 182A	London Parachuting Ltd	
G-ATNV	PA-24 Comanche 260	A. J. Bradley & P. A. Jay	
G-ATNX	Cessna F.150F	B. R. Walker/Halfpenny Green	
G-ATOA	PA-23 Apache 160	S. J. Green & P. N. Tilney	
G-ATOD	Cessna F.150F	Cornwall Flying Club Ltd/Bodmin	
G-ATOE	Cessna F.150F	T. J. Butler/Shoreham	
G-ATOH	D.62B Condor	E. D. Burke/Staverton	
G-ATOI	PA-28 Cherokee 140	O. & E. Flying Ltd/Stapleford	
G-ATOJ	PA-28 Cherokee 140	Liteflight Ltd	
G-ATOK	PA-28 Cherokee 140	J. D. & M. Cheetham	
G-ATOL	PA-28 Cherokee 140	Mike Bennett Ltd/Bodmin	
G-ATOM	PA-28 Cherokee 140	A. Reynard/Kidlington	
G-ATON	PA-28 Cherokee 140	R. G. Walters	
G-ATOO	PA-28 Cherokee 140	P. J. Stead/Cark	
G-ATOP	PA-28 Cherokee 140	Aero 80 Flying Group/Lasham	
G-ATOR	PA-28 Cherokee 140	T. A. J. Morgan & ptnrs/Shobdon	
G-ATOS	PA-28 Cherokee 140	AFT Craft Ltd/Halfpenny Green	
G-ATOT	PA-28 Cherokee 180	J. H. Parker/Halfpenny Green	
G-ATOU	Mooney M.20E Super 21	B. C. Dietrich & C. V. Margrane-Jones	
G-ATOY	PA-24 Comanche 260 ★	Museum of Flight/E. Fortune	
G-ATOZ	Bensen B.8M	J. Jordan	
G-ATPD	H.S.125 Srs 1B	Euroguard Ltd	
G-ATPE	H.S.125 Srs 1B	Colt Executive Aviation Ltd/Staverton	
G-ATPJ	BAC One-Eleven 301	Dan-Air Services Ltd/Gatwick	
G-ATPK	BAC One-Eleven 301	Dan-Air Services Ltd/Gatwick	
G-ATPL	BAC One-Eleven 301	Dan-Air Services Ltd/Gatwick	
G-ATPM	Cessna F.150F	Dan-Air Flying Club/Lasham	
G-ATPN	PA-28 Cherokee 140	D. A. Thompson & L. Martin/Southend	
G-ATPT	Cessna 182J Skylane	Western Models Ltd/Lydd	

Notes	Reg.	Type	Owner or Operator
	G-ATPV	JB.01 Minicab	S. Russell
	G-ATRC	Beech B.95A Travel Air	Bulldog Aviation Ltd/Ipswich
	G-ATRG	PA-18 Super Cub 150	Lasham Gliding Soc Ltd
	G-ATRI	Bo 208C Junior	W. H. Jones/Dunkeswell
	G-ATRK	Cessna F.150F	A. B. Mills
	G-ATRL	Cessna F.150F	Loganair Ltd/Glasgow
	G-ATRM	Cessna F.150F	J. W. C. A. Coulcutt/Sandown
	G-ATRN	Cessna F.150F	Westair Flying Services Ltd
	G-ATRO	PA-28 Cherokee 140	390th Flying Group
	G-ATRR	PA-28 Cherokee 140	S. Richardson
	G-ATRU	PA-28 Cherokee 180	Britannia Airways Ltd/Luton
	G-ATRW	PA-32 Cherokee Six 260	Eastern Enterprises/Ipswich
	G-ATRX	PA-32 Cherokee Six 260	R. F. Gibbs/Panshanger
	G-ATRY	Alon A-2 Aircoupe	B. W. Swann/Sandown
	G-ATSI	Bo 208C Junior	T. M. H. Paterson/Shoreham
	G-ATSL	Cessna F.172G	H. G. Le Cheminant/Guernsey
	G-ATSM	Cessna 337A	Anglo-European Trustees Ltd
	G-ATSR	Beech M.35 Bonanza	Alstan Aviation Ltd
	G-ATST	M.S.893A Rallye Commodore	Severnside International Aviation/ Cardiff
	G-ATSU	Jodel D.140B	J. Gillham/Redhill
	G-ATSX	Bo 208C Junior	N. M. G. Pearson/Bristol
	G-ATSY	Wassmer WA41 Super Baladou IV	Horizon Flying Group
	G-ATTB	Wallis WA.116-1	D. A. Wallis
	G-ATTD	Cessna 182J Skylane	Hanro Aviation Ltd/Leicester
	G-ATTF	PA-28 Cherokee 140	S. J. & H. Y. George
	G-ATTG	PA-28 Cherokee 140	Arrow Air Services Engineering Ltd/ Shipdham
	G-ATTI	PA-28 Cherokee 140	D. Newman & C. Babb
	G-ATTK	PA-28 Cherokee 140	P. J. Smyth
	G-ATTM	Jodel DR.250-160	R. W. Tomkinson
	G-ATTP	BAC One-Eleven 207	Dan-Air Services Ltd/Gatwick
	G-ATTR	Bo 208C Junior 3	S. Luck
	G-ATTU	PA-28 Cherokee 140	Leith Air Ltd/Elstree
	G-ATTV	PA-28 Cherokee 140	W. J. Brogan
	G-ATTX	PA-28 Cherokee 180	A. Gray & ptnrs/Andrewsfield
	G-ATTY	PA-32 Cherokee Six 260	L. A. Dingemans & D. J. Everett/ Stapleford
	G-ATUB	PA-28 Cherokee 140	R. H. Partington & M. J. Porter
	G-ATUC	PA-28 Cherokee 140	Executive Communications Ltd/Booker
	G-ATUD	PA-28 Cherokee 140	E. J. Clempson
	G-ATUF	Cessna F.150F	C. J. Lynn/Langar
	G-ATUG	D.62B Condor	C. B. Marsh & D. J. R. Williams
	G-ATUH	T.66 Nipper	G. L. Winterbourne
	G-ATUI	Bo 208C Junior	A. J. H. Martin/Southampton
	G-ATUL	PA-28 Cherokee 180	R. F. W. Warner
	G-ATVF	D.H.C.1 Chipmunk 22	RAFGSA/Dishforth
	G-ATVH	BAC One-Eleven 207	Dan-Air Services Ltd *City of Newcastle-upon-Tyne*/Gatwick
	G-ATVK	PA-28 Cherokee 140	JRB Aviation Ltd/Southend
	G-ATVL	PA-28 Cherokee 140	West London Aero Services/White Waltham
	G-ATVO	PA-28 Cherokee 140	R. J. Midgley/Stapleford
	G-ATVP	F.B.5 Gunbus (2345) ★	RAF Museum
	G-ATVS	PA-28 Cherokee 180	Marshalls Woodflakes Ltd/Bristol
	G-ATVW	D.62B Condor	J. P. Coulter & J. Chidley/Panshanger
	G-ATVX	Bo 208C Junior	G. W. Stanmore
	G-ATWA	Jodel DR.1050	R. S. Arbuthnot & ptnrs/Thruxton
	G-ATWB	Jodel D.117	A. R. Mangham
	G-ATWE	M.S.892A Rallye Commodore	D. I. Murray
	G-ATWJ	Cessna F.172F	C. J. & J. Freeman/Headcorn
	G-ATWP	Alon A-2 Aircoupe	H. Dodd & I. Wilson
	G-ATWR	PA-30 Twin Comanche 160	Lubair (Transport Services) Ltd E. Midlands
	G-ATWZ	M.S.892 Rallye Commodore	N. A. Hall & ptnrs/Bodmin
	G-ATXA	PA-22 Tri-Pacer 150	B. R. Gaunt/Blackpool
	G-ATXD	PA-30 Twin Comanche 160	Alphameric Systems Ltd
	G-ATXF	GY-80 Horizon 150	A. I. Milne/Swanton Morley
	G-ATXM	PA-28 Cherokee 180	J. Khan/Stapleford
	G-ATXN	Mitchell-Proctor Kittiwake	D. W. Kent/Lasham
	G-ATXO	SIPA 903	M. Hillam/Sherburn
	G-ATXR	AFB 1 gas balloon	Mrs C. M. Bulmer *Omega One*
	G-ATXZ	Bo 208C Junior	B. Flintoff

Reg.	Type	Owner or Operator	Notes
G-ATYM	Cessna F.150G	J. F. Perry & Co	
G-ATYN	Cessna F.150G	Skegness Air Taxi Services Ltd	
G-ATYS	PA-28 Cherokee 180	R. V. Waite	
G-ATZA	Bo 208C Junior	W. C. Roberts	
G-ATZG	AFB2 gas balloon	Flt Lt S. Cameron *Aeolis*	
G-ATZK	PA-28 Cherokee 180	P. L. Williams/Shoreham	
G-ATZM	Piper J-3C-65 Cub	R. W. Davison	
G-ATZS	Wassmer WA41 Super Baladou IV	J. R. MacAlpine-Downie & P. A. May	
G-ATZY	Cessna F.150G	P. P. D. Howard-Johnstone/Edinburgh	
G-ATZZ	Cessna F.150G	C. Wren/Southend	
G-AUTO	Cessna 441 Conquest	Mediplan International PLC	
G-AVAA	Cessna F.150G	A. C. Garrett	
G-AVAJ	Hiller UH-12B	Bristow Helicopters Ltd	
G-AVAK	M.S.893A Rallye Commodore 180	W. K. Anderson/Perth	
G-AVAO	PA-30 Twin Comanche 160	Ghan-Air	
G-AVAP	Cessna F.150G	Seawing Flying Club Ltd/Southend	
G-AVAR	Cessna F.150G	J. Tapley	
G-AVAS	Cessna F.172H	Birmingham Aviation Ltd/ Halfpenny Green	
G-AVAU	PA-30 Twin Comanche 160	L. Batin/Fairoaks	
G-AVAW	D.62B Condor	Avato Flying Group	
G-AVAX	PA-28 Cherokee 180	Three Counties Flying Group	
G-AVAZ	PA-28 Cherokee 180	Intermark Ltd/Redhill	
G-AVBG	PA-28 Cherokee 180	S. S. Padam/Booker	
G-AVBH	PA-28 Cherokee 180	P. Crisp	
G-AVBP	PA-28 Cherokee 140	Bristol & Wessex Aeroplane Club Ltd/Bristol	
G-AVBS	PA-28 Cherokee 180	F. B. Miles/St Just	
G-AVBT	PA-28 Cherokee 180	P. O. Hire & D. J. Spicer	
G-AVBZ	Cessna F.172H	J. Seville	
G-AVCA	Brantly B.2B	M. J. & Mrs G. M. Page/Norwich	
G-AVCC	Cessna F.172H	Mercia Estates Ltd/Coventry	
G-AVCE	Cessna F.172H	Astrashine Ltd	
G-AVCM	PA-24 Comanche 260	F. Smith & Sons Ltd/Stapleford	
G-AVCS	A.61 Terrier 1	L. M. Farrell & A. R. C. Hunter	
G-AVCT	Cessna F.150G	Sierra Aviation Services/(*stored*)	
G-AVCU	Cessna F.150G	P. R. Moss/Alderney	
G-AVCV	Cessna 182J Skylane	University of Manchester Institute of Science & Technology/Barton	
G-AVCX	PA-30 Twin Comanche 160	F. J. Stevens/Leicester	
G-AVCY	PA-30 Twin Comanche 160	Thornhurst & Co/Halfpenny Green	
G-AVDA	Cessna 182K Skylane	J. W. Grant	
G-AVDE	Turner Gyroglider Mk 1	J. S. Smith	
G-AVDF	Beagle Pup 100 ★	Brighton Transport Musuem	
G-AVDG	Wallis WA-116 Srs 1	K. H. Wallis	
G-AVDR	Beech B80 Queen Air	Shoreham Flight Simulation/ Bournemouth	
G-AVDS	Beech B80 Queen Air	Shoreham Flight Simulation/ Bournemouth	
G-AVDT	Aeronca 7AC Champion	D. Cheney & J. G. Woods	
G-AVDV	PA-22 Tri-Pacer 150 (modified to Pacer)	S. C. Brooks/Slinfold	
G-AVDW	D.62B Condor	Essex Aviation/Andrewsfield	
G-AVDY	Luton LA-4A Minor	D. E. Evans & ptnrs	
G-AVDZ	PA-25 Pawnee 235	Apple Aviation/Sibson	
G-AVEB	Morane MS 230 (1076)	Hon P. Lindsay/Booker	
G-AVEC	Cessna F.172H	W. H. Ekin (Engineering) Co Ltd	
G-AVEF	Jodel D.150	Tiger Club Ltd/Redhill	
G-AVEH	SIAI-Marchetti S.205	P. D. Winborne	
G-AVEM	Cessna F.150G	C. P. Osbourne	
G-AVEN	Cessna F.150G	N. J. Rudd/Aberdeen	
G-AVEO	Cessna F.150G	Computaplane Ltd/Glasgow	
G-AVER	Cessna F.150G	B. I. Chapman/Ipswich	
G-AVET	Beech 95-C55A Baron	Bob Crowe Aircraft Sales Ltd	
G-AVEU	Wassmer WA.41 Baladou	Baladou Flying Group/Aberdeen	
G-AVEX	D.62B Condor	Cotswold Roller Hire Ltd/Long Marston	
G-AVEY	Currie Super Wot	A. Eastelow/Dunkeswell	
G-AVEZ	HPR-7 Herald 210 ★	Norwich Aviation Museum	
G-AVFB	H.S.121 Trident 2E ★	Imperial War Museum/Duxford	
G-AVFE	H.S.121 Trident 2E ★	Belfast Airport Authority	

Notes	Reg.	Type	Owner or Operator
	G-AVFG	H.S.121 Trident 2E ★	British Airways/Heathrow
	G-AVFH	H.S.121 Trident 2E ★	Mosquito Aircraft Museum (Fuselage only)
	G-AVFM	H.S.121 Trident 2E ★	Brunel Technical College/Bristol
	G-AVFP	PA-28 Cherokee 140	H. D. Vince Ltd/Woodvale
	G-AVFR	PA-28 Cherokee 140	J. J. Ballagh/Newtownards
	G-AVFS	PA-32 Cherokee Six 300	Headcorn Parachute Club Ltd
	G-AVFU	PA-32 Cherokee Six 300	Couesnon Ltd/Biggin Hill
	G-AVFX	PA-28 Cherokee 140	J. D. Palfreman/Ipswich
	G-AVFY	PA-28 Cherokee 140	Vectaphone Manufacturing Ltd
	G-AVFZ	PA-28 Cherokee 140	R. S. Littlechild & V. B. G. Childs/Exeter
	G-AVGA	PA-24 Comanche 260	W. B. Baillie/Tees-side
	G-AVGB	PA-28 Cherokee 140	D. Jenkins & ptnrs/Swansea
	G-AVGC	PA-28 Cherokee 140	B. A. Bennett/Redhill
	G-AVGD	PA-28 Cherokee 140	A. D. Wren/Southend
	G-AVGE	PA-28 Cherokee 140	A. Dobson/Southend
	G-AVGH	PA-28 Cherokee 140	Avon Aviation Services Ltd/Bristol
	G-AVGI	PA-28 Cherokee 140	F. Cooper
	G-AVGJ	Jodel DR.1050	L. G. J. R. Wallis & J. S. Palmer/ Headcorn
	G-AVGK	PA-28 Cherokee 180	Liverpool Aero Club Ltd
	G-AVGP	BAC One-Eleven 408	British Airways *County of Nottinghamshire*/Birmingham
	G-AVGU	Cessna F.150G	G. R. W. Brown
	G-AVGV	Cessna F.150G	J. P. Lassey
	G-AVGY	Cessna 182K Skylane	K. A. Ellis/Fenland
	G-AVGZ	Jodel DR.1050	D. C. Webb & S. P. Johnson
	G-AVHF	Beech A.23 Musketeer	R. W. Neale/Coventry
	G-AVHH	Cessna F.172H	V. W. Wharton & ptnrs/Goodwood
	G-AVHJ	Wassmer WA.41 Baladou	D. G. Pickering & ptnrs
	G-AVHL	Jodel DR.105A	G. L. Winterbourne/Redhill
	G-AVHM	Cessna F.150G	M. Tosh/Panshanger
	G-AVHN	Cessna F.150G	RGR High Service Ltd
	G-AVHT	Auster AOP.9 (WZ711)	M. Somerton-Rayner/Middle Wallop
	G-AVHY	Fournier RF.4D	R. Swinn & J. Conolly
	G-AVIA	Cessna F.150G	P. N. Voltzenlogel
	G-AVIB	Cessna F.150G	Gt. Consall Copper Mine Co Ltd
	G-AVIC	Cessna F.172H	Pembrokeshire Air/Haverfordwest
	G-AVID	Cessna 182J	T. D. Boyle/Prestwick
	G-AVIE	Cessna F.172H	Red Fir Aviation Ltd/Clacton
	G-AVII	A-B 206A JetRanger	Bristow Helicopters Ltd
	G-AVIL	Alon A.2 Aircoupe	D. W. Vernon/Woodvale
	G-AVIN	M.S.880B Rallye Club	D. R. F. Sapte/Elstree
	G-AVIO	M.S.880B Rallye Club	A. R. Johnston/Popham
	G-AVIP	Brantly B.2B	Cosworth Engineering Ltd
	G-AVIR	Cessna F.172H	F. E. Gooding/Biggin Hill
	G-AVIS	Cessna F.172H	Jon Paul Photography Ltd/Rochester
	G-AVIT	Cessna F.150G	Shropshire Aero Club Ltd/Sleap
	G-AVIZ	Scheibe SF.25A Motorfalke	D. C. Pattison & D. A. Wilson
	G-AVJB	V.815 Viscount	British Air Ferries/Southend
	G-AVJE	Cessna F.150G	P. R. Green & ptnrs/Booker
	G-AVJF	Cessna F.172H	J. A. & G. M. Rees/Haverfordwest
	G-AVJG	Cessna 337B	P. R. Moss/Bournemouth
	G-AVJH	D.62 Condor	Lleyn Flying Group/Mona
	G-AVJI	Cessna F.172H	J. R. Pearce/Compton Abbas
	G-AVJJ	PA-30 Twin Comanche 160	A. H. Manser Ltd/Staverton
	G-AVJK	Jodel DR.1050 M.1	G. Wylde/Manchester
	G-AVJN	Brantly B.2B	B. J. & G. A. Finch
	G-AVJO	Fokker E.III Replica (422-15)	Personal Plane Services Ltd/Booker
	G-AVJU	PA-24 Comanche 260	Syd Ward (South Normanton) Ltd
	G-AVJV	Wallis WA-117 Srs 1	K. H. Wallis (G-ATCV)
	G-AVJW	Wallis WA-118 Srs 2	K. H. Wallis (G-ATPW)
	G-AVKB	MB.50 Pipistrelle	R. A. Fairclough
	G-AVKD	Fournier RF.4D	Lasham RF4 Group
	G-AVKE	Gadfly HDW.1 ★	British Rotorcraft Museum
	G-AVKG	Cessna F.172H	W. Lancs Aero Club Ltd/Woodvale
	G-AVKI	Nipper T.66 Srs 3	J. P. Tribe & K. D. G. Courtney/ Swansea
	G-AVKJ	Nipper T.66 Srs 3	N. M. Yeo/Barton
	G-AVKK	Nipper T.66 Srs 3	C. Watson
	G-AVKN	Cessna 401	Strand Furniture Ltd/E. Midlands
	G-AVKP	A.109 Airedale	H. F. Igoe (*stored*)
	G-AVKR	Bo 208C Junior	P. E. Hinkley/Redhill
	G-AVKY	Hiller UH-12E	Agricopters Ltd/Chilbolton

Reg.	Type	Owner or Operator	Notes
G-AVKZ	PA-23 Aztec 250	Volvo BM (UK) Ltd/Stansted	
G-AVLA	PA-28 Cherokee 140	C. Walker/Perth	
G-AVLB	PA-28 Cherokee 140	J. A. Overton Ltd/Andrewsfield	
G-AVLC	PA-28 Cherokee 140	F. C. V. Hopkins/Swansea	
G-AVLD	PA-28 Cherokee 140	M&E Machinery Ltd/Elstree	
G-AVLE	PA-28 Cherokee 140	P. A. Johnstone & E. J. Morgan	
G-AVLF	PA-28 Cherokee 140	W. London Aero Services Ltd/ White Waltham	
G-AVLG	PA-28 Cherokee 140	R. W. Biggs	
G-AVLH	PA-28 Cherokee 140	T. L. Wilkinson	
G-AVLI	PA-28 Cherokee 140	A. J. Molle & R. P. I. Scott/Ipswich	
G-AVLJ	PA-28 Cherokee 140	E. Berks Boat Company Ltd/White Waltham	
G-AVLM	B.121 Pup 2	R. Towle	
G-AVLN	B.121 Pup 2	R. Watson	
G-AVLO	Bo 208C Junior	J. A. Webb & K. F. Barnard/Popham	
G-AVLP	PA-23 Aztec 250	Survey Data Ltd	
G-AVLR	PA-28 Cherokee 140	Peter Dolan & Co Ltd/Sibson	
G-AVLS	PA-28 Cherokee 140	G. C. Stewart/Humberside	
G-AVLT	PA-28 Cherokee 140	E. A. Clack & M. T. Pritchard/Southend	
G-AVLU	PA-28 Cherokee 140	I. K. George/Fairoaks	
G-AVLW	Fournier RF 4D	P. J. Sellar & B. M. O'Brien/Redhill	
G-AVLY	Jodel D.120A	J. S. Parlour & ptnrs	
G-AVMA	GY.80 Horizon 180	B. R. & S. Hildick	
G-AVMB	D.62B Condor	A. J. Starkey/Fairoaks	
G-AVMD	Cessna 150G	K. J. Jarvis/Southend	
G-AVMF	Cessna F. 150G	J. F. Marsh & M. J. Oliver	
G-AVMH	BAC One-Eleven 510	British Airways *County of Cheshire*/ Manchester	
G-AVMI	BAC One-Eleven 510	British Airways *County of Avon*/ Manchester	
G-AVMJ	BAC One-Eleven 510	British Airways *Strathclyde Region*/ Manchester	
G-AVMK	BAC One-Eleven 510	British Airways *County of Kent*/ Manchester	
G-AVML	BAC One-Eleven 510	British Airways *County of Surrey*/ Manchester	
G-AVMM	BAC One-Eleven 510	British Airways *County of Antrim*/ Manchester	
G-AVMN	BAC One-Eleven 510	British Airways *County of Essex*/ Manchester	
G-AVMO	BAC One-Eleven 510	British Airways *Lothian Region*/ Manchester	
G-AVMP	BAC One-Eleven 510	British Airways *Bailiwick of Jersey*/ Manchester	
G-AVMR	BAC One-Eleven 510	British Airways *County of Tyne & Wear*/ Manchester	
G-AVMS	BAC One-Eleven 510	British Airways *County of West Sussex*/ Manchester	
G-AVMT	BAC One-Eleven 510	British Airways *County of Berkshire*/ Manchester	
G-AVMU	BAC One-Eleven 510	British Airways *County of Dorset*/ Manchester	
G-AVMV	BAC One-Eleven 510	British Airways *Greater Manchester County*/Manchester	
G-AVMW	BAC One-Eleven 510	British Airways *Grampian Region*/ Manchester	
G-AVMX	BAC One-Eleven 510	British Airways *County of East Sussex*/ Manchester	
G-AVMY	BAC One-Eleven 510	British Airways *County of Derbyshire*/ Manchester	
G-AVMZ	BAC One-Eleven 510	British Airways *County of Lancashire*/ Manchester	
G-AVNB	Cessna F.150G	G. A. J. Bowles/Elstree	
G-AVNC	Cessna F.150G	J. Turner	
G-AVNI	PA-30 Twin Comanche 160	D.P. Aviation/Coventry	
G-AVNL	PA-23 Aztec 250	Nalson Aviation Ltd/Biggin Hill	
G-AVNM	PA-28 Cherokee 180	Brands Hatch Circuit Ltd/Biggin Hill	
G-AVNN	PA-28 Cherokee 180	Barum Alloys Ltd	
G-AVNO	PA-28 Cherokee 180	Ninasky Ltd/Southend	
G-AVNP	PA-28 Cherokee 180	W. Easton/Glasgow	
G-AVNR	PA-28 Cherokee 180	L. R. Davies/Biggin Hill	
G-AVNS	PA-28 Cherokee 180	B. J. Haylor & ptnrs/Southampton	
G-AVNU	PA-28 Cherokee 180	F. E. Gooding/Biggin Hill	

Notes	Reg.	Type	Owner or Operator
	G-AVNW	PA-28 Cherokee 180	Len Smith's School & Sports Ltd
	G-AVNX	Fournier RF-4D	O. C. Harris & C. G. Masterman
	G-AVNY	Fournier RF-4D	A. N. Mavrogordato/Biggin Hill
	G-AVNZ	Fournier RF-4D	Aviation Special Developments (ASD) (Redated) Ltd/Biggin Hill
	G-AVOA	Jodel DR.1050	D. A. Willies
	G-AVOD	Beagle D5/180 Husky	D. Bonsall & ptnrs
	G-AVOH	D.62B Condor	J. E. Hobbs/Sandown
	G-AVOM	Jodel DR.221	M. A. Mountford/Headcorn
	G-AVON	Luton LA-5A Major	G. R. Mee
	G-AVOO	PA-18-150 Super Cub	London Gliding Club Ltd/ Dunstable
	G-AVOR	Lockspeiser LDA-01	D. Lockspeiser (*stored*)
	G-AVOZ	PA-28 Cherokee 180	Downley Garages Ltd/Booker
	G-AVPC	D.31 Turbulent	J. Sharp (*stored*)
	G-AVPD	D.9 Bebe	S. W. McKay (*stored*)
	G-AVPE	H.S.125 Srs 3B	British Aerospace/Filton
	G-AVPH	Cessna F.150G	W. Lancashire Aero Club/Woodvale
	G-AVPI	Cessna F.172H	R. Jones & ptnrs/Exeter
	G-AVPJ	D.H.82A Tiger Moth	R. W. Livett/Sywell
	G-AVPK	M.S.892A Rallye Commodore	J. F. & J. M. Gosling
	G-AVPM	Jodel D.117	J. Houghton/Breighton
	G-AVPN	HPR.7 Herald 213	—
	G-AVPO	Hindustan HAL-26 Pushpak	A. & R. Rimington
	G-AVPR	PA-30 Twin Comanche 160	Cold Storage (Jersey) Ltd
	G-AVPS	PA-30 Twin Comanche 160	Russell Foster Holdings Ltd
	G-AVPT	PA-18 Super Cub 150	Tiger Club Ltd/Redhill
	G-AVPV	PA-28 Cherokee 180	A. W. Chapman
	G-AVPX	Taylor JT.1 Monoplane	S. M. Smith/Redhill
	G-AVRF	H.S.125 Srs 3B	Bridgelink Finance Ltd
	G-AVRG	H.S.125 Srs 3B	Hatfield Executive Aviation
	G-AVRK	PA-28 Cherokee 180	S. R. Culley Developments Ltd
	G-AVRN	Boeing 737-204	Britannia Airways Ltd *Capt James Cook*/Luton
	G-AVRP	PA-28 Cherokee 140	K. Cooper & N. D. Douglas/ Halfpenny Green
	G-AVRS	GY.80 Horizon 180	Horizon Flyers Ltd/Booker
	G-AVRT	PA-28 Cherokee 140	F. Clarke/Stapleford
	G-AVRU	PA-28 Cherokee 180	H. B. Holden & ptnrs
	G-AVRW	GY-20 Minicab	R. B. Pybus
	G-AVRY	PA-28 Cherokee 180	Roses Flying Group/Barton
	G-AVRZ	PA-28 Cherokee 180	Briglea Engineering Ltd/Guernsey
	G-AVSA	PA-28 Cherokee 180	Alliance Aviation Ltd/Barton
	G-AVSB	PA-28 Cherokee 180	White House Garage Ashford Ltd
	G-AVSC	PA-28 Cherokee 180	W. London Aero Services Ltd/ White Waltham
	G-AVSD	PA-28 Cherokee 180	Landmate Ltd
	G-AVSE	PA-28 Cherokee 180	Yorkshire Light Aircraft Ltd/Leeds
	G-AVSF	PA-28 Cherokee 180	Goodwood Terrena Ltd
	G-AVSH	PA-28 Cherokee 180	Elken Ltd/Blackpool
	G-AVSI	PA-28 Cherokee 140	W. London Aero Services/White Waltham
	G-AVSP	PA-28 Cherokee 180	Trig Engineering Ltd
	G-AVSR	Beagle D 5/180 Husky	A. L. Young
	G-AVTB	Nipper T.66 Srs 3	N. J. Smith & B. A. Wright
	G-AVTC	Nipper T.66 Srs 3	M. K. Field
	G-AVTJ	PA-32 Cherokee Six 260	K. A. Goodchild/Southend
	G-AVTK	PA-32 Cherokee Six 260	Mannix Aviation Ltd
	G-AVTP	Cessna F.172H	J. H. A. Clarke & ptnrs
	G-AVTT	Ercoupe 415D	Wright's Farm Eggs Ltd/Andrewsfield
	G-AVTV	M.S.893A Rallye Commodore	Crowland Flying Group
	G-AVUA	Cessna F.172H	Recreational Flying Centre (Popham) Ltd
	G-AVUD	PA-30 Twin Comanche 160B	F. M. Aviation/Biggin Hill
	G-AVUG	Cessna F.150H	Skyways Flying Group/Netherthorpe
	G-AVUH	Cessna F.150H	D. M. Leonard/Tees-side
	G-AVUL	Cessna F.172H	D. H. Stephens & D. J. Reason/Elstree
	G-AVUS	PA-28 Cherokee 140	M. J. Bishop
	G-AVUT	PA-28 Cherokee 140	Bencray Ltd/Blackpool
	G-AVUU	PA-28 Cherokee 140	C. H. Dennis/Rochester
	G-AVUZ	PA-32 Cherokee Six 300	Ceesix Ltd/Jersey
	G-AVVB	H.S. 125 Srs 3B	Hitchens (Hatfield) Ltd
	G-AVVC	Cessna F.172H	Kestrel Air Ltd/Swansea
	G-AVVE	Cessna F.150H	R. Windley

Reg.	Type	Owner or Operator	Notes
G-AVVF	D.H.104 Dove 8	Sparline Agencies Ltd	
G-AVVI	PA-30 Twin Comanche 160B	Express Flight Ltd/Southend	
G-AVVJ	M.S.893A Rallye Commodore	D. E. Schofield	
G-AVVL	Cessna F.150H	Osprey Air Services Ltd/Cranfield	
G-AVVN	D.62C Condor	Avato Flying Group	
G-AVVO	Avro 652A Anson 19 (VL348) ★	Newark Air Museum	
G-AVVS	Hughes 269B	W. Holmes	
G-AVVT	PA-23 Aztec 250	Kayglynn Investment Holdings Ltd	
G-AVVV	PA-28 Cherokee 180	D. F. Field/Goodwood	
G-AVVX	Cessna F.150H	Hatfield Flying Club	
G-AVVY	Cessna F.150H	Scotia Safari Ltd/Prestwick	
G-AVWA	PA-28 Cherokee 140	W. London Aero Services Ltd/ White Waltham	
G-AVWD	PA-28 Cherokee 140	Telepoint Ltd	
G-AVWE	PA-28 Cherokee 140	W. C. C. Meyer/Biggin Hill	
G-AVWG	PA-28 Cherokee 140	Bencray Ltd/Blackpool	
G-AVWH	PA-28 Cherokee 140	P. Elliott/Biggin Hill	
G-AVWI	PA-28 Cherokee 140	L. M. Veitch	
G-AVWJ	PA-28 Cherokee 140	M. J. Steer/Biggin Hill	
G-AVWL	PA-28 Cherokee 140	Bristol & Wessex Aeroplane Club Ltd/ Bristol	
G-AVWM	PA-28 Cherokee 140	Southend Flying Club	
G-AVWN	PA-28R Cherokee Arrow 180	M. J. White & ptnrs	
G-AVWO	PA-28R Cherokee Arrow 180	C & S Controls Ltd/Biggin Hill	
G-AVWR	PA-28R Cherokee Arrow 180	D. J. Cooper/Netherthorpe	
G-AVWT	PA-28R Cherokee Arrow 180	G. W. Barker & ptnrs/Leeds	
G-AVWU	PA-28R Cherokee Arrow 180	Horizon Flyers Ltd/Denham	
G-AVWV	PA-28R Cherokee Arrow 180	Mapair Ltd/Birmingham	
G-AVWY	Fournier RF-4D	ASD Formaero Ltd	
G-AVXA	PA-25 Pawnee 235	Agricultural Aerial Services Ltd	
G-AVXB	Lovegrove PL-1 gyrocopter	A. Stone	
G-AVXC	Nipper T.66 Srs 3	W. G. Wells & ptnrs	
G-AVXD	Nipper T.66 Srs 3	C. Watson	
G-AVXF	PA-28R Cherokee-Arrow 180	J. G. Stewart & I. M. S. Ferriman/ Cranfield	
G-AVXI	H.S.748 Srs 2A	Civil Aviation Authority/Stansted	
G-AVXJ	H.S.748 Srs 2A	Civil Aviation Authority/Stansted	
G-AVXV	Bleriot XI (BAPC 104) ★	Museum/RAF St Athan	
G-AVXW	D.62B Condor	M. D. Bailey/Rochester	
G-AVXX	Cessna FR.172E	Hadrian Flying Group/Newcastle	
G-AVXY	Auster AOP.9 (XK417)	R. Windley	
G-AVXZ	PA-28 Cherokee 140 ★	ATC Hayle (*instructional airframe*)	
G-AVYE	H.S.121 Trident 1E-140 ★	Science Museum/Wroughton	
G-AVYK	A.61 Terrier 3	G. J. Busby	
G-AVYL	PA-28 Cherokee 180	N. J. Allcoat	
G-AVYM	PA-28 Cherokee 180	Carlisle Aviation Co Ltd/Crosby	
G-AVYP	PA-28 Cherokee 140	T. D. Reid (Braids) Ltd	
G-AVYR	PA-28 Cherokee 140	D.R. Flying Club Ltd/Staverton	
G-AVYS	PA-28R Cherokee Arrow 180	E. W. Passmore	
G-AVYT	PA-28R Cherokee Arrow 180	H. Stephenson/Tees-side	
G-AVYV	Jodel D.120	Long Mountain Aero Group	
G-AVYX	AB-206A JetRanger	S.W. Electricity Board/Bristol	
G-AVZA	IMCO Callair A-9	Arable & Bulb Chemicals Ltd	
G-AVZB	Aero Z-37 Cmelak	ADS (Aerial) Ltd/Southend	
G-AVZC	Hughes 269B	Lake Farms	
G-AVZE	D.62B Condor	A. J. M. Trowbridge & J. Harris	
G-AVZI	Bo 208C Junior	C. F. Rogers	
G-AVZM	Beagle B.121 Pup 1	ARAZ Group/Elstree	
G-AVZN	Beagle B.121 Pup 1	W. E. Cro & Sons Ltd/Shoreham	
G-AVZP	Beagle B.121 Pup 1	T. A. White	
G-AVZR	PA-28 Cherokee 180	W. E. Lowe/Shobdon	
G-AVZU	Cessna F.150H	E. J. R. McDowell	
G-AVZV	Cessna F.172H	Bencray Ltd/Blackpool	
G-AVZW	EAA Model P Biplane	R. G. Maidment & G. R. Edmundson/ Goodwood	
G-AVZX	M.S.880B Rallye Club	M. Atkins & G. Harris	
G-AWAA	M.S.880B Rallye Club	P. A. Cairns/Dunkeswell	
G-AWAC	GY-80 Horizon 180	Applied Signs Ltd	
G-AWAD	Beech D 55 Baron	Aero Lease Ltd/Bournemouth	
G-AWAH	Beech D 55 Baron	B. J. S. Grey/Booker	
G-AWAI	Beech D 55 Baron	Alibear Ltd/Booker	
G-AWAJ	Beech D 55 Baron	Standard Hose Ltd	
G-AWAT	D.62B Condor	Tarwood Ltd/Redhill	

Notes	*Reg.*	*Type*	*Owner or Operator*
	G-AWAU	Vickers F.B.27A Vimy (replica) (F8614) ★	Bomber Command Museum/Hendon
	G-AWAV	Cessna F.150F	Ipswich School of Flying
	G-AWAW	Cessna F.150F	G. J. Charlton
	G-AWAX	Cessna 150D	Cambridge Technical Developments (Leasing) Ltd
	G-AWAZ	PA-28R Cherokee Arrow 180	C. J. Chapman
	G-AWBA	PA-28R Cherokee Arrow 180	March Flying Group/Andrewsfield
	G-AWBB	PA-28R Cherokee Arrow 180	Brian Neale Ltd/Sibson
	G-AWBC	PA-28R Cherokee Arrow 180	G. K. Furneaux/Fairoaks
	G-AWBE	PA-28 Cherokee 140	Gloucester & Cheltenham Aviation Services Flying Club Ltd/Staverton
	G-AWBG	PA-28 Cherokee 140	C. S. & B. A. Frost
	G-AWBH	PA-28 Cherokee 140	R. C. A. Mackworth
	G-AWBJ	Fournier RF-4D	The BJ Group/Plymouth
	G-AWBL	BAC One-Eleven 416	British Airways *County of Leicestershire*/Birmingham
	G-AWBM	D.31 Turbulent	J. T. S. Lewis
	G-AWBN	PA-30 Twin Comanche 160	Stourfield Investments Ltd/Jersey
	G-AWBP	Cessna 182L Skylane	N. Y. Souster/Southampton
	G-AWBS	PA-28 Cherokee 140	W. London Aero Services Ltd/ White Waltham
	G-AWBT	PA-30 Twin Comanche 160	R. M. English & Son Ltd
	G-AWBU	Morane-Saulnier N (replica) (M.S.50)	Personal Plane Services Ltd/Booker
	G-AWBV	Cessna 182L Skylane	Hunting Surveys & Consultants Ltd/ Manchester
	G-AWBW	Cessna F.172H ★	Brunel Technical College/Bristol
	G-AWBX	Cessna F.150H	D. F. Ranger & ptnrs
	G-AWCD	CEA DR.253	D. H. Smith
	G-AWCH	Cessna F.172H	M. Bua/Bournemouth
	G-AWCJ	Cessna F.150H	Transknight Ltd/Hinton-in-the-Hedges
	G-AWCL	Cessna F.150H	Signtest Ltd
	G-AWCM	Cessna F.150H	R. J. Jackson
	G-AWCN	Cessna FR.172E	LEC Refrigeration Ltd
	G-AWCP	Cessna F.150H (tailwheel)	Herefordshire Aero Club Ltd/Shobdon
	G-AWCW	Beech E.95 Travel Air	H. W. Astor/White Waltham
	G-AWCY	PA-32 Cherokee Six 260	J. Harvey
	G-AWDA	Nipper T.66 Srs. 3	D. F. Lea
	G-AWDD	Nipper T.66 Srs. 3	T. D. G. Roberts/Inverness
	G-AWDI	PA-23 Aztec 250	Air Foyle Ltd/Luton
	G-AWDO	D.31 Turbulent	R. Watling-Greenwood
	G-AWDP	PA-28 Cherokee 180	Brian Ilston Ltd/Shipdham
	G-AWDR	Cessna FR.172E	Levendene Ltd
	G-AWDU	Brantly B.2B	S. N. Cole
	G-AWEF	Stampe SV-4B	Tiger Club Ltd/Redhill
	G-AWEG	Cessna 172G	H. Lawson
	G-AWEI	D.62B Condor	M. A. Pearce/Redhill
	G-AWEL	Fournier RF.4D	A. B. Clymo/Halfpenny Green
	G-AWEM	Fournier RF.4D	B. J. Griffin/Wickenby
	G-AWEN	Jodel DR.1050	L. G. Earnshaw & ptnrs
	G-AWEO	Cessna F.150H	Banbury Plant Hire Ltd/Wellesbourne
	G-AWEP	JB-01 Minicab	S. E. Bond/Barton
	G-AWER	PA-23 Aztec 250	Woodgate Air Services (IOM) Ltd
	G-AWES	Cessna 150H	Ralair Ltd/Blackpool
	G-AWET	PA-28 Cherokee 180	Broadland Flying Group Ltd/ Swanton Morley
	G-AWEV	PA-28 Cherokee 140	Smith Engineering/Shipdham
	G-AWEX	PA-28 Cherokee 140	R. V. Bowles/Coventry
	G-AWEZ	PA-28R Cherokee Arrow 180	BTG Plant Hire & Repairs Ltd/Stapleford
	G-AWFB	PA-28R Cherokee Arrow 180	Luke Aviation Ltd/Bristol
	G-AWFC	PA-28R Cherokee Arrow 180	K. A. Goodchild/Southend
	G-AWFD	PA-28R Cherokee Arrow 180	B. Knight & A. J. Harvey
	G-AWFF	Cessna F.150H	Keatlord Ltd
	G-AWFJ	PA-28R Cherokee Arrow 180	R. Watt
	G-AWFK	PA-28R Cherokee Arrow 180	J. A. Rundle (Holdings) Ltd/Kidlington
	G-AWFN	D.62B Condor	A. F. Bullock
	G-AWFO	D.62B Condor	T. A. Major
	G-AWFP	D.62B Condor	Blackbushe Flying Club
	G-AWFT	Jodel D.9 Bebe	W. H. Cole
	G-AWFW	Jodel D.117	F. H. Greenwell
	G-AWFX	Sikorsky S-61N	British Airways Helicopters Ltd/ Aberdeen
	G-AWFZ	Beech A23 Musketeer	R. Sweet & B D. Corbett

Reg.	Type	Owner or Operator	Notes
G-AWGA	A.109 Airedale	RAFGSA/Bicester	
G-AWGD	Cessna F.172H	B. J. M. Vermilio	
G-AWGJ	Cessna F.172H	J. & C. J. Freeman/Headcorn	
G-AWGK	Cessna F.150H	Aerial Flying Group/Shoreham	
G-AWGL	Bensen B.8	M. H. J. Goldring	
G-AWGM	Mitchell-Procter Kittiwake 2	RAF Halton Flying Club Ltd	
G-AWGN	Fournier RF.4D	R. H. Ashforth/Staverton	
G-AWGR	Cessna F.172H	P. Bushell/Liverpool	
G-AWGZ	Taylor JT.1 Monoplane	A. Hill	
G-AWHB	CASA 2.111 (6J+PR) ★	P. Raymond	
G-AWHV	Rollason Beta B.2A	Tiger Club Ltd/Redhill	
G-AWHW	Rollason Beta B.2A	C. E. Bellhouse/Redhill	
G-AWHX	Rollason Beta B.2	J. J. Cooke/White Waltham	
G-AWII	V.S.349 Spitfire VC (AR501)	Shuttleworth Trust/Duxford	
G-AWIN	Campbell-Bensen B.8MC	M. J. Cuttel & J. Deane	
G-AWIO	Brantly B.2B	G. J. Ward & ptnrs	
G-AWIP	Luton LA-4A Minor	J. Houghton/Kinloss	
G-AWIR	Midget Mustang	K. E. Sword/Leicester	
G-AWIT	PA-28 Cherokee 180	H.M. Air Ltd	
G-AWIV	Storey TSR.3	C. J. Jesson/Redhill	
G-AWIW	Stampe SV.4B ★	Aerospace Museum/Cosford	
G-AWIY	PA-23 Aztec 250	Queen's University of Belfast	
G-AWJA	Cessna 182L Skylane	D. Penny	
G-AWJC	Brighton gas balloon	P. D. Furlong *Slippery William*	
G-AWJE	Nipper T.66 Srs. 3	N. McArthur & T. Mosedale	
G-AWJF	Nipper T.66 Srs. 3	R. Wilcock/Shoreham	
G-AWJI	M.S.880B Rallye Club	D. V. Tyler/Ipswich	
G-AWJO	Tigercraft Tiger Mk. II	K. Aziz	
G-AWJV	D.H.98 Mosquito TT Mk.35 (TA634) ★	Mosquito Aircraft Museum	
G-AWJX	Z.526 Akrobat	Aerobatics International Ltd	
G-AWJY	Z.526 Akrobat	Elco Manufacturing Co/Kidlington	
G-AWKB	M.J.5 Sirocco F2/39	G. D. Claxton	
G-AWKD	PA-17 Vagabond	A. T. & Mrs M. R. Dowie/ White Waltham	
G-AWKM	B.121 Pup 1	D. M. G. Jenkins/Swansea	
G-AWKO	B.121 Pup 1	Golden Lion Flying Group/Shoreham	
G-AWKP	Jodel DR.253	R. C. Chandless	
G-AWKT	M.S.880B Rallye Club	D. C. Strain	
G-AWKX	Beech A65 Queen Air	Westgate Shipping Ltd	
G-AWLA	Cessna F.150H	S. R. Cameron/Cranfield	
G-AWLE	Cessna F.172H	H. Mendelssohn & H. I. Shott	
G-AWLF	Cessna F.172H	Burbage Farms Ltd/Coventry	
G-AWLG	SIPA 903	S. W. Markham	
G-AWLI	PA-22 Tri-Pacer 150	S. J. Saggers/Biggin Hill	
G-AWLJ	Cessna F.150H	D. S. Watts/Southend	
G-AWLL	AB-206B JetRanger 2	Helicare Ltd/Liverpool	
G-AWLM	Bensen B.8MS	C. J. E. Ashby	
G-AWLO	Boeing N2S-5 Kaydet	Warbirds of GB	
G-AWLP	Mooney M.20F	Siminco Ltd	
G-AWLR	Nipper T. 66 Srs. 3	J. D. Lawther	
G-AWLS	Nipper T. 66 Srs. 3	Stapleford Nipper Group	
G-AWLY	Cessna F.150H	Banbury Plant Hire Ltd/Wellesbourne	
G-AWLZ	Fournier RF.4D	E. V. Goodwin & C. R. Williamson	
G-AWMC	Campbell-Bensen B.8MS	M. E. Sykes-Hankinson	
G-AWMD	Jodel D.11	K. Dawson	
G-AWMF	PA-18-150 Super Cub	Airways Aero Associations Ltd/Booker	
G-AWMI	Glos-Airtourer 115	Red Dragon Aviation Ltd/Cardiff	
G-AWMK	AB-206B JetRanger	Bristow Helicopters Ltd	
G-AWMM	M.S.893A Rallye Commodore 180	Tug 83 Group/Perranporth	
G-AWMN	Luton LA-4A Minor	R. E. R. Wilks	
G-AWMP	Cessna F.172H	W. Rennie-Roberts/Ipswich	
G-AWMR	D.31 Turbulent	P. R. M. Bowlan	
G-AWMT	Cessna F.150H	R. V. Grocott/Sleap	
G-AWMU	Cessna F.172H	H. M. Jackson & B. D. Jones/ Coventry	
G-AWMZ	Cessna F.172H ★	*Parachute jump trainer*/Cark	
G-AWNA	Boeing 747-136	British Airways *City of Peterborough*/ Heathrow	
G-AWNB	Boeing 747-136	British Airways *City of Newcastle*/ Heathrow	
G-AWNC	Boeing 747-136	British Airways *City of Belfast*/ Heathrow	

Notes	Reg.	Type	Owner or Operator
	G-AWND	Boeing 747-136	British Airways *City of Leeds*/ Heathrow
	G-AWNE	Boeing 747-136	British Airways *City of Southampton*/ Heathrow
	G-AWNF	Boeing 747-136	British Airways *City of Westminster*/ Heathrow
	G-AWNG	Boeing 747-136	British Airways *City of London*/ Heathrow
	G-AWNH	Boeing 747-136	British Airways *City of Inverness*/ Heathrow
	G-AWNJ	Boeing 747-136	British Airways *City of Sheffield*/ Heathrow
	G-AWNL	Boeing 747-136	British Airways *City of Nottingham*/ Heathrow
	G-AWNM	Boeing 747-136	British Airways *City of Bristol*/Heathrow
	G-AWNN	Boeing 747-136	British Airways *City of Leicester*/ Heathrow
	G-AWNO	Boeing 747-136	British Airways *City of Durham*/ Heathrow
	G-AWNP	Boeing 747-136	British Airways *City of Portsmouth*/ Heathrow
	G-AWNT	BN-2A Islander	Peterborough Parachute Centre/Sibson
	G-AWOA	M.S.880B Rallye Club	W. E. Taylor & Son Ltd
	G-AWOE	Aero Commander 680E	J. M. Houlder/Elstree
	G-AWOF	PA-15 Vagabond	R. S. O. B. Evans
	G-AWOH	PA-17 Vagabond	K. M Bowen
	G-AWOL	Bell 206B JetRanger 2	Tayshaw Ltd
	G-AWOT	Cessna F.150H	J. M. Montgomerie & J. Ferguson
	G-AWOU	Cessna 170B	Parker Airways Ltd/Denham
	G-AWPH	P.56 Provost T.1	J. A. D. Bradshaw
	G-AWPJ	Cessna F.150H	W. J. Greenfield
	G-AWPN	Shield Xyla	T. Worrall/Finmere
	G-AWPP	Cessna F.150H	D. Williams
	G-AWPS	PA-28 Cherokee 140	Jakecourt Ltd/Bournemouth
	G-AWPU	Cessna F.150J	Light Planes (Lancashire) Ltd/Barton
	G-AWPW	PA-12 Super Cruiser	T. S. Warren & ptnrs/Sandown
	G-AWPX	Cessna 150E	W. R. Emberton/Southampton
	G-AWPY	Bensen B.8M	J. Jordan
	G-AWPZ	Andreasson BA-4B	J. M. Venine
	G-AWRK	Cessna F.150J	V. C. Pierce & A. R. Clark/Shoreham
	G-AWRL	Cessna F.172H	R. D. Snow
	G-AWRS	Avro 19 Srs. 2 ★	N. E. Aircraft Museum
	G-AWRY	P.56 Provost T.1 (XF836)	Slymar Aviation & Services Ltd/ Thruxton
	G-AWRZ	Bell 47G-5	Hammond Aerial Spraying Ltd
	G-AWSA	Avro 652A Anson 19 (VL349) ★	Norfolk & Suffolk Aviation Museum
	G-AWSD	Cessna F.150J	A. C. Byrne/Norwich
	G-AWSL	PA-28 Cherokee 180D	Fascia Ltd/Southend
	G-AWSM	PA-28 Cherokee 235	Firecrest Aviation Ltd/Leavesden
	G-AWSN	D.62B Condor	J. Leader
	G-AWSP	D.62B Condor	R. Q. & A. S. Bond/Enstone
	G-AWSS	D.62C Condor	G. Bruce
	G-AWST	D.62B Condor	Humberside Aviation/Cranwell North
	G-AWSV	Skeeter 12 (XM553)	Maj. M. Somerton-Rayner/ Middle Wallop
	G-AWSY	Boeing 737-204	Britannia Airways Ltd *General James Wolfe*/Luton
	G-AWSZ	M.S.894A Rallye Minerva 220	D. Quinn & J. McCloskey
	G-AWTA	Cessna E.310N	A. H. Wiltshire/Fairoaks
	G-AWTJ	Cessna F.150J	Metropolitan Police Flying Club/ Biggin Hill
	G-AWTL	PA-28 Cherokee 180D	S. J. Lambden/Southend
	G-AWTM	PA-28 Cherokee 140	Keenair Services Ltd/Liverpool
	G-AWTR	Beech A.23 Musketeer	J. & P. Donoher
	G-AWTS	Beech A.23 Musketeer	M. Corbett/Blackbushe
	G-AWTV	Beech A.23 Musketeer	A. Johnston
	G-AWTW	Beech B.55 Baron	Ingham Aviation Ltd/Bristol
	G-AWTX	Cessna F.150J	Tango X-Ray Flying Club/Andrewsfield
	G-AWUA	Cessna P.206D	Balmar Aviation/Thruxton
	G-AWUB	GY.201-Minicab	H. P. Burrill
	G-AWUE	Jodel DR.1050	S. Bichan
	G-AWUG	Cessna F.150H	P. P. D. Howard-Johnstone/Edinburgh
	G-AWUH	Cessna F.150H	M. J. Passingham
	G-AWUJ	Cessna F.150H	W. Lawton

Reg.	Type	Owner or Operator	Notes
G-AWUL	Cessna F.150H	Scotia Safari Ltd/Prestwick	
G-AWUN	Cessna F.150H	Northamptonshire School of Flying Ltd/ Sywell	
G-AWUO	Cessna F.150H	K. A. Ellis/Sibson	
G-AWUP	Cessna F.150H	R. H. Timmis	
G-AWUS	Cessna F.150J	Recreational Flying Centre (Popham) Ltd	
G-AWUT	Cessna F.150J	L. Salmon/Tollerton	
G-AWUW	Cessna F.172H	T. A. Holding/Clacton	
G-AWUX	Cessna F.172H	J. D. A. Shields & ptnrs/Lydd	
G-AWUZ	Cessna F.172H	March Microwave Ltd	
G-AWVA	Cessna F.172H	C. F. Bishop	
G-AWVB	Jodel D.117	C. M. & T. R. C. Griffin	
G-AWVC	B.121 Pup 1	S. W. Bates	
G-AWVE	Jodel DR.1050M.I	E. A. Taylor/Southend	
G-AWVF	P.56 Provost T.1 (XF877)	J. Harper/White Waltham	
G-AWVG	AESL Airtourer T.2	P. Fabish/Cambridge	
G-AWVK	H.P.137 Jetstream	Decca Navigator Co Ltd/Biggin Hill	
G-AWVN	Aeronca 7AC Champion	Bowker Air Services Ltd/Rush Green	
G-AWVZ	Jodel D.112	D. C. Stokes/Dunkeswell	
G-AWWE	B.121 Pup 2	J. M. Randle/Coventry	
G-AWWF	B.121 Pup 1	Air South/Shoreham	
G-AWWI	Jodel D.117	R. L. Sambell/Coventry	
G-AWWM	GY-201 Minicab	J. S. Brayshaw	
G-AWWN	Jodel DR.1051	T. W. M. Beck & ptnrs/Shoreham	
G-AWWO	Jodel DR.1050	Whiskey Oscar Group/Barton	
G-AWWP	Woody Pusher III	M. S. Bird & Mrs R. D. Bird	
G-AWWT	D.31 Turbulent	M. A. Sherry & J. Tring/Redhill	
G-AWWU	Cessna FR.172F	B. W. Wells/Coventry	
G-AWWW	Cessna 401	Westair Flying Services Ltd/Blackpool	
G-AWWX	BAC One-Eleven 509	Dan-Air Services Ltd/Gatwick	
G-AWWZ	BAC One-Eleven 509	British Island Airways Ltd/Gatwick	
G-AWXO	H.S.125 Srs. 400B	Alkharafi Aviation Ltd/Kuwait	
G-AWXR	PA-28 Cherokee 180D	J. D. Williams	
G-AWXS	PA-28 Cherokee 180D	Rayhenro Flying Group/Shobdon	
G-AWXU	Cessna F.150J	B. B. Burtenshaw & ptnrs	
G-AWXV	Cessna F.172H	G. R. V. Haynes/Blackpool	
G-AWXX	Wessex Mk, 60 Srs. 1	Sykes Group Co Ltd/Bournemouth	
G-AWXY	M.S.885 Super Rallye	Arrow Group	
G-AWXZ	SNCAN SV-4C (D88)	Personal Plane Services Ltd/Booker	
G-AWYB	Cessna FR.172F	C. W. Larkin/Southend	
G-AWYE	H.S.125 Srs 1B	Rolls-Royce Ltd/Filton	
G-AWYF	G.159 Gulfstream 1	Ford Motor Co Ltd/Stansted	
G-AWYJ	B.121 Pup 2	H. C. Taylor/Southampton	
G-AWYL	Jodel DR.253B	Clarville Ltd/Headcorn	
G-AWYO	B.121 Pup 1	B. R. C. Wild/Popham	
G-AWYR	BAC One-Eleven 501	British Caledonian Airways *Isle of Tiree*/ Gatwick	
G-AWYS	BAC One-Eleven 501	British Caledonian Airways *Isle of Bute*/ Gatwick	
G-AWYT	BAC One-Eleven 501	British Caledonian Airways *Isle of Barra*/ Gatwick	
G-AWYU	BAC One-Eleven 501	British Caledonian Airways *Isle of Colonsay*/Gatwick	
G-AWYV	BAC One-Eleven 501	British Caledonian Airways *Isle of Harris*/Gatwick	
G-AWYX	M.S.880B Rallye Club	Joy M. L. Edwards/Thruxton	
G-AWYY	T.57 Camel replica (C1701)	Fleet Air Arm Museum/Yeovilton	
G-AWZE	H.S.121 Trident 3B ★	*Instructional airframe*/Heathrow	
G-AWZH	H.S.121 Trident 3B	*Withdrawn*/Heathrow	
G-AWZI	H.S.121 Trident 3B	*Withdrawn*/Heathrow	
G-AWZJ	H.S.121 Trident 3B	*Withdrawn*/Heathrow	
G-AWZK	H.S.121 Trident 3B	*Withdrawn*/Heathrow	
G-AWZM	H.S.121 Trident 3B	*Withdrawn*/Heathrow	
G-AWZN	H.S.121 Trident 3B	*Withdrawn*/Heathrow	
G-AWZO	H.S.121 Trident 3B	*Withdrawn*/Heathrow	
G-AWZP	H.S.121 Trident 3B	*Withdrawn*/Heathrow	
G-AWZR	H.S.121 Trident 3B	*Withdrawn*/Heathrow	
G-AWZS	H.S.121 Trident 3B	*Withdrawn*/Heathrow	
G-AWZU	H.S.121 Trident 3B	*Withdrawn*/Heathrow	
G-AWZV	H.S.121 Trident 3B	*Withdrawn*/Heathrow	
G-AWZX	H.S.121 Trident 3B ★	BAA Fire Services/Gatwick	
G-AWZZ	H.S.121 Trident 3B ★	Airport Fire Services/Birmingham	

Notes	Reg.	Type	Owner or Operator
	G-AXAB	PA-28 Cherokee 140	Bencray Ltd/Blackpool
	G-AXAK	M.S.880B Rallye Club	R. L. & Mrs C. Stewart
	G-AXAN	D.H.82A Tiger Moth (EM720)	Comarket Aviation Ltd
	G-AXAO	Omega 56 balloon	P. D. Furlong
	G-AXAS	Wallis WA-116T	K. H. Wallis (G-AVDH)
	G-AXAT	Jodel D.117A	J. F. Barber/Southend
	G-AXAU	PA-30 Twin Comanche 160C	Bartcourt Ltd/Southampton
	G-AXAV	PA-30 Twin Comanche 160C	P. S. King/Guernsey
	G-AXAW	Cessna 421A	Translift (UK) Ltd/Stansted
	G-AXAX	PA-23 Aztec 250D	M. J. Deedman/Shoreham
	G-AXBB	BAC One-Eleven 409	British Island Airways *Island Entente*/ Gatwick
	G-AXBF	Beagle D.5/180 Husky	C. H. Barnes
	G-AXBH	Cessna F.172H	P. H. Joinson/Shoreham
	G-AXBJ	Cessna F.172H	K. M. Brennan & C. Mackay/Leicester
	G-AXBW	DH.82A Tiger Moth	R. Venning
	G-AXBZ	D.H.82A Tiger Moth	D. H. McWhir
	G-AXCA	PA-28R Cherokee Arrow 200	J. S. Malcolm/Halfpenny Green
	G-AXCC	Bell 47G-2	P. Lancaster
	G-AXCG	Jodel D.117	P. H. A. Dupon
	G-AXCI	Bensen B.8M ★	Loughborough & Leicester Aircraft Museum
	G-AXCL	M.S.880B Rallye Club	Long Marston Flying Group
	G-AXCM	M.S.880B Rallye Club	Avcom Developments Ltd/Denham
	G-AXCN	M.S.880B Rallye Club	A. Chester
	G-AXCP	BAC One-Eleven 401	Dan Air Services Ltd/Gatwick
	G-AXCX	B.121 Pup 2	R. S. Blackman/Elstree
	G-AXCY	Jodel D.117	J. Gillespie
	G-AXDB	Piper L-4B Cub	N. D. Norman/Bembridge
	G-AXDC	PA-23 Aztec 250D	Trago Mills (South Devon) Ltd/Bodmin
	G-AXDE	Bensen B.8	T. J. Hartwell
	G-AXDH	BN-2A Islander	Parachute Regiment Freefall Club
	G-AXDI	Cessna F.172H	Klingair Ltd/Conington
	G-AXDK	Jodel DR.315	P. J. Checketts & T. J. Thomas
	G-AXDL	PA-30 Twin Comanche 160C	Northern Executive Aviation Ltd/ Manchester
	G-AXDM	H.S.125 Srs 400B	Ferranti Ltd/Edinburgh
	G-AXDN	BAC-Sud Concorde 01 ★	Duxford Aviation Soc
	G-AXDU	B.121 Pup 2	R. Wilson/Cambridge
	G-AXDV	B.121 Pup 1	C. N. G. Hobbs & J. J. Teagle
	G-AXDW	B.121 Pup 1	Cranfield Institute of Technology
	G-AXDY	Falconar F-II	J. Nunn
	G-AXDZ	Cassutt Racer Srs IIIM	A. Chadwick/Little Staughton
	G-AXEB	Cassutt Racer Srs IIIM	G. E. Horder/Denham
	G-AXEC	Cessna 182M	H. S. Mulligan
	G-AXED	PA-25 Pawnee 235	T. G. R. Oglesby
	G-AXEG	H.S.125 Srs 3B	British Aerospace PLC/Hatfield
	G-AXEH	B.121 Pup ★	Museum of Flight/E. Fortune
	G-AXEI	Ward Gnome ★	Lincolnshire Aviation Museum
	G-AXEO	Scheibe SF.25B Falke	D. Collinson
	G-AXES	B.121 Pup 2	D. A. Lowe & A. Molesworth/Nairobi
	G-AXEV	B.121 Pup 2	B. Richardson/Southend
	G-AXEX	B.121 Pup 1	Lubair (Transport Services) Ltd/ E. Midlands
	G-AXFD	PA-25 Pawnee 235	J.E.F. Aviation Ltd
	G-AXFE	Beech B.90 King Air	T. S. Grimshaw Ltd/Cardiff
	G-AXFH	D.H.114 Heron 1B/C	Topflight Aviation Co Ltd/Fairoaks
	G-AXFN	Jodel D.119	D. C. Barber & B. Sleddon
	G-AXGA	PA-19 Super Cub 95	Felthorpe Flying Group Ltd
	G-AXGC	M.S.880B Rallye Club	Ian Richard Transport Services Ltd
	G-AXGE	M.S.880B Rallye Club	R. P. Loxton
	G-AXGG	Cessna F.150J	A. R. Nicholls
	G-AXGP	Piper L-4B Cub	W. K. Butler
	G-AXGR	Luton LA-4A Minor	T. M. W. Webster/Long Marston
	G-AXGS	D.62B Condor	Tiger Club Ltd/Redhill
	G-AXGT	D.62B Condor	P. J. Simpson & ptnrs
	G-AXGU	D.62B Condor	Tiger Club Ltd/Redhill
	G-AXGV	D.62B Condor	R. J. Wrixon/Dunkeswell
	G-AXGZ	D.62B Condor	Lincoln Condor Group/Sturgate
	G-AXHA	Cessna 337A	S. E. Fellows/Southampton
	G-AXHC	Stampe SV-4C	BLS Aviation Ltd/Denham
	G-AXHE	BN-2A Islander	NW Parachute Centre/Blackpool
	G-AXHG	M.S.880B Rallye Club	The Rallye Groupe/Tees-side
	G-AXHI	M.S.880B Rallye Club	J. M. Whittard

Reg.	Type	Owner or Operator	Notes
G-AXHO	B.121 Pup 2	L. W. Grundy/Clacton	
G-AXHP	Piper L-4H Cub	R. Giles	
G-AXHR	Piper L-4H Cub (329601)	M. G. H. Slater	
G-AXHS	M.S.880B Rallye Club	D. Horne & ptnrs	
G-AXHT	M.S.880B Rallye Club	J. L. Osbourne & A. M. Sutton	
G-AXHV	Jodel D.117A	D. M. Cashmore & K. R. Payne	
G-AXHX	M.S.892A Rallye Commodore	D. W. Weever	
G-AXIA	B.121 Pup 1	Cranfield Institute of Technology	
G-AXIE	B.121 Pup 2	I. J. Ross & D. I. McBride	
G-AXIF	B.121 Pup 2	T. G. Hiscock/Elstree	
G-AXIG	B.125 Bulldog 104	George House (Holdings) Ltd	
G-AXIO	PA-28 Cherokee 140B	W. London Aero Services Ltd/ White Waltham	
G-AXIR	PA-28 Cherokee 140B	A. H. Canvin/Weston Zoyland	
G-AXIT	M.S.893A Rallye Commodore 180	South Wales Gliding Club Ltd	
G-AXIW	Scheibe SF.25B Falke	Kent Motor Gliding & Soaring Centre/ Manston	
G-AXIX	Glos-Airtourer 150	G. Thomas & M. Mann/Shotteswell	
G-AXIY	Bird Gyrocopter	E. N. Grace	
G-AXJB	Omega 84 balloon	Hot-Air Group *Jester*	
G-AXJH	B.121 Pup 2	J. S. Chillingworth	
G-AXJI	B.121 Pup 2	Cole Aviation Ltd	
G-AXJJ	B.121 Pup 2	The B.U.M.P.F. Group/Crosland Moor	
G-AXJK	BAC One-Eleven 501	British Caledonian Airways *Isle of Staffa*/Gatwick	
G-AXJM	BAC One-Eleven 501	British Caledonian Airways *Isle of Islay*/ Gatwick	
G-AXJN	B.121 Pup 2	J. B. Goodrich & D. M. Jenkins/ Shoreham	
G-AXJO	B.121 Pup 2	J. A. D. Bradshaw	
G-AXJR	Scheibe SF.25B Falke	D. R. Chatterton/Staverton	
G-AXJV	PA-28 Cherokee 140B	Mona Aviation Ltd	
G-AXJW	PA-28 Cherokee 140B	D. L. Claydon	
G-AXJX	PA-28 Cherokee 140B	MSF Aviation Ltd/Manchester	
G-AXJY	Cessna U-206D	Hereford Parachute Club Ltd/Shobdon	
G-AXKD	PA-23 Aztec 250D	Jones & Bailey Contractors Ltd/ Glenrothes	
G-AXKH	Luton LA-4A Minor	M. E. Vaisey	
G-AXKI	Jodel D.9 Bebe	A. F. Cashin	
G-AXKJ	Jodel D.9 Bebe	J. R. Surbey	
G-AXKK	Westland Bell 47G-4A	Bristow Helicopters Ltd	
G-AXKL	Westland Bell 47G-4A	Helicare Ltd/Liverpool	
G-AXKN	Westland Bell 47G-4A	Helicare Ltd/Liverpool	
G-AXKO	Westland Bell 47G-4A	Bristow Helicopters Ltd	
G-AXKR	Westland Bell 47G-4A	Robert & Hewett Agriculture Ltd/ Liverpool	
G-AXKS	Westland Bell 47G-4A	Museum of Army Flying/Middle Wallop	
G-AXKU	Westland Bell 47G-4A	Bristow Helicopters Ltd	
G-AXKX	Westland Bell 47G-4A	Bristow Helicopters Ltd	
G-AXKY	Westland Bell 47G-4A	Bristow Helicopters Ltd	
G-AXKZ	Westland Bell 47G-4A	Helicare Ltd/Liverpool	
G-AXLA	Westland Bell 47G-4A	Helicare Ltd/Liverpool	
G-AXLG	Cessna 310K	Smiths (Outdrives) Ltd	
G-AXLI	Nipper T.66 Srs 3	N. J. Arthur	
G-AXLL	BAC One-Eleven 523FJ	British Caledonian Airways/Gatwick	
G-AXLN	BAC One-Eleven 523FJ	British Island Airways *Island Envoy*/Gatwick	
G-AXLS	Jodel DR.105A	T. L. Giles/Doncaster	
G-AXLZ	PA-19 Super Cub 95	J. C. Quantrell/Ludham	
G-AXMA	PA-24 Comanche 180	Tegrel Products Ltd/Newcastle	
G-AXMB	Slingsby T.7 Motor Cadet	I. G. Smith/Langar	
G-AXMD	Omega 20 balloon ★	British Balloon Museum	
G-AXME	SNCAN SV-4C	D. W. Hawthorne/Oporto	
G-AXMG	BAC One-Eleven 518	British Island Airways Ltd/Gatwick	
G-AXMN	J/5B Autocar	A. Phillips	
G-AXMP	PA-28 Cherokee 180	Midland Battery Distribution Ltd/Sywell	
G-AXMS	PA-30 Twin Comanche 160C	Ernest Green International Ltd	
G-AXMT	Bu 133 Jungmeister	D. J. Berry/Shoreham	
G-AXMU	BAC One-Eleven 432	Air UK Ltd/Norwich	
G-AXMW	B.121 Pup 1	DJP Engineering (Knebworth) Ltd/ Cambridge	
G-AXMX	B.121 Pup 2	Susan A. Jones/Cannes	

Notes	Reg.	Type	Owner or Operator
	G-AXNA	Boeing 737-204C	Britannia Airways Ltd *Robert Clive of India*/Luton
	G-AXNB	Boeing 737-204C	Britannia Airways Ltd *Charles Darwin*/ Luton
	G-AXNC	Boeing 737-204	Britannia Airways Ltd *Isambard Kingdom Brunel*/Luton
	G-AXNJ	Wassmer Jodel D.120	Clive Flying Group/Sleap
	G-AXNK	Cessna F.150J	N. Abram
	G-AXNL	B.121 Pup 1	D. C. Barber/St Just
	G-AXNM	B.121 Pup 1	Edgley Aircraft Flying Group/ Old Sarum
	G-AXNN	B.121 Pup 2	Romney Marsh Flying Group/Lydd
	G-AXNP	B.121 Pup 2	J. W. & K. E. Ellis/Chester
	G-AXNR	B.121 Pup 2	Specialised Mouldings Ltd & ptnrs
	G-AXNS	B.121 Pup 2	S. J. Figures & N. Fields
	G-AXNW	SNCAN SV-4C	E. N. Grace
	G-AXNX	Cessna 182M	P. Reid
	G-AXNZ	Pitts S.1C Special	W. A. Jordan
	G-AXOG	PA-23 Aztec 250D	R. W. Diggens/Denham
	G-AXOH	M.S.894 Rallye Minerva	Bristol Cars Ltd/White Waltham
	G-AXOI	Jodel D.9 Bebe	P. R. Underhill
	G-AXOJ	B.121 Pup 2	TM Air Ltd/Rochester
	G-AXOL	Currie Wot	A. Kennedy & ptnrs
	G-AXOR	PA-28 Cherokee 180D	Axor Aviation Ltd/Compton Abbas
	G-AXOS	M.S.894A Rallye Minerva	T. E. H. Simmons & S. G. Busby
	G-AXOT	M.S.893 Rallye Commodore 180	P. Evans & D. Riley/Elstree
	G-AXOV	Beech B55A Baron	S. Brod/Elstree
	G-AXOW	PA-23 Aztec 250	G. Costello/Dublin
	G-AXOX	BAC One-Eleven 432	Air UK Ltd/Norwich
	G-AXOZ	B.121 Pup 1	Arrow Air Services (Engineering) Ltd/ Shipdham
	G-AXPA	B.121 Pup 1	H. J. Bateman/Stapleford
	G-AXPB	B.121 Pup 1	D. J. Sage/Cardiff
	G-AXPD	B.121 Pup 1	C. A. Thorpe
	G-AXPF	Cessna F.150K	Y. Newell/Booker
	G-AXPG	Mignet HM-293	W. H. Cole
	G-AXPM	B.121 Pup 1	D. Taylor
	G-AXPN	B.121 Pup 2	Isohigh Ltd/Elstree
	G-AXPZ	Campbell Cricket	W. R. Partridge
	G-AXRA	Campbell Cricket	L. E. Schnurr
	G-AXRC	Campbell Cricket	K. W. Hayr/(*Stored*)
	G-AXRK	Practavia Sprite 115	E. G. Thale
	G-AXRL	PA-28 Cherokee 160	T. W. Clark/Headcorn
	G-AXRO	PA-30 Twin Comanche 160C	Express Flight Ltd/Southend
	G-AXRP	SNCAN SV-4C	C. C. Manning
	G-AXRR	Auster AOP.9 (XR241)	British Aerial Museum/Duxford
	G-AXRT	Cessna FA.150K (tailwheel)	W. H. Milner/Brough
	G-AXRU	Cessna FA.150K	Arnval Enterprises Ltd
	G-AXSC	B.121 Pup 1	T. R. Golding & C. Spencer
	G-AXSD	B.121 Pup 1	Surrey & Kent Flying Club Ltd/ Biggin Hill
	G-AXSF	Nash Petrel	Nash Aircraft Ltd/Lasham
	G-AXSG	PA-28 Cherokee 180	Shropshire Aero Club Ltd/Sleap
	G-AXSJ	Cessna FA.150K	C. N. Peate & E. Maggs/Fairoaks
	G-AXSM	Jodel DR.1051	C. Cousten/White Waltham
	G-AXSV	Jodel DR.340	Leonard F. Jollye Ltd/Panshanger
	G-AXSW	Cessna FA.150K	Furness Aviation Ltd/Cark
	G-AXSZ	PA-28 Cherokee 140B	N. Cureton & R. B. Cheek/Sandown
	G-AXTA	PA-28 Cherokee 140B	P. Barry & I. Cameron
	G-AXTC	PA-28 Cherokee 140B	S. R. Baugh & D. Rose/Booker
	G-AXTD	PA-28 Cherokee 140B	G. R. Walker/Southend
	G-AXTH	PA-28 Cherokee 140B	W. London Aero Services Ltd/ White Waltham
	G-AXTI	PA-28 Cherokee 140B	I. K. George/Fairoaks
	G-AXTJ	PA-28 Cherokee 140B	E. Kelk
	G-AXTL	PA-28 Cherokee 140B	P. Murphy & R. Flower/Blackbushe
	G-AXTO	PA-24 Comanche 260	J. L. Wright
	G-AXTP	PA-28 Cherokee 180	E. R. Moore/Elstree
	G-AXTX	Jodel D.112	J. J. Penney
	G-AXUA	B.121 Pup 1	F. R. Blennerhassett & C. Wedlake/ Tees-side
	G-AXUB	BN-2A Islander	Headcorn Parachute Club
	G-AXUC	PA-12 Super Cruiser	V. N. Mukaloff/Manston

Reg.	Type	Owner or Operator	Notes
G-AXUE	Jodel DR.105A	Six Group/Netherthorpe	
G-AXUF	Cessna FA.150K	Turnhouse Flying Club Ltd	
G-AXUI	H.P.137 Jetstream 1	Cranfield Institute of Technology	
G-AXUJ	J/I Autocrat	R. G. Earp & J. W. H. Lee/Sibson	
G-AXUK	Jodel DR.1050	PERME Westcott Social & Sports Club Ltd	
G-AXUM	H.P.137 Jetstream 1	Cranfield Institute of Technology	
G-AXUW	Cessna FA.150K	Coventry Air Training School	
G-AXUX	Beech B95 Travel Air	Melbren Air Ltd/Liverpool	
G-AXVB	Cessna F.172H	C. Gabbitas/Staverton	
G-AXVC	Cessna FA.150K	V. F. Lynn/Sibson	
G-AXVK	Campbell Cricket	Campbell Gyroplanes Ltd	
G-AXVM	Campbell Cricket	D. M. Organ/Staverton	
G-AXVN	McCandless M.4	W. R. Partridge	
G-AXVS	Jodel DR.1050	F. W. Tilley/Rochester	
G-AXVU	Omega 84 balloon	Brede Balloons Ltd *Henry VIII*	
G-AXVV	Piper L-4H Cub	J. MacCarthy	
G-AXVW	Cessna F.150K	R. Nichols/Elstree	
G-AXVX	Cessna F.172H	Staverton Flying Services Ltd	
G-AXWA	Auster AOP.9 (XN437)	T. Platt/Biggin Hill	
G-AXWB	Omega 65 balloon	A. Robinson & M. J. Moore *Ezekiel*	
G-AXWD	Jurca MJ 10	F.P.A. Group	
G-AXWE	Cessna F.150K	Light Planes (Lancashire) Ltd/Barton	
G-AXWF	Cessna F.172H	Red Fir Aviation Ltd/Clacton	
G-AXWH	BN-2A Islander	Midland Parachute Centre Ltd	
G-AXWP	BN-2A Islander	Aurigny Air Services/Guernsey	
G-AXWR	BN-2A Islander	Aurigny Air Services/Guernsey	
G-AXWT	Jodel D.11	R. Owen/Shoreham	
G-AXWV	Jodel DR.253	Murray Motors/Southend	
G-AXWZ	PA-28R Cherokee Arrow 200	R. F. Grute & M. K. Taylor	
G-AXXC	CP.301B Emeraude	J. R. R. Gale & J. Tetley	
G-AXXG	BN-2A Islander	Avon Aviation Services Ltd/Bristol	
G-AXXJ	BN-2A Islander	Air Wight Ltd/Bembridge	
G-AXXM	CP.301A Emeraude	Mrs P. H. Wren	
G-AXXV	D.H.82A Tiger Moth (DE992)	N. Smith/Wellesbourne	
G-AXXW	Jodel D.117	D. J. & M. Watson/Sherburn	
G-AXYA	PA-31-300 Navajo	W. R. M. C. Foyle/Luton	
G-AXYD	BAC One-Eleven 509	Dan-Air Services Ltd/Gatwick	
G-AXYK	Taylor JT.1 Monoplane	C. Oakins	
G-AXYM	BN-2A Islander	Balmar Aviation/Thruxton	
G-AXYU	Jodel D.9 Bebe	D. P. Jones/Panshanger	
G-AXYX	WHE Airbuggy	W. B. Lumb	
G-AXYY	WHE Airbuggy	M. P. Chetwynd-Talbot	
G-AXYZ	WHE Airbuggy	W. H. Ekin	
G-AXZA	WHE Airbuggy	P. H. Dyson & W. B. Lumb	
G-AXZB	WHE Airbuggy	D. R. C. Pugh	
G-AXZD	PA-28 Cherokee 180E	Damair/Goodwood	
G-AXZF	PA-28 Cherokee 180E	E. P. C. & W. R. Rabson/Southampton	
G-AXZM	Slingsby Nipper T.66 Mk III	G. R. Harlow	
G-AXZO	Cessna 180	A. R. Brett	
G-AXZP	PA-23 Aztec 250	White House Garage, Ashford Ltd/ Denham	
G-AXZT	Jodel D.117	H. W. Baines	
G-AXZU	Cessna 182N	Yorkshire Light Aircraft Ltd/Leeds	
G-AYAA	PA-28 Cherokee 180E	Briskloom Ltd/Manchester	
G-AYAB	PA-28 Cherokee 180E	J. A. & J. C. Cunningham	
G-AYAC	PA-28R Cherokee Arrow 200	Fersfield Flying Group	
G-AYAI	Fournier RF-5	Exeter RF Group	
G-AYAJ	Cameron O-84 balloon	E. T. Hall *Flaming Pearl*	
G-AYAL	Omega 56 balloon	British Balloon Museum	
G-AYAN	Slingsby Motor Cadet Mk III	I. Stevenson/Sherburn	
G-AYAP	PA-28 Cherokee 180E	Gala Air Holidays Ltd	
G-AYAR	PA-28 Cherokee 180E	C. H. Campbell/Leavesden	
G-AYAS	PA-28 Cherokee 180E	Aviation Enterprises/Fairoaks	
G-AYAT	PA-28 Cherokee 180E	P. J. Messervy/Manston	
G-AYAU	PA-28 Cherokee 180E	Tiarco Ltd	
G-AYAV	PA-28 Cherokee 180E	Cavendish Aviation Ltd/Sandtoft	
G-AYAW	PA-28 Cherokee 180E	Wizard Air Services Ltd/Coventry	
G-AYBD	Cessna F.150K	D. G. & W. B. Adams	
G-AYBG	Scheibe SF.25B Falke	D. J. Rickman	
G-AYBK	PA-28 Cherokee 180E	C. Lotetsky	
G-AYBO	PA-23 Aztec 250D	Twinguard Aviation Ltd/Elstree	
G-AYBP	Jodel D.112	Fairwood Flying Group/Swansea	

Notes	Reg.	Type	Owner or Operator
	G-AYBU	Western 84 balloon	D. R. Gibbons
	G-AYBV	Chasle Tourbillon	B. A. Mills
	G-AYCC	Campbell Cricket	K. W. E. Denson
	G-AYCE	CP.301C Emeraude	R. A. Austin/Bodmin
	G-AYCF	Cessna FA.150K	E. J. Atkins/Popham
	G-AYCG	SNCAN SV-4C	N. Bignall/Booker
	G-AYCJ	Cessna TP.206D	H. O. Holm
	G-AYCM	Bell 206A JetRanger	W.R. Finance Ltd
	G-AYCN	Piper L-4H Cub	W. R. & B. M. Young
	G-AYCO	CEA DR.360	L. M. Gould/Jersey
	G-AYCP	Jodel D.112	W. Hutchings
	G-AYCT	Cessna F.172H	Fife Airport Management & Kontrox Ltd/Glenrothes
	G-AYDG	M.S.894A Rallye Minerva	R. Cushing & A. Levitt
	G-AYDI	D.H.82A Tiger Moth	R. B. Woods & ptnrs
	G-AYDR	SNCAN SV-4C	R. A. Phillips
	G-AYDU	AJEP W.8 Tailwind (nosewheel)	AJEP Development Ltd
	G-AYDV	Coates SA.11-1 Swalesong	J. R. Coates/Rush Green
	G-AYDW	A.61 Terrier 2	J. S. Harwood
	G-AYDX	A.61 Terrier 2	Homelink Services Ltd/Tollerton
	G-AYDY	Luton LA-4A Minor	R. M. Hill-Venning/White Waltham
	G-AYDZ	Jodel DR.200	F. M. Ward
	G-AYEB	Jodel D.112	B. Ibbott
	G-AYEC	CP.301A Emeraude	A. P. Docherty/Redhill
	G-AYED	PA-24 Comanche 260	C. A. Saville
	G-AYEE	PA-28 Cherokee 180E	D. J. Beale
	G-AYEF	PA-28 Cherokee 180E	Axholme Aviation Ltd
	G-AYEG	Falconar F-9	A. G. Thelwall
	G-AYEH	Jodel DR.1050	R. O. F. Harper & P. R. Skeels/Barton
	G-AYEI	PA-31-300 Navajo	Hubbard Air Ltd/Norwich
	G-AYEJ	Jodel DR.1050	G. Weaver
	G-AYEK	Jodel DR.1050	I. Shaw & B. Hanson/Sherburn
	G-AYEN	Piper L-4H Cub	P. Warde & C. F. Morris
	G-AYES	M.S.892A Rallye Commodore 150	Waveney Flying Group/Seething
	G-AYET	M.S.892A Rallye Commodore 150	Lands End Flying Club/St Just
	G-AYEU	Brookland Hornet	M. G. Reilly
	G-AYEV	Jodel DR.1050	L. G. Evans/Headcorn
	G-AYEW	Jodel DR.1051	Taildragger Group/Halfpenny Green
	G-AYEY	Cessna F.150K	B&M Motors/Goodwood
	G-AYFA	SA Twin Pioneer 3	Flight One Ltd/Shobdon
	G-AYFC	D.62B Condor	R. A. Smith/Redhill
	G-AYFD	D.62B Condor	Tiger Club Ltd/Redhill
	G-AYFE	D.62C Condor	R. R. Harris
	G-AYFF	D.62B Condor	A. F. S. Caldecourt/Fairoaks
	G-AYFG	D.62C Condor	Wolds Gliding Club
	G-AYFJ	M.S.880B Rallye Club	S. Hackett & R. P. H. Lake
	G-AYFP	Jodel D.140	S. K. Minocha/Sherburn
	G-AYFT	PA-39 Twin Comanche C/R	Kirby Oldham Ltd/Manchester
	G-AYFV	Crosby BA-4B	A. N. R. Houghton/Leicester
	G-AYFX	AA-I Yankee	P. A. Ellway & R. M. Bainbridge
	G-AYFZ	PA-31-300 Navajo	Air Luton Ltd
	G-AYGA	Jodel D.117	E. J. Baxter & ptnrs/Stapleford
	G-AYGB	Cessna 310Q	Airwork Services Ltd/Perth
	G-AYGC	Cessna F.150K	J. M. Hill & ptnrs
	G-AYGD	Jodel DR.1051	I. S. Walsh
	G-AYGE	SNCAN SV-4C	The Hon A. M. J. Rothschild/Booker
	G-AYGG	Jodel D.120	R. F. Sothcott
	G-AYGK	BN-2A Islander	Prop-Jet Ltd/Stansted
	G-AYGN	Cessna 210K	J. W. O'Sullivan/Jersey
	G-AYGX	Cessna FR.172G	J. A. Edwards/Blackpool
	G-AYGZ	Beech 58 Baron	General Engineering Co (Ilford) Ltd/ Stansted
	G-AYHA	AA-1 Yankee	D. L. Harrisberg & I. J. Widger/ Elstree
	G-AYHI	Campbell Cricket	J. F. MacKay
	G-AYHX	Jodel D.117A	L. J. E. Goldfinch
	G-AYHY	Fournier RF-4D	Tiger Club Ltd/Redhill
	G-AYIA	Hughes 369HS	G. D. E. Bilton/Sywell
	G-AYIB	Cessna 182N Skylane	R. M. Clarke
	G-AYIF	PA-28 Cherokee 140C	The Hare Flying Group/Elstree
	G-AYIG	PA-28 Cherokee 140C	Klingair Ltd/Conington
	G-AYIH	PA-28 Cherokee 140C	B. Lince/Elstree

Reg.	Type	Owner or Operator	Notes
G-AYII	PA-28R Cherokee Arrow 200	Devon Growers Ltd & A. L. Bacon/ Exeter	
G-AYIJ	SNCAN SV-4B	A. J. Ditheridge/Ipswich	
G-AYIL	Scheibe SF.25B Falke	J. M. Salt & ptnrs	
G-AYIO	PA-28 Cherokee 140C	Pettigrew Supermarkets (Lytham) Ltd/ Blackpool	
G-AYIT	D.H.82A Tiger Moth	R. L. H. Alexander & ptnrs/Barton	
G-AYJA	Jodel DR.1050	A. A. Alderdice & ptnrs	
G-AYJB	SNCAN SV-4C	F. J. M. & J. P. Esson/Middle Wallop	
G-AYJD	Alpavia-Fournier RF-3	C. Wren/Southend	
G-AYJE	BN-2A-6 Islander	Headcorn Parachute Club Ltd	
G-AYJP	PA-28 Cherokee 140C	RAF Brize Norton Flying Club Ltd	
G-AYJR	PA-28 Cherokee 140C	RAF Brize Norton Flying Club Ltd	
G-AYJS	PA-28 Cherokee 140C	W. R. Griffiths & Sons (Office Furnishers) Ltd	
G-AYJT	PA-28 Cherokee 140C	Jennifer M. Lesslie/Leicester	
G-AYJU	Cessna TP-206A	J. E. Ball & R. W. F. Marsh/Thruxton	
G-AYJW	Cessna FR.172G	Rogers Aviation Ltd/Cranfield	
G-AYJY	Isaacs Fury II	A. V. Francis	
G-AYKA	Beech 95-B55A Baron	Sasha Fashions (International) Ltd Biggin Hill	
G-AYKD	Jodel DR.1050	J. J. Collis	
G-AYKF	M.S.880B Rallye Club	R. V. Screen & ptnrs	
G-AYKJ	Jodel D.117A	G. R. W. Monksfield/Stapleford	
G-AYKK	Jodel D.117	D. M. Whitham	
G-AYKL	Cessna F.150L	Aero Group 78/Netherthorpe	
G-AYKS	Leopoldoff L-7	C. E. & W. B. Cooper	
G-AYKT	Jodel D.117	G. Wright/Sherburn	
G-AYKV	PA-28 Cherokee 140C	A. Wright/Liverpool	
G-AYKW	PA-28 Cherokee 140C	T. P. Sheff/Southend	
G-AYKX	PA-28 Cherokee 140C	M. J. Garland & ptnrs/Woodford	
G-AYKZ	SAI KZ-8	R. E. Mitchell/Coventry	
G-AYLA	Glos-Airtourer 115	Vagabond Flying Group	
G-AYLB	PA-39 Twin Comanche C/R	Mercia Aviation/Coventry	
G-AYLE	M.S.880B Rallye Club	B. W. Hogger	
G-AYLF	Jodel DR.1051	A. C. Frost	
G-AYLG	H.S.125 Srs 400B	British Aerospace PLC	
G-AYLK	Stampe SV-4C	R. W. & P. R. Budge/Southend	
G-AYLL	Jodel DR.1050	Firefly Aviation Ltd/Glasgow	
G-AYLO	AA-1 Yankee	J. A. & A. J. Boyd/Cardiff	
G-AYLP	AA-1 Yankee	D. Nairn & E. Y. Hawkins/Bournemouth	
G-AYLU	Pitts S-ID Special	I. M. G. Senior & J. G. Harper	
G-AYLV	Jodel D.120	R. E. Wray/Stapleford	
G-AYLX	Hughes 269C	Feastlight Ltd/Sywell	
G-AYLY	PA-23 Aztec 250	British Island Airways/Shoreham	
G-AYLZ	Super Aero 45 Srs 2	A. Topen	
G-AYMA	Stolp Starduster Too	R. W. Davies & J. C. Wood/Staverton	
G-AYME	Fournier RF.5	R. D. Goodger/Biggin Hill	
G-AYMG	HPR-7 Herald 213	Securicor Ltd/Birmingham	
G-AYMK	PA-28 Cherokee 140C	The Piper Flying Group	
G-AYML	PA-28 Cherokee 140C	J. M. Bendle/Elstree	
G-AYMM	Cessna 421B	Rogers Aviation Sales Ltd/Cranfield	
G-AYMN	PA-28 Cherokee 140C	F. R. Aviation Ltd/Coventry	
G-AYMO	PA-23 Aztec 250	Beechwood Properties Ltd	
G-AYMP	Currie Wot Special	H. F. Moffatt/Shobdon	
G-AYMR	Lederlin 380L Ladybug	J. S. Brayshaw	
G-AYMT	Jodel DR.1050	Merlin Flying Club Ltd/Hucknall	
G-AYMU	Jodel D.112	Shoreham Aero Club	
G-AYMV	Western 20 balloon	G. F. Turnbull & ptnrs *Tinkerbelle*	
G-AYMW	Bell 206A JetRanger 2	Dollar Air Services Ltd/Coventry	
G-AYMX	Bell 206A JetRanger	W. Holmes	
G-AYMZ	PA-28 Cherokee 140C	D. Tuke/Crosby	
G-AYNA	Currie Wot	D. G. Crew/Biggin Hill	
G-AYNC	Wessex Mk 60 Srs 1	Sykes Group Co Ltd/Bournemouth	
G-AYND	Cessna 310Q	Source Premium & Promotional Consultants Ltd/Fairoaks	
G-AYNF	PA-28 Cherokee 140C	P. Weston	
G-AYNJ	PA-28 Cherokee 140C	T. L. Deamer/Elstree	
G-AYNN	Cessna 185B Skywagon	Bencray Ltd/Blackpool	
G-AYNP	Westland S.55 Srs 3	Bristow Helicopters Ltd	
G-AYNS	Airmaster H2-B1	D. J. Fry/Blackbushe	
G-AYOC	BN-2A-8 Islander	Harvest Aviation Ltd/Southend	
G-AYOD	Cessna 172	R. A. Bowes & ptnrs	

Notes	Reg.	Type	Owner or Operator
	G-AYOM	Sikorsky S-61N Mk 2	British Airways Helicopters Ltd/ Aberdeen
	G-AYOP	BAC One-Eleven 530	British Caledonian Airways *Isle of Hoy*/ Gatwick
	G-AYOW	Cessna 182N Skylane	C. H. Royal
	G-AYOX	V.814 Viscount	British Midland Airways Ltd/E. Midlands
	G-AYOY	Sikorsky S-61N Mk 2	British Airways Helicopters Ltd/ Aberdeen
	G-AYOZ	Cessna FA.150L	Exeter Flying Club Ltd
	G-AYPB	Beech C-23 Musketeer	Wrighton Homes Ltd
	G-AYPD	Beech 95 B.55 Baron	Sir W. S. Dugdale/Birmingham
	G-AYPE	Bo 209 Monsun	Papa Echo Ltd/Biggin Hill
	G-AYPF	Cessna F.177RG	H.W. Structures Ltd/Cranfield
	G-AYPG	Cessna F.177RG	D. Davies
	G-AYPH	Cessna F.177RG	D. Hewerdine & J. G. Collins
	G-AYPI	Cessna F.177RG	Cardinal Aviation Ltd/Guernsey
	G-AYPJ	PA-28 Cherokee 180	Snowdon Aviation Ltd/Mona
	G-AYPM	PA-19 Super Cub 95	C. H. A. Bott
	G-AYPO	PA-19 Super Cub 95	R. T. Love
	G-AYPP	PA-19 Super Cub 95	M. Kirk
	G-AYPR	PA-19 Super Cub 95	J. W. Hollingsworth
	G-AYPS	PA-19 Super Cub 95	Tony Dyer Television
	G-AYPT	PA-19 Super Cub 95	Laarbruch Flying Club
	G-AYPU	PA-28R Cherokee Arrow 200	Alpine Ltd/Jersey
	G-AYPV	PA-28 Cherokee 140D	R. J. Steed/Elstree
	G-AYPZ	Campbell Cricket	A. Melody/Lasham
	G-AYRF	Cessna F.150L	Northern Auto Salvage/Inverness
	G-AYRG	Cessna F.172K	G. H. Aviation Ltd
	G-AYRH	M.S.892A Rallye Commodore 150	J. D. Watt
	G-AYRI	PA-28R Cherokee Arrow 200	E. P. Van Mechelen & Delta Motor Co (Windsor) Sales Ltd/White Waltham
	G-AYRK	Cessna 150J	K. A. Learmonth/Southend
	G-AYRL	Fournier SFS.31 Milan	P. A. Moorehead
	G-AYRM	PA-28 Cherokee 140D	E. S. Dignam/Biggin Hill
	G-AYRN	Schleicher ASK-14	V. J. F. Falconer/Dunstable
	G-AYRO	Cessna FA.150L Aerobat	Thruxton Flight Centre
	G-AYRP	Cessna FA.150L Aerobat	Andrewsfield Flying Club Ltd
	G-AYRS	Jodel D.120A	J. H. Tetley & G. C. Smith/Sherburn
	G-AYRT	Cessna F.172K	W. Gibson/Bournemouth
	G-AYRU	BN-2A-6 Islander	Joint Services Parachute Centre/ Netheravon
	G-AYSA	PA-23 Aztec 250C	J. M. Yendall & R. Greenhill
	G-AYSB	PA-30 Twin Comanche 160C	Sandcliffe Aviation/Biggin Hill
	G-AYSD	Slingsby T.67A Falke	J. Conolly/Tees-side
	G-AYSH	Taylor JT.1 Monoplane	C. J. Lodge
	G-AYSK	Luton L.A.4A Minor	P. F. Bennison & ptnrs/Barton
	G-AYSX	Cessna F.177RG	Nasaire Ltd/Liverpool
	G-AYSY	Cessna F.177RG	Sierra Yankee Flying Group/Elstree
	G-AYTA	M.S.880B Rallye Club	N. Butcher
	G-AYTC	PA-E23 Aztec 250C	New Guarantee Trust Finance Ltd/ E. Midlands
	G-AYTF	Bell 206B JetRanger 2	N. Hutchings
	G-AYTJ	Cessna 207 Super Skywagon	Foxair/Glenrothes
	G-AYTN	Cameron O-65 balloon	P. G. Hall & R. F. Jessett *Prometheus*
	G-AYTR	CP.301A Emeraude	D. J. Knight & M. E. Sain
	G-AYTT	Phoenix PM-3 Duet	Gp Capt A. S. Knowles/Fairoaks
	G-AYTV	MJ.2A Tempete	P. Russell/Netherthorpe
	G-AYTY	Bensen B.8	J. H. Wood (*Stored*)
	G-AYUB	CEA DR.253B	D. J. Brook/Shoreham
	G-AYUH	PA-28 Cherokee 180F	M. S. Bayliss/Coventry
	G-AYUI	PA-28 Cherokee 180	Routair Aviation Services Ltd/Southend
	G-AYUJ	Evans VP.1 Volksplane	J. A. Wills
	G-AYUL	PA-23 Aztec 250E	ADM Air Services Ltd
	G-AYUM	Slingsby T-61A Falke	The Burn Gliding Club
	G-AYUN	Slingsby T-61A Falke	C. W. Vigar & R. J. Watts
	G-AYUP	Slingsby T-61A Falke	Cranwell Gliding Club
	G-AYUR	Slingsby T-61A Falke	W. A Urwin
	G-AYUS	Taylor JT.1 Monoplane	R. R. McKinnon & A. D. Lincoln/ Southampton
	G-AYUT	Jodel DR.1050	R. Norris
	G-AYUV	Cessna F.172H	J. R. Wheeler & M. Cresswell
	G-AYUX	D.H.82A Tiger Moth (PG651)	Ardentland Ltd/Booker
	G-AYUY	Cessna FA.150L Aerobat	J. A. Wills/Dundee

Reg.	Type	Owner or Operator	Notes
G-AYVA	Cameron O-84 balloon	A. Kirk *April Fool*	
G-AYVI	Cessna T.210H	Trident Marine Ltd/Edinburgh	
G-AYVO	Wallis WA120 Srs 1	K. H. Wallis	
G-AYVP	Woody Pusher	J. R. Wraight	
G-AYVT	Brochet MB.84	Dunelm Flying Group	
G-AYVU	Cameron O-56 balloon	Shell-Mex & B.P. Ltd *Hot Potato*	
G-AYVY	D.H.82A Tiger Moth (PG617)	G. Smith/Ronaldsway	
G-AYWA	Avro 19 Srs 2 ★	Strathallan Aircraft Collection	
G-AYWB	BAC One-Eleven 531FS	British Island Airways Ltd *Island Enterprise*/Gatwick	
G-AYWD	Cessna 182N	Trans Para Aviation Ltd/Barton	
G-AYWG	PA-E23 Aztec 250C	Telepoint Ltd/Manchester	
G-AYWH	Jodel D.117A	J. M. Knapp & ptnrs	
G-AYWI	BN-2A Mk III-I Trislander	Aurigny Air Services/Guernsey	
G-AYWM	Glos-Airtourer Super 150	F. B. Miles/Staverton	
G-AYWT	Stampe SV-4B	B. K. Lecomber/Denham	
G-AYWW	PA-28R Cherokee Arrow 200D	Acorn Aviation Ltd/Leicester	
G-AYXO	Luton LA-5A Major	A. C. T. Broomcroft	
G-AYXP	Jodel D.117A	G. N. Davies/Shobdon	
G-AYXS	SIAI-Marchetti S205-18R	J. C. Scott/Liverpool	
G-AYXT	Westland Sikorsky S-55 Srs 2	J. E. Wilkie/Blackpool	
G-AYXU	Champion 7KCAB Citabria	H. Fould & ptnrs	
G-AYXV	Cessna FA.150L	*Wreck*/Popham	
G-AYXW	Evans VP.1 Volksplane	J. S. Penny/Doncaster	
G-AYYD	M.S.894A Rallye Minerva	G. Lyons	
G-AYYF	Cessna F.150L	Falcon Aero Club/Swansea	
G-AYYK	Slingsby T-61A Falke	Cornish Gliding & Flying Club Ltd/ Perranporth	
G-AYYL	Slingsby T.61A Falke	C. Wood	
G-AYYO	Jodel DR.1050/M1	Bustard Flying Club Ltd/Old Sarum	
G-AYYT	Jodel DR.1050/M1	Sicile Flying Group/Sandown	
G-AYYU	Beech C23 Musketeer	R. L. C. Appleton/Staverton	
G-AYYW	BN-2A Islander	Foster Yeoman Ltd	
G-AYYX	M.S.880B Rallye Club	S. &. J. Cattle/Newcastle	
G-AYYY	M.S.880B Rallye Club	T. W. Heffer/Panshanger	
G-AYYZ	M.S.880B Rallye Club	R. J. Napp/Southend	
G-AYZC	PA-E23 Aztec 250D	N. W. Aero Services Ltd	
G-AYZE	PA-39 Twin Comanche 160 C/R	Alger, Brownless & Court Ltd	
G-AYZH	Taylor JT-2 Titch	K. J. Munro	
G-AYZI	Stampe SV-4C	W. H. Smout & F. A. R. de Lavergne	
G-AYZJ	Westland Sikorsky S-55 (XM685) ★	Newark Air Museum	
G-AYZK	Jodel DR.1050/M1	G. S. Claybourn/Doncaster	
G-AYZN	PA-E23 Aztec 250	International Institute of Tropical Agriculture	
G-AYZS	D.62B Condor	R. Simpson/Staverton	
G-AYZT	D.62B Condor	Essex & Suffolk Gliding Club	
G-AYZU	Slingsby T.61A Falke	The Falcon Gliding Group/Enstone	
G-AYZW	Slingsby T.61A Falke	J. A. Dandie & R. J. M. Clement	
G-AZAB	PA-30 Twin Comanche 160	T. W. P. Sheffield/Humberside	
G-AZAD	Jodel DR.1051	I. C. Young & J. S. Paget/Bodmin	
G-AZAJ	PA-28R Cherokee Arrow 200B	Driscoll Tyres Ltd & J. McHugh & Son (Civil Engineers) Ltd/Stapleford	
G-AZAV	Cessna 337F	W. T. Johnson & Sons (Huddersfield) Ltd	
G-AZAW	GY-80 Horizon 160	G. Barlow	
G-AZAZ	Bensen B.8M	FAA Museum/Yeovilton	
G-AZBA	T.66 Nipper 3	I. McKenzie	
G-AZBB	MBB Bo 209 Monsun 160FV	C.E. Aviation Ltd/Staverton	
G-AZBC	PA-39 Twin Comanche 160 C/R	E. Shipley/Clacton	
G-AZBE	Glos-Airtourer Super 150	S. A. Warwick/Popham	
G-AZBI	Jodel D.150	T. A. Rawson & C. R. Warcup	
G-AZBK	PA-E23 Aztec 250E	Qualitair Engineering Ltd/Blackbushe	
G-AZBL	Jodel D.9 Bebe	West Midlands Flying Group/ Halfpenny Green	
G-AZBN	AT-16 Harvard 2B (FT391)	Ashbon Associates Ltd	
G-AZBT	Western O-65 balloon	D. J. Harris *Hermes*	
G-AZBU	Auster AOP.9	K. H. Wallis	
G-AZBY	Westland Wessex 60 Srs 1	Westland Helicopters Ltd/Yeovil	
G-AZBZ	Westland Wessex 60 Srs 1	Westland Helicopters Ltd/Yeovil	
G-AZCB	Stampe SV-4B	M. J. Coburn	
G-AZCI	Cessna 320A Skyknight	Landsurcon (Air Survey) Ltd/Staverton	

Notes	Reg.	Type	Owner or Operator
	G-AZCK	B.121 Pup 2	Wickenby Flying Club Ltd
	G-AZCL	B.121 Pup 2	Cameron Rainwear Ltd/Lympne
	G-AZCP	B.121 Pup 1	R. I. Scadeng
	G-AZCT	B.121 Pup 1	Northamptonshire School of Flying/ Sywell
	G-AZCU	B.121 Pup 1	Leyline Aviation Ltd/Tees-side
	G-AZCZ	B.121 Pup 2	P. R. Moorehead/Booker
	G-AZDA	B.121 Pup 1	G. H. G. Bishop & K. E. Fehrenbach/ Shoreham
	G-AZDC	Sikorsky S-61N	Bristow Helicopters Ltd
	G-AZDD	MBB Bo 209 Monsun 150FF	Double Delta Flying Group/Elstree
	G-AZDE	PA-28R Cherokee Arrow 200B	Electro-Motion UK (Export) Ltd/ E. Midlands
	G-AZDF	Cameron O-84 balloon	K. L. C. M. Busemeyer
	G-AZDG	B.121 Pup 2	D. J. Sage
	G-AZDK	Beech B55 Baron	Forbury Foods Ltd
	G-AZDX	PA-28 Cherokee 180F	S. D. Quigley & K. J. Gallagher/Glasgow
	G-AZDY	D.H.82A Tiger Moth	J. B. Mills
	G-AZEA	Cessna 182N	G. P. Grant-Suttie
	G-AZED	BAC One-Eleven 414	Dan-Air Services Ltd/Gatwick
	G-AZEE	M.S.880B Rallye Club	P. L. Clements
	G-AZEF	Jodel D.120	J. R. Legge
	G-AZEG	PA-28 Cherokee 140D	J. W. Simmons/Barrow
	G-AZER	Cameron O-42 balloon	M. P. Dokk-Olsen & P. L. Jaye *Shy Tot*
	G-AZEU	B.121 Pup 2	J. N. Russell/Aberdeen
	G-AZEV	B.121 Pup 2	G. P. Martin/Shoreham
	G-AZEW	B.121 Pup 2	Deltair Ltd/Chester
	G-AZEY	B.121 Pup 2	M. A. Pearce
	G-AZFA	B.121 Pup 2	K. F. Plummer/Sywell
	G-AZFC	PA-28 Cherokee 140D	A. H. Lavender/Biggin Hill
	G-AZFF	Jodel D.112	J. M. Newbold/Enstone
	G-AZFI	PA-28R Cherokee Arrow 200B	Hawksworth Garage Ltd
	G-AZFM	PA-28R Cherokee Arrow 200B	Linco Poultry Machinery Ltd/Biggin Hill
	G-AZFO	PA-39 Twin Comanche 160 C/R	Handhorn Ltd/Blackpool
	G-AZFP	Cessna F.177RG	G. A. Stead/Bournemouth
	G-AZFR	Cessna 401B	Westair Flying Services Ltd/Blackpool
	G-AZFZ	Cessna D.120	Redapple Ltd/Fairoaks
	G-AZGA	Jodel D.120	G. B. Morris
	G-AZGB	PA-E23 Aztec 250D	Qualitair Engineering Ltd/Blackbushe
	G-AZGC	Stampe SV-4C (No 120)	The Hon Patrick Lindsay/Booker
	G-AZGE	Stampe SV-4A	M. R. L. Astor/Booker
	G-AZGF	B.121 Pup 2	K. Singh
	G-AZGH	M.S.880B Rallye Club	R. G. Moore
	G-AZGI	M.S.880B Rallye Club	G. E. M. Hallett & ptnrs/Newcastle
	G-AZGJ	M.S.880B Rallye Club	P. Rose
	G-AZGL	M.S.894A Rallye Minerva	The Cambridge Aero Club Ltd
	G-AZGY	CP.301B Emeraude	Rodingair Flying Group/Stapleford
	G-AZGZ	D.H.82A Tiger Moth (NM181)	F. R. Manning
	G-AZHB	Robin HR 100-200	C. & P. P. Scarlett/Lydd
	G-AZHC	Jodel D.112	J. A. Summer & A. Burton/Netherthorpe
	G-AZHD	Slingsby T.61A Falke	J. Sentance
	G-AZHH	SA 102.5 Cavalier	D. W. Buckle *Time*
	G-AZHI	Glos-Airtourer Super 150	H. J. Douglas & ptnrs/Biggin Hill
	G-AZHJ	S.A. Twin Pioneer Srs 3	Flight One Ltd/Shobdon
	G-AZHK	Robin HR 100-200	S. V. Swallow
	G-AZHL	PA-31-300 Navajo	BAC Aviation Ltd/Stansted
	G-AZHM	Cassutt Racer	J. A. H. Chadwick/Redhill
	G-AZHO	Jodel DR.1050	S. Alexander
	G-AZHR	Piccard Ax6 balloon	J. W. Moss *Happiness*
	G-AZHT	Glos-Airtourer T.3	D. C. Giles/Southend
	G-AZHU	Luton LA-4A Minor	F. Didsbury/Netherthorpe
	G-AZIB	ST-10 Diplomate	Wilmslow Audio Ltd/Tollerton
	G-AZID	Cessna FA.150L	Oldment Ltd/Brough
	G-AZII	Jodel D.117A	J. S. Brayshaw
	G-AZIJ	Jodel DR.360	K. H. Tostevin/Guernsey
	G-AZIK	PA-34-200 Seneca	C.S.E. (Aircraft Services) Ltd/Kidlington
	G-AZIL	Slingsby T.61A Falke	I. Jamieson
	G-AZIO	SNCAN SV-4C	Rollason Aircraft & Engines Ltd/(*Stored*)
	G-AZIP	Cameron O-65 balloon	Dante Balloon Group *Dante*
	G-AZIR	Stampe SV-4C	Rollason Aircraft & Engines Ltd
	G-AZJB	PA-34-200 Seneca	Air Charter Scotland & Phoenix Hydrocarbons/Glasgow
	G-AZJC	Fournier RF-5	J. J. Butler/Biggin Hill

Reg.	Type	Owner or Operator	Notes
G-AZJE	JB-01 Minicab	J. B. Evans/Sandown	
G-AZJI	Western O-65 balloon	W. Davison *Peek-a-Boo*	
G-AZJN	Robin DR 300/140	Wright Farm Eggs Ltd	
G-AZJV	Cessna F.172L	J. A. & A. J. Boyd/Denham	
G-AZJW	Cessna F.150L	A. J. Fletcher/Shoreham	
G-AZJY	Cessna FRA.150L	Shropshire Aero Club Ltd/Sleap	
G-AZJZ	PA-23 Aztec 250E	Encee Services Ltd/Cardiff	
G-AZKC	M.S.880B Rallye Club	L. J. Martin/Redhill	
G-AZKD	M.S.880B Rallye Club	P. Feeney/Kidlington	
G-AZKE	M.S.880B Rallye Club	B. S. Rowden & W. L. Rogers	
G-AZKG	Cessna F.172L	Multree Ltd	
G-AZKK	Cameron O-56 balloon	Gemini Balloon Group *Gemini*	
G-AZKN	Robin HR.100/200	K. Marriott/Tollerton	
G-AZKO	Cessna F.337F	Crispair Aviation Services Ltd	
G-AZKP	Jodel D.117	J. Lowe/Tollerton	
G-AZKR	PA-24 Comanche 180	A. W. Crawford	
G-AZKS	AA-1A Trainer	J. G. Hill	
G-AZKV	Cessna FRA.150L	Penguin Flight/Bodmin	
G-AZKW	Cessna F.172L	Banbury Plant Hire Ltd/ Hinton-in-the-Hedges	
G-AZKZ	Cessna F.172L	Sprowston Engineering Ltd/Norwich	
G-AZLE	Boeing N2S-5 Kaydet	A. E. Poulson	
G-AZLF	Jodel D.120	J. Brooks	
G-AZLH	Cessna F.150L	Skegness Air Taxi Service Ltd/Boston	
G-AZLJ	BN-2A-1 Mk III Trislander	Avon Aviation Services Ltd/Bristol	
G-AZLL	Cessna FRA.150L	Airwork Ltd/Perth	
G-AZLM	Cessna F.172L	J. F. Davis & D. M. Boddy	
G-AZLN	PA-28 Cherokee 180F	D. H. L. Wigan	
G-AZLO	Cessna F.337F	Leasetec Ltd	
G-AZLP	V.813 Viscount	British Midland Airways Ltd/E. Midlands	
G-AZLR	V.813 Viscount	British Midland Airways Ltd/E. Midlands	
G-AZLS	V.813 Viscount	British Midland Airways Ltd/E. Midlands	
G-AZLV	Cessna 172K	D. S. McKay	
G-AZLY	Cessna F.150L	Cleveland Flying School Ltd/Tees-side	
G-AZLZ	Cessna F.150L	Exeter Flying Club Ltd	
G-AZMA	Jodel D.140B	I. J. Bishop & M. Kirk	
G-AZMB	Bell 47G-3B	Helicopter Farming Ltd	
G-AZMC	Slingsby T.61A Falke	Essex Gliding Club Ltd	
G-AZMD	Slingsby T.61C Falke	P. J. Moss & ptnrs/Long Marston	
G-AZMF	BAC One-Eleven 530	British Caledonian Airways/Gatwick	
G-AZMH	Morane-Saulnier M.S.500 (7A+WN)	Earl of Suffolk & Berkshire & B. D. Woodford	
G-AZMJ	AA-5 Traveler	R. T. Love/Bodmin	
G-AZMK	PA-23 Aztec 250	Andrew Edie Aviation/Shoreham	
G-AZMN	Glos-Airtourer T.5	R. G. Lowerson & T. Ellefson	
G-AZMO	PA-32 Cherokee Six 260	A. S. Bamrah	
G-AZMV	D.62C Condor	Ouse Gliding Club Ltd/Rufforth	
G-AZMX	PA-28 Cherokee 140 ★	Kelsterton College (*instructional airframe*)/Deeside	
G-AZMZ	M.S.893A Rallye Commodore 150	W. A. L. Mitchell	
G-AZNA	V.813 Viscount	British Midland Airways Ltd/E. Midlands	
G-AZNB	V.813 Viscount	British Midland Airways Ltd/E. Midlands	
G-AZNC	V.813 Viscount	British Midland Airways Ltd/E. Midlands	
G-AZNF	Stampe SV-4C	H. J. Smith/Booker	
G-AZNI	S.A.315B Lama	Dollar Air Services Ltd (G-AWLC)/ Coventry	
G-AZNK	Stampe SV-4A	A. E. Hutton/Duxford	
G-AZNL	PA-28R Cherokee Arrow 200D	Medical-Assist Ltd	
G-AZNO	Cessna 182P	M&D Aviation/Bournemouth	
G-AZNT	Cameron O-84 balloon	N. Tasker	
G-AZOA	MBB Bo 209 Monsun 150FF	R. J. O. Walker	
G-AZOB	MBB Bo 209 Monsun 150FF	G. N. Richardson	
G-AZOD	PA-23 Aztec 250D	Peregrine Air Services Ltd/Inverness	
G-AZOE	Glos-Airtourer 115	E. G. Whelan	
G-AZOF	Glos-Airtourer Super 150	Lands End Flying Club/St Just	
G-AZOG	PA-28R Cherokee Arrow 200D	Winchfield Enterprises Ltd	
G-AZOH	Beech 65-B90 Queen Air	Clyde Surveys Ltd/White Waltham	
G-AZOL	PA-34-200 Seneca	MTV Design Ltd/Jersey	
G-AZOM	MBB Bo 105D	Bristow Helicopters Ltd/Aberdeen	
G-AZON	PA-34-200-2 Seneca	Willowvale Electronics Ltd/Blackbushe	
G-AZOO	Western O-65 balloon	Southern Balloon Group *Carousel*	
G-AZOS	MJ.5-F1 Sirocco	R. Wells	
G-AZOT	PA-34-200-2 Seneca	L. G. Payne/Elstree	

Notes	Reg.	Type	Owner or Operator
	G-AZOU	Jodel DR.1051	T. W. Jones & ptnrs/Slinfold
	G-AZOZ	Cessna FRA.150L	L. C. Cole
	G-AZPA	PA-25 Pawnee 235	Black Mountain Gliding Co Ltd
	G-AZPC	Slingsby T.61C Falke	B. C. Dixon
	G-AZPF	Fournier RF-5	R. Pye/Blackpool
	G-AZPH	Craft-Pitts S-1S Special	Aerobatics International Ltd/ Farnborough
	G-AZPV	Luton LA-4A Minor	J. Scott/(*Stored*)
	G-AZPX	Western O-31 balloon	E. R. McCosh
	G-AZPZ	BAC One-Eleven 515	British Caledonian Airways Ltd/Gatwick
	G-AZRA	MBB Bo 209 Monsun	Alpha Flying Ltd/Denham
	G-AZRD	Cessna 401B	Spring Valley Trading Co Ltd/ Ronaldsway
	G-AZRF	Sikorsky S-61N	Bristow Helicopters *Pitcaple*
	G-AZRG	PA-23 Aztec 250D	Woodgate Aviation (IOM) Ltd/ Ronaldsway
	G-AZRH	PA-28 Cherokee 140D	Newcastle-upon-Tyne Aero Club Ltd
	G-AZRI	Payne balloon	G. F. Payne *Shoestring*
	G-AZRK	Fournier RF-5	Strathtay Flying Group/Perth
	G-AZRL	PA-19 Super Cub 95	S. J. & L. M. Harmer/Middle Wallop
	G-AZRM	Fournier RF-5	R. S. A. Lloyd-Bostock/Shoreham
	G-AZRN	Cameron O-84 balloon	M. Yarrow *Gravida II*
	G-AZRP	Glos-Airtourer 115	Tudor Flying School/Wellesbourne
	G-AZRR	Cessna 310Q	Ames Company (Transport) Ltd/ Blackpool
	G-AZRS	PA-22 Tri-Pacer 150	E. A. Harrhy/Shoreham
	G-AZRU	AB-206B JetRanger 2	Dollar Air Services Ltd/Coventry
	G-AZRV	PA-28R Cherokee Arrow 200B	Design for Sound Ltd
	G-AZRW	Cessna T.337C	R. C. Frazle/Southend
	G-AZRX	GY-80 Horizon 160	J. B. McBride/Aldergrove
	G-AZRZ	Cessna U-206F	Army Parachute Association/ Netheravon
	G-AZSA	Stampe SV-4B	J. K. Faulkner/Biggin Hill
	G-AZSC	AT-16 Harvard IIB (FT323)	Machine Music Ltd/Redhill
	G-AZSD	Slingsby T.29B Motor Tutor	R. G. Boynton
	G-AZSE	PA-28R Cherokee Arrow 200D	Northlink Storage Systems Ltd/ Aberdeen
	G-AZSF	PA-28R Cherokee Arrow 200D	W. T. Northorpe & R. J. Mills/Coventry
	G-AZSG	PA-28 Cherokee 180E	Cherokee Flying Group/Netherthorpe
	G-AZSH	PA-28R Cherokee Arrow 180	C. R. Hayward
	G-AZSK	Taylor JT.1 Monoplane	R. R. Lockwood
	G-AZSM	PA-28R Cherokee Arrow 180	Van Globe Ltd/Stapleford
	G-AZSN	PA-28R Cherokee Arrow 200	Northern Printing Machinery International Ltd
	G-AZSU	H.S.748 Srs 2A	Dan-Air Services Ltd/Gatwick
	G-AZSW	Beagle 121 Pup 1	Northamptonshire School of Flying Ltd/Sywell
	G-AZSX	Beagle 121 Pup 1	P. W. Hunter/Elstree
	G-AZSZ	PA-23 Aztec 250	Air Kilroe/Manchester
	G-AZTA	MBB Bo 209 Monsun 150FF	R. S. Perks/Elstree
	G-AZTD	PA-32 Cherokee Six 300	Presshouse Publications Ltd/Enstone
	G-AZTF	Cessna F.177RG	J. Bolson & Son Ltd/Bournemouth
	G-AZTK	Cessna F.172F	Y. R. Stewart/Andrewsfield
	G-AZTM	Glos-Airtourer 115	I. J. Smith
	G-AZTO	PA-34-200-2 Seneca	E. W. Noakes
	G-AZTR	SNCAN SV.4C	P. G. Palumbo/Booker
	G-AZTS	Cessna F.172L	Transgap Ltd/Manchester
	G-AZTV	Stolp SA.500 Starlet	The Stolp Group/Old Warden
	G-AZTW	Cessna F.177RG	R. M. Clarke/Leicester
	G-AZUM	Cessna F.172L	Shetland Flying Club Ltd
	G-AZUO	Cessna F.177RG	Newbury Sand and Gravel Co Ltd
	G-AZUP	Cameron O-65 balloon	R. S. Bailey & ptnrs
	G-AZUT	M.S.893A Rallye Commodore 180	Rallye Flying Group
	G-AZUV	Cameron O-65 balloon ★	British Balloon Museum
	G-AZUX	Western O-56 balloon	H. C. J. & Mrs S. L. G. Williams *Slow Djinn*
	G-AZUY	Cessna E.310L	Fountain Forestry Ltd
	G-AZUZ	Cessna FRA.150L	D. J. Parker/Netherthorpe
	G-AZVA	MBB Bo 209 Monsun 150FF	K. H. Wallis
	G-AZVB	MBB Bo 209 Monsun 150FF	P. C. Logsdon/Dunkeswell
	G-AZVE	AA-5 Traveler	R. N. Morant
	G-AZVF	M.S.894A Rallye Minerva	R. J. Cole & W. G. Gregory/Cardiff
	G-AZVG	AA-5 Traveler	J. N. C. Shields & ptnrs

Reg.	Type	Owner or Operator	Notes
G-AZVH	M.S.894A Rallye Minerva	Bristol Cars Ltd/White Waltham	
G-AZVI	M.S.892A Rallye Commodore	W. A. McCartney & T. A. Pugh	
G-AZVJ	PA-34-200-2 Seneca	Business Air Travel Ltd/Lydd	
G-AZVL	Jodel D.119	Forest Flying Group/Stapleford	
G-AZVM	Hughes 369HS	Diagnostic Reagents Ltd	
G-AZVP	Cessna F.177RG	R. W. Martin & R. G. Saunders/ Biggin Hill	
G-AZVR	Cessna F.150L	Suffolk Aero Club Ltd/Ipswich	
G-AZVT	Cameron O-84 balloon	Sky Soarer Ltd *Jules Verne*	
G-AZVV	PA-28 Cherokee 180G	H. M. Clarke	
G-AZWB	PA-28 Cherokee 140	J. Bowers/White Waltham	
G-AZWD	PA-28 Cherokee 140	Airways Aero Associations Ltd/Booker	
G-AZWE	PA-28 Cherokee 140	Airways Aero Associations Ltd/Booker	
G-AZWF	SAN Jodel DR.1050	R. A. Jarvis/Perth	
G-AZWS	PA-28R Cherokee Arrow 180	A. W. Gibbs (Holdings)/Manchester	
G-AZWT	Westland Lysander III (V9441)	Strathallan Aircraft Collection	
G-AZWW	PA-23 Aztec 250E	Phoenix Aviation (Bedford) Ltd/ Cranfield	
G-AZWY	PA-24 Comanche 260	Keymer Son & Co Ltd/Biggin Hill	
G-AZXA	Beechcraft 95-C55 Baron	Flight Refuelling Ltd/Bournemouth	
G-AZXB	Cameron O-65 balloon	London Balloon Club Ltd *London Pride II*	
G-AZXC	Cessna F.150L	S. J. Lawrence	
G-AZXD	Cessna F.172L	Birdlake Ltd/Wellesbourne	
G-AZXE	Jodel D.120A	Kestrel Flying Group/Hucknall	
G-AZXG	PA-23 Aztec 250	K. J. Le Fevre/Norwich	
G-AZXH	PA-34-200-2 Seneca	Carentals Ltd/Coventry	
G-AZXR	BN-2A-9 Islander	Stanton Aircraft Management Ltd	
G-AZYA	GY-80 Horizon 160	T. Poole & ptnrs/Sywell	
G-AZYD	M.S.893A Rallye Commodore	Deeside Gliding Club	
G-AZYF	PA-28 Cherokee 180	J. C. Glynn/E. Midlands	
G-AZYK	Cessna 310Q	Johnston Aviation	
G-AZYM	Cessna E-310Q	Kingswinford Engineering Co Ltd	
G-AZYS	CP.301C-1 Emeraude	J. R. Hughes/Stapleford	
G-AZYU	PA-E23 Aztec 250	L. J. Martin/Redhill	
G-AZYV	Burns O-77 balloon	B. F. G. Ribbans *Contrary Mary*	
G-AZYX	M.S.893A Rallye Commodore	Black Mountain Gliding Co Ltd	
G-AZYY	Slingsby T.61A Falke	J. A. Towers	
G-AZYZ	WA.51A Pacific	A. E. O'Broin	
G-AZZA	PA-E23 Aztec 250	Air Charter (Scotland) Ltd/Glasgow	
G-AZZB	AB-206B JetRanger 2	Gleneagle Helicopter Services (Scotland) Ltd	
G-AZZF	M.S.880B Rallye Club	I. C. Davies/Swansea	
G-AZZG	Cessna 188 Agwagon	*Wreck*/Southend	
G-AZZH	Practavia Pilot Sprite 115	K. G. Stewart	
G-AZZK	Cessna 414	Unifix Air Ltd/Stansted	
G-AZZO	PA-28 Cherokee 140	R. Hewitt & ptnrs/Stapleford	
G-AZZP	Cessna F.172H	Michael Gardner Ltd	
G-AZZR	Cessna F.150L	Herefordshire Aero Club Ltd/Shobdon	
G-AZZS	PA-34-200-2 Seneca	Medical Express Ltd/Shoreham	
G-AZZT	PA-28 Cherokee 180 ★	*Ground instruction airframe*/Cranfield	
G-AZZV	Cessna F.172L	Cleveland Flying School Ltd/Tees-side	
G-AZZW	Fournier RF-5	Aviation Special Developments	
G-AZZX	Cessna FRA.150L	J. E. Uprichard & ptnrs/Newtownards	
G-AZZZ	D.H.82A Tiger Moth	S. W. McKay	
G-BAAD	Evans Super VP-1	R. W. Husband/Netherthorpe	
G-BAAF	Manning-Flanders MF1 replica	Aviation Film Services Ltd/Booker	
G-BAAH	Coates SA.III Swalesong	J. R. Coates	
G-BAAI	M.S.893A Rallye Commodore	A. F. Butcher	
G-BAAK	Cessna 207	Sunderland Parachute Centre Ltd	
G-BAAL	Cessna 172A	B. Earl/Bodmin	
G-BAAP	PA-28R Cherokee Arrow 200	Shirley A. Shelley	
G-BAAT	Cessna 182P Skylane	J. B. Anderson/Newtownards	
G-BAAU	Enstrom F-28C-UK	Norman Bailey Helicopters Ltd/ Southampton	
G-BAAW	Jodel D.119	S. W. Ward/Ipswich	
G-BAAX	Cameron O-84 balloon	The New Holker Estate Co Ltd *Holker Hall*	
G-BAAY	Valtion Viima II (VI-3)	P. H. McConnell/White Waltham	
G-BAAZ	PA-28A Cherokee Arrow 200D	A. W. Rix/Guernsey	
G-BABA	D.H.82A Tiger Moth	S. W. McKay	
G-BABB	Cessna F.150L	DEH Designs Ltd/Southampton	
G-BABC	Cessna F.150L	Suffolk Aero Club Ltd/Ipswich	

Notes	Reg.	Type	Owner or Operator
	G-BABD	Cessna FRA.150L	Phoenix Aviation (Bedford) Ltd/ Cranfield
	G-BABE	Taylor JT.2 Titch	P. D. G. Grist
	G-BABG	PA-28 Cherokee 180	P. D. Foster
	G-BABH	Cessna F.150L	B. McIntyre & N. Simpson
	G-BABK	PA-34-200-2 Seneca	D. F. J. & N. R. Flashman/Biggin Hill
	G-BABY	Taylor JT.2 Titch	J. R. D. Bygraves/O. Warden
	G-BACA	BAC Petrel	British Aircraft Corporation Ltd/Warton
	G-BACB	PA-34-200-2 Seneca	Business Air Travel Ltd/Lydd
	G-BACC	Cessna FRA.150L	Chainrose Ltd/Little Staughton
	G-BACE	Fournier RF-5	R. W. K. Stead/Perranporth
	G-BACH	Enstrom F.28A	Finload Ltd/Southampton
	G-BACJ	Jodel D.120	Wearside Flying Association
	G-BACK	D.H.82A Tiger Moth (DF130)	G. R. French & ptnrs
	G-BACL	Jodel D.150	G. R. French
	G-BACM	Cessna FRA.150L	W. Vinten Ltd/Ipswich
	G-BACN	Cessna FRA.150L	Airwork Ltd/Perth
	G-BACO	Cessna FRA.150L	H. G. & V. Fawkes/Bodmin
	G-BACP	Cessna FRA.150L	B. A. Mills
	G-BADC	Luton Beta B.2A	H. M. Mackenzie
	G-BADE	PA-23 Aztec 250	RCR Aviation Ltd
	G-BADH	Slingsby T.61A Falke	E. M. Andrew & ptnrs/Old Sarum
	G-BADI	PA-E23 Aztec 250	W. London Aero Services Ltd/ White Waltham
	G-BADJ	PA-E23 Aztec 250	J. G. Hogg/Rochester
	G-BADK	BN-2A-8 Islandser	Harvest Air Ltd/Southend
	G-BADL	PA-34-200-2 Seneca	Cartographical Services (Southampton) Ltd/Birmingham
	G-BADO	PA-32 Cherokee Six 300	B. J. Haylor & ptnrs/Southampton
	G-BADP	Boeing 737-204	Britannia Airways Ltd *Sir Arthur Whitten Brown*/Luton
	G-BADR	Boeing 737-204	Britannia Airways Ltd *Capt Robert Falconer Scott*/Luton
	G-BADT	Cessna 402B	British Aircraft Corp Ltd/Warton
	G-BADU	Cameron O-56 balloon	J. Philp *Dream Machine*
	G-BADV	Brochet MB-50	P. A. Cairns/Dunkeswell
	G-BADW	Pitts S-2A Special	Aerospace Museum/Cosford
	G-BADZ	Pitts S-2A Special	A. L. Brown & ptnrs/Bourn
	G-BAEB	Robin DR.400/160	Bracknell Refrigeration Services Ltd/ Fairoaks
	G-BAEC	Robin HR.100/210	Autographics Ltd/White Waltham
	G-BAED	PA-E23 Aztec 250	Dairo Air Services (Jersey) Ltd
	G-BAEE	Jodel DR.1050/M1	J. B. Randle
	G-BAEF	Boeing 727-46	Dan-Air Services Ltd/Gatwick
	G-BAEM	Robin DR.400/125	Store Equipment (London) Ltd
	G-BAEN	Robin DR.400/180	Trans Europe Air Charter Ltd/Biggin Hill
	G-BAEP	Cessna FRA.150L	RFC (Bourn) Ltd
	G-BAER	Cosmic Wind	R. S. Voice/Redhill
	G-BAES	Cessna 337A	Page & Moy Ltd & High Voltage Applications Ltd/Leicester
	G-BAET	Piper L-4H Cub	C. M. G. Ellis
	G-BAEU	Cessna F.150L	Skyviews & General Ltd
	G-BAEV	Cessna FRA.150L	P. H. A. Dupon
	G-BAEW	Cessna F.172M	Northamptonshire School of Flying Ltd/ Sywell
	G-BAEX	Cessna F.172M	Wycombe Air Centre Ltd
	G-BAEY	Cessna F.172M	R. Fursman/Southampton
	G-BAEZ	Cessna FRA.150L	J. C. Glynn/E. Midlands
	G-BAFA	AA-5 Traveler	C. F. Mackley
	G-BAFD	MBB Bo 105D	Gleneagle Helicopter Services (Scotland) Ltd
	G-BAFG	D.H.82A Tiger Moth	C. D. Cyster
	G-BAFH	Evans VP-1 Volksplane	R. H. W. Beath
	G-BAFI	Cessna F.177RG	W. J. Jenkins & M. Giles/Swansea
	G-BAFL	Cessna 182P	Ingham Aviation Ltd/Lulsgate
	G-BAFM	AT-16 Harvard IIB (FS728)	J. Parks/Southampton
	G-BAFP	Robin DR.400/160	P. R. Ashley & M. Gauna/ Netherthorpe
	G-BAFS	PA-18 Super Cub 150	Burns Gliding Club Ltd
	G-BAFT	PA-18 Super Cub 150	Cambridge University Gliding Trust Ltd/ Duxford
	G-BAFU	PA-28 Cherokee 140	Goshawk Aviation Ltd/Southend
	G-BAFV	PA-18 Super Cub 95	P. Elliott/Biggin Hill

Reg.	Type	Owner or Operator	Notes
G-BAFW	PA-28 Cherokee 140	H. T. Boal & ptnrs/Cambridge	
G-BAFX	Robin DR.400/140	D. C. R. Writer/Headcorn	
G-BAFZ	Boeing 727-46	Dan-Air Services Ltd/Gatwick	
G-BAGB	SIAI-Marchetti SF.260	British Midland Airways Ltd/ E. Midlands	
G-BAGC	Robin DR.400/140	Hempalm Ltd/Headcorn	
G-BAGF	Jodel D.92 Bebe	G. R. French & J. D. Watt	
G-BAGG	PA-32 Cherokee Six 300E	Hornair Ltd/Jersey	
G-BAGI	Cameron O-31 balloon	Cameron Balloons Ltd *Vital Spark*	
G-BAGL	SA.341 G. Gazelle Srs 1	Westland Helicopters Ltd/Yeovil	
G-BAGN	Cessna F.177RG	M. L. Rhodes/Halfpenny Green	
G-BAGO	Cessna 421B	Donington Aviation Ltd/E. Midlands	
G-BAGR	Robin DR.400/140	F. C. Aris & ptnrs/Mona	
G-BAGS	Robin DR.400/180 2+2	Headcorn Flying School Ltd	
G-BAGT	Helio H.295 Courier	B. J. C. Woodall Ltd	
G-BAGU	Luton LA-5A Major	J. Gawley	
G-BAGV	Cessna U.206F	Scottish Parachute Club/Strathallan	
G-BAGX	PA-28 Cherokee 140	Klingair Ltd/Conington	
G-BAGY	Cameron O-84 balloon	P. G. Dunnington *Beatrice*	
G-BAHD	Cessna 182P Skylane	S. Brunt (Silverdale Staffs) Ltd/Sleap	
G-BAHE	PA-28 Cherokee 140	A. H. Evans & A. O. Jones	
G-BAHF	PA-28 Cherokee 140	Blackberry Lane Garage Ltd & Midland Aircraft Maintenance Ltd	
G-BAHG	PA-24 Comanche 260	Friendly Aviation (Jersey) Ltd	
G-BAHH	Wallis WA.121	K. H. Wallis	
G-BAHI	Cessna F.150H	M. D. Barrow	
G-BAHJ	PA-24 Comanche 250	M. D. Faiers	
G-BAHL	Robin DR.400/160	Norvett Electronics Ltd	
G-BAHN	Beech 58 Baron	British Midland Airways/E. Midlands	
G-BAHO	Beech C.23 Sundowner	G-ATJG Private Aircraft Syndicate Ltd/Denham	
G-BAHP	Volmer VJ.22 Sportsman	M. T. Moore & E. P. Beck	
G-BAHS	PA-28R Cherokee Arrow 200-II	A. A. Wild & ptnrs	
G-BAHU	Enstrom F-28A	Rassler Aero Services/Booker	
G-BAHX	Cessna 182P	S. Shorrock/Blackpool	
G-BAHZ	PA-28R Cherokee Arrow 200-II	Tenison Air Ltd	
G-BAIA	PA-32 Cherokee Six 300E	Langham International (Aircraft) Ltd/ Southend	
G-BAIB	Enstrom F-28A	C.S.E. Aviation Ltd/Kidlington	
G-BAIH	PA-28R Cherokee Arrow 200-II	J. Pemberton/Cambridge	
G-BAII	Cessna FRA.150L	Airwork Ltd/Perth	
G-BAIK	Cessna F.150L	Wickenby Aviation Ltd	
G-BAIL	Cessna FR.172J	London Parachuting Ltd/Pampisford	
G-BAIM	Cessna 310Q	Airwork Ltd/Perth	
G-BAIN	Cessna FRA.150L	Airwork Ltd/Perth	
G-BAIP	Cessna F.150L	W. D. Cliffe/Wellesbourne	
G-BAIR	Thunder Ax7-77 balloon	P. A. & Mrs M. Hutchins	
G-BAIS	Cessna F.177RG	I. H. Bewley	
G-BAIU	Hiller UH-12E (Soloy)	Heliwork Finance Ltd/Thruxton	
G-BAIW	Cessna F.172M	Humber Aviation Ltd	
G-BAIX	Cessna F.172M	John Cordery Aviation Ltd/Elstree	
G-BAIY	Cameron O-65 balloon	Budget Rent A Car (UK) Ltd *Lady Budget*	
G-BAIZ	Slingsby T.61A Falke	W. L. C. O'Neill & ptnrs	
G-BAJA	Cessna F.177RG	Don Ward Productions Ltd/Biggin Hill	
G-BAJB	Cessna F.177RG	Brittfish (Hull) Ltd	
G-BAJC	Evans VP-1	J. R. Clements/Headcorn	
G-BAJE	Cessna 177 Cardinal	J. E. Cull	
G-BAJN	AA-5 Traveler	Janacrew Ltd/Sherburn	
G-BAJO	AA-5 Traveler	A. Townson/Blackpool	
G-BAJR	PA-28 Cherokee 180	K. Foster & Son/Humberside	
G-BAJT	PA-28R Cherokee Arrow 200-II	J. M. Giles & B. R. Rossiter	
G-BAJW	Boeing 727-46	Dan-Air Services Ltd/Gatwick	
G-BAJY	Robin DR.400/180	F. Birch & K. J. Pike/Sturgate	
G-BAJZ	Robin DR.400/125	Readwell Aviation/Rochester	
G-BAKA	Sikorsky S-61N	Bristow Helicopters Ltd *West Sole*	
G-BAKB	Sikorsky S-61N	Bristow Helicopters Ltd *Montrose*	
G-BAKC	Sikorsky S-61N	Bristow Helicopters Ltd *Forties*	
G-BAKD	PA-34-200-2 Seneca	Andrews Professional Colour Laboratories/Elstree	
G-BAKF	Bell 206B JetRanger 2	M. J. K. Belmont/Coventry	
G-BAKG	Hughes 269C	W. K. MacGillivray	

Notes	Reg.	Type	Owner or Operator
	G-BAKH	PA-28 Cherokee 140	Woodgate Air Services (IoM) Ltd/ Ronaldsway
	G-BAKJ	PA-30 Twin Comanche 160	M. F. Fisher
	G-BAKK	Cessna F.172H	C. M. Hampson/Coventry
	G-BAKL	F.27 Friendship 200	Air UK/Norwich
	G-BAKM	Robin DR.400/140	M. N. King & J. D. Spencer
	G-BAKN	SNCAN SV-4C	M. Holloway
	G-BAKO	Cameron O-84 balloon	D. C. Dokk-Olsen *Pied Piper*
	G-BAKP	PA-E23 Aztec 250	J. J. Woodhouse
	G-BAKR	Jodel D.117	A. B. Bailey/White Waltham
	G-BAKS	A-B 206B JetRanger 2	Kilroe Helicopters Ltd/Manchester
	G-BAKT	A-B 206B JetRanger 2	Gleneagle Helicopter Services (Scotland) Ltd
	G-BAKV	PA-18 Super Cub 150	Pounds Marine Shipping Ltd/ Goodwood
	G-BAKW	B.121 Pup 2	John Hicks Group/Shoreham
	G-BAKY	Slingsby T.61C Falke	G. Hill & P. Shepherd
	G-BALC	Bell 206B JetRanger 2	Dollar Air Services Ltd/Coventry
	G-BALE	Enstrom F.28A	Reynard Racing Cars Ltd
	G-BALF	Robin DR.400/140	F. A. Spear/Panshanger
	G-BALG	Robin DR.400/180	R. Jones
	G-BALH	Robin DR.400/140B	R. Pidcock & ptnrs
	G-BALI	Robin DR.400 2+2	R. A. Gridley
	G-BALJ	Robin DR.400/180	D. Batt & ptnrs/Headcorn
	G-BALK	SNCAN SV-4C	J. C. Brierley
	G-BALM	Cessna 340	Manro Transport Ltd/Manchester
	G-BALN	Cessna T.310Q	O'Brien Properties Ltd/Shoreham
	G-BALP	PA-39 Twin Comanche 160 C/R	N. K. Thomas
	G-BALS	Nipper T.66 Mk 3	L. W. Shaw
	G-BALT	Enstrom F28A	W. Gray/Cambridge
	G-BALW	PA-28R Cherokee Arrow 200-II	H. R. Fenwick/Glasgow
	G-BALX	D.H.82A Tiger Moth (N6848)	Toadair
	G-BALY	Practavia Pilot Sprite 150	A. L. Young
	G-BALZ	Bell 212	B.E.A.S. Ltd/Redhill
	G-BAMB	Slingsby T.61C Falke	Universities of Glasgow & Strathclyde Gliding Club/Strathaven
	G-BAMC	Cessna F.150L	D. H. Pattison
	G-BAME	Volmer VJ-22 Sportsman	V. H. Bellamy/St Just
	G-BAMF	MBB Bo 105D	Bond Helicopters Ltd/Bourn
	G-BAMG	Avions Lobet Ganagobie	J. A. Brompton
	G-BAMJ	Cessna 182P	Melbourns Brewery Ltd
	G-BAMK	Cameron D.96 hot-air airship	Cameron Balloons Ltd
	G-BAML	Bell 206A JetRanger	Blue Star Ship Management Ltd
	G-BAMM	PA-28 Cherokee 235	Holmfield Wakefield Ltd
	G-BAMR	PA-16 Clipper	H. Royce
	G-BAMS	Robin DR.400/160	G-BAMS Ltd/Headcorn
	G-BAMU	Robin DR.400/160	The Alternative Flying Group/Sywell
	G-BAMV	Robin DR.400/180	W. J. Gooding
	G-BAMY	PA-28R Cherokee Arrow 200-II	B. A. W. Gittins & ptnrs/Birmingham
	G-BANA	Robin DR.221	G. T. Pryor
	G-BANB	Robin DR.400/180	Time Electronics Ltd/Headcorn
	G-BANC	GY-201 Minicab	D. H. Pattison
	G-BAND	Cameron O-84 balloon	Mid-Bucks Farmers Balloon Group *Clover*
	G-BANE	Cessna FRA.150L	D. J. Park & A. P. Clarke/Blackpool
	G-BANF	Luton LA-4A Minor	D. W. Bosworth/Shobdon
	G-BANK	PA-34-200-2 Seneca	C. E. Branfield & D. Phillips
	G-BANL	BN-2A-8 Islander	Loganair Ltd/Glasgow
	G-BANS	PA-34-200-2 Seneca	G. Knowles/Halfpenny Green
	G-BANT	Cameron O-65 balloon	M. A. Dworski & R. M. Bishop
	G-BANU	Wassmer Jodel D.120	C. E. McKinney
	G-BANV	Phoenix Currie Wot	K. Knight/Shobdon
	G-BANW	CP-1330 Super Emeraude	J. D. McCracker & ptnrs/E. Fortune
	G-BANX	Cessna F.172M	I. R. March
	G-BAOB	Cessna F.172M	Gordon King (Aviation) Ltd/Biggin Hill
	G-BAOG	M.S.880B Rallye Club	J. L. Quinlan
	G-BAOH	M.S.880B Rallye Club	S. P. Bryant & ptnrs/Wellesbourne
	G-BAOJ	M.S.880B Rallye Club	D. W. Busby & J. L. Howard
	G-BAOM	M.S.880B Rallye Club	D. W. Brown/Seething
	G-BAOP	Cessna FRA.150L	Business Air Travel Ltd/Lydd
	G-BAOS	Cessna F.172M	M. E. Aldridge
	G-BAOT	M.S.880B Rallye Club	H. F. Hambling
	G-BAOU	AA-5 Traveler	W. H. Ingram/St Just
	G-BAOV	AA-5A Cheetah	G. K. Ellerker

Reg.	Type	Owner or Operator	Notes
G-BAOW	Cameron O-65 balloon	P. I. White *Winslow Boy*	
G-BAOY	Cameron S-31 balloon	Shell-Mex BP Ltd *New Potato*	
G-BAPA	Fournier RF-5B Sperber	R. Pasold & D. Steynor/Booker	
G-BAPB	DHC-1 Chipmunk 22	R. C. P. Brookhouse/Redhill	
G-BAPC	Luton LA-4A Minor	Midland Aircraft Preservation Soc	
G-BAPF	V.814 Viscount	British Midland Airways Ltd/E. Midlands	
G-BAPG	V.814 Viscount	Philstone International Ltd/Exeter	
G-BAPH	Cessna FRA.150L	E. Shipley	
G-BAPI	Cessna FRA.150L	Industrial Supplies (Peterborough) Ltd	
G-BAPJ	Cessna FRA.150L	M. D. Page/Manston	
G-BAPK	Cessna F.150L	Andrewsfield Flying Club Ltd	
G-BAPL	PA-23 Aztec 250E	Scottish Malt Distillers Ltd/Lossiemouth	
G-BAPM	Fuji FA.200-160	Papa Mike Group/Swansea	
G-BAPN	PA-28 Cherokee 180	H. Clayton & J. O. Carlisle/E. Midlands	
G-BAPP	Evans VP-1	N. Crow	
G-BAPR	Jodel D.11	Gt. Consall Copper Mines Co Ltd	
G-BAPS	Campbell Cougar ★	British Rotorcraft Museum	
G-BAPT	Fuji FA.200-180	J. F. Thurlow & J. H. Pickering/Ipswich	
G-BAPV	Robin DR.400/160	J. D. Millne & ptnrs	
G-BAPW	PA-28R Cherokee Arrow 180	G. & R. Consultants Ltd	
G-BAPX	Robin DR.400/160	R. R. Hall & R. H. Richards	
G-BAPY	Robin HR.100/210	Engineering Appliances Ltd/Booker	
G-BARB	PA-34-200-2 Seneca	S. J. Dixon	
G-BARC	Cessna FR.172J	C. Porter & ptnrs/Defford	
G-BARD	Cessna 337C	P. L. Aviation Consultants	
G-BARF	Jodel D.112 Club	J. E. Shepherd & ptnrs	
G-BARG	Cessna E.310Q	Nottingham Building Soc Ltd/ Tollerton	
G-BARH	Beech C.23 Sundowner	L. J. Fisher	
G-BARJ	Bell 212	Autair International Ltd/Panshanger	
G-BARN	Taylor JT.2 Titch	R. G. W. Newton	
G-BARP	Bell 206B JetRanger 2	S.W. Electricity Board/Bristol	
G-BARS	D.H.C.1. Chipmunk 22	T. I. Sutton/Chester	
G-BARV	Cessna 310Q	Old England Watches Ltd/Elstree	
G-BARX	Bell 206B JetRanger 2	W. R. Finance Ltd	
G-BARY	CP.301A Emeraude	J. J. Butler	
G-BARZ	Scheibe SF.28A Tandem Falke	J. A. Fox & ptnrs/Dishforth	
G-BASB	Enstrom F-28A	Blades Helicopters Ltd/Goodwood	
G-BASD	B.121 Pup 2	C. C. Brown/Leicester	
G-BASE	Bell 206B JetRanger 2	Air Hanson Ltd/Brooklands	
G-BASG	AA-5 Traveler	R. Brooks	
G-BASH	AA-5 Traveler	M. J. Metham/Blackbushe	
G-BASI	PA-28 Cherokee 140	Telepoint Ltd/Manchester	
G-BASJ	PA-28 Cherokee 180	Guildhaven Ltd	
G-BASL	PA-28 Cherokee 140	Air Navigation & Trading Ltd/Blackpool	
G-BASM	PA-34-200-2 Seneca	Eastern Enterprises/Southend	
G-BASN	Beech C.23 Sundowner	M. F. Fisher	
G-BASO	Lake LA-4 Amphibian	P. B. W. Spearing & H. W. A. Deacon	
G-BASP	B.121 Pup 1	Northamtonshire School of Flying Ltd/ Sywell	
G-BASU	PA-31-350 Navajo Chieftain	Streamline Aviation	
G-BASX	PA-34-200-2 Seneca	Willowvale Electronics Ltd/Elstree	
G-BASY	Jodel D.9 Bebe	R. L. Sambell	
G-BATC	MBB Bo 105D	Bond Helicopters Ltd/Bourn	
G-BATH	Cessna F.337G	DJF Group Management	
G-BATJ	Jodel D.119	E. G. Waite	
G-BATM	PA-32 Cherokee Six 300	Patgrove Ltd/Bolney	
G-BATN	PA-E23 Aztec 250	Marshall of Cambridge Ltd	
G-BATR	PA-34-200-2 Seneca	Midland Air Taxis Ltd/Halfpenny Green	
G-BATS	Taylor JT.1 Monoplane	J. Jennings	
G-BATT	Hughes 269C	Keith Sutcliffe & Sons Ltd	
G-BATU	Enstrom F.28A-UK	Executive Helicopters Ltd	
G-BATV	PA-28 Cherokee 180D	The Scoreby Flying Group/Sherburn	
G-BATW	PA-28 Cherokee 140	J. E. Shepherd	
G-BATX	PA-23 Aztec 250E	Tayside Aviation Ltd/Dundee	
G-BAUA	PA-E23 Aztec 250	David Parr & Associates Ltd/Shobdon	
G-BAUC	PA-25 Pawnee 235	P. M. Charles/Little Snoring	
G-BAUD	Robin DR.400/160	R. E. Delvis/Shoreham	
G-BAUE	Cessna 310Q	A. J. Dyer/Elstree	
G-BAUH	Jodel D.112	G. A. & D. Shepherd	
G-BAUI	PA-E23 Aztec 250	SFT Aviation Ltd/Bournemouth	
G-BAUJ	PA-E23 Aztec 250	Express Aviation Services Ltd/ Biggin Hill	
G-BAUK	Hughes 269C	Curtis Engineering (Frome) Ltd	

Notes	Reg.	Type	Owner or Operator
	G-BAUR	F.27 Friendship Mk 200	Air UK Ltd/Norwich
	G-BAUV	Cessna F.150L	Cooper Airmotive (UK) Ltd/Kidlington
	G-BAUW	PA-E23 Aztec 250	R. E. Myson/Stapleford
	G-BAUY	Cessna FRA.150L	Glencair (Aero Services) Ltd
	G-BAUZ	Nord NC.854S	W. A. Ashley & D. Horne
	G-BAVB	Cessna F.172M	Hudson Bell Aviation/Southend
	G-BAVC	Cessna F.150L	Elles Aviation/Biggin Hill
	G-BAVE	Beech A.100 King Air	Vernair Transport Services/Liverpool
	G-BAVF	Beech 58 Baron	I. K. I. Stewart/Elstree
	G-BAVH	D.H.C.1 Chipmunk 22	Portsmouth Naval Gliding Club/ Lee-on-Solent
	G-BAVL	PA-E23 Aztec 250	Hampshire Consultancy Ltd
	G-BAVO	Boeing Stearman N2S (26)	Keenair Services Ltd/Liverpool
	G-BAVR	AA-5 Traveler	W. Midlands Flying Centre
	G-BAVS	AA-5 Traveler	V. J. Peake
	G-BAVU	Cameron A-105 balloon	J. D. Michaelis
	G-BAVX	HPR-7 Herald 214	British Air Ferries Ltd/Southend
	G-BAVZ	PA-E23 Aztec 250	Merseyside Air Charter Ltd/Liverpool
	G-BAWB	PA-E23 Aztec 250	Sutaberry Ltd
	G-BAWG	PA-28R-200-2 Cherokee Arrow	South Coast Aero Club/Goodwood
	G-BAWI	Enstrom F-28A-UK	Swiftshield Ltd/Staverton
	G-BAWK	PA-28 Cherokee 140	Newcastle-Upon-Tyne Aero Club Ltd
	G-BAWL	Airborne Industries gas airship	A. F. J. Smith *The Santos-Dumont*
	G-BAWN	PA-30C Twin Comanche 160	J. & Y. Plastics (Aviation) Ltd/ Manchester
	G-BAWR	Robin HR.100/210	Nelson Air/Shoreham
	G-BAWU	PA-30 Twin Comanche 160	Kwikflite Ltd/Fairoaks
	G-BAWV	PA-E23 Aztec 250	Merseyside Commercial Sales & Service Ltd/Manchester
	G-BAWW	Thunder Ax7-77 balloon	M. L. C. Hutchins *Taurus*/Holland
	G-BAWX	PA-28 Cherokee 180	Bawxair Ltd/Leeds
	G-BAWZ	Cessna 402B	Beavergrain Ltd/Biggin Hill
	G-BAXD	BN-2A Mk III Trislander	Aurigny Air Services/Guernsey
	G-BAXE	Hughes 269A	Reethorpe Engineering Ltd
	G-BAXF	Cameron O-77 balloon	R. D. Sargeant & M. F. Lasson
	G-BAXH	Cessna 310Q	D. A. Williamson
	G-BAXJ	PA-32 Cherokee Six 300	UK Parachute Services/Ipswich
	G-BAXK	Thunder Ax7-77 balloon	Newbury Balloon Group *Jack O'Newbury*
	G-BAXL	H.S.125 Srs 3B	Dennis Vanguard International (Switchgear) Ltd/Coventry
	G-BAXM	Beech B.24R Sierra	McGovern PVC Windows
	G-BAXP	PA-E23 Aztec 250	Prestwick Aircraft Leasing
	G-BAXS	Bell 47G-5	Helicopter Supplies & Engineering Ltd/ Bournemouth
	G-BAXT	PA-28R-200 Cherokee Arrow	Williams & Griffin Ltd
	G-BAXU	Cessna F.150L	W. Lancs Aero Club Ltd/Woodvale
	G-BAXY	Cessna F.172M	The Bearing Mart Ltd/Manchester
	G-BAXZ	PA-28 Cherokee 140	H. Martin & D. Norris
	G-BAYC	Cameron O-65 balloon	D. Whitlock & R. T. F. Mitchell
	G-BAYL	Nord 1203/III Norecrin	D. M. Fincham/Bodmin
	G-BAYO	Cessna 150L	Cheshire Air Training School Ltd/ Liverpool
	G-BAYP	Cessna 150L	Three Counties Aero Club Ltd/ Blackbushe
	G-BAYR	Robin HR.100/210	Gilbey Warren Co Ltd
	G-BAYV	SNCAN 1101 Noralpha ★	Booker Aircraft Museum
	G-BAYX	Bell 47G-5	John Holbord (Farm Helicopters) Ltd
	G-BAYY	Cessna 310C	Specialist Flying Training Ltd/Carlisle
	G-BAYZ	Bellanca 7GC BC Citabria	Cambridge University Gliding Trust Ltd/ Duxford
	G-BAZB	H.S.125 Srs 400B	Short Bros Ltd/Sydenham
	G-BAZC	Robin DR.400/160	Sherburn Robin Group
	G-BAZF	AA-5 Traveler	N. London Flying Club/Elstree
	G-BAZG	Boeing 737-204	Britannia Airways Ltd *Florence Nightingale*/Luton
	G-BAZI	Boeing 737-204	Airways International Cymru Ltd/ Cardiff
	G-BAZJ	HPR-7 Herald 209 ★	Guernsey Airport Fire Services
	G-BAZM	Jodel D.11	Bingley Flying Group/Leeds
	G-BAZN	Bell 206B JetRanger 2	Blue Star Ship Management Ltd
	G-BAZS	Cessna F.150L	Sherburn Aero Club Ltd
	G-BAZT	Cessna F.172M	M. Fraser/Exeter
	G-BAZU	PA-28R-200 Cherokee Arrow	Andytruc Ltd/White Waltham

Reg.	Type	Owner or Operator	Notes
G-BBAE	L.1011-385 TriStar	British Airtours *Torbay*/Gatwick	
G-BBAF	L.1011-385 TriStar	British Airways *Babbacombe Bay*/ Heathrow	
G-BBAG	L.1011-385 TriStar	British Airways *Bridgwater Bay*/ Heathrow	
G-BBAH	L.1011-385 TriStar	British Airways *Lyme Bay*/Heathrow	
G-BBAI	L.1011-385 TriStar	British Airtours *St Brides Bay*/ Gatwick	
G-BBAJ	L.1011-385 TriStar	British Airtours Ltd *The Elizabeth Harkness Rose*/Gatwick	
G-BBAK	M.S.894A Rallye Minerva	W. G. Henderson/Glenrothes	
G-BBAR	Jodel D.117	J. F. Wright	
G-BBAW	Robin HR.100/210	Scoba Ltd/Goodwood	
G-BBAX	Robin DR.400/140	S. R. Young	
G-BBAY	Robin DR.400/140	G. A. Pentelow & D. B. Roadnight	
G-BBAZ	Hiller UH-12E	Bond Helicopters Ltd/Bourn	
G-BBBA	Hiller UH-12E	Bond Helicopters Ltd/Bourn	
G-BBBB	Taylor JT.1 Monoplane	S. A. MacConnacher	
G-BBBC	Cessna F.150L	K. Firth	
G-BBBI	AA-5 Traveler	C. B. Dew	
G-BBBK	PA-28 Cherokee 140	Bencray Ltd/Blackpool	
G-BBBM	Bell 206B JetRanger 2	D. M. Leasing Co	
G-BBBN	PA-28 Cherokee 180	B. R. Rossiter/White Waltham	
G-BBBO	SIPA 903	J. S. Hemmings/Lydd	
G-BBBW	FRED Series 2	D. L. Webster/Sherburn	
G-BBBX	Cessna E310L	Air Atlantique Ltd/Jersey	
G-BBBY	PA-28 Cherokee 140	R. A. E. Tremlett/Guernsey	
G-BBCA	Bell 206B JetRanger 2	Helicopter Hire Ltd/Southend	
G-BBCB	Western O-65 balloon	M. Westwood *Cee Bee*	
G-BBCC	PA-E23 Aztec 250	Dorglen Ltd	
G-BBCD	Beech 95-B55 Baron	L. M. Tulloch	
G-BBCF	Cessna FRA.150L	Yorkshire Light Aircraft Ltd/Leeds	
G-BBCG	Robin DR.400/2+2	Headcorn Flying School Ltd	
G-BBCH	Robin DR. 400/2+2	Headcorn Flying School Ltd	
G-BBCI	Cessna 150H	N. R. Windley	
G-BBCJ	Cessna 150J	P. Channon	
G-BBCK	Cameron O-77 balloon	R. J. Leathart *The Mary Gloster*	
G-BBCM	PA-E23 Aztec 250	W. R. C. M. Foyle	
G-BBCN	Robin HR.100/210	K. T. G. Atkins	
G-BBCP	Thunder Ax6-56 balloon	J. M. Robinson *Jack Frost*	
G-BBCS	Robin DR.400/140	J. A. Thomas	
G-BBCW	PA-E23 Aztec 250	Deborah Services Ltd/Sturgate	
G-BBCY	Luton LA-4A Minor	C. H. Difford/Dunkeswell	
G-BBCZ	AA-5 Traveler	Stronghill Flying Group/Bournemouth	
G-BBDA	AA-5 Traveler	David Burke Marine Ltd	
G-BBDB	PA-28 Cherokee 180	T. D. Strange/Newtownards	
G-BBDC	PA-28 Cherokee 140	P. E. Quick	
G-BBDD	PA-28 Cherokee 140	FR Aviation	
G-BBDE	PA-28R-200-2 Cherokee Arrow	R. L. Coleman	
G-BBDG	Concorde 100	British Aerospace PLC/Filton	
G-BBDH	Cessna F.172M	A. E. & G. R. Garner Ltd/Mona	
G-BBDI	PA-18-150 Super Cub	Scottish Gliding Union Ltd	
G-BBDJ	Thunder Ax6-56 balloon	S. W. D. & H. B. Ashby *Jack Tar*	
G-BBDK	V.808F Viscount Freightmaster	British Air Ferries *Viscount Lindley*/ Southend	
G-BBDL	AA-5 Traveler	J. Jones/Halfpenny Green	
G-BBDM	AA-5 Traveler	E. M. Pettit Construction Ltd/Stapleford	
G-BBDN	Taylor JT.1 Monoplane	J. A. Bennett	
G-BBDO	PA-E23 Aztec 250	R. Long/Bristol	
G-BBDP	Robin DR.400/160	Jarrett & Plumb Aviation (Rochester) Ltd	
G-BBDS	PA-31 Navajo	Broad Oak Air Services/Rochester	
G-BBDT	Cessna 150H	J. M. McCloy/Sherburn	
G-BBDU	PA-31 Navajo	ITT Components Ltd/Stansted	
G-BBDV	SIPA S.903	A. W. Webster	
G-BBEA	Luton LA-4A Minor	D. J. Wells & ptnrs/Fenland	
G-BBEB	PA-28R-200-2 Cherokee Arrow	R. D. Rippingale/Thruxton	
G-BBEC	PA-28 Cherokee 180	J. H. Kimber	
G-BBED	M.S.894B Rallye Minerva	Trago Mills Ltd/Bodmin	
G-BBEF	PA-28 Cherokee 140	Air Navigation & Trading Co Ltd/ Blackpool	
G-BBEI	PA-31 Navajo	BKS Surveys Ltd/Exeter	
G-BBEL	PA-28R Cherokee Arrow 180	J. M. McRitchie/Newtownards	

Notes	Reg.	Type	Owner or Operator
	G-BBEN	Bellanca 7GCBC Citabria	Ulster Gliding Club Ltd
	G-BBEO	Cessna FRA.150L	Granair Ltd
	G-BBEU	Bell 206B JetRanger 2	Air Hanson Ltd/Brooklands
	G-BBEV	PA-28 Cherokee 140	J. C. McVey/Liverpool
	G-BBEW	PA-E23 Aztec 250	Air Furness
	G-BBEX	Cessna 185A Skywagon	R. G. Brooks & D. E. Wilson/Dunkeswell
	G-BBEY	PA-E23 Aztec 250	Berrard Aviation/Blackbushe
	G-BBFC	AA-1B Trainer	R. C. Gillingham & G. Mobey/Lydd
	G-BBFD	PA-28R-200-2 Cherokee Arrow	Delta Sound Services Ltd
	G-BBFE	Bell 206 JetRanger	Heliwork (Finance) Ltd/Thruxton
	G-BBFL	GY-201 Minicab	M. F. Coy/Wellesbourne
	G-BBFS	Van Den Bemden gas balloon	A. J. F. Smith *Le Tomate*
	G-BBFU	PA-E23 Aztec 250	Reedtrend Ltd
	G-BBFV	PA-32 Cherokee Six 260	Southend Securities Ltd
	G-BBFW	PA-E23 Aztec 250B	T. Bartlett/Stapleford
	G-BBFZ	PA-28R-200-2 Cherokee Arrow	Larkfield Garage (Chepstow) Ltd
	G-BBGB	PA-E23 Aztec 250	Lassair International Ltd/Bristol
	G-BBGC	M.S.893E Rallye Commodore 180	A. Somerville/Alderney
	G-BBGE	PA-E23 Aztec 250	Dollar Air Services Ltd/Coventry
	G-BBGF	Cessna 340	Hillair Ltd & Lawrence Wilson & Son Ltd/Leeds
	G-BBGH	AA-5 Traveler	J. W. Skene
	G-BBGI	Fuji FA.200-160	J. J. Young
	G-BBGJ	Cessna 180	Med-Co Hospital Supplies Ltd
	G-BBGL	Baby Great Lakes	P. W. Thomas/Barton
	G-BBGR	Cameron O-65 balloon	M. L. & L. P. Willoughby
	G-BBGS	Sikorsky S-61N	Bristow Helicopters Ltd *Indefatigable*
	G-BBGX	Cessna 182P Skylane	H. I. Williams & ptnrs/Sleap
	G-BBGZ	CHABA 42 balloon	British Balloon Museum *Phlogiston*
	G-BBHB	PA-31-300 Navajo	Kondair/Stansted
	G-BBHC	Enstrom F-28A	S. Warwick
	G-BBHD	Enstrom F-28A	Stott Demolition Ltd
	G-BBHE	Enstrom F-28A	C. Leonard
	G-BBHF	PA-23 Aztec 250E	Bevan Lynch Aviation Ltd/Birmingham
	G-BBHG	Cessna E-310Q	Airwork Services Ltd/Perth
	G-BBHI	Cessna 177RG	Independent Tape Duplicators Ltd
	G-BBHJ	Piper J-3C-65 Cub	R. V. Miller & R. H. Heath
	G-BBHK	AT-16 Harvard IIB	Bob Warner Aviation/Exeter
	G-BBHL	Sikorsky S-61N Mk II	Bristow Helicopters Ltd *Glamis*
	G-BBHM	Sikorsky S-61N Mk II	Bristow Helicopters Ltd *Braemar*
	G-BBHU	SA.341G Gazelle 1	Specialist Flying Training Ltd/Carlisle
	G-BBHW	SA.341G Gazelle 1	McAlpine Aviation Ltd/Hayes
	G-BBHX	M.S.893E Rallye Commodore	Exeter Flying Club Ltd
	G-BBHY	PA-28 Cherokee 180	Air Operations Ltd/Guernsey
	G-BBIA	PA-28R-200 Cherokee Arrow	A. G. (Commodities) Ltd/Stapleford
	G-BBIC	Cessna 310Q	Angus Motor Co/Dundee
	G-BBID	PA-28 Cherokee 140	R. K. Tamlinson
	G-BBIF	PA-E23 Aztec 250	Northern Executive Aviation Ltd/ Manchester
	G-BBIH	Enstrom F-28A	Dolphin Property (Management) Ltd
	G-BBII	Fiat G-46-3B	The Hon Patrick Lindsay/Booker
	G-BBIL	PA-28 Cherokee 140	D. Arlette
	G-BBIN	Enstrom F28A	Jarogate Ltd
	G-BBIO	Robin HR.100/210	R. A. King/Headcorn
	G-BBIT	Hughes 269B	Contract Development & Projects (Leeds) Ltd
	G-BBIV	Hughes 269C	W. R. Finance Ltd
	G-BBIX	PA-28 Cherokee 140	R. J. Hill/Biggin Hill
	G-BBJB	Thunder Ax7-77 balloon	St Crispin Balloon Group *Dick Darby*
	G-BBJI	Isaacs Spitfire	A. N. R. Houghton & ptnrs
	G-BBJT	Robin HR.200/100	Headcorn Flying School Ltd
	G-BBJU	Robin DR.400/140	J. C. Lister
	G-BBJV	Cessna F.177RG	Pilot Magazine/Biggin Hill
	G-BBJW	Cessna FRA.150L	Coventry School of Flying Ltd
	G-BBJX	Cessna F.150L	Yorkshire Flying Services Ltd/Leeds
	G-BBJY	Cessna F.172M	J. Lucketti/Barton
	G-BBJZ	Cessna F.172M	Burks, Green & ptnrs
	G-BBKA	Cessna F.150L	Sherburn Aero Club Ltd
	G-BBKB	Cessna F.150L	Shoreham Flight Simulation/ Bournemouth
	G-BBKC	Cessna F.172M	D. E. H. Designs Ltd/Bournemouth
	G-BBKE	Cessna F.150L	Wickenby Aviation Ltd
	G-BBKF	Cessna FRA.150L	Compton Abbas Airfield Ltd

Reg.	Type	Owner or Operator	Notes
G-BBKG	Cessna FR.172J	Exmine Ltd/Lydd	
G-BBKI	Cessna F.172M	B. C. Lemon & D. Godfrey/ Fenland	
G-BBKJ	Cessna FT.337G	Carter Aviation Ltd/E. Midlands	
G-BBKL	CP.301A Emeraude	W. J. Walker	
G-BBKR	Scheibe SF.24A Motorspatz	P. I. Morgans	
G-BBKU	Cessna FRA.150L	Balgin Ltd/Bourn	
G-BBKV	Cessna FRA.150L	Skegness Air Taxi Service Ltd	
G-BBKX	PA-28 Cherokee 180	P. E. Eglington	
G-BBKY	Cessna F.150L	W. of Scotland Flying Club Ltd/Glasgow	
G-BBKZ	Cessna 172M	Exeter Flying Club Ltd	
G-BBLA	PA-28 Cherokee 140	Woodgate Aviation Co Ltd/Woodvale	
G-BBLC	Hiller UH-12E	Agricopters Ltd/Chilbolton	
G-BBLE	Hiller UH-12E	Agricopters Ltd/Chilbolton	
G-BBLG	Hiller UH-12E	*Stored*/Sywell	
G-BBLH	Piper O-59A Grasshopper	Shipping & Airlines Ltd	
G-BBLL	Cameron O-84 balloon	University of East Anglia Hot-Air Ballooning Club *Boadicea*	
G-BBLM	MS.880 Rallye 100 Sport	R. J. Lewis & P. Walker	
G-BBLP	PA-E23 Aztec 250D	Euro-Drill Equipment (UK) Ltd/ E. Midlands	
G-BBLS	AA-5 Traveler	Turnhouse Flying Club/Edinburgh	
G-BBLU	PA-34-200-2 Seneca	F. Tranter/Manchester	
G-BBMB	Robin DR.400/180	W. B. Wright & Sons & G. K. Hare	
G-BBME	BAC One-Eleven 401	British Airways *County of Shropshire* (G-AZMI)/Birmingham	
G-BBMF	BAC One-Eleven 401	British Airways *County of Worcestershire* (G-ATVU)/ Birmingham	
G-BBMG	BAC One-Eleven 408	British Airways *County of* Gloucestershire (G-AWEJ)/ Birmingham	
G-BBMH	E.A.A. Sports Biplane Model P.1.	K. Dawson/Andrewsfield	
G-BBMJ	PA-E23 Aztec 250	Expressflight Ltd/Southend	
G-BBMK	PA-31-300 Navajo	Steer Aviation Ltd/Biggin Hill	
G-BBMN	D.H.C.l Chipmunk 22	R. Steiner/Panshanger	
G-BBMO	D.H.C.l Chipmunk 22	A. J. Hurst/Holland	
G-BBMR	D.H.C.l Chipmunk T.10 ★ (WB763)	Southall Technical College	
G-BBMT	D.H.C.l Chipmunk 22	A. T. Letts & ptnrs/Dunstable	
G-BBMV	D.H.C.l Chipmunk 22 (WG348)	B. Rossiter/Booker	
G-BBMW	D.H.C.l Chipmunk 22	M. S. Evans	
G-BBMX	D.H.C.l Chipmunk 22	A. L. Brown & P. S. Murchison	
G-BBMZ	D.H.C.l Chipmunk 22	Wycombe Gliding School Syndicate/ Booker	
G-BBNA	D.H.C.l Chipmunk 22 (Lycoming)	Coventry Gliding Club Ltd/Husbands Bosworth	
G-BBNC	D.H.C.l Chipmunk T.10 ★ (WP790)	Mosquito Aircraft Museum	
G-BBND	D.H.C.l Chipmunk 22	A. J. Organ	
G-BBNG	Bell 206B JetRanger 2	M. Burgin & K. Rix/Netherthorpe	
G-BBNH	PA-34-200-2 Seneca	Lawrence Goodwin Machine Tools Ltd Coventry	
G-BBNI	PA-34-200-2 Seneca	Colnenay Ltd/Guernsey	
G-BBNJ	Cessna F.150L	Sherburn Aero Club	
G-BBNN	PA-E23 Aztec 250D	British Caledonian Airways Ltd/Gatwick	
G-BBNO	PA-E23 Aztec 250E	Andrew Edie Aviation Ltd/Shoreham	
G-BBNR	Cessna 340	J. Lipton/Elstree	
G-BBNT	PA-31-350 Navajo Chieftain	Northern Executive Aviation Ltd/ Manchester	
G-BBNV	Fuji FA.200-160	C.S.E. Aviation Ltd/Kidlington	
G-BBNX	Cessna FRA.150L	Three Counties Aero Club Ltd/ Blackbushe	
G-BBNY	Cessna FRA.150L	Air Tows Ltd/Lasham	
G-BBNZ	Cessna F.172M	Lunnons Commercials/Stapleford	
G-BBOA	Cessna F.172M	Michael Gardner Ltd	
G-BBOB	Cessna 421B	D. J. Parry & G. A. Macfarlane	
G-BBOC	Cameron O-77 balloon	Bacchus Balloons *Bacchus*	
G-BBOE	Robin HR.200/100	Aberdeen Flying Group	
G-BBOH	Pitts S-1S Special	P. Meeson	
G-BBOI	Bede BD-5B	Heather V. B. Wheeler/Greenland	
G-BBOL	PA-18-150 Super Cub	Lakes Gliding Club Ltd/Barrow	
G-BBOO	Thunder Ax6-56 balloon	K. Meehan *Tigerjack*	
G-BBOR	Bell 206B JetRanger 2	Helicrops Ltd	

Notes	Reg.	Type	Owner or Operator
	G-BBOX	Thunder Ax7-77 balloon	R. C. Weyda *Rocinante*
	G-BBOY	Thunder Ax6-56A balloon	N. C. Faithfull *Eric of Titchfield*
	G-BBPJ	Cessna F.172M	Simmette Ltd/Exeter
	G-BBPK	Evans VP-1	P. D. Kelsey
	G-BBPM	Enstrom F-28A	Source Promotions & Premium Consultants Ltd
	G-BBPN	Enstrom F-28A	C.S.E. Aviation Ltd/Kidlington
	G-BBPO	Enstrom F-28A	M. Page
	G-BBPP	PA-28 Cherokee 180	Hartmann Ltd
	G-BBPS	Jodel D.117	A. Appleby/Redhill
	G-BBPU	Boeing 747-136	British Airways *Henry Hudson*/ Heathrow
	G-BBPW	Robin HR.100/210	Parlway Ltd/Biggin Hill
	G-BBPX	PA-34-200-2 Seneca	Richel Investments Ltd/Guernsey
	G-BBPY	PA-28 Cherokee 180	George Hill (Oldham) Ltd
	G-BBRA	PA-E23 Aztec 250E	W. London Aviation Services/ White Waltham
	G-BBRB	D.H.82A Tiger Moth (DF198)	R. Barham/Biggin Hill
	G-BBRC	Fuji FA.200-180	W. & L. Installations & Co Ltd/Fairoaks
	G-BBRH	Bell 47G-5A	Helicopter Supplies & Engineering Ltd
	G-BBRI	Bell 47G-5A	Camlet Helicopters Ltd/Fairoaks
	G-BBRN	Procter Kittiwake	Vari-Prop (GB) Ltd
	G-BBRV	D.H.C.I Chipmunk 22	HSA (Chester) Sports & Social Club
	G-BBRW	PA-28 Cherokee 140	March Flying Group/Stapleford
	G-BBRX	SIAI-Marchetti S.205-18F	W. Chrystal
	G-BBRZ	AA-5 Traveler	M. J. Coleman/Exeter
	G-BBSA	AA-5 Traveler	K. Lynn
	G-BBSB	Beech C23 Sundowner	Sundowner Group/Manchester
	G-BBSC	Beech B24R Sierra	Beechcombers Flying Group
	G-BBSE	D.H.C. I Chipmunk 22	Hartley House Investments Ltd/Elstree
	G-BBSM	PA-32 Cherokee Six 300	R. Robson/Biggin Hill
	G-BBSO	PA-28 Cherokee 140	K. Edwards/Coventry
	G-BBSS	D.H.C.IA Chipmunk 22	Northumbria Tug Group
	G-BBSU	Cessna 421B	Fisons Ltd/E. Midlands
	G-BBSV	Cessna 421B	B. A. & M. L. M. Langevad
	G-BBSW	Pietenpol Air Camper	J. K. S. Wills
	G-BBTB	Cessna FRA.150L	Compton Abbas Airfield Ltd
	G-BBTG	Cessna F.172M	D. H. Laws
	G-BBTH	Cessna F.172M	M. F. Calvert & J. Henderson
	G-BBTJ	PA-E23 Aztec 250E	Midas Aviation Ltd/Cambridge
	G-BBTK	Cessna FRA.150L	Airwork Ltd/Perth
	G-BBTL	PA-E23 Aztec 250C	Air Navigation & Trading Co Ltd/ Blackpool
	G-BBTS	Beech V35B Bonanza	Charles Lock Motors Ltd/Elstree
	G-BBTU	ST-10 Diplomate	P. Campion/Stapleford
	G-BBTW	PA-31P Navajo	M. G. Tyrell & Co Ltd
	G-BBTX	Beech C23 Sundowner	Celahurst Ltd/Stapleford
	G-BBTY	Beech C23 Sundowner	Torlid Ltd/Biggin Hill
	G-BBTZ	Cessna F.150L	Woodgate Air Services Ltd
	G-BBUD	Sikorsky S-61N Mk II	British Airways Helicopters Ltd/ Aberdeen
	G-BBUE	AA-5 Traveler	C. A. Arnold/Southend
	G-BBUF	AA-5 Traveler	W. J. Brogan
	G-BBUG	PA-16 Clipper	C. C. & J. M. Lovell
	G-BBUJ	Cessna 421B	Hawtal Whiting Design & Engineering Co Ltd/Southend
	G-BBUL	Mitchell-Procter Kittiwake 1	R. Bull
	G-BBUO	Cessna 150L	Exeter Flying Club Ltd
	G-BBUT	Western O-65 balloon	Wg. Cdr. G. F. Turnbull & Mrs K. Turnbull *Christabelle II*
	G-BBUU	Piper L-4B Cub	Cooper Bros/Little Snoring
	G-BBUY	Bell 206B JetRanger 2	Helicopter Hire Ltd/Southend
	G-BBVA	Sikorsky S-61N Mk. II	Bristow Helicopters Ltd *Vega*
	G-BBVE	Cessna 340	R. M. Cox Ltd/Biggin Hill
	G-BBVF	SA Twin Pioneer III ★	Museum of Flight/E. Fortune
	G-BBVG	PA-E23 Aztec 250D	R. F. Wanbon & P. G. Warmerdan/ Panshanger
	G-BBVH	V.807 Viscount	GB Airways Ltd/Gibraltar
	G-BBVI	Enstrom F-28A ★	*Ground trainer*/Kidlington
	G-BBVJ	Beech B24R Sierra	Beech Sierra Group/Gamston
	G-BBVM	Beech A.100 King Air	Vernair Transport Services/Liverpool
	G-BBVO	Isaacs Fury II	D. B. Wilson/Jersey
	G-BBVP	Westland-Bell 47G-3BI	Freemans of Bewdley (Aviation) Ltd
	G-BBWM	PA-E23 Aztec 250E	Guernsey Air Search Ltd

Reg.	Type	Owner or Operator	Notes
G-BBWN	D.H.C.I Chipmunk 22	G. R. Tait & J. Ripley/Elstree	
G-BBWZ	AA-1B Trainer	Hornet Aviation Ltd/Breighton	
G-BBXB	Cessna FRA.150L	M. L. Swain/Bourn	
G-BBXG	PA-34-200-2 Seneca	London Flight Centre (Stansted) Ltd	
G-BBXH	Cessna FR.172F	H. H. Metal Finishing (Wales) Ltd	
G-BBXK	PA-34-200-2 Seneca	Dashback Aviation Ltd	
G-BBXL	Cessna E310Q	G. R. Patrick	
G-BBXO	Enstrom F-28A	C.S.E. Aviation Ltd/Kidlington	
G-BBXR	PA-31-350 Navajo Chieftain	Maldocrest Ltd/Bristol	
G-BBXS	Piper J-3C-65 Cub	M. J. Butler (G-ALMA)/Leicester	
G-BBXT	Cessna F.172M	Northair Aviation Ltd/Leeds	
G-BBXU	Beech B24R Sierra	Glidegold Ltd/Coventry	
G-BBXV	PA-28-151 Warrior	Beechwood Marine Ltd/Staverton	
G-BBXW	PA-28-151 Warrior	3D Aluminium Ltd/Kidlington	
G-BBXX	PA-31-350 Navajo Chieftain	Natural Environment Research Council	
G-BBXY	Bellanca 7GC BC Citabria	Peter Hilton (Wickham) Ltd/ Lee-on-Solent	
G-BBXZ	Evans VP-1	P. J. Egan	
G-BBYB	PA-18 Super Cub 95	Modeller's World (Leeds) Ltd/Breighton	
G-BBYE	Cessna 195	R. J. Willies	
G-BBYH	Cessna 182P	Sanderson (Forklifts) Ltd	
G-BBYK	PA-E23 Aztec 250	Kraken Air/Cardiff	
G-BBYL	Cameron O-77 balloon	Buckingham Balloon Club *Jammy*	
G-BBYM	H.P.137 Jetstream 200	British Aerospace PLC (G-AYWR)/ Woodford	
G-BBYN	PA-30 Twin Comanche 160	Express Aviation Services Ltd	
G-BBYO	BN-2A Mk. III Trislander	Aurigny Air Services (G-BBWR)/ Guernsey	
G-BBYP	PA-28 Cherokee 140	A. J. Bamrah/Biggin Hill	
G-BBYR	Cameron O-65 balloon	D. M. Winder *Phoenix*	
G-BBYS	Cessna 182P Skylane	Forth Engineering Ltd/Birmingham	
G-BBYU	Cameron O-56 balloon	C. J. T. Davey *Chieftain*	
G-BBYW	PA-28 Cherokee 140	C.S.E. (Aircraft Services) Ltd/Kidlington	
G-BBZF	PA-28 Cherokee 140	A. G. Lewis & B. R. Davis	
G-BBZH	PA-28R-200 Cherokee Arrow	Creedair Ltd	
G-BBZI	PA-31-310 Navajo	Incat (Jersey) Ltd	
G-BBZJ	PA-34-200-2 Seneca	C.S.E. (Aircraft Services) Ltd/Kidlington	
G-BBZK	Westland-Bell 47G-3B1	Helicopter Supplies & Engineering Ltd	
G-BBZN	Fuji FA.200-180	J. Westwood & P. D. Wedd	
G-BBZO	Fuji FA.200-160	Seal Executive Aircraft Ltd	
G-BBZS	Enstrom F-28A	Spooner Aviation (Enstrom Helicopters) Ltd/Shoreham	
G-BBZV	PA-28R-200-2 Cherokee Arrow	Unicol Engineering/Kidlington	
G-BCAC	M.S.894A Rallye Minerva 220	R. S. Rogers/Cardiff	
G-BCAD	M.S.894A Rallye Minerva 220	V. S. Williams	
G-BCAN	Thunder Ax7-77 balloon	D. & L. Cole *Billboard*	
G-BCAR	Thunder Ax7-77 balloon	T. J. Woodbridge/Australia	
G-BCAS	Thunder Ax7-77 balloon	Zebedee Balloon Service *Drifter*	
G-BCAT	PA-31-310 Turbo Navajo	David Martin Couriers Ltd	
G-BCAZ	PA-12 Super Cruiser	A. D. Williams	
G-BCBD	Bede BD-5B	Brockmore-Bede Aircraft (UK) Ltd/ Shobdon	
G-BCBG	PA-E23 Aztec 250	M. J. L. Batt/Booker	
G-BCBH	Fairchild 24R-46A Argus III	Bluegale Ltd/Biggin Hill	
G-BCBJ	PA-25 Pawnee 235	CKS Air Ltd/Southend	
G-BCBK	Cessna 421B	Sullivan Management Control Ltd	
G-BCBL	Fairchild 24R-46A Argus III (HB751)	W. A. Jordan	
G-BCBM	PA-23 Aztec 250	J. Turnbull/Blackpool	
G-BCBP	M.S.880B Rallye 100S Sport	A. A. Thomas/Dunkeswell	
G-BCBR	AJEP/Wittman W.8 Tailwind	G. McMillan	
G-BCBW	Cessna 182P	E. Reed	
G-BCBX	Cessna F.150L	Woodgate Air Services Ltd/Aldergrove	
G-BCBY	Cessna F.150L	Scottish Airways Flyers (Prestwick) Ltd	
G-BCBZ	Cessna 337C	H. Tempest Ltd	
G-BCCB	Robin HR.200/100	Tradebase Ltd	
G-BCCC	Cessna F.150L	J. C. Glynn	
G-BCCD	Cessna F.172M	Personal Service Travel (UK) Ltd	
G-BCCE	PA-E23 Aztec 250	Gilbert Doyle Associates Ltd/Shoreham	
G-BCCF	PA-28 Cherokee 180	J. T. Friskney Ltd/Skegness	
G-BCCG	Thunder Ax7-65 balloon	N. H. Ponsford	
G-BCCK	AA-5 Traveler	Prospect Air Ltd	

Notes	Reg.	Type	Owner or Operator
	G-BCCP	Robin HR. 200/100	Northampton School of Flying Ltd/ Sywell
	G-BCCR	CP.301B Emeraude	A. B. Fisher
	G-BCCU	BN-2A Mk III-1 Trislander	Kondair/Stansted
	G-BCCX	D.H.C.I Chipmunk 22 (Lycoming)	RAFGSA/Bicester
	G-BCCY	Robin HR.200/100	D. S. Farler/Lulsgate
	G-BCDA	Boeing 727-46	Dan-Air Services Ltd/Gatwick
	G-BCDB	PA-34-200-2 Seneca	C.S.E. (Aircraft Services) Ltd/Kidlington
	G-BCDC	PA-18 Super Cub 95	ALY Aviation Ltd
	G-BCDH	MBB Bo.105D	Bond Helicopters Ltd/Bourn
	G-BCDJ	PA-28 Cherokee 140	J. A. Renelt/Southend
	G-BCDK	Partenavia P.68B	Nottingham Offshore Marine
	G-BCDL	Cameron O-42 balloon	D. P. & Mrs B. O. Turner *Chums*
	G-BCDN	F.27 Friendship Mk. 200	Air UK/Norwich
	G-BCDO	F.27 Friendship Mk. 200	Air UK/Norwich
	G-BCDR	Thunder Ax7-77 balloon	W. G. Johnston & ptnrs *Obelix*
	G-BCDS	PA-E23 Aztec 250	Hamilton Aviation Ltd
	G-BCDY	Cessna FRA.150L	Airwork Ltd/Perth
	G-BCDZ	H.S.748 Srs 2A	British Aerospace/Woodford
	G-BCEA	Sikorsky S-61N Mk. II	British Airways Helicopters Ltd/ Aberdeen
	G-BCEB	Sikorsky S-61N Mk. II	British Airways Helicopters Ltd/ Penzance
	G-BCEC	Cessna F.172M	Great Consall Copper Mine Co Ltd
	G-BCEE	AA-5 Traveler	Echo Echo Ltd/Bournemouth
	G-BCEF	AA-5 Traveler	Echo Fox Ltd/Jersey
	G-BCEN	BN-2A Islander	Westward Aviation Co Ltd
	G-BCEO	AA-5 Traveler	E. Bartholomew
	G-BCEP	AA-5 Traveler	Nottingham Industrial Cleaners Ltd/ Tollerton
	G-BCER	GY-201 Minicab	D. Beaumont
	G-BCEU	Cameron O-42 balloon	Entertainment Services Ltd *Harlequin*
	G-BCEX	PA-E23 Aztec 250	Weekes Bros (Welling) Ltd/Biggin Hill
	G-BCEY	D.H.C.1 Chipmunk 22	D. O. Wallis
	G-BCEZ	Cameron O-84 balloon	Anglia Aeronauts Ascension Association *Stars and Bars*
	G-BCFB	Cameron O-77 balloon	J. J. Harris & P. Pryce-Jones *Teutonic Turkey*
	G-BCFC	Cameron O-65 balloon	B. H. Mead *Candy Twist*
	G-BCFD	West balloon	British Balloon Museum *Hellfire*
	G-BCFF	Fuji FA-200-160	Flight Preparations Ltd/Southampton
	G-BCFN	Cameron O-65 balloon	W. G. Johnson & H. M. Savage
	G-BCFO	PA-18-150 Super Cub	Bristol & Gloucestershire Gliding Club (Pty) Ltd/Nympsfield
	G-BCFP	Enstrom F-28A	Downland Construction Ltd/Goodwood
	G-BCFR	Cessna FRA.150L	J. J. Baumhardt Associates Ltd/ Southend
	G-BCFW	Saab 91D Safir	D. R. Williams
	G-BCFY	Luton LA-4A Minor	D. V. Magee
	G-BCGB	Bensen B.8	A. Melody
	G-BCGC	D.H.C.I Chipmunk 22	Culdrose Gliding Club
	G-BCGD	PA-28R-200-2 Cherokee Arrow	BLA Ltd/Birmingham
	G-BCGG	Jodel DR.250 Srs 160	C. G. Gray (G-ATZL)
	G-BCGH	Nord NC.854S	G. T. Roberts & A. W. Hughes
	G-BCGI	PA-28 Cherokee 140	G. W. Peace
	G-BCGJ	PA-28 Cherokee 140	I. T. D. Hall & B. R. Sedgeman Tees-side
	G-BCGK	PA-28 Cherokee 140	CSE (Aircraft Services) Ltd/Kidlington
	G-BCGL	Jodel D.112	R. A. Lock
	G-BCGM	Jodel D.120	I. E. Fisher/Wick
	G-BCGN	PA-28 Cherokee 140	Oxford Flyers Ltd/Kidlington
	G-BCGS	PA-28R-200 Cherokee Arrow	Atholhurst Ltd/Southampton
	G-BCGT	PA-28 Cherokee 140	I. M. Fieldsend/Cranfield
	G-BCGW	Jodel D.11	G. H. & M. D. Chittenden/Panshanger
	G-BCGX	Bede BD-5A/B	R. Hodgson
	G-BCHK	Cessna F.172H	N. Yorks Aviation Ltd/Grindale
	G-BCHL	D.H.C.I Chipmunk 22A (WP788)	B. D. Bate & R. Rutherford/Sleap
	G-BCHM	SA.341G Gazelle	Bristol Helicopters Ltd/Yeovil
	G-BCHP	CP.1310C-3 Super Emeraude	H. Swift (G-JOSI)
	G-BCHT	Schleicher ASK.16	K. M. Barton & ptnrs/Dunstable
	G-BCHU	Dawes VP-2	G. Dawes
	G-BCHV	D.H.C.I Chipmunk 22	N. F. Charles/Sywell
	G-BCHX	SF.23A Sperling	R. L. McClean/Netherthorpe

Reg.	Type	Owner or Operator	Notes
G-BCID	PA-34-200-2 Seneca	Comanche Air Services Ltd/Lydd	
G-BCIE	PA-28-151 Warrior	Channel Islands Aero Holdings (Jersey) Ltd	
G-BCIF	PA-28 Cherokee 140	Fryer-Robins Aviation Ltd/E. Midlands	
G-BCIH	D.H.C.I Chipmunk 22 (WD363)	J. M. Hosey & R. A. Schofield/Stansted	
G-BCIJ	AA-5 Traveler	Crosby Agents & Brokers Ltd	
G-BCIK	AA-5 Traveler	W. Nutt & Son Ltd	
G-BCIL	AA-1B Trainer	Aerospares (IoM) Ltd/Ronaldsway	
G-BCIM	AA-1B Trainer	R. H. Partington & R. J. Bourner/Strubby	
G-BCIN	Thunder Ax7-77 balloon	P. G & R. A. Vale	
G-BCIR	PA-28-151 Warrior	Call Sign Aviation/Leavesden	
G-BCIT	CIT/AI Srs 1	Cranfield Institute of Technology	
G-BCIW	D.H.C.I Chipmunk 22	R. K. J. Hadlow & ptnrs/Old Warden	
G-BCJF	Beagle B.206 Srs 1	A. A. Mattacks/Biggin Hill	
G-BCJH	Mooney M.20F	S. R. Cannell/Panshanger	
G-BCJM	PA-28 Cherokee 140	Pearson Charlton Engineers Ltd	
G-BCJN	PA-28 Cherokee 140	A. J. Steed/Goodwood	
G-BCJO	PA-28R-200 Cherokee Arrow	G. I. Cooper	
G-BCJP	PA-28 Cherokee 140	G. C. Smith/Bourn	
G-BCJR	PA-E23 Aztec 250	Creedair Ltd	
G-BCJS	PA-E23 Aztec 250	Woodgate Air Services (IOM) Ltd/ Ronaldsway	
G-BCKD	PA-28R-200-2 Cherokee Arrow	A. B. Plant (Aviation) Ltd/Bristol	
G-BCKF	SA.102.5 Cavalier	K. Fairness	
G-BCKN	D.H.C.I A Chipmunk 22	RAFGSA/Cranwell	
G-BCKO	PA-E23 Aztec 250	W. R. M. C. Foyle/Luton	
G-BCKP	Luton LA-5A Major	J. R. Callow	
G-BCKS	Fuji FA.200-180	J. T. Hicks/Goodwood	
G-BCKT	Fuji FA.200-180	Littlewick Green Service Station Ltd/ Booker	
G-BCKU	Cessna FRA.150L	Airwork Ltd/Perth	
G-BCKV	Cessna FRA.150L	Airwork Ltd/Perth	
G-BCLA	Sikorsky S-61N	Bristow Helicopters Ltd	
G-BCLC	Sikorsky S-61N	Bristow Helicopters Ltd *Craigievar*	
G-BCLD	Sikorsky S-61N	Bristow Helicopters Ltd *Slains*	
G-BCLI	AA-5 Traveler	R. A. Williams/Elstree	
G-BCLJ	AA-5 Traveler	C. P. W. Villa & ptnrs/Shoreham	
G-BCLL	PA-28 Cherokee 180	Stu Davidson & Son Plant Hire Ltd	
G-BCLS	Cessna 170B	C. W. Proffitt-White/Shotteswell	
G-BCLU	Jodel D.117	N. A. Wallace	
G-BCLV	Bede BD-5A	R. A. Gardiner	
G-BCLW	AA-1B Trainer	G. W. Cartledge/Manchester	
G-BCMB	Partenavia P.68B	DK Aviation Ltd/Birmingham	
G-BCMC	Bell 212	Bristow Helicopters Ltd	
G-BCMD	PA-19 Super Cub 95	R. G. Brooks/Dunkeswell	
G-BCMJ	SA.102.5 Cavalier (tailwheel)	R. G. Sykes/Shoreham	
G-BCMR	Robin HR.100/285	D. Betts	
G-BCMT	Isaacs Fury II	M. H. Turner	
G-BCNC	GY.201 Minicab	J. R. Wraight	
G-BCNP	Cameron O-77 balloon	B. A. Nathan & ptnrs	
G-BCNR	Thunder Ax7-77A balloon	S. J. Miliken & ptnrs *Howdy*	
G-BCNS	Cameron O-77 balloon	Cathay Pacific Airways Ltd *Cathay I*	
G-BCNT	Partenavia P.68B	Welsh Airways Ltd	
G-BCNX	Piper J.3C-65 Cub	N. J. R. Empson & ptnrs	
G-BCNZ	Fuji FA.200-160	J. Bruton	
G-BCOB	Piper J-3C-65 Cub	R. W. & Mrs J. W. Marjoram	
G-BCOE	HS.748 Srs 2B	British Airways *Glen Livet*/Glasgow	
G-BCOF	HS.748 Srs 2B	British Airways *Glen Fiddich*/Glasgow	
G-BCOG	Jodel D.112	B. A. Bower & ptnrs	
G-BCOH	Avro 683 Lancaster 10 (KB976)	Strathallan Aircraft Collection	
G-BCOI	D.H.C.I Chipmunk 22	D. S. McGregor & A. T. Letham	
G-BCOJ	Cameron O-56 balloon	T. J. Knott & M. J. Webber	
G-BCOL	Cessna F.172M	J. Birkett/Wickenby	
G-BCOM	Piper J-3C-65 Cub	P. M. Whitlock & J. P. Whitham/ Sywell	
G-BCOO	D.H.C.I Chipmunk 22	T. G. Fielding & M. S. Morton/Blackpool	
G-BCOP	PA-28R-200 Cherokee Arrow	E. A. Saunders/Halfpenny Green	
G-BCOR	SOCATA Rallye 100ST	H. J. Pincombe/Dunkeswell	
G-BCOU	D.H.C.I Chipmunk 22	P. J. Loweth	
G-BCOX	Bede BD-5A	H. J. Cox	
G-BCOY	D.H.C.I Chipmunk 22	Coventry Gliding Club Ltd/ Husbands Bosworth	
G-BCPB	Howes radio-controlled model free balloon	R. B. & Mrs C. Howes *Posbee 1*	

Notes	Reg.	Type	Owner or Operator
	G-BCPD	GY-201 Minicab	A. H. K. Denniss/Halfpenny Green
	G-BCPE	Cessna F.150M	Channel Islands Aero Holdings Ltd/ Jersey
	G-BCPF	PA-23 Aztec 250	M. A. Bonsall/E. Midlands
	G-BCPG	PA-28R-200 Cherokee Arrow	Progressive Business Services Ltd/ Birmingham
	G-BCPH	Piper J-3C-65 Cub (329934)	J. J. Anziani/Booker
	G-BCPJ	Piper J-3C-65 Cub	M. C. Barraclough & T. M. Storey
	G-BCPK	Cessna F.172M	Skegness Air Taxi Services Ltd
	G-BCPN	AA-5 Traveler	B.W. Agricultural Equipments Ltd
	G-BCPO	Partenavia P.68B	A. Jolliffe/Booker
	G-BCPU	D.H.C.I Chipmunk T.10	P. Waller/Booker
	G-BCPX	Szep HFC.125	A. Szep/Netherthorpe
	G-BCRA	Cessna F.150M	Three Counties Aero Club/Blackbushe
	G-BCRB	Cessna F.172M	F. D. & M. D. Forbes/Biggin Hill
	G-BCRE	Cameron O-77 balloon	A. R. Langton
	G-BCRH	Alaparma Baldo B.75	A. L. Scadding/(*stored*)
	G-BCRI	Cameron O-65 balloon	V. J. Thorne *Joseph*
	G-BCRJ	Taylor JT.1 Monoplane	Canary Flying Group
	G-BCRK	SA.102.5 Cavalier	T. Barlow/Barton
	G-BCRL	PA-28-151 Warrior	F. N. Garland/Biggin Hill
	G-BCRN	Cessna FRA.150L	Airwork Ltd/Perth
	G-BCRP	PA-E23 Aztec 250	LEC Refrigeration Ltd/Bognor
	G-BCRR	AA-5B Tiger	Travelworth Ltd
	G-BCRT	Cessna F.150M	Suffolk Aero Club Ltd/Ipswich
	G-BCRX	D.H.C.I Chipmunk 22	J. P. V. Hunt & P. G. H. Tory/Denham
	G-BCSA	D.H.C.I Chipmunk 22	RAFGSA/Bicester
	G-BCSB	D.H.C.I Chipmunk 22	RAFGSA/Cosford
	G-BCSL	D.H.C.I Chipmunk 22	Jalawain Ltd/Barton
	G-BCSM	Bellanca 8GC BC Scout	Buckminster Gliding Club
	G-BCST	M.S.893A Rallye Commodore 180	P. J. Wilcox/Cranfield
	G-BCSX	Thunder Ax7-77 balloon	A. T. Wood *Whoopski*
	G-BCSY	Taylor JT.2 Titch	T. Hartwell & D. Wilkinson
	G-BCSZ	PA-28R-200 Cherokee Arrow	P. J. C. Crispe
	G-BCTA	PA-28-151 Warrior	T. G. Aviation Ltd/Manston
	G-BCTF	PA-28-151 Warrior	D. R. Stanley
	G-BCTI	Schleicher ASK.16	R. J. Steward
	G-BCTJ	Cessna 310Q	Airwork Ltd/Perth
	G-BCTK	Cessna FR.172J	Kernow Caravans & Transport Ltd
	G-BCTR	Taylor JT.2 Titch	T. Reagan/Redhill
	G-BCTT	Evans VP-1	B. J. Boughton
	G-BCTU	Cessna FRA.150M	Mona Aviation Ltd
	G-BCTV	Cessna F.150M	Andrewsfield Flying Club Ltd
	G-BCTW	Cessna F.150M	Woodgate Air Services Ltd/Aldergrove
	G-BCUB	Piper J-3C-65 Cub	A. L. Brown & G. Attwell/Bourn
	G-BCUF	Cessna F.172M	G. H. Kirke Ltd
	G-BCUH	Cessna F.150M	Gordon King (Aviation) Ltd/Biggin Hill
	G-BCUI	Cessna F.172M	Hillhouse Estates Ltd
	G-BCUJ	Cessna F.150M	G. Stewart/Humberside
	G-BCUK	Cessna F.172M	Wycombe Air Centre Ltd/Booker
	G-BCUL	SOCATA Rallye 100ST	J. J. Strain
	G-BCUM	Stinson HW-75	J. P. Turner/Shoreham
	G-BCUY	Cessna FRA.150M	S. R. Cameron
	G-BCUZ	Beech A200 Super King Air	Hawk Aviation Ltd/Aberdeen
	G-BCVA	Cameron O-65 balloon	J. C. Bass & ptnrs *Crepe Suzette*
	G-BCVB	PA-17 Vagabond	A. T. Nowak/Popham
	G-BCVC	SOCATA Rallye 100ST	Brettshire Ltd/Southend
	G-BCVE	Evans VP-2	D. Masterson & D. B. Winstanley
	G-BCVF	Practavia Pilot Sprite	G. B. Castle/Elstree
	G-BCVG	Cessna FRA.150L	Airwork Ltd/Perth
	G-BCVH	Cessna FRA.150L	W. Lancs Aero Club Ltd/Woodvale
	G-BCVI	Cessna FR.172J	R. M. Savage & O. Bayliss
	G-BCVJ	Cessna F.172M	D. S. Newland & J. Rothwell/ Manchester
	G-BCVW	GY-80 Horizon 180	P. M. A. Parrett/Dunkeswell
	G-BCVX	Jodel DR.1050	G. Hopkins & J. R. Heaton
	G-BCVY	PA-34-200T Seneca	C.S.E. Aviation Ltd/Kidlington
	G-BCWA	BAC One-Eleven 518	Dan-Air Services Ltd (G-AXMK)/ Gatwick
	G-BCWB	Cessna 182P	British Car Auctions (Aviation) Ltd/ Blackbushe
	G-BCWE	HPR-7 Herald 206	British Air Ferries *Herald Angel*/ Southend

Reg.	Type	Owner or Operator	Notes
G-BCWF	S.A. Twin Pioneer 1	Flight One Ltd (G-APRS)/Shobdon	
G-BCWH	Practavia Pilot Sprite	R. Tasker/Warton	
G-BCWI	Bensen B.8M	C. J. Blundell	
G-BCWK	Alpavia Fournier RF-3	D. I. Nickolls & ptnrs	
G-BCWL	Westland Lysander III (V9281)	Wessex Aviation & Transport Ltd	
G-BCWM	AB-206B JetRanger 2	Helicare Ltd/Liverpool	
G-BCXB	SOCATA Rallye 100ST	Tindon Ltd	
G-BCXE	Robin DR.400/2+2	Headcorn Flying School Ltd	
G-BCXF	H.S.125 Srs 600B	Beecham International Aviation Ltd/ Heathrow	
G-BCXH	PA-28 Cherokee 140F	Sandwell Scaffold Co Ltd/Manchester	
G-BCXJ	Piper J-3C-65 Cub (413048)	W. F. Stockdale/Compton Abbas	
G-BCXN	D.H.C.I Chipmunk 22	J. D. Scott/Swanton Morley	
G-BCXO	MBB Bo 105D	Bond Helicopters Ltd/Bourn	
G-BCXR	BAC One-Eleven 517	Dan-Air Services Ltd (G-BCCV)/Gatwick	
G-BCXZ	Cameron O-56 balloon	Olives from Spain Ltd *Olives from Spain*	
G-BCYE	D.H.C.I Chipmunk 22 (WG350)	Pulsegrove Ltd/Shoreham	
G-BCYF	Dassault Mystère 20	Nidiva Services (UK) Ltd/Heathrow	
G-BCYH	DAW Privateer Mk. 2	D. B. Limbert	
G-BCYI	Schleicher ASK-16	J. Fox & J. Harding/Lasham	
G-BCYJ	D.H.C.I Chipmunk 22 (WG307)	R. A. L. Falconer	
G-BCYK	Avro CF.100 Mk 4 Canuck (18393) ★	Imperial War Museum/Duxford	
G-BCYL	D.H.C.I Chipmunk 22	R. F. Bowman/Southend	
G-BCYM	D.H.C.I Chipmunk 22	C. R. R. Eagleton/Headcorn	
G-BCYR	Cessna F.172M	Laxtonbridge Ltd/St Just	
G-BCYZ	Westland-Bell 47G-3B1	Helicrops Ltd	
G-BCZF	PA-28 Cherokee 180	Wycombe Air Centre/Booker	
G-BCZH	D.H.C.I Chipmunk 22	A. C. Byrne & H. Marshall/Norwich	
G-BCZI	Thunder Ax7-77 balloon	Motor Tyres & Accessories *Motorway for Tyres*	
G-BCZM	Cessna F.172M	London Flight Centre (Stansted) Ltd	
G-BCZN	Cessna F.150M	Mona Aviation Ltd	
G-BCZO	Cameron O-77 balloon	W. O. T. Holmes *Leo*	
G-BDAB	SA.102.5 Cavalier	A. H. Brown	
G-BDAC	Cameron O-77 balloon	D. Fowler & J. Goody *Chocolate Ripple*	
G-BDAD	Taylor JT.1 Monoplane	Rochester Flying Group	
G-BDAE	BAC One-Eleven 518	Dan-Air Services Ltd (G-AXMI)/Gatwick	
G-BDAG	Taylor JT.1 Monoplane	R. S. Basinger	
G-BDAH	Evans VP-1	R. W. Lowe/Popham	
G-BDAI	Cessna FRA.150M	Scotia Safari Ltd/Prestwick	
G-BDAJ	R. Commander 112A	Ortonport Ltd	
G-BDAK	R. Commander 112A	T. E. Abell	
G-BDAL	R. 500S Shrike Commander	Micro Consultants Ltd/Biggin Hill	
G-BDAM	AT-16 Harvard IIB (FE992)	M. V. Gauntlett & ptnrs/Goodwood	
G-BDAP	AJEP Tailwind	J. Whiting	
G-BDAR	Evans VP-1	S. C. Foggin & M. T. Dugmore	
G-BDAS	BAC One-Eleven 518	Dan-Air Services Ltd (G-AXMH)/ Gatwick	
G-BDAT	BAC One-Eleven 518	Dan-Air Services Ltd (G-AYOR)/ Gatwick	
G-BDAU	Cessna FRA.150M	Airwork Ltd/Perth	
G-BDAV	PA-23 Aztec 250	Air Ipswich	
G-BDAW	Enstrom F-28A	R. Windley	
G-BDAX	PA-E23 Aztec 250	R. Phillips/Biggin Hill	
G-BDAY	Thunder Ax5-42A balloon	T. M. Donnelly *Meconium*	
G-BDBD	Wittman W.8 Tailwind	R. W. L. Breckell/Woodvale	
G-BDBF	FRED Srs 2	W. T. Morrell	
G-BDBH	Bellanca 7GCBC Citabria	Inkpen Gliding Club Ltd/Thruxton	
G-BDBI	Cameron O-77 balloon	R. L. Harbord	
G-BDBJ	Cessna 182P	H. C. Wilson	
G-BDBP	D.H.C.I Chipmunk 22	Sherwood Flying Club Ltd/Tollerton	
G-BDBR	AB-206B JetRanger 2	P.L.M. Helicopters Ltd/Inverness	
G-BDBS	Short SD3-30	Short Bros Ltd/Sydenham	
G-BDBU	Cessna F.150M	Channel Islands Aero Holdings Ltd/ Jersey	
G-BDBV	Jodel D.11A	J. P. de Hevingham	
G-BDBW	Heintz Zenith 100 A18	D. B. Winstanley	
G-BDBX	Evans VP-1	Montgomeryshire Ultra-Light Flying Club	
G-BDBZ	WS.55 Whirlwind Srs 2 ★	*Ground instruction airframe*/Kidlington	
G-BDCA	SOCATA Rallye 150ST	B. W. J. Pring & ptnrs/Dunkeswell	
G-BDCB	D.H.C.I Chipmunk 22	P. Rees/Halfpenny Green	

Notes	Reg.	Type	Owner or Operator
	G-BDCC	D.H.C.I Chipmunk 22 (WD321)	Coventry Gliding Club Ltd/ Husbands Bosworth
	G-BDCD	Piper J-3C-65 Cub (480133)	Suzanne C. Brooks/Slinfold
	G-BDCE	Cessna F.172H	H. A. Baillie/Blackbushe
	G-BDCI	CP.301A Emeraude	H. A. R. Haresign
	G-BDCK	AA-5 Traveler	Sequoia Air Ltd/Aberdeen
	G-BDCM	Cessna F.177RG	P. R. Gunnel
	G-BDCO	B.121 Pup 1	Dr R. D. H. & Mrs K. N. Maxwell/Leeds
	G-BDCP	AB-206B JetRanger 2	Ben Turner & Sons (Helicopters) Ltd
	G-BDCS	Cessna 421B	Shorrock Security Systems Ltd/ Blackpool
	G-BDCT	PA-25 Pawnee 235C	Apple Aviation Ltd/Sibson
	G-BDCU	Cameron O-77 balloon	Bristol Balloons
	G-BDDD	D.H.C.I Chipmunk 22	RAE Aero Club Ltd/Farnborough
	G-BDDF	Jodel D.120	Sywell Skyriders Flying Group
	G-BDDG	Jodel D.112	A. C. Watt & T. C. Greig/Perth
	G-BDDH	F.27 Friendship Mk 200	Air UK Ltd/Norwich
	G-BDDJ	Luton LA-4A Minor	D. D. Johnson
	G-BDDS	PA-25 Pawnee 235	Lasham Gliding Soc Ltd
	G-BDDT	PA-25 Pawnee 235	Farm Aviation Services Ltd/Enstone
	G-BDDV	BN-2A Islander	Prop-Jet Ltd/Stansted
	G-BDDX	Whittaker MW.2B Excalibur ★	Cornwall Aero Park/Helston
	G-BDDZ	CP.301A Emeraude	D. L. Sentence
	G-BDEA	Boeing 707-338C	Anglo Cargo Airlines Ltd/Gatwick
	G-BDEB	SOCATA Rallye 100ST	W. G. Dunn & ptnrs/Exeter
	G-BDEC	SOCATA Rallye 100ST	Cambridge Chemical Co Ltd
	G-BDEF	PA-34-200T-2 Seneca	European Paper Sales Ltd/Biggin Hill
	G-BDEH	Jodel D.120A	R. E. Figg/Headcorn
	G-BDEI	Jodel D.9 Bebe	P. J. Griggs
	G-BDEJ	R. Commander 112	R. W. Fairless
	G-BDEN	SIAI-Marchetti SF.260	Micro Consultants Ltd/Biggin Hill
	G-BDES	Sikorsky S-61N Mk II	British Airways Helicopters Ltd/ Aberdeen
	G-BDEU	D.H.C.I Chipmunk 22 (WP808)	A. Taylor
	G-BDEV	Taylor JT.1 Monoplane	D. A. Bass
	G-BDEW	Cessna FRA.150M	Compton Abbas Airfield Ltd
	G-BDEX	Cessna FRA.150M	Compton Abbas Airfield Ltd
	G-BDEY	Piper J-3C-65 Cub	Ducksworth Flying Club
	G-BDEZ	Piper J-3C-65 Cub	G. V. Wallis
	G-BDFB	Currie Wot	D. F. Faulkner-Bryant/Shoreham
	G-BDFC	R. Commander 112A	Lemrest Ltd/Biggin Hill
	G-BDFF	Supermarine S.5 replica	W. G. Hosie & ptnrs
	G-BDFG	Cameron O-65 balloon	N. A. Robertson *Golly II*
	G-BDFH	Auster AOP.9 (XR240)	R. O. Holden/Booker
	G-BDFI	Cessna F.150M	Coventry Civil Aviation Ltd
	G-BDFJ	Cessna F.150M	Coventry Civil Aviation Ltd
	G-BDFM	Caudron C.270 Luciole	G. V. Gower
	G-BDFO	Hiller UH-12E	Heliscot Ltd & F. F. Chamberlain/ Inverness
	G-BDFR	Fuji FA.200-160	C.S.E. (Aircraft Services) Ltd/Kidlington
	G-BDFS	Fuji FA.200-160	D. J. Carding/Halfpenny Green
	G-BDFU	Dragonfly MPA Mk 1 ★	Museum of Flight/E. Fortune
	G-BDFW	R. Commander 112A	DTR Audio Visual Ltd
	G-BDFX	Auster 5	K. E. Ballington
	G-BDFY	AA-5 Traveler	E. O. Liebert/Jersey
	G-BDFZ	Cessna F.150M	Skyviews & General Ltd/Exeter
	G-BDGA	Bushby-Long Midget Mustang	J. R. Owen
	G-BDGB	GY-20 Minicab	D. G. Burden
	G-BDGH	Thunder Ax7-77 balloon	The London Balloon Club Ltd *London Pride III*
	G-BDGK	Beechcraft D.17S	P. M. J. Wolf/Biggin Hill
	G-BDGM	PA-28-151 Warrior	A. J. Breakspear & ptnrs
	G-BDGN	AA-5B Tiger	C. S. Wilkin
	G-BDGO	Thunder Ax7-77 balloon	International Distillers & Vintners Ltd *J. & B. Rare*
	G-BDGY	PA-28 Cherokee 140	R. E. Woolridge/Staverton
	G-BDHB	Isaacs Fury II	D. H. Berry
	G-BDHJ	Pazmany PL.1	C. T. Millner
	G-BDHK	Piper J-3C-65 Cub (329417)	A. Liddiard
	G-BDHL	PA-E23 Aztec 250E	Air Northwest Ltd/Manchester
	G-BDHM	SA.102.5 Cavalier	D. H. Mitchell
	G-BDIC	D.H.C.1 Chipmunk 22	T. Bibby & D. Halliwell/Blackpool
	G-BDID	D.H.C.1 Chipmunk 22	Coventry Gliding Club Ltd/ Husbands Bosworth

Reg.	Type	Owner or Operator	Notes
G-BDIE	R. Commander 112A	Time Out (Ashby) Ltd/E. Midlands	
G-BDIG	Cessna 182P	A. W. Saxon/Panshanger	
G-BDIH	Jodel D.117	J. Chisholm	
G-BDII	Sikorsky S-61N	Bristow Helicopters Ltd	
G-BDIJ	Sikorsky S-61N	Bristow Helicopters Ltd	
G-BDIM	D.H.C.1 Chipmunk 22	Protechnic Computers Ltd/Cambridge	
G-BDIW	D.H.106 Comet 4C ★	Air Classik/Dusseldorf	
G-BDIX	D.H.106 Comet 4C ★	Museum of Flight/E. Fortune	
G-BDIY	Luton LA-4A Minor	M. A. Musselwhite	
G-BDJB	Taylor JT.1 Monoplane	J. F. Barber	
G-BDJC	AJEP W.8 Tailwind	A. Whiting	
G-BDJD	Jodel D.112	C. Davidson	
G-BDJF	Bensen B.8MV	R. P. White	
G-BDJN	Robin HR.200/100	Northampton School of Flying Co Ltd/ Sywell	
G-BDJP	Piper J-3C-65 Cub	Mrs J. M. Pothecary/Slinfold	
G-BDJR	Nord NC.858	R. F. M. Marson & ptnrs/Redhill	
G-BDKC	Cessna A185F	Bridge of Tilt Co Ltd	
G-BDKD	Enstrom F-28A	D. Philp/Shoreham	
G-BDKH	CP.301A Emeraude	R. F. Bridge/Goodwood	
G-BDKI	Sikorsky S-61N MK II	British Airways Helicopters Ltd/ Aberdeen	
G-BDKJ	SA.102.5 Cavalier	H. B. Yardley	
G-BDKK	Bede BD-5B	A. W. Odell	
G-BDKM	SIPA 903	S. W. Markham/White Waltham	
G-BDKS	Pitts S-2A Special	Airmiles Ltd/Cardiff	
G-BDKU	Taylor JT.1 Monoplane	A. C. Dove	
G-BDKV	PA-28R-200-2 Cherokee Arrow	Jacqui-Toni Smithies	
G-BDKW	R. Commander 112A	Denny Bros Printing Ltd	
G-BDLO	AA-5A Cheetah	J. A. Gordon/Doncaster	
G-BDLR	AA-5B Tiger	Coldmix Ltd	
G-BDLS	AA-1B Trainer	M. Brown/Andrewsfield	
G-BDLT	R. Commander 112A	Wintergrain Ltd/Exeter	
G-BDLY	SA.102.5 Cavalier	J. A. Espin/Popham	
G-BDMB	Robin HR.100/210	R. J. Hitchman & Son	
G-BDMC	MBB Bo 105D	Bond Helicopters Ltd/Bourn	
G-BDME	Robin DR.400/140B	Miss S. A. Pound/Fairoaks	
G-BDMM	Jodel D.11	D. M. Metcalf	
G-BDMO	Thunder Ax7-77A balloon	H. G. Twilley Ltd *Flash Harry*	
G-BDMS	Piper J-3C-65 Cub	A. T. H. Martin & K. G. Harris	
G-BDMW	Jodel DR.100	J. T. Nixon/Blackpool	
G-BDNC	Taylor JT.1 Monoplane	T. A. Hodges & M. Sargeant	
G-BDNF	Bensen B.8M	W. F. O'Brien	
G-BDNG	Taylor JT.1 Monoplane	D. J. Phillips/Lasham	
G-BDNO	Taylor JT.1 Monoplane	W. R. Partridge/Bodmin	
G-BDNP	BN-2A Islander ★	*Ground parachute trainer*/Headcorn	
G-BDNR	Cessna FRA.150M	Cheshire Air Training School Ltd/ Liverpool	
G-BDNT	Jodel D.92	E. R. Gould	
G-BDNU	Cessna F.172M	Brencham Air Charter Ltd	
G-BDNW	AA-1B Trainer	I. J. Turner/Doncaster	
G-BDNX	AA-1B Trainer	R. M. North/Manchester	
G-BDNY	AA-1B Trainer	M. R. Langford/Doncaster	
G-BDNZ	Cameron O-77 balloon	N. R. Page *Winston Churchill*	
G-BDOC	Sikorsky S-61N Mk II	Bristow Helicopters Ltd *Tolquhoun*	
G-BDOD	Cessna F.150M	Latharp Ltd/Booker	
G-BDOE	Cessna FR.172J	R. C. Harding/St. Just	
G-BDOF	Cameron O-56 balloon	New Holker Estates Co *Fred Cavendish*	
G-BDOG	SA Bullfinch Srs 2100	I. Drake/Netherthorpe	
G-BDOI	Hiller UH-12E	Heli-Spray Ltd	
G-BDOL	Piper J-3C-65 Cub	Allman & Son (Redhill) Ltd/Shoreham	
G-BDON	Thunder Ax7-77A balloon	J. R. Henderson & ptnrs	
G-BDOR	Thunder Ax6-56A balloon	M. S. Drinkwater & G. Fitzpatrick	
G-BDOS	BN-2A Mk III-2 Trislander	Kondair/Stansted	
G-BDOW	Cessna FRA.150M	Stornoway Flying Club	
G-BDOY	Hughes 369HS	Cosworth Engineering Ltd	
G-BDPA	PA-28-151 Warrior	Noon (Aircraft Leasing) Ltd/ Shoreham	
G-BDPB	Falconar F-II-3	A. E. Pritchard	
G-BDPC	Bede BD-5A	P. R. Cremer	
G-BDPF	Cessna F.172M	Huntara Ltd/Shoreham	
G-BDPK	Cameron O-56 balloon	R. L. Rummery	
G-BDPL	Falconar F-II	P. J. Shone & D. L. Scott/Popham	

Notes	Reg.	Type	Owner or Operator
	G-BDPV	Boeing 747-136	British Airways *City of Aberdeen*/ Heathrow
	G-BDRB	AA-5B Tiger	R. Ginn/Leeds
	G-BDRC	V.724 Viscount ★	Fire School/Manston
	G-BDRD	Cessna FRA.150M	Airwork Ltd/Perth
	G-BDRE	AA-1B Trainer	M. C. Hastings & ptnrs/Elstree
	G-BDRF	Taylor JT.1 Monoplane	D. G. Hannan
	G-BDRG	Taylor JT.2 Titch	D. R. Gray
	G-BDRH	Sikorsky S-61N	Bristow Helicopters Ltd
	G-BDRI	PA-34-200T-2 Seneca	Video Vision Air
	G-BDRJ	D.H.C.I Chipmunk 22 (WP857)	J. C. Schooling
	G-BDRK	Cameron O-65 balloon	D. L. Smith *Smirk*
	G-BDRL	Stitts SA-3 Playboy	D. L. MacLean/Rochester
	G-BDRO	BN-2A-21 Islander	Aerogulf Services (UK) Ltd/ Goodwood
	G-BDSB	PA-28-181 Archer II	Santacane Ltd/Fairoaks
	G-BDSC	Cessna F.150M	Suffolk Aero Club Ltd/Ipswich
	G-BDSD	Evans VP-1	J. E. Worthington
	G-BDSE	Cameron O-77 balloon	British Airways *Concorde*
	G-BDSF	Cameron O-56 balloon	A. R. Greensides & B. H. Osbourne
	G-BDSH	PA-28 Cherokee 140	Bamberhurst Ltd/Tollerton
	G-BDSK	Cameron O-65 balloon	Southern Balloon Group *Carousel II*
	G-BDSL	Cessna F.150M	Cleveland Flying School Ltd/Tees-side
	G-BDSM	Slingsby/Kirby Cadet Mk 3	D. W. Savage
	G-BDSN	Wassmer WA.52 Europa	E. A. L. Glover & ptnrs (G-BADN)
	G-BDSO	Cameron O-31 balloon	Budget Rent-a-Car *Baby Budget*
	G-BDSP	Cessna U.206F Stationair	White, Morgan & Co Ltd/Biggin Hill
	G-BDSZ	BN-2A Islander	Fletcher & Stewart Ltd/Tollerton
	G-BDTB	Evans VP-1	T. E. Boyes
	G-BDTL	Evans VP-1	A. K. Lang/Dunkeswell
	G-BDTP	BN-2A Mk III-2 Trislander	Kondair/Stansted
	G-BDTU	Omega III gas balloon	Mrs K. E. Turnbull *Omega II*
	G-BDTV	Mooney M.20F	J. P. McDermott & ptnrs/Biggin Hill
	G-BDTW	Cassutt Racer	B. E. Smith & C. S. Thompson/Redhill
	G-BDTX	Cessna F.150M	A. A. & R. N. Croxford/Southend
	G-BDUI	Cameron V-56 balloon	D. C. Johnson
	G-BDUJ	PA-31-310 Navajo	Frantham Property Ltd
	G-BDUL	Evans VP-1	D. Beevers & ptnrs
	G-BDUM	Cessna F.150M	Swanton Morley Flying Group
	G-BDUN	PA-34-200T-2 Seneca	R. Paris/White Waltham
	G-BDUO	Cessna F.150M	Sandown Aero Club
	G-BDUX	Slingsby T.31B motor glider	J. C. Anderson/Southend
	G-BDUY	Robin DR.400/140B	Waveney Flying Group/Seething
	G-BDUZ	Cameron V-56 balloon	Balloon Stable Ltd *Hot Lips*
	G-BDVA	PA-17 Vagabond	Mrs H. S. & I. M. Callier/ White Waltham
	G-BDVB	PA-15 (PA-17) Vagabond	B. P. Gardner
	G-BDVC	PA-17 Vagabond	A. R. Caveen
	G-BDVG	Thunder Ax6-56A balloon	R. F. Pollard *Argonaut*
	G-BDVH	H.S.748 Srs 2A	British Aerospace/Woodford
	G-BDVS	F.27 Friendship 200	Air UK/Norwich
	G-BDVT	F.27 Friendship 200	Air UK/Norwich
	G-BDVU	Mooney M.20F	Uplands Video Ltd
	G-BDVW	BN-2A Islander	Loganair Ltd/Glasgow
	G-BDWA	SOCATA Rallye 150ST	H. Cowan/Newtownards
	G-BDWB	SOCATA Rallye 150ST	P. H. Johnson
	G-BDWE	Flaglor Scooter	D. W. Evernden
	G-BDWG	BN-2A Islander	Peterborough Parachute Centre/Sibson
	G-BDWH	SOCATA Rallye 150ST	J. Scott/Kirkwall
	G-BDWJ	SE-5A Replica (F8010)	S. M. Smith/Booker
	G-BDWK	Beech 95-B58 Baron	David Huggett Motor Factors Ltd/ Stapleford
	G-BDWL	PA-25 Pawnee 235	J. E. F. Aviation
	G-BDWM	Mustang replica	D. C. Bonsall
	G-BDWO	Howes Ax6 balloon	R. B. & Mrs C. Howes *Griffin*
	G-BDWP	PA-32R-300 Cherokee Lance	Le Maitre Fireworks Ltd/Sibson
	G-BDWV	BN-2A Mk III-2 Trislander	Aurigny Air Services Ltd/Guernsey
	G-BDWX	Jodel D.120A	J. P. Lassey
	G-BDWY	PA-28 Cherokee 140	D. M. Leonard
	G-BDXA	Boeing 747-236B	British Airways *City of Cardiff*/Heathrow
	G-BDXB	Boeing 747-236B	British Airways *City of Liverpool*/ Heathrow
	G-BDXC	Boeing 747-236B	British Airways *City of Manchester*/ Heathrow

Reg.	Type	Owner or Operator	Notes
G-BDXD	Boeing 747-236B	British Airways *City of Plymouth*/Heathrow	
G-BDXE	Boeing 747-236B	British Airways *City of Glasgow*/Heathrow	
G-BDXF	Boeing 747-236B	British Airways *City of York*/Heathrow	
G-BDXG	Boeing 747-236B	British Airways *City of Oxford*/Heathrow	
G-BDXH	Boeing 747-236B	British Airways *City of Edinburgh*/Heathrow	
G-BDXI	Boeing 747-236B	British Airways *City of Cambridge*/Heathrow	
G-BDXJ	Boeing 747-236B	British Airways *City of Birmingham*/Heathrow	
G-BDXK	Boeing 747-236B	British Airways *City of Canterbury*/Heathrow	
G-BDXL	Boeing 747-236B	British Airways *City of Winchester*/Heathrow	
G-BDXW	PA-28R-200 Cherokee Arrow	Bebecar (UK) Ltd	
G-BDXX	Nord NC.858S	S. E. Elvins	
G-BDXY	Auster AOP.9 (XR269)	B. A. Webster	
G-BDXZ	Pitts S-1S Special	P. Meeson	
G-BDYB	AA-5B Tiger	S. Lowe/Elstree	
G-BDYC	AA-1B Trainer	T. D. Saveker	
G-BDYD	R. Commander 114	SRJ Aviation	
G-BDYF	Cessna 421C	Nullifire Ltd/Coventry	
G-BDYG	Percival Provost T.1 (WV493)	Museum of Flight/E. Fortune	
G-BDYH	Cameron V-56 balloon	I. S. & S. W. Matthews	
G-BDYL	Beech C23 Sundowner	Gordon-Air Ltd/Goodwood	
G-BDYM	Skysales S-31 balloon	Miss A. I. Smith & M. J. Moore *Cheeky Devil*	
G-BDYY	Hiller UH-12E	Agricopters Ltd	
G-BDYZ	MBB Bo 105D	Bond Helicopters Ltd/Bourn	
G-BDZA	Scheibe SF.25E Super Falke	Norfolk Gliding Club Ltd/Tibenham	
G-BDZB	Cameron S-31 balloon	Kenning Motor Group Ltd *Kenning*	
G-BDZC	Cessna F.150M	Air Tows Ltd/Blackbushe	
G-BDZD	Cessna F.172M	M. T. Hodges/Blackbushe	
G-BDZF	G.164 Ag-Cat B	Miller Aerial Spraying Ltd/Wickenby	
G-BDZS	Scheibe SF.25E Super Falke	A. D. Gubbay/Panshanger	
G-BDZU	Cessna 421C	Page & Moy Ltd & ptnrs/Leicester	
G-BDZW	PA-28 Cherokee 140	Oldbus Ltd/Shoreham	
G-BDZX	PA-28-151 Warrior	Lyndair Aviation Ltd/Biggin Hill	
G-BDZY	Phoenix LA-4A Minor	P. J. Dalby	
G-BEAA	Taylor JT.1 Monoplane	R. C. Hobbs/Bembridge	
G-BEAB	Jodel DR.1051	C. Fitton	
G-BEAC	PA-28 Cherokee 140	Eileen R. Purfield/Biggin Hill	
G-BEAE	PA-25 Pawnee 235	R. V. Miller & J. F. P. Burn/Slinfold	
G-BEAG	PA-34-200T-2 Seneca	C.S.E. Aviation Ltd/Kidlington	
G-BEAH	J/2 Arrow	W. J. & Mrs M. D. Horler	
G-BEAK	L-1011-385 TriStar	British Airways *Carmarthen Bay*/Heathrow	
G-BEAL	L-1011-385 TriStar	British Airtours Ltd *Cardigan Bay*/Gatwick	
G-BEAM	L-1011-385 TriStar	British Airtours Ltd *Swansea Bay*/Gatwick	
G-BEAR	Viscount V.5 balloon	B. Hargraves & B. King	
G-BEAU	Pazmany PL.4A	B. H. R. Smith	
G-BEBB	HPR-7 Herald 214	Channel Express/Bournemouth	
G-BEBC	WS.55 Whirlwind 3 (XP355) ★	Norwich Aviation Museum	
G-BEBE	AA-5A Cheetah	G. N. Smith/Netherthorpe	
G-BEBF	Auster AOP.9	M. D. N. & Mrs A. C. Fisher	
G-BEBG	WSK-PZL SDZ-45A Ogar	D. S. McKay & ptnrs/Booker	
G-BEBI	Cessna F.172M	Calder Equipment Ltd/Hatfield	
G-BEBK	PA-31-300 Turbo Navajo	Vange Scaffolding & Engineering Co Ltd	
G-BEBL	Douglas DC-10-30	British Caledonian Airways *Sir Alexander Flemming-The Scottish Challenger*/Gatwick	
G-BEBM	Douglas DC-10-30	British Caledonian Airways *Robert Burns-The Scottish Bard*/Gatwick	
G-BEBN	Cessna 177B	M. A. Berriman	
G-BEBO	Turner TSW-2 Wot	E. Newsham & ptnrs	
G-BEBR	GY-201 Minicab	A. S. Jones & D. R. Upton	
G-BEBS	Andreasson BA-4B	D. M. Fenton/Breighton	

Notes	Reg.	Type	Owner or Operator
	G-BEBU	R. Commander 112A	M. Rowland & J. K. Woodford
	G-BEBZ	PA-28-151 Warrior	Goodwood Terrena Ltd/Goodwood
	G-BECA	SOCATA Rallye 100ST	Goricstar Ltd
	G-BECB	SOCATA Rallye 100ST	A. J. Trible/Bodmin
	G-BECC	SOCATA Rallye 150ST	Lapwing Flying Group Ltd
	G-BECD	SOCATA Rallye 150ST	J. F. O'Hara & E. M. Walmsley/ Bournemouth
	G-BECF	Scheibe SF.25A Falke	D. A. Wilson & ptnrs
	G-BECG	Boeing 737-204ADV	Britannia Airways Ltd *Amy Johnson*/ Luton
	G-BECH	Boeing 737-204ADV	Britannia Airways Ltd *Viscount Montgomery of Alamein*/Luton
	G-BECJ	Partenavia P.68B	Hereford Parachute Club Ltd/ Shobdon
	G-BECK	Cameron V-56 balloon	K. H. Greenaway
	G-BECL	C.A.S.A. C.352L (N9+AA)	Junkers Ju.52/3m Flight Ltd
	G-BECN	Piper J-3C-65 Cub	Harvest Air Ltd/Southend
	G-BECO	Beech A.36 Bonanza	Thorney Machinery Co Ltd
	G-BECT	C.A.S.A.1.131 Jungmann	Rendermere Ltd/Shoreham
	G-BECW	C.A.S.A.1.131 Jungmann	N. C. Jensen/Redhill
	G-BECZ	CAARP CAP.10B	Aerobatic Associates Ltd/Booker
	G-BEDA	C.A.S.A.1.131 Jungmann	M. G. Kates & D. J. Berry
	G-BEDB	Nord 1203 Norecrin	B. F. G. Lister
	G-BEDD	Jodel D.117A	A. T. Croy/Kirkwall
	G-BEDE	Bede BD-5A	Biggin Hill BD5 Syndicate
	G-BEDF	Boeing B-17G-105-VE (485784)	B-17 Preservation Ltd/Duxford
	G-BEDG	R. Commander 112A	L. E. Blackburn
	G-BEDJ	Piper J-3C-65 Cub (44-80594)	D. J. Elliott
	G-BEDK	Hiller UH-12E	T. C. Jay/Chilbolton
	G-BEDL	Cessna T.337D	John Bisco (Cheltenham) Ltd
	G-BEDU	Scheibe SF.23C Sperling	Burns Gliding Club Ltd
	G-BEDV	V.668 Varsity T.1 (WJ945)	D. S. Selway/Duxford
	G-BEDZ	BN-2A Islander	Loganair Ltd/Glasgow
	G-BEEE	Thunder Ax6-56A balloon	I. R. M. Jacobs *Avia*
	G-BEEG	BN-2A Islander	Loganair Ltd/Glasgow
	G-BEEH	Cameron V-56 balloon	J. M. Langley *Kaleidoscope*
	G-BEEI	Cameron N-77 balloon	Hedgehoppers Balloon Group
	G-BEEJ	Cameron O-77 balloon	DAL (Builders Merchants) Ltd *Dal's Pal*
	G-BEEL	Enstrom F-280C-UK-2 Shark	K. E. Wills
	G-BEEN	Cameron O-56 balloon	Swire Bottlers Ltd *Coke*/Hong Kong
	G-BEEO	Short SD3-30	Jersey European Airways Ltd
	G-BEEP	Thunder Ax5-42 balloon	Mrs B. C. Faithful *Also Kenneth*
	G-BEER	Isaacs Fury II	M. J. Clark/Southampton
	G-BEEU	PA-28 Cherokee 140E	Berkshire Aviation Services Ltd/Fairoaks
	G-BEEV	PA-28 Cherokee 140E	V. M. Lambeth/Dunkeswell
	G-BEEW	Taylor JT.1 Monoplane	K. Wigglesworth/Breighton
	G-BEFA	PA-28-151 Warrior	Firmbeam Ltd/Booker
	G-BEFC	AA-5B Tiger	A. G. McLeod/Shobdon
	G-BEFF	PA-28 Cherokee 140	Sherwood Flying Club Ltd/Tollerton
	G-BEFH	Nord 3202	William Tomkins Ltd/Sibson
	G-BEFR	Fokker DR.I Replica (1425/17)	R. A. Bowes & P. A. Crawford
	G-BEFT	Cessna 421C	Eclipsol Oil Co Ltd
	G-BEFV	Evans VP-2	Yeadon Aeroplane Group/Leeds
	G-BEFX	Hiller UH-12E	Agricopters Ltd/Chilbolton
	G-BEFY	Hiller UH-12E	Peter Scott Agriculture Ltd
	G-BEGA	Westland Bell 47G-3BI	P. Pilkington & K. M. Armitage/ Coventry
	G-BEGG	Scheibe SF.25E Super Falke	T. Miller/Dunstable
	G-BEGV	PA-23 Aztec 250F	Advance Air Charter Ltd/Manchester
	G-BEHG	AB-206B JetRanger 2	Compass Helicopters/Bristol
	G-BEHH	PA-32R-300 Cherokee Lance	SMK Engineering Ltd
	G-BEHJ	Evans VP-1	K. Heath
	G-BEHK	Agusta-Bell 47G-3BI (Soloy)	Dollar Air Services Ltd/Coventry
	G-BEHM	Taylor JT-1 Monoplane	H. McGovern
	G-BEHN	Westland Bell 47G-3BI (Soloy)	Dollar Air Services Ltd/Coventry
	G-BEHS	PA-25 Pawnee 260C	Farm Aviation Services Ltd/Enstone
	G-BEHU	PA-34-200T-2 Seneca	Thomas Long & Sons Ltd/Tollerton
	G-BEHV	Cessna F.172N	L. Mitchell
	G-BEHW	Cessna F.150M	S. R. Taylor/Leeds
	G-BEHX	Evans VP-2	G. S Adams
	G-BEHY	PA-28-181 Archer II	Q. P. Cope/Dubai
	G-BEIA	Cessna FRA.150M	Airwork Ltd/Perth

Reg.	Type	Owner or Operator	Notes
G-BEIB	Cessna F.172N	R. L. Orsborn & Son Ltd/Sywell	
G-BEIC	Sikorsky S-61N	British Airways Helicopters Ltd/ Aberdeen	
G-BEID	Sikorsky S-61N	British Airways Helicopters Ltd/ Aberdeen	
G-BEIE	Evans VP-2	F. G. Morris	
G-BEIF	Cameron O-65 balloon	C. Vening	
G-BEIG	Cessna F.150M	Gordon King (Aviation) Ltd/Biggin Hill	
G-BEIH	PA-25 Pawnee 235D	Crop Protection Services Ltd	
G-BEII	PA-25 Pawnee 235D	Agricola Aerial Work Ltd	
G-BEIJ	G-164B Ag Cat	Miller Aircraft Hire Ltd/Wickenby	
G-BEIK	Beech A.36 Bonanza	Scot-Stock Ltd/Inverness	
G-BEIL	SOCATA Rallye 150T	D. H. Tonkin/Bodmin	
G-BEIP	PA-28-181 Archer II	M. Ferguson Ltd/Newtownards	
G-BEIS	Evans VP-1	J. A. Joynes	
G-BEIZ	Cessna 500 Citation	Solid State Logic Ltd/Bristol	
G-BEJA	Thunder Ax6-56A balloon	P. A. Hutchins *Jackson*	
G-BEJB	Thunder Ax6-56A balloon	International Distillers & Vinters Ltd *Baby J. & B.*	
G-BEJD	H.S.748 Srs 1	Dan-Air Services Ltd/Gatwick	
G-BEJE	H.S.748 Srs 1	Dan-Air Services Ltd/Gatwick	
G-BEJK	Cameron S-31 balloon	Esso Petroleum Ltd	
G-BEJL	Sikorsky S-61N	British Airways Helicopters/ Aberdeen	
G-BEJM	BAC One-Eleven 423	Ford Motor Co Ltd/Stansted	
G-BEJO	Saffrey S.250 free balloon	Cupro Sapphire Ltd *Firefly*	
G-BEJP	D.H.C.-6 Twin Otter 310	Manx Airlines Ltd/Ronaldsway	
G-BEJT	PA-23 Aztec 250F	Distance No Object Ltd/Stansted	
G-BEJV	PA-34-200T-2 Seneca	C.S.E. Aviation Ltd/Kidlington	
G-BEJW	BAC One-Eleven 423	Ford Motor Co Ltd/Stansted	
G-BEKA	BAC One-Eleven 520	Dan-Air Services Ltd/Gatwick	
G-BEKB	PA-23 Aztec 250	Alidair Ltd/E. Midlands	
G-BEKC	H.S.748 Srs 1	Dan-Air Services Ltd/Gatwick	
G-BEKE	H.S.748 Srs 1	Dan-Air Services Ltd/Gatwick	
G-BEKH	AB-206B JetRanger 2	W. R. Finance Ltd	
G-BEKL	Bede BD-4E	G. A. Hodges	
G-BEKM	Evans VP-1	G. J. McDill	
G-BEKN	Cessna FRA.150M	RFC (Bourn) Ltd	
G-BEKO	Cessna F.182Q	Tyler International	
G-BEKR	Rand KR-2	K. B. Raven	
G-BELP	PA-28-151 Warrior	Coventry Civil Aviation Ltd	
G-BELR	PA-28 Cherokee 140	H. M. Clarke	
G-BELT	Cessna F.150J	Yorkshire Light Aircraft Ltd (G-AWUV)/ Leeds	
G-BELX	Cameron V-56 balloon	W.H. & Mrs J. P. Morgan *Topsy*	
G-BEMB	Cessna F.172M	P. B. Hollands Associates Ltd	
G-BEMD	Beech 95-B55 Baron	Vaux (Aviation) Ltd	
G-BEMF	Taylor JT.1 Monoplane	J. Simpson	
G-BEMM	Slingsby T.31B Motor Cadet	M. N. Martin	
G-BEMU	Thunder Ax5-42 balloon	Zebedee Balloon Service	
G-BEMV	PA-28 Cherokee 140	Worldair Sales Ltd	
G-BEMW	PA-28-181 Archer II	Morgan Laboratories Ltd/Elstree	
G-BEMY	Cessna FRA.150M	L. G. Sawyer/Lasham	
G-BEND	Cameron V-56 balloon	Dante Balloon Group *Le Billet*	
G-BENE	Cessna 402B	Greenline Refrigerated Transport Ltd/ Liverpool	
G-BENH	Phoenix LA-5A Major	C. D. Macartney/Newtownards	
G-BENJ	R. Commander 112B	F. T. Arnold/Elstree	
G-BENK	Cessna F.172M	Capeston Aviation Ltd	
G-BENM	PA-31-325 Navajo	Gull Air Ltd	
G-BENN	Cameron V-56 balloon	S. L. G. & H. C. J. Williams/Ipswich	
G-BENO	Enstrom F-280C Shark	Red Baron Property Ltd/Shoreham	
G-BENS	Saffrey S.330 balloon	D. Whitlock *Hot Plastic*	
G-BENT	Cameron N-77 balloon	N. Tasker	
G-BEOD	Cessna 180	N. London Parachuting Ltd/Cranfield	
G-BEOE	Cessna FRA.150M	J. R. Nicholls & M. A. Paine	
G-BEOH	PA-28R-201T Turbo Arrow III	Priorvale Contractors Ltd	
G-BEOI	PA-18-150 Super Cub	S. Down Gliding Club Ltd	
G-BEOK	Cessna F.150M	Gordon King (Aviation) Ltd/Biggin Hill	
G-BEOO	Sikorsky S-61N Mk. II	British Airways Helicopters Ltd/ Hong Kong	
G-BEOT	PA-25 Pawnee 235D	Moonraker Aviation Ltd/Thruxton	
G-BEOW	PA-28 Cherokee 140	H. Slater/Dubai	
G-BEOX	L-414 Hudson IV (A16-199) ★	RAF Museum	

Notes	Reg.	Type	Owner or Operator
	G-BEOY	Cessna FRA.150L	Fowler Aviation Ltd
	G-BEOZ	A.W.650 Argosy 101	Elan International/E. Midlands
	G-BEPB	Pereira Osprey II	J. J. & A. J. C. Zwetsloot
	G-BEPC	SNCAN SV-4C	M. Harbron/Bodmin
	G-BEPD	SA.102.5 Cavalier	P. & Mrs E. A. Donaldson
	G-BEPE	SC.5 Belfast	HeavyLift Cargo Airlines Ltd (G-ASKE)/ Southend
	G-BEPF	SNCAN SV-4A	M. Stelfox & ptnrs
	G-BEPI	BN-2A Mk. III-2 Trislander	Aurigny Air Services Ltd/Guernsey
	G-BEPO	Cameron N-77 balloon	P. C. C. Clarke
	G-BEPP	AB-206B JetRanger 2	Trent Valley Grain Ltd
	G-BEPS	SC.5 Belfast	HeavyLift Cargo Airlines Ltd/Stansted
	G-BEPV	Fokker S.11-I Instructor	Strathallan Aircraft Collection
	G-BEPY	R. Commander 112B	D. S. Thomas/Booker
	G-BEPZ	Cameron D-96 hot-air airship	IAZ International (UK) Ltd
	G-BERA	SOCATA Rallye 150ST	B. J. Durrant-Peatfield/Biggin Hill
	G-BERC	SOCATA Rallye 150ST	Severn Valley Aero Group
	G-BERD	Thunder Ax6-56A balloon	H. G. Twilley Ltd
	G-BERI	R. Commander 114	M. L. Farrar/Leeds
	G-BERJ	Bell 47G-4A	Land Air Services Ltd
	G-BERN	Saffrey S-330 balloon	B. Martin *Beeze*
	G-BERT	Cameron V-56 balloon	Southern Balloon Group *Bert*
	G-BERW	R. Commander 114	C. D. Allison
	G-BERY	AA-1B Trainer	P. R. Botterill/Brough
	G-BESO	BN-2A Islander	John Jolly (1978) Ltd/Kirkwall
	G-BESS	Hughes 369D	Rassler Aero Services/Booker
	G-BETD	Robin HR.200/100	R. A. Parsons/Bourn
	G-BETE	Rollason B.2A Beta	T. M. Jones/Tollerton
	G-BETF	Cameron 'Champion' balloon	Balloon Stable Ltd *Champion*
	G-BETG	Cessna 180K Skywagon	Euro-Drill Equipment (UK) Ltd/ E. Midlands
	G-BETH	Thunder Ax6-56A balloon	Debenhams Ltd *Debenhams I*
	G-BETI	Pitts S-1D Special	P. Metcalfe/Tees-side
	G-BETJ	Douglas DC-8-33 ★	*Derelict at Stansted*
	G-BETL	PA-25 Pawnee 235D	Crop Aviation (UK) Ltd/Wyberton
	G-BETM	PA-25 Pawnee 235D	Crop Aviation (UK) Ltd/Wyberton
	G-BETO	MS.885 Super Rallye	A. Somerville/Lasham
	G-BETP	Cameron O-65 balloon	J. R. Rix & Sons Ltd
	G-BETS	Cessna A.188B Ag Truck	J. H. Farrar
	G-BETT	PA-34-200-2 Seneca	Andrews Professional Colour Laboratories Ltd/Headcorn
	G-BETU	Piper J-3C-65 Cub	M. J. Curran & R. G. Fitton
	G-BETV	HS.125 Srs 600B	Tenneco Aviation Ltd/Luton
	G-BETW	Rand KR-2	T. A. Wiffen
	G-BEUA	PA-18-150 Super Cub	London Gliding Club (Pty) Ltd/ Dunstable
	G-BEUC	PA-28-161 Warrior II	Bailey Aviation Ltd/Fairoaks
	G-BEUD	Robin HR.100/285R	E. A. & L. M. C. Payton
	G-BEUI	Piper J-3C-65 Cub	M. J. Whatley
	G-BEUK	Fuji FA-200-160	C.S.E Aviation Ltd/Kidlington
	G-BEUL	Beech 95-58 Baron	Basic Metal Co Ltd
	G-BEUM	Taylor JT.1 Monoplane	M. T. Taylor
	G-BEUN	Cassutt Racer IIIm	R. S. Voice/Redhill
	G-BEUP	Robin DR.400/180	J. Button/Rochester
	G-BEUR	Cessna F.172M	B. V. Ruckstuhl
	G-BEUS	SNCAN SV-4C	G. V. Gower
	G-BEUU	PA-19 Super Cub 95	F. Sharples/Sandown
	G-BEUV	Thunder Ax6-56A balloon	Thunder Balloons Ltd
	G-BEUX	Cessna F.172N	Light Planes (Lancashire) Ltd/Barton
	G-BEUY	Cameron N-31 balloon	Southern Balloon Group
	G-BEVA	SOCATA Rallye 150ST	The Rallye Group/Tees-side
	G-BEVB	SOCATA Rallye 150ST	T. R. Sinclair
	G-BEVC	SOCATA Rallye 150ST	B. W. Walpole
	G-BEVG	PA-34-200T-2 Seneca	T. Chalu & Stratton Motor Co (Norwich) Ltd
	G-BEVH	Holland D.700 balloon	D. I. Holland *Sally*
	G-BEVI	Thunder Ax7-77A balloon	The Painted Clouds Balloon Co Ltd
	G-BEVO	Sportavia-Pützer RF-5	D. Lister & R. F. Bradshaw
	G-BEVP	Evans VP-2	C. F. Bloyce
	G-BEVS	Taylor JT.1 Monoplane	D. Hunter
	G-BEVT	BN-2A Mk. III-2 Trislander	Aurigny Air Services Ltd/ Guernsey
	G-BEVV	BN-2A Mk. III-2 Trislander	Avon Aviation Services Ltd/Bristol
	G-BEVW	SOCATA Rallye 150ST	P. C. Goodwin & M. G. Wiltshire

Reg.	Type	Owner or Operator	Notes
G-BEWJ	Westland-Bell 47G-3B1	Lamplight Music Ltd & K. Jones	
G-BEWL	Sikorsky S-61N Mk. II	British Airways Helicopters Ltd/ Aberdeen	
G-BEWM	Sikorsky S-61N Mk. II	British Airways Helicopters Ltd/ Hong Kong	
G-BEWN	D.H.82A Tiger Moth	H. D. Labouchere	
G-BEWO	Zlin 326 Trener Master	R. C. Poolman & K. D. Ballinger/ Staverton	
G-BEWR	Cessna F.172N	Cheshire Air Training School Ltd/ Liverpool	
G-BEWS	Cameron 56 'Lamp Bulb' hot-air balloon	Osram (GEC) Ltd *Osram*	
G-BEWX	PA-28R-201 Arrow III	I. Farini & A. P. J. Lavelle	
G-BEWY	Bell 206B JetRanger 2	John Holborn (Farm Helicopters) Ltd	
G-BEXK	PA-25 Pawnee 235D	Howard Avis (Aviation) Ltd	
G-BEXL	PA-25 Pawnee 235D	Aerial Farm Assistance Ltd/Breighton	
G-BEXN	AA-1C Lynx	Scotia Safari Ltd/Prestwick	
G-BEXO	PA-23 Apache 160	B. Burton/Bournemouth	
G-BEXR	Mudry/CAARP CAP-10B	R. P. Lewis/Booker	
G-BEXS	Cessna F.150M	Coventry School of Flying	
G-BEXW	PA-28-181 Archer II	Acorn Press (Hove) Ltd/Shoreham	
G-BEXX	Cameron V-56 balloon	A. Tyler & ptnrs *Rupert of Rutland*	
G-BEXY	PA-28 Cherokee 140	W. M. Coupar Ltd/Perth	
G-BEXZ	Cameron N-56 balloon	D. C. Eager & G. C. Clark	
G-BEYA	Enstrom F-280C Shark	Guy Morton & Sons Ltd	
G-BEYB	Fairey Flycatcher (replica) (S1287)	John S. Fairey/Yeovilton	
G-BEYD	HPR-7 Herald 401	Panavia Air Cargo/Southend	
G-BEYF	HPR-7 Herald 401	Elan International/E. Midlands	
G-BEYK	HPR-7 Herald 401	—	
G-BEYL	PA-28 Cherokee 180	B. G. & G. Airlines Ltd/Jersey	
G-BEYM	Cessna F.150M	Anglian Flight Training Ltd/Seething	
G-BEYN	Evans VP-2	C. D. Denham	
G-BEYO	PA-28 Cherokee 140	Solid State Logic Ltd/Kidlington	
G-BEYP	Fuji FA-200-180AO	A. C. Pritchard/Booker	
G-BEYV	Cessna T.210M	Valley Motors/Bournemouth	
G-BEYW	Taylor JT.I Monoplane	R. A. Abrahams	
G-BEYY	PA-31-310 Turbo Navajo	Auxili-Air Aviation Ltd/Stansted	
G-BEYZ	Jodel DR.1051/MI	M. J. McCarthy & S. Aarons/ Biggin Hill	
G-BEZA	Zlin 226T Trener	L. Bezak	
G-BEZB	HPR-7 Herald 209	Channel Express/Bournemouth	
G-BEZC	AA-5 Traveler	R. M. Gosling & P. J. Schwind	
G-BEZE	Rutan VariEze	M. F. Sharples & ptnrs	
G-BEZF	AA-5 Traveler	KAL Aviation/Coventry	
G-BEZG	AA-5 Traveler	D. L. Wood	
G-BEZH	AA-5 Traveler	H. & L. Sims Ltd	
G-BEZI	AA-5 Traveler	D. Boyd & ptnrs/Cranfield	
G-BEZJ	MBB Bo 105D	Bond Helicopters Ltd/Bourn	
G-BEZK	Cessna F.172H	R. J. Lock & ptnrs/Andrewsfield	
G-BEZL	PA-31-310 Navajo	A. J. Hows	
G-BEZM	Cessna F.182Q	Walton Summit Truck Centre Ltd/ Blackpool	
G-BEZO	Cessna F.172M	Rogers Aviation Ltd/Cranfield	
G-BEZP	PA-32-300D Cherokee Six	Falcon Styles Ltd/Kidlington	
G-BEZR	Cessna F.172M	Kirmington Aviation Ltd	
G-BEZS	Cessna FR.172J	R. E. Beeton & M. R. Cavinder	
G-BEZU	PA-31-350 Navajo Chieftain	Commercial & Capital Leasing Ltd	
G-BEZV	Cessna F.172M	S. G. Brady/Aberdeen	
G-BEZW	Practavia Pilot Sprite	T. S. Wilkins & ptnrs	
G-BEZY	Rutan VariEze	R. J. Jones	
G-BEZZ	Jodel D.112	A. J. Stevens & ptnrs/Barton	
G-BFAA	GY-80 Horizon 160	Mary Poppins Ltd	
G-BFAB	Cameron N-56 balloon	Phonogram Ltd *Phonogram* Southend	
G-BFAD	PA-28-161 Warrior II	Elso Properties Ltd/Newcastle	
G-BFAF	Aeronca 7BCM (7797)	D. C. W. Harper/Finmere	
G-BFAH	Phoenix Currie Wot	J. F. Dowe	
G-BFAI	R. Commander 114	D. S. Innes/Guernsey	
G-BFAK	MS.892A Rallye Commodore 150	R. Jennings & ptnrs/Alderney	
G-BFAM	PA-31P Navajo	Video Unlimited Motion Pictures	
G-BFAN	H.S.125 Srs 600F	British Aerospace (G-AZHS)/Hatfield	

Notes	Reg.	Type	Owner or Operator
	G-BFAO	PA-20 Pacer 135	J. Days & ptnrs/Denham
	G-BFAP	SIAI-Marchetti S.205-20R	Miss M. A. Eccles
	G-BFAR	Cessna 550 Citation	Bermuda Aviation Associates Ltd
	G-BFAS	Evans VP-1	A. I. Sutherland
	G-BFAV	Orion model free balloon	D. C. Boxall
	G-BFAW	D.H.C.1 Chipmunk 22	R. V. Bowles
	G-BFAX	D.H.C.1 Chipmunk 22 (WG422)	P. G. D. Bell
	G-BFBA	Jodel DR.100A	Wasp Flying Group/Redhill
	G-BFBB	PA-23 Aztec 250E	Donington Aviation Ltd/E. Midlands
	G-BFBC	Taylor JT.1 Monoplane	A. Brooks
	G-BFBD	Partenavia P.68B	Pegasus Aviation Ltd/Aberdeen
	G-BFBE	Robin HR.200/100	Charles Major Ltd/Blackpool
	G-BFBF	PA-28 Cherokee 140	J. J. Donnelly
	G-BFBH	PA-31-325 Turbo Navajo	Civil Aviation Authority/Stansted
	G-BFBM	Saffery S.330 balloon	B. Martin *Beeze II*
	G-BFBR	PA-28-161 Warrior II	Lowery Holdings Ltd/Wellesbourne
	G-BFBU	Partenavia P.68B	Nordic Oil Services Ltd/Bournemouth
	G-BFBV	Brügger Colibri M.B.2	J. D. Hutton & ptnrs/Netherthorpe
	G-BFBW	PA-25 Pawnee 235D	A & A Aviation Services Ltd/Staverton
	G-BFBX	PA-25 Pawnee 235D	Bowker Aircraft Services Ltd/ Rush Green
	G-BFBY	Piper J-3C-65 Cub	L. W. Usherwood
	G-BFCT	Cessna TU-206F	Cecil Aviation Ltd/Cambridge
	G-BFCX	BN-2A Islander	Loganair Ltd/Glasgow
	G-BFCY	AB-206B JetRanger 2	Window Boxes Ltd/Cardiff
	G-BFCZ	Sopwith Camel (B7270) ★	FAA Museum/Yeovilton
	G-BFDA	PA-31-350 Navajo Chieftain	Avon Aviation Services Ltd/Bristol
	G-BFDC	D.H.C.1 Chipmunk 22	Aldhaven Ltd/Newtownards
	G-BFDE	Sopwith Tabloid (replica) (168) ★	Bomber Command Museum/Hendon
	G-BFDF	SOCATA Rallye 235E	J. H. Atkinson/Skegness
	G-BFDG	PA-28R-201T Turbo-Arrow III	Noblair Ltd
	G-BFDI	PA-28-181 Archer II	Reedtrend Ltd/Birmingham
	G-BFDK	PA-28-161 Warrior II	C.S.E. Aviation Ltd/Kidlington
	G-BFDL	Piper L-4J Cub (454537)	P. F. Craven & J. H. Shearer
	G-BFDM	Jodel D.120	Worcestershire Gliding Ltd
	G-BFDN	PA-31-350 Navajo Chieftain	Topflight Aviation Ltd/Blackbushe
	G-BFDO	PA-28R-201T Turbo-Arrow III	Peter Scott Agriculture Ltd
	G-BFDP	CP.301A Emeraude	J. P. McGrath & J. Robson/Barton
	G-BFDZ	Taylor JT.1 Monoplane	D. C. Barber/Woodvale
	G-BFEB	Jodel D.150	D. Aldersea & ptnrs/Sherburn
	G-BFEC	PA-23 Aztec 250F	Midas Aviation Ltd/Cambridge
	G-BFEE	Beech 95-E55 Baron	MLP Aviation Ltd/Elstree
	G-BFEF	Agusta-Bell 47G-3B1	Dollar Air Services Ltd/Coventry
	G-BFEG	Westland-Bell 47G-3B1	Heliwork Ltd/Thruxton
	G-BFEH	Jodel D.117A	C. V. & S. J. Philpott
	G-BFEI	Westland-Bell 47G-3B1	Trent Air Services Ltd/Cranfield
	G-BFEJ	Agusta-Bell 47G-3B1	Copley Farms Ltd/Cambridge
	G-BFEK	Cessna F.152	Staverton Flying Services Ltd
	G-BFEO	Boeing 707-323C	*Stored*
	G-BFER	Bell 212	Bristow Helicopters Ltd
	G-BFES	Bell 212	Bristow Helicopters Ltd
	G-BFEV	PA-25 Pawnee 235	Bowker Aircraft Services Ltd/ Rush Green
	G-BFEW	PA-25 Pawnee 235	Agricola Aerial Work Ltd
	G-BFEX	PA-25 Pawnee 235	CKS Air Ltd/Southend
	G-BFEY	PA-25 Pawnee 235	Howard Avis Aviation Ltd
	G-BFFB	Evans VP-2	D. Bradley
	G-BFFC	Cessna F.152-II	Yorkshire Flying Services Ltd/Leeds
	G-BFFE	Cessna F.152-II	Doncaster Aero Club
	G-BFFF	Cessna 188B Ag Truck	Plains Aerial Spraying Ltd/ Compton Abbas
	G-BFFG	Beech 95-B55 Baron	Beaucette Holdings Ltd
	G-BFFJ	Sikorsky S-61N Mk II	British Airways Helicopters Ltd/ Aberdeen
	G-BFFK	Sikorsky S.61N Mk II	British Airways Helicopters Ltd/ Aberdeen
	G-BFFP	PA-18-150 Super Cub	Airways Aero Associations Ltd/Booker
	G-BFFT	Cameron V-56 balloon	R. I. M. Kerr & D. C. Boxall
	G-BFFW	Cessna F.152	Midland Aircraft Leasing Ltd
	G-BFFY	Cessna F.150M	Pegasus Aviation Ltd
	G-BFFZ	Cessna FR.172 Hawk XP	B. W. Wells
	G-BFGA	SOCATA Rallye 150ST	The BBC Club/Denham
	G-BFGC	Cessna F.150M	Northair Aviation Ltd/Perth

Reg.	Type	Owner or Operator	Notes
G-BFGD	Cessna F.172N-II	Reedtrend Ltd/Birmingham	
G-BFGF	Cessna F.177RG	Victree (V.M.) Ltd/Birmingham	
G-BFGG	Cessna FRA.150M	Airwork Ltd/Perth	
G-BFGH	Cessna F.337G	T. Perkins/Leeds	
G-BFGI	Douglas DC-10-30	British Caledonian Airways *David Livingstone — The Scottish Explorer*/Gatwick	
G-BFGK	Jodel D.117	P. Cawkwell	
G-BFGL	Cessna FA.152	Yorkshire Flying Services Ltd/Leeds	
G-BFGM	Boeing 727-095	Dan-Air Services Ltd/Gatwick	
G-BFGO	Fuji FA.200-160	M. C. Wroe	
G-BFGP	D.H.C.6 Twin Otter 310	Jersey European Airways Ltd/Blackpool	
G-BFGS	MS.893E Rallye 180GT	P. A. Cairns/Dunkeswell	
G-BFGW	Cessna F.150H	J. F. Morgan	
G-BFGX	Cessna FRA.150M	Airwork Ltd/Perth	
G-BFGY	Cessna F.182P	Oxford Aviation Co Ltd/Kidlington	
G-BFGZ	Cessna FRA.150M	Airwork Ltd/Perth	
G-BFHD	C.A.S.A. C.352L (N8+AA)	Wessex Aviation & Transport Ltd	
G-BFHF	C.A.S.A. C.352L	Junkers Ju.52/3m Flight Ltd	
G-BFHG	C.A.S.A. C.352L	Aces High Ltd/Duxford	
G-BFHH	D.H.82A Tiger Moth	P. Harrison & M. J. Gambrell	
G-BFHI	Piper J-3C-65 Cub	J. M. Robinson	
G-BFHK	Cessna F.177RG-II	Facet Group Holdings Ltd/Southend	
G-BFHM	Steen Skybolt	N. M. Bloom	
G-BFHN	Scheibe SF.25E Super Falke	W. J. Dyer	
G-BFHP	Champion 7GCAA Citabria	R. J. Fray/Sibson	
G-BFHR	Jodel DR.220/2+2	J. B. Keith & J. Bugg/Fenland	
G-BFHT	Cessna F.152-II	Riger Ltd/Luton	
G-BFHU	Cessna F.152-II	Ingham Aviation Ltd	
G-BFHV	Cessna F.152-II	Angelsword Ltd/Halfpenny Green	
G-BFHX	Evans VP-1	P. Johnson	
G-BFIB	PA-31-310 Turbo Navajo	Mann Aviation Ltd/Fairoaks	
G-BFID	Taylor JT.2 Titch Mk III	W. F. Adams	
G-BFIE	Cessna FRA.150M	RFC (Bourn) Ltd	
G-BFIF	Cessna FR.172K XPII	D. H. R. Jenkins	
G-BFIG	Cessna FR.172K XPII	Tenair Ltd	
G-BFII	PA-23 Aztec 250E	Tenza Tapes Ltd	
G-BFIJ	AA-5A Cheetah	J. H. Wise/Redhill	
G-BFIL	AA-5A Cheetah	Paine Electrics/Elstree	
G-BFIN	AA-5A Cheetah	M. F. D. Bartley	
G-BFIP	Wallbro Monoplane 1909 replica	K. H. Wallis	
G-BFIR	Avro 652A Anson 21 (WD413)	G. M. K. Fraser/Bournemouth	
G-BFIT	Thunder Ax6-56Z balloon	J. A. G. Tyson	
G-BFIU	Cessna FR.172K XP	P. Fletcher & ptnrs/Netherthorpe	
G-BFIV	Cessna F.177RG	C. Fisher/Denham	
G-BFIX	Thunder Ax7-77A balloon	E. Sowden Ltd	
G-BFIY	Cessna F.150M	Yorkshire Light Aircraft Ltd/Leeds	
G-BFJA	AA-5B Tiger	G. W. Hind/Perth	
G-BFJH	SA.102-5 Cavalier	B. F. J. Hope	
G-BFJI	Robin HR.100/250	MPW Aviation Ltd/Booker	
G-BFJJ	Evans VP-1	P. R. Pykett & B. J. Dyke/Thruxton	
G-BFJK	PA-23 Aztec 250E	Drive Petroleum Co Ltd/Shoreham	
G-BFJM	Cessna F.152	Pegasus Aviation Ltd/Aberdeen	
G-BFJN	Westland-Bell 47G-3B1	Dollar Air Services Ltd/Coventry	
G-BFJO	G.164B Ag-Cat 450	Miller Aerial Spraying Ltd/Wickenby	
G-BFJP	G.164B Ag-Cat	E. S. Axford	
G-BFJR	Cessna F.337G	C. J. Harling/Cranfield	
G-BFJV	Cessna F.172H	Herefordshire Aero Club Ltd/Shobdon	
G-BFJW	AB-206B JetRanger	Dollar Air Services Ltd/Coventry	
G-BFJZ	Robin DR.400/140B	Forge House Restaurant Ltd/Biggin Hill	
G-BFKA	Cessna F.172N	D. J. A. Seagram	
G-BFKB	Cessna F.172N	Eastern Helicopters Ltd/Norwich	
G-BFKC	Rand KR.2	K. K. Cutt	
G-BFKD	R. Commander 114B	C. W. Ford/Leicester	
G-BFKF	Cessna FA.152	Klingair Ltd/Conington	
G-BFKG	Cessna F.152	W. R. C. Foyle/Luton	
G-BFKH	Cessna F.152	B. W. Wells & Burbage Farms Ltd	
G-BFKL	Cameron N-56 balloon	Merrythought Toys Ltd *Merrythought*	
G-BFKM	Westland-Bell 47G-3B1	Heliwork Ltd/Thruxton	
G-BFKN	PA-23 Aztec 250F	Air Envoy Ltd/Birmingham	
G-BFKP	Partenavia P.68B	Curry & Pennick (Builders) Ltd	
G-BFKV	PA-25 Pawnee 235D	Moonraker Aviation Co Ltd/Thruxton	
G-BFKY	PA-34-200 Seneca	S.L.H. Construction Ltd/Biggin Hill	

Notes	Reg.	Type	Owner or Operator
	G-BFLC	Cessna 210L	Save The Children Fund
	G-BFLD	Boeing 707-338C	British Midland Airways Ltd/E. Midlands
	G-BFLE	Boeing 707-338C	British Midland Airways Ltd/E. Midlands
	G-BFLH	PA-34-200T-2 Seneca	C.S.E. Aviation Ltd/Kidlington
	G-BFLI	PA-28R-201T Turbo Arrow III	Peter Walker (Heritage) Ltd
	G-BFLK	Cessna F.152	Gordon King (Aviation) Ltd/Biggin Hill
	G-BFLL	H.S.748 Srs 2A	Dan-Air Services Ltd/Gatwick
	G-BFLM	Cessna 150M	Cornwall Flying Club Ltd/Bodmin
	G-BFLN	Cessna 150M	Sherburn Aero Club Ltd
	G-BFLO	Cessna F.172M	W. A. Cook & ptnrs/Sherburn
	G-BFLP	Amethyst Ax6 balloon	K. J. Hendry *Amethyst*
	G-BFLR	Hiller UH-12E	Fountain Forestry Ltd/Perth
	G-BFLU	Cessna F.152	Inverness Flying Services Ltd
	G-BFLV	Cessna F.172N	C. J. Williams/Elstree
	G-BFLX	AA-5A Cheetah	Sheffield Auto Hire/Barton
	G-BFLZ	Beech 95-A55 Baron	K. K. Demel Ltd/Kidlington
	G-BFMC	BAC One-Eleven 414	Ford Motor Co Ltd/Stansted
	G-BFME	Cameron V-56 balloon	Warwick Balloons Ltd
	G-BFMF	Cassutt Racer Mk III	P. H. Lewis
	G-BFMG	PA-28-161 Warrior II	R. E. & D. Coldham/Biggin Hill
	G-BFMH	Cessna 177B	Span Aviation
	G-BFMK	Cessna FA.152	RAF Halton Aeroplane Club Ltd
	G-BFMM	PA-28-181 Archer II	Bristol & Wessex Aeroplane Club Ltd/ Bristol
	G-BFMR	PA-20 Pacer 125	B. C. & J. I. Cooper
	G-BFMS	MS.893E Rallye 180GT	Air Space Advertising Ltd/Barton
	G-BFMT	Robin HR.200/100	Eagle Forms (GB) Ltd/Biggin Hill
	G-BFMU	AA-5A Cheetah	A. Bazar/Elstree
	G-BFMW	V.735 Viscount	(*Derelict*)/E. Midlands
	G-BFMX	Cessna F.172N	Phoenix Aviation (Bedford) Ltd/ Cranfield
	G-BFMY	Sikorsky S-61N	Bristow Helicopters Ltd
	G-BFMZ	Payne Ax6 balloon	G. F. Payne
	G-BFNC	AS.350B Ecureuil	Dollar Air Services Ltd/Coventry
	G-BFNG	Jodel D.112	M. J. Sanders
	G-BFNH	Cameron V-77 balloon	P. O. Atkins & R. Emms *Red Pepper*
	G-BFNI	PA-28-161 Warrior II	C.S.E. Aviation Ltd/Kidlington
	G-BFNJ	PA-28-161 Warrior II	C.S.E. Aviation Ltd/Kidlington
	G-BFNK	PA-28-161 Warrior II	C.S.E. Aviation Ltd/Kidlington
	G-BFNM	Globe GC.1 Swift	Nottingham Flying Group/E. Midlands
	G-BFNU	BN-2B Islander	Isle of Scilly Sky Bus Ltd/St Just
	G-BFNV	BN-2A Islander	Loganair Ltd/Glasgow
	G-BFOD	Cessna F.182Q	Graphiking Publicity Ltd/Staverton
	G-BFOE	Cessna F.152	Armstrong Whitworth Flying Group/ Coventry
	G-BFOF	Cessna F.152	Staverton Flying School Ltd
	G-BFOG	Cessna 150M	D. E. Tisdale
	G-BFOH	Westland-Bell 47G-3B1	Helicopter Hire Ltd/Southend
	G-BFOI	Westland-Bell 47G-3B1	CKS Air Ltd/Southend
	G-BFOJ	AA-1 Yankee	A. J. Morton/Bournemouth
	G-BFOM	PA-31-325 Navajo	Air Atlantique Ltd/Coventry
	G-BFON	PA-31-310 Navajo	Air Kilroe/Manchester
	G-BFOP	Jodel D.120	H. Cope/Stapleford
	G-BFOS	Thunder Ax6-56A balloon	N. T. Petty
	G-BFOT	Thunder Ax6-56A balloon	Thunder Balloons Ltd
	G-BFOU	Taylor JT.1 Monoplane	G. Bee
	G-BFOV	Cessna F.172N	Gooda Walker Ltd/Shoreham
	G-BFOX	D.H.83 Fox Moth Replica	R. K. J. Hadlow
	G-BFOZ	Thunder Ax6-56 balloon	Motorway Tyres & Accessories Ltd
	G-BFPA	Scheibe SF.25B Super Falke	Yorkshire Gliding Club (Pty) Ltd
	G-BFPB	AA-5B Tiger	Aero Stratus Ltd
	G-BFPF	Sikorsky S-61N	British Caledonian Helicopters Ltd
	G-BFPH	Cessna F.172K	Red Fir Aviation Ltd/Clacton
	G-BFPJ	Procter Petrel	S. G. Craggs
	G-BFPK	Beech A.23 Musketeer	A. Anoufriou
	G-BFPL	Fokker D.VII Replica (4253/18)	Coys of Kensington (Petrol Sales) Ltd
	G-BFPM	Cessna F.172M	Abbey Windows Ltd
	G-BFPO	R. Commander 112B	J. G. Hale Ltd
	G-BFPP	Bell 47J-2	D. Fordham
	G-BFPS	PA-25 Pawnee 235D	Bowker Aviation Services Ltd/ Rush Green
	G-BFPX	Taylor JT.1 Monoplane	E. A. Taylor
	G-BFPZ	Cessna F.177RG	S. R. Cherry-Downes

Reg.	Type	Owner or Operator	Notes
G-BFRA	R. Commander 114	Sabre Engines Ltd/Bournemouth	
G-BFRC	AA-5A Cheetah	Hamble Aeroplane Co	
G-BFRD	Bowers Flybaby 1A	F. R. Donaldson	
G-BFRF	Taylor JT.1 Monoplane	E. R. Bailey	
G-BFRI	Sikorsky S-61N	Bristow Helicopters Ltd	
G-BFRL	Cessna F.152	J. J. Baumhardt Associates Ltd/ Southend	
G-BFRM	Cessna 550 Citation II	Marshall of Cambridge (Engineering) Ltd	
G-BFRO	Cessna F.150M	Skyviews & General Ltd/Carlisle	
G-BFRR	Cessna FRA.150M	Rolim Ltd/Aberdeen	
G-BFRS	Cessna F.172N	Poplar Toys Ltd	
G-BFRT	Cessna FR.172K XP II	B. J. Sharpe/Booker	
G-BFRV	Cessna FA.152	Rogers Aviation Ltd/Cranfield	
G-BFRX	PA-25 Pawnee 260	Howard Avis Aviation Ltd	
G-BFRY	PA-25 Pawnee 260	C. J. Pearce/Goodwood	
G-BFSA	Cessna F.182Q	Clark Masts Ltd/Sandown	
G-BFSB	Cessna F.152	Seal Executive Aircraft Ltd	
G-BFSC	PA-25 Pawnee 235D	Farm Aviation Services Ltd/Enstone	
G-BFSD	PA-25 Pawnee 235D	A. W. Evans	
G-BFSJ	Westland-Bell 47G-3B1	R.K.B. Leasing Services/Shobdon	
G-BFSK	PA-23 Apache 160 ★	Oxford Air Training School/Kidlington	
G-BFSL	Cessna U.206F Stationair	Range Air/Nairobi	
G-BFSO	H.S.125 Srs 700B	Brown & Root (UK) Ltd	
G-BFSR	Cessna F.150J	C. Simms	
G-BFSS	Cessna FR.172G	Minerva Services	
G-BFST	Partenavia P.68B	Autofarm Ltd	
G-BFSY	PA-28-181 Archer II	Downland Aviation/Goodwood	
G-BFTC	PA-28R-201T Turbo Arrow II	Noor Investments Ltd	
G-BFTE	AA.5A Cheetah	B. Refson/Conington	
G-BFTF	AA-5B Tiger	F. C. Burrow Ltd/Leeds	
G-BFTG	AA-5B Tiger	R. A. House	
G-BFTH	Cessna F.172N	F. G. Taylor & Sons/Ipswich	
G-BFTR	Bell 206L Long Ranger	Air Hanson Ltd/Brooklands	
G-BFTT	Cessna 421C	P&B Metal Components Ltd/Manston	
G-BFTW	PA-23 Aztec 250F	Southampton Airport Ltd	
G-BFTX	Cessna F.172N	J. N. Collins/Manston	
G-BFTY	Cameron V-77 balloon	Regal Motors (Bilston) Ltd *Regal Motors*	
G-BFTZ	MS.880B Rallye Club	R. & B. Legge Ltd	
G-BFUB	PA-32RT-300 Turbo Lance II	Bumbles Ltd/Jersey	
G-BFUD	Scheibe SF.25E Super Falke	S. H. Hart	
G-BFUG	Cameron N-77 balloon	Headland Services Ltd	
G-BFUZ	Cameron V-77 balloon	Skysales Ltd	
G-BFVA	Boeing 737-204ADV	Britannia Airways Ltd *Sir John Alcock*/ Luton	
G-BFVF	PA-38-112 Tomahawk	Ipswich School of Flying	
G-BFVG	PA-28-181 Archer II	P. A. Cornah & S. Reed/Blackpool	
G-BFVH	D.H.2 Replica (5964)	Russavia Collection/Duxford	
G-BFVI	H.S.125 Srs 700B	Bristow Helicopters Ltd	
G-BFVM	Westland-Bell 47G-3B1	Pilotmoor Ltd	
G-BFVO	Partenavia P.68B	P. Meeson	
G-BFVP	PA-23 Aztec 250	B. J. Eastwood/Newtownards	
G-BFVS	AA-5B Tiger	S. W. Biroth & ptnrs/Denham	
G-BFVU	Cessna 150L	Thruxton Flight Centre Ltd	
G-BFVV	SA.365 Dauphin 2	Bond Helicopters Ltd/Bourn	
G-BFVW	SA.365 Dauphin 2	Bond Helicopters Ltd/Bourn	
G-BFVX	Beech C90 King Air	Vernair Transport Services/Liverpool	
G-BFVY	Beech C90 King Air	Vernair Transport Services/Liverpool	
G-BFWB	PA-28-161 Warrior II	C.S.E. Aviation Ltd/Kidlington	
G-BFWD	Currie Wot	F. E. Nuthall	
G-BFWE	PA-23 Aztec 250	Air Navigation & Trading Co Ltd/ Blackpool	
G-BFWF	Cessna 421B	Alcon Oil Ltd	
G-BFWK	PA-28-161 Warrior II	Woodgate Air Services (IOM) Ltd	
G-BFWL	Cessna F.150L	J. Dolan/Eglinton	
G-BFWM	Hiller UH-12D	Management Aviation Ltd/Bourn	
G-BFWW	Robin HR.100/210	Willingair Ltd	
G-BFXD	PA-28-161 Warrior II	C.S.E. Aviation Ltd/Kidlington	
G-BFXE	PA-28-161 Warrior II	C.S.E. Aviation Ltd/Kidlington	
G-BFXF	Andreasson BA.4B	A. Brown/Sherburn	
G-BFXG	D.31 Turbulent	S. Griffin	
G-BFXH	Cessna F.152	Angelsword Ltd	
G-BFXI	Cessna F.172M	Thanet Electronics	

Notes	Reg.	Type	Owner or Operator
	G-BFXK	PA-28 Cherokee 140	G. S. & Mrs M. T. Pritchard/Southend
	G-BFXL	Albatross D.5A (D5397/17)	Fleet Air Arm Museum, Yeovilton
	G-BFXM	Jurca MJ.5 Sirocco	R. Bradbury & A. R. Greenfield
	G-BFXO	Taylor JT.1 Monoplane	A. S. Nixon
	G-BFXR	Jodel D.112	R. E. Walker & M. Riddin/Netherthorpe
	G-BFXS	R. Commander 114	Niglon Ltd/Birmingham
	G-BFXT	H.S.125 Srs 700B	Coca Cola Export Corporation
	G-BFXU	American Beta Z Airship	G. Turnbull/USA
	G-BFXW	AA-5B Tiger	Crosswind Aviation Ltd/Leeds
	G-BFXX	AA-5B Tiger	Stanton Aircraft Management Ltd/ Biggin Hill
	G-BFXY	AA-5A Cheetah	Kaal Electrics Ltd/Elstree
	G-BFXZ	PA-28-181 Archer II	Cleanacres Ltd
	G-BFYA	MBB Bo 105D	Helicopter Hire Ltd/Southend
	G-BFYB	PA-28-161 Warrior II	C.S.E. Aviation Ltd/Kidlington
	G-BFYC	PA-32RT-300 Lance II	Peter Lang International Ltd
	G-BFYE	Robin HR.100/285	J. Hackett/Jersey
	G-BFYI	Westland-Bell 47G-3B1	Dollar Air Services Ltd/Coventry
	G-BFYJ	Hughes 369HE	Wilford Aviation Ltd/Fairoaks
	G-BFYL	Evans VP.2	A. G. Wilford
	G-BFYM	PA-28-161 Warrior II	C.S.E. Aviation Ltd/Kidlington
	G-BFYN	Cessna FA.152	Phoenix Flying Services Ltd/Glasgow
	G-BFYO	Spad XIII (replica) (53398)	Peter Long International Ltd
	G-BFYP	Bensen B.7	A. J. Philpotts
	G-BFYU	SC.5 Belfast	HeavyLift Cargo Airlines Ltd/Stansted
	G-BFZA	Alpavia Fournier RF-3	T. J. Hartwell & D. R. Wilkinson
	G-BFZB	Piper J-3C-85 Cub	Zebedee Flying Group/Shoreham
	G-BFZD	Cessna FR.182RG	R. B. Lewis & Co/Sleap
	G-BFZF	Boeing 707 321C	*Stored*/Lasham
	G-BFZG	PA-28-161 Warrior II	C.S.E. Aviation Ltd/Kidlington
	G-BFZH	PA-28R-200 Cherokee Arrow	W. J. Olaney
	G-BFZL	V.836 Viscount	Manx Airlines Ltd/Ronaldsway
	G-BFZM	R. Commander 112TC	Rolls-Royce Ltd/Filton
	G-BFZN	Cessna FA.152	Leicestershire Aero Club Ltd
	G-BFZO	AA-5A Cheetah	Heald Air Ltd/Manchester
	G-BFZP	AA-5B Tiger	J. Hodson-Klein/Edinburgh
	G-BFZS	Cessna F.152	R. T. Haddow
	G-BFZT	Cessna FA.152	One Zero One Three Ltd/Guernsey
	G-BFZU	Cessna FA.152	Reedtrend Ltd/Stapleford
	G-BFZV	Cessna F.172M	W. J. Kavanagh
	G-BGAA	Cessna 152 II	Louth Equipment Hire Services
	G-BGAB	Cessna F.152 II	TG Aviation Ltd/Manston
	G-BGAD	Cessna F.152 II	W. J. Overhead
	G-BGAE	Cessna F.152 II	Northfield Garage (Cowdenbeath) Ltd
	G-BGAF	Cessna FA.152	Suffolk Aero Club Ltd/Ipswich
	G-BGAG	Cessna F.172N	Adifer Ltd
	G-BGAH	FRED Srs 2	G. A. Harris
	G-BGAJ	Cessna F.182Q II	Ground Airport Services Ltd/Guernsey
	G-BGAK	Cessna F.182Q II	Safari World Services Ltd
	G-BGAT	Douglas DC-10-30	British Caledonian Airways *James Watt — The Scottish Engineer*/Gatwick
	G-BGAU	Rearwin 9000L	Shipping & Airlines Ltd/Biggin Hill
	G-BGAX	PA-28 Cherokee 140	Hillvine Ltd
	G-BGAY	Cameron O-77 balloon	Dante Balloon Group *Antonia*
	G-BGAZ	Cameron V-77 balloon	Cameron Balloons Ltd *Silicon Chip*
	G-BGBA	Robin R.2100A	D. Faulkner/Redhill
	G-BGBB	L.1011-385 TriStar 200	British Airways *The Lakeland Rose*/ Heathrow
	G-BGBC	L.1011-385 TriStar 200	British Airways *The Shot Silk Rose*/ Heathrow
	G-BGBE	Jodel DR.1050	P. B. Turner/Stapleford
	G-BGBF	D.31A Turbulent	S. Haye
	G-BGBG	PA-28-181 Archer II	Harlow Printing Ltd/Newcastle
	G-BGBI	Cessna F.150L	Air Fenland Ltd
	G-BGBK	PA-38-112 Tomahawk	Sandtoft Air Services Ltd
	G-BGBN	PA-38-112 Tomahawk	Leavesden Flight Centre Ltd
	G-BGBP	Cessna F.152	Solo Leasing/Guernsey
	G-BGBR	Cessna F172N	Stanton Aircraft Management Ltd/ Biggin Hill
	G-BGBU	Auster AOP.9	P. Neilson
	G-BGBW	PA-38-112 Tomahawk	Spatial Air Brokers & Forwarders Ltd/ E. Midlands
	G-BGBX	PA-38-112 Tomahawk	Ipswich School of Flying

Reg.	Type	Owner or Operator	Notes
G-BGBY	PA-38-112 Tomahawk	Cheshire Flying Services Ltd/ Manchester	
G-BGBZ	R. Commander 114	R. S. Fenwick/Biggin Hill	
G-BGCC	PA-31-325 Navajo	Denham School of Flying	
G-BGCG	Douglas C-47A	Datran Holdings/Thruxton	
G-BGCH	PA-38-112 Tomahawk	B. W. Wells/Wellesbourne	
G-BGCL	AA-5A Cheetah	Kestrel Air Services	
G-BGCM	AA-5A Cheetah	Flight Preparations Ltd/Southampton	
G-BGCO	PA-44-180 Seminole	J. R. Henderson	
G-BGCX	Taylor JT.1 Monoplane	G. M. R. Walters	
G-BGCY	Taylor JT.1 Monoplane	R. L. A. Davies	
G-BGDA	Boeing 737-236	British Airways *River Tamar*/Heathrow	
G-BGDB	Boeing 737-236	British Airways *River Tweed*/Heathrow	
G-BGDC	Boeing 737-236	British Airways *River Humber*/Heathrow	
G-BGDD	Boeing 737-236	British Airways *River Tees*/Heathrow	
G-BGDE	Boeing 737-236	British Airways *River Avon*/Heathrow	
G-BGDF	Boeing 737-236	British Airways *River Thames*/Heathrow	
G-BGDG	Boeing 737-236	British Airways *River Medway*/ Heathrow	
G-BGDH	Boeing 737-236	British Airways *River Clyde*/Heathrow	
G-BGDI	Boeing 737-236	British Airways *River Ouse*/Heathrow	
G-BGDJ	Boeing 737-236	British Airways *River Trent*/Heathrow	
G-BGDK	Boeing 737-236	British Airways *River Mersey*/Heathrow	
G-BGDL	Boeing 737-236	British Airways *River Don*/Heathrow	
G-BGDN	Boeing 737-236	British Airways *River Tyne*/Heathrow	
G-BGDO	Boeing 737-236	British Airways *River Usk*/Heathrow	
G-BGDP	Boeing 737-236	British Airways *River Taff*/Heathrow	
G-BGDR	Boeing 737-236	British Airways *River Bann*/Heathrow	
G-BGDS	Boeing 737-236	British Airways *River Severn*/Heathrow	
G-BGDT	Boeing 737-236	British Airways *River Forth*/Heathrow	
G-BGDU	Boeing 737-236	British Airways *River Dee*/Heathrow	
G-BGEA	Cessna F.150M	R. L. Beverley/Bournemouth	
G-BGED	Cessna U.206F	Elecwind (Clay Cross) Ltd/E. Midlands	
G-BGEE	Evans VP-1	B. H. D. H. Frere	
G-BGEF	Jodel D.112	G. G. Johnson & S. J. Davies	
G-BGEH	Monnet Sonerai II	R. E. Finlay/Barton	
G-BGEI	Baby Great Lakes	D. J. Wright	
G-BGEK	PA-38-112 Tomahawk	Cheshire Flying Services Ltd/ Manchester	
G-BGEL	PA-38-112 Tomahawk	Cheshire Flying Services Ltd/ Manchester	
G-BGEM	Partenavia P.68B	AEW Engineering Co Ltd	
G-BGEN	D.H.C.-6 Twin Otter 310	Loganair Ltd/Glasgow	
G-BGEO	PA-31-350 Navajo Chieftain	Oxford Aerocharter Ltd/Kidlington	
G-BGEP	Cameron D-38 balloon	Cameron Balloons Ltd	
G-BGES	Currie Wot	K. E. Ballington	
G-BGET	PA-38-112 Tomahawk	Goodwood Terrena Ltd	
G-BGEV	PA-38-112 Tomahawk	Cheshire Flying Services Ltd/ Manchester	
G-BGEW	Nord NC.854S	R. A. Yates	
G-BGEX	Brookland Mosquito 2	D. R. C. Pugh/Shobdon	
G-BGFC	Evans VP-2	J. A. Jones	
G-BGFF	FRED Srs 2	G. R. G. Smith	
G-BGFG	AA-5A Cheetah	Fletcher Aviation Ltd	
G-BGFH	Cessna F.182Q	Mindon Engineering (Nottingham) Ltd/ Tollerton	
G-BGFI	AA-5A Cheetah	Maston Property Holdings Ltd	
G-BGFJ	Jodel D.9 Bebe	C. M. Fitton	
G-BGFN	PA-25 Pawnee 235	Farmwork Services (Eastern) Ltd	
G-BGFS	Westland-Bell 47G-3B1	G. S. Mason	
G-BGFT	PA-34-200T-2 Seneca	C.S.E. Aviation Ltd/Kidlington	
G-BGFX	Cessna F.152	A. W. Fay/Biggin Hill	
G-BGGA	Bellanca 7GCBC Citabria	A. G. Forshaw/Barton	
G-BGGB	Bellanca 7GCBC Citabria	R. J. W. Wood	
G-BGGC	Bellanca 7GCBC Citabria	R. P. Ashfield & B. A. Jesty	
G-BGGD	Bellanca 8GCBC Scout	Bristol & Gloucestershire Gliding Club/Nympsfield	
G-BGGE	PA-38-112 Tomahawk	C.S.E. (Aircraft Services) Ltd/Kidlington	
G-BGGF	PA-38-112 Tomahawk	C.S.E. (Aircraft Services) Ltd/Kidlington	
G-BGGG	PA-38-112 Tomahawk	C.S.E. (Aircraft Services) Ltd/Kidlington	
G-BGGI	PA-38-112 Tomahawk	C.S.E. (Aircraft Services) Ltd/Kidlington	
G-BGGJ	PA-38-112 Tomahawk	C.S.E. (Aircraft Services) Ltd/Kidlington	
G-BGGL	PA-38-112 Tomahawk	C.S.E. (Aircraft Services) Ltd/Kidlington	
G-BGGM	PA-38-112 Tomahawk	C.S.E. Aviation Ltd/Kidlington	

Notes	Reg.	Type	Owner or Operator
	G-BGGN	PA-38-112 Tomahawk	C.S.E. (Aircraft Services) Ltd/Kidlington
	G-BGGO	Cessna F.152	E. Midlands Flying School Ltd
	G-BGGP	Cessna F.152	E. Midlands Flying School Ltd
	G-BGGT	Zenith CH.200	P. R. M. Nind
	G-BGGU	Wallis WA-116R-R	K. H. Wallis
	G-BGGV	Wallis WA-120 Srs 2	K. H. Wallis
	G-BGGW	Wallis WA-112	K. H. Wallis
	G-BGGY	AB-206B JetRanger 3	Mann Aviation Ltd/Fairoaks
	G-BGHA	Cessna F.152	Wickwell (UK) Ltd
	G-BGHC	Saffery Hot Pants Firefly balloon	H. C. Saffery *Petuniga*
	G-BGHD	Saffery Helios Blister balloon	H. C. Saffery
	G-BGHE	Convair L-13A	J. Davis/USA
	G-BGHF	Westland WG.30 ★	*Instructional airframe*/Yeovil
	G-BGHI	Cessna F.152	Taxon Ltd/Shoreham
	G-BGHJ	Cessna F.172N	Timelion Ltd
	G-BGHK	Cessna F.152	Wilson Leasing/Leeds
	G-BGHM	Robin R.1180T	P. T. Bolton
	G-BGHP	Beech 76 Duchess	J. J. Baumhardt Associates Ltd
	G-BGHS	Cameron N-31 balloon	Balloon Stable Ltd
	G-BGHT	Falconar F-12	T. Kerr-Baillie
	G-BGHU	T-6G Harvard (115042)	S. M. & P. S. Warner/Wellesbourne
	G-BGHV	Cameron V-77 balloon	E. Davies
	G-BGHW	Thunder Ax8-90 balloon	Edinburgh University Balloon Group *James Tytler*
	G-BGHX	Chasle YC-12 Tourbillon	C. Clark
	G-BGHY	Taylor JT.1 Monoplane	J. Prowse
	G-BGIB	Cessna 152 II	Mona Aviation Ltd
	G-BGIC	Cessna 172N	T. R. Sinclair
	G-BGID	Westland-Bell 47G-3B1	A. E. & B. G. Brown
	G-BGIG	PA-38-112 Tomahawk	Scotia Safari Ltd/Prestwick
	G-BGIH	Rand KR-2	G. & D. G. Park
	G-BGII	PA-32-300 Cherokee Six	Rosefair Electronics Ltd/Elstree
	G-BGIK	Taylor JT.1 Monoplane	J. H. Medforth
	G-BGIM	AS.350B Ecureuil	Lord Glendyne/Hayes
	G-BGIO	Bensen B.8M	C. G. Johns
	G-BGIP	Colt 56A balloon	Capitol Balloon Club
	G-BGIU	Cessna F.172H	Metro Equipment (Chesham) Ltd/ Panshanger
	G-BGIV	Bell 47G-5	Helitech (Luton) Ltd
	G-BGIW	Bell 47G-2	Autair Ltd/Panshanger
	G-BGIX	H.295 Super Courier	Nordic Oil Services Ltd/Edinburgh
	G-BGIY	Cessna F.172N	Glasgow Flying Club
	G-BGIZ	Cessna F.152	Creaton Aviation Services
	G-BGJA	Cessna FA.152	Flairhire Ltd
	G-BGJB	PA-44-180 Seminole	Reedtrend Ltd/Birmingham
	G-BGJE	Boeing 737-236	British Airways *Sandpiper*/Gatwick
	G-BGJF	Boeing 737-236	British Airways *River Axe*/Gatwick
	G-BGJG	Boeing 737-236	British Airways *River Arun*/Gatwick
	G-BGJH	Boeing 737-236	British Airways *River Lyne*/Gatwick
	G-BGJI	Boeing 737-236	British Airways *River Wey*/Gatwick
	G-BGJJ	Boeing 737-236	British Airways *River Swale*/Gatwick
	G-BGJK	Boeing 737-236	British Airways *River Cherwell*/Gatwick
	G-BGJM	Boeing 737-236	British Airways *River Ribble*/Gatwick
	G-BGJU	Cameron V-65 Balloon	D. T. Watkins *Spoils*
	G-BGJV	H.S.748 Srs 2B	British Airways/Glasgow
	G-BGJW	GA-7 Cougar	Trent Air Services Ltd/Cranfield
	G-BGKA	P.56 Provost T.1 (XF690)	D. W. Mickleburgh
	G-BGKC	SOCATA Rallye 110ST	Martin Ltd/Biggin Hill
	G-BGKD	SOCATA Rallye 110ST	Air Westward Co Ltd/Dunkeswell
	G-BGKE	BAC One Eleven 539	British Airways *County of Gwynedd*/ Birmingham
	G-BGKF	BAC One-Eleven 539	British Airways *County of Warwick*/ Birmingham
	G-BGKG	BAC One-Eleven 539	British Airways *County of Stafford*/ Birmingham
	G-BGKJ	MBB Bo 105C	Bond Helicopters Ltd/Bourn
	G-BGKM	SA.365C Dauphin	Bond Helicopters Ltd/Bourn
	G-BGKO	GY-20 Minicab	R. B. Webber
	G-BGKP	MBB Bo 105C	Bond Helicopters Ltd/Bourn
	G-BGKS	PA-28-161 Warrior II	Woodgate Air Services (IOM) Ltd
	G-BGKT	Auster AOP.9 (XN441)	K. H. Wallis
	G-BGKU	PA-28R-201 Arrow III	Farr (Metal Fabrications) Ltd/Coventry
	G-BGKV	PA-28R-201 Arrow III	J. Lloyd
	G-BGKW	Evans VP-1	I. W. Black

Reg.	Type	Owner or Operator	Notes
G-BGKY	PA-38-112 Tomahawk	J. J. Cook Aviation & Sandwell Scaffold Co Ltd/Manchester	
G-BGKZ	J/5F Aiglet Trainer	R. C. H. Hibberd/Dunkeswell	
G-BGLA	PA-38-112 Tomahawk	Scotia Safari Ltd/Prestwick	
G-BGLB	Bede BD-5B	W. Sawney	
G-BGLD	Beech 76 Duchess	Roy Hall Travel Ltd/Manchester	
G-BGLE	Saffrey S.330 Balloon	C. J. Dodd & ptnrs	
G-BGLF	Evans VP-1	M. P. Edwards/Barton	
G-BGLG	Cessna 152	Skyviews & General Ltd/Stapleford	
G-BGLH	Cessna 152	Deltair Ltd/Chester	
G-BGLI	Cessna 152	Knack Air Ltd	
G-BGLK	Monnet Sonerai II	G. L. Kemp & J. Beck	
G-BGLN	Cessna FA.152	Shoreham Flight Simulation Ltd	
G-BGLO	Cessna F.172N	A. H. Slaughter/Southend	
G-BGLR	Cessna F.152	Peter Roberts Associates	
G-BGLS	Super Baby Great Lakes	D. S. Morgan/Lasham	
G-BGLW	PA-34-200 Seneca	Andrew Edie Aviation Ltd/Shoreham	
G-BGLX	Cameron N-56 balloon	Sara A. G. Williams	
G-BGLZ	Stits SA-3A Playboy	P. E. Barker	
G-BGMA	D.31 Turbulent	G. C. Masterson	
G-BGMB	Taylor JT.2 Titch	E. M. Bourne	
G-BGMC	D.H.C.-6 Twin Otter 310	Loganair Ltd/Glasgow	
G-BGMD	D.H.C.-6 Twin Otter 310	Hubbardair Ltd	
G-BGME	SIPA S.903	M. Emery (G-BCML)/Redhill	
G-BGMJ	GY-201 Minicab	H. P. Burrill	
G-BGMP	Cessna F.172G	Norvic Racing Engines Ltd/ Cranfield	
G-BGMR	GY-201 Minicab	T. J. D. Hodge & A. B. Holloway/ Southend	
G-BGMS	Taylor JT.2 Titch	M. A. J. Spice	
G-BGMT	MS.894E Rallye 235GT	M. E. Taylor	
G-BGMU	Westland Bell 47G-3B1	Carskiey Ltd	
G-BGMV	Scheibe SF.25B Falke	Wolds Gliding Club Ltd/Pocklington	
G-BGMW	Edgley EA-7 Optica	Edgley Aircraft Ltd/Old Sarum	
G-BGMX	Enstrom F-280C-UK-2 Shark	Nelson Helicopter Services Ltd/ Shoreham	
G-BGND	Cessna F.172N	Stansted Fluid Power Products Ltd	
G-BGNM	SA.365C1 Dauphin	Bond Helicopters Ltd/Bourn	
G-BGNR	Cessna F.172N	Bevan Lynch Aviation Ltd/Birmingham	
G-BGNS	Cessna F.172N	Reedtrend Ltd/Shoreham	
G-BGNT	Cessna F.152	Klingair Ltd/Conington	
G-BGNU	Beech E90 King Air	Norwich Union Fire Insurance Ltd/ Norwich	
G-BGNV	GA-7 Cougar	H. Snelson/Manchester	
G-BGNW	Boeing 737-219ADV	Britannia Airways Ltd *George Stephenson*/Luton	
G-BGNZ	Cessna FRA.150L	Kingsmetal Ltd/Lydd	
G-BGOA	Cessna FR.182RG	The Forestry Commission/Fairoaks	
G-BGOC	Cessna F.152	Elliot Forbes Ltd/Aberdeen	
G-BGOD	Colt 77A balloon	J. R. Gore	
G-BGOE	Beech 76 Duchess	H. Nelson (Engineers) Ltd/Manchester	
G-BGOF	Cessna F.152	Kingsmetal Ltd/Lydd	
G-BGOG	PA-28-161 Warrior II	M. J. Cowham	
G-BGOH	Cessna F.182Q	Zonex Ltd/Blackpool	
G-BGOI	Cameron O-56 balloon	Balloon Stable Ltd *Skymaster*	
G-BGOL	PA-28R-201T Turbo Arrow IV	P.A.C.K. Enterprise (Sussex) Ltd/ Shoreham	
G-BGOM	PA-31-310 Navajo	Oxford Aero Charter Ltd/Kidlington	
G-BGON	GA-7 Cougar	Wendexim Trading Co Ltd/Denham	
G-BGOO	Colt 56 SS balloon	British Gas Corporation	
G-BGOP	Dassault Falcon 20F	Nissan (UK) Ltd/Heathrow	
G-BGOR	AT-6D Harvard III	M. L. Sargeant	
G-BGOX	PA-31-350 Navajo Chieftain	Berrard Ltd	
G-BGOY	PA-31-350 Navajo Chieftain	Berrard Ltd	
G-BGPA	Cessna 182Q	R. A. Robinson	
G-BGPB	AT-16 Harvard IV (385)	A. G. Walker & R. Lamplough/Duxford	
G-BGPC	D.H.C.-6 Twin Otter 310	Nordic Oil Services Ltd/Glasgow	
G-BGPD	Piper L-4H Cub	P. D. Whiteman	
G-BGPE	Thunder Ax6-56 balloon	C. Wolstenholme *Sergeant Pepper*	
G-BGPF	Thunder Ax6-56Z balloon	Thunder Balloons Ltd *Pepsi*	
G-BGPG	AA-5B Tiger	Tentergate Aviation/Elstree	
G-BGPH	AA-5B Tiger	Peter Turnbull (York) Ltd/Sherburn	
G-BGPI	Plumb BGP-1	B. G. Plumb	

Notes	Reg.	Type	Owner or Operator
	G-BGPJ	PA-28-161 Warrior II	R. P. Maughan & A. E. Hart/ Biggin Hill
	G-BGPK	AA-5B Tiger	Ann Green Manufacturing Co Ltd/ Elstree
	G-BGPL	PA-28-161 Warrior II	Cormack (Aircraft Services) Ltd/ Glasgow
	G-BGPM	Evans VP-2	T. G. Painter
	G-BGPN	PA-18-150 Super Cub	Roy Moore Ltd/Blackpool
	G-BGPP	PA-25 Pawnee 235	Miller Aerial Spraying Ltd/Wickenby
	G-BGPS	Aero Commander 200D	G. Jones & K. Abbott/Cardiff
	G-BGPT	Parker Teenie Two	K. Atkinson
	G-BGPU	PA-28 Cherokee 140	Air Navigation & Trading Co Ltd/ Blackpool
	G-BGPZ	MS.890A Rallye Commodore	J. A. Espin/Popham
	G-BGRA	Taylor JT.2 Titch	J. R. C. Thompson
	G-BGRC	PA-28 Cherokee 140	Falcon Flying Services/Biggin Hill
	G-BGRE	Beech A200 Super King Air	Martin-Baker (Engineering) Ltd
	G-BGRG	Beech 76 Duchess	Arrows Ltd/Manchester
	G-BGRH	Robin DR.400/2+2	T.M.A. Associates Ltd/Biggin Hill
	G-BGRI	Jodel DR.1051	C. R. Warcup
	G-BGRJ	Cessna T.310R	Gledhill Water Storage Ltd/Blackpool
	G-BGRK	PA-38-112 Tomahawk	Goodwood Terrena Ltd
	G-BGRL	PA-38-112 Tomahawk	Goodwood Terrena Ltd
	G-BGRM	PA-38-112 Tomahawk	Goodwood Terrena Ltd
	G-BGRN	PA-38-112 Tomahawk	Goodwood Terrena Ltd
	G-BGRO	Cessna F.172M	M. Hmaidatou/Humberside
	G-BGRR	PA-38-112 Tomahawk	Cambrian Flying Club/Cardiff
	G-BGRS	Thunder Ax7-77Z balloon	P. Hassall Ltd
	G-BGRT	Steen Skybolt	R. C. Teverson
	G-BGRX	PA-38-112 Tomahawk	T. Millar/Elstree
	G-BGSA	MS.892E Rallye 150GT	G. A. Schulz & ptnrs/Leicester
	G-BGSC	Ayres S2R-T34 Turbo Thrush 500	Shoreham Flight Simulation/ Bournemouth
	G-BGSE	Pitts S-2A Special	P. H. Meeson
	G-BGSG	PA-44-180 Seminole	D. J. McSorley
	G-BGSH	PA-38-112 Tomahawk	Scotia Safari Ltd/Prestwick
	G-BGSI	PA-38-112 Tomahawk	Sandwell Scaffold Co Ltd/Manchester
	G-BGSJ	Piper J-3C-65 Cub	W. J. Higgins/Dunkeswell
	G-BGSM	MS.892E Rallye 150GT	Tyre & Tune Service Station
	G-BGSN	Enstrom F-28C-UK-2	Nicelynn Ltd
	G-BGSO	PA-31-310 Navajo	Reedtrend Ltd
	G-BGST	Thunder Ax7-65 balloon	L. H. T. Large & ptnrs *Eclipse*
	G-BGSV	Cessna F.172N	Wickenby Flying Club Ltd
	G-BGSW	Beech F33 Debonair	D. J. Shires/Stapleford
	G-BGSX	Cessna F.152	RJS Aviation Ltd/Coventry
	G-BGSY	GA-7 Cougar	Van Allen Ltd/Guernsey
	G-BGTA	Firebird Bunce B.500 balloon	S. J. Bunce
	G-BGTB	SOCATA TB.10 Tobago ★	S. Yorks Aviation Soc
	G-BGTC	Auster AOP.9 (XP282)	A. C. Byrne
	G-BGTF	PA-44-180 Seminole	New Guarantee Trust Ltd/Jersey
	G-BGTG	PA-23 Aztec 250	R. J. Howard
	G-BGTI	Piper J-3C-65 Cub	The Grasshopper Group
	G-BGTJ	PA-28 Cherokee 180	Serendipity Aviation/Staverton
	G-BGTK	Cessna FR.182RG	Kestrel Air Services Ltd
	G-BGTL	GY-20 Minicab	A. K. Lang
	G-BGTP	Robin HR.100/210	B. G. Graham
	G-BGTR	PA-28 Cherokee 140	Keenair Services Ltd/Liverpool
	G-BGTS	PA-28 Cherokee 140	Keenair Services Ltd/Liverpool
	G-BGTT	Cessna 310R	Air Atlantique/Jersey
	G-BGTU	BAC One-Eleven 409	Turbo Union Ltd/Filton
	G-BGTX	Jodel D.117	Madley Flying Group/Shobdon
	G-BGTY	Boeing 737-2Q8	Orion Airways Ltd/E. Midlands
	G-BGUA	PA-38-112 Tomahawk	Truman Aviation Ltd/Tollerton
	G-BGUB	PA-32-300 Cherokee Six	J. Beckers & ptnrs

NOTE: The G-BGUx sequence will not be issued unless specifically requested

Notes	Reg.	Type	Owner or Operator
	G-BGUY	Cameron V-56 balloon	G. V. Beckwith
	G-BGVA	Cessna 414A	Clymsil Holdings Ltd
	G-BGVB	Robin DR.315	J. R. D. Bygraves/O. Warden
	G-BGVE	CP1310-C3 Super Emeraude	E. J. A. Woolnough
	G-BGVF	Colt 77A balloon	Hot Air Balloon Co Ltd
	G-BGVH	Beech 76 Duchess	Velco Marketing

Reg.	Type	Owner or Operator	Notes
G-BGVI	Cessna F.152	Farr (Metal Fabrications) Ltd/ Stapleford	
G-BGVK	PA-28-161 Warrior II	W. E. B. Wordsworth	
G-BGVL	PA-38-112 Tomahawk	Moore House Freight/Edinburgh	
G-BGVM	Wilson Cassutt 3M	J. T. Mirley/Halfpenny Green	
G-BGVN	PA-28RT-201 Arrow IV	Essex Aviation Ltd/Stapleford	
G-BGVP	Thunder Ax6-56Z balloon	A. Bolger	
G-BGVR	Thunder Ax6-56Z balloon	A. N. G. Howie	
G-BGVS	Cessna F.172M	P. D. A. Aviation Ltd/Tollerton	
G-BGVT	Cessna R.182RG	Barnes Plastics Group Ltd	
G-BGVU	PA-28 Cherokee 180	Cheshire Flying Services Ltd/ Manchester	
G-BGVV	AA-5A Cheetah	R. M. Messenger	
G-BGVW	AA-5A Cheetah	Matthew Associates Ltd/Elstree	
G-BGVX	Cessna P.210N	Cheshire Scaffold Co Ltd/Liverpool	
G-BGVY	AA-5B Tiger	Porter Bell Ltd/Bournemouth	
G-BGVZ	PA-28-181 Archer II	Robinson-Wyllie Ltd	
G-BGWA	GA-7 Cougar	Lough Erne Aviation Ltd	
G-BGWC	Robin DR.400/180	E. F. Braddon & D. C. Shepherd/ Rochester	
G-BGWD	Robin HR.100/285	Hordell Engineering Ltd/Fairoaks	
G-BGWF	PA-18-150 Super Cub	E. D. Burke/Staverton	
G-BGWH	PA-18-150 Super Cub	Earnshaw-Brown (Battle) Ltd	
G-BGWI	Cameron V-65 balloon	Army Balloon Club/W. Germany	
G-BGWJ	Sikorsky S-61N	Bristow Helicopters Ltd	
G-BGWK	Sikorsky S-61N	British Executive Air Services Ltd	
G-BGWM	PA-28-181 Archer II	Zitair Flying Club Ltd/Booker	
G-BGWN	PA-38-112 Tomahawk	Scotia Safari Ltd/Prestwick	
G-BGWO	Jodel D.112	A. J. Court	
G-BGWP	MBB Bo 105C	Bond Helicopters Ltd/Bourn	
G-BGWS	Enstrom F-280C Shark	G. Firbank & N. M. Grimshaw	
G-BGWT	WS-58 Wessex 60 Srs 1	Bristow Helicopters Ltd	
G-BGWU	PA-38-112 Tomahawk	Sandwell Scaffold Co Ltd/Manchester	
G-BGWV	Aeronca 7AC Champion	RFC Flying Group/Popham	
G-BGWW	PA-23 Aztec 250E	Ski Air Ltd/Biggin Hill	
G-BGWY	Thunder Ax6-56Z balloon	S. L. G. Williams	
G-BGWZ	Eclipse Super Eagle ★	FAA Museum/Yeovilton	
G-BGXA	Piper J-3C-65 Cub	E. F. Fryer	
G-BGXB	PA-38-112 Tomahawk	Keenair Services Ltd/Liverpool	
G-BGXC	SOCATA TB.10 Tobago	A. J. Halliday/Shoreham	
G-BGXD	SOCATA TB.10 Tobago	Selles Dispensing Chemists Ltd	
G-BGXJ	Partenavia P.68B	Cecil Aviation Ltd/Cambridge	
G-BGXK	Cessna 310R	Garage Micro-Systems Ltd/Exeter	
G-BGXL	Bensen B.8MV	B. P. Triefus	
G-BGXN	PA-38-112 Tomahawk	Keats Web Offset Ltd/Denham	
G-BGXO	PA-38-112 Tomahawk	C.S.E. Aviation Ltd/Kidlington	
G-BGXP	Westland-Bell 47G-3B1	B. A. Hogan & ptnrs	
G-BGXR	Robin HR.200/100	Graham Churchill (Plant Hire) Ltd	
G-BGXS	PA-28-236 Dakota	Debian Car Hire Ltd/Jersey	
G-BGXT	SOCATA TB.10 Tobago	County Aviation Ltd/Halfpenny Green	
G-BGXU	WMB-1 balloon	C. J. Dodd & ptnrs	
G-BGXX	Jodel DR.1051M1	C. Evans	
G-BGXZ	Cessna FA.152	Kingsmetal Ltd/Lydd	
G-BGYG	PA-28-161 Warrior II	C.S.E. Aviation Ltd/Kidlington	
G-BGYH	PA-28-161 Warrior II	C.S.E. Aviation Ltd/Kidlington	
G-BGYJ	Boeing 737-204	Britannia Airways Ltd *Sir Barnes Wallis*/Luton	
G-BGYK	Boeing 737-204	Britannia Airways Ltd *R. J. Mitchell*/ Luton	
G-BGYL	Boeing 737-204	Britannia Airways Ltd *Jean Batten*/ Luton	
G-BGYN	PA-18-150 Super Cub	A. G. Walker	
G-BGYR	H.S.125 Srs 600B	British Aerospace/Warton	
G-BGYT	EMB-110P1 Bandeirante	Jersey European Airways	
G-BGYV	EMB-110P1 Bandeirante	Jersey European Airways	
G-BGZC	C.A.S.A. 1.131 Jungmann	J. E. Douglas	
G-BGZE	PA-38-112 Tomahawk	G. C. J. Moffatt & Co Ltd/Cardiff	
G-BGZF	PA-38-112 Tomahawk	Shirlster Container Transport Ltd/ Swansea	
G-BGZH	PA-38-112 Tomahawk	G. C. J. Moffatt & Co Ltd/Cardiff	
G-BGZJ	PA-38-112 Tomahawk	W. R. C. Foyle/Luton	
G-BGZK	Westland-Bell 47G-3B1	Sensehover Ltd	
G-BGZL	Eiri PIK-20E	D. I. Liddell-Grainger	
G-BGZN	WMB.2 Windtracker balloon	S. R. Woolfries	

Notes	Reg.	Type	Owner or Operator
	G-BGZO	M.S.880B Rallye Club	W. G. R. Wunderlich/Biggin Hill
	G-BGZP	D.H.C.6 Twin Otter 310	Hubbardair Ltd
	G-BGZR	Meagher Model balloon Mk.1	S. C. Meagher
	G-BGZS	Keirs Heated Air Tube	M. N. J. Kirby
	G-BGZW	PA-38-112 Tomahawk	Angelsword Ltd/Wellesbourne
	G-BGZX	PA-32 Cherokee Six 260	CEL Electronics Ltd/Stapleford
	G-BGZY	Jodel D.120	P. J. Sebastian/Popham
	G-BGZZ	Thunder Ax6-56 balloon	J. M. Robinson
	G-BHAA	Cessna 152	Herefordshire Aero Club Ltd/Shobdon
	G-BHAB	Cessna 152	Herefordshire Aero Club Ltd/Shobdon
	G-BHAC	Cessna A.152	Herefordshire Aero Club Ltd/Shobdon
	G-BHAD	Cessna A.152	Shropshire Aero Club Ltd/Sleap
	G-BHAF	PA-38-112 Tomahawk	R. Colin Snow/Humberside
	G-BHAG	Scheibe SF.25E Super Falke	British Gliding Association/Lasham
	G-BHAI	Cessna F.152	Channel Islands Aero Holdings Ltd/ Jersey
	G-BHAJ	Robin DR.400/160	Crocker Aviation Services/Rochester
	G-BHAL	Rango Saffery S.200 SS	A. M. Lindsay *Anneky Panky*
	G-BHAM	Thunder Ax6-56 balloon	D. Sampson
	G-BHAR	Westland-Bell 47G-3B1	E. A. L. Sturmer
	G-BHAT	Thunder Ax7-77 balloon	C. P. Witter Ltd *Witter*
	G-BHAV	Cessna F.152	Iceni Leasing
	G-BHAW	Cessna F.172N	W. Lancs Aero Club Ltd/Woodvale
	G-BHAX	Enstrom F-28C-UK-2	Southern Air Ltd/Shoreham
	G-BHAY	PA-28RT-201 Arrow IV	V. J. Holden/Newcastle
	G-BHBA	Campbell Cricket	S. M. Irwin
	G-BHBB	Colt 77A balloon	S. D. Bellew/USA
	G-BHBE	Westland Bell 47G-3B1 (Soloy)	Dollar Air Services Ltd/Coventry
	G-BHBF	Sikorsky S-76A	Bristow Helicopters Ltd
	G-BHBG	PA-32R-300 Lance	D. A. Stewart/Birmingham
	G-BHBI	Mooney M.20J	B. K. Arthur/Exeter
	G-BHBK	Viscount V-5 balloon	B. Hargraves & B. King
	G-BHBL	L.1011-385 TriStar 200	British Airways *Largs Bay*/Heathrow
	G-BHBM	L.1011-385 TriStar 200	British Airways *Poole Bay*/Heathrow
	G-BHBN	L.1011-385 TriStar 200	British Airways *Bideford Bay*/Heathrow
	G-BHBO	L.1011-385 TriStar 200	British Airways *St Magnus Bay*/ Heathrow
	G-BHBP	L.1011-385 TriStar 200	British Airways *Whitesand Bay*/ Heathrow
	G-BHBR	L.1011-385 TriStar 200	British Airtours *Bude Bay*/Gatwick
	G-BHBS	PA-28RT-201T Turbo Arrow IV	Zipmaster Ltd/Elstree
	G-BHBT	MA.5 Charger	R. G. & C. J. Maidment/Shoreham
	G-BHBW	Westland Bell 47G-3BI	Heliwork Ltd/Thruxton
	G-BHBZ	Partenavia P.68B	Alpine Press Ltd/Leavesden
	G-BHCC	Cessna 172M	C. I. McAndrew/Bournemouth
	G-BHCE	Jodel D.112	G. F. M. Garner
	G-BHCF	WMB.2 Windtracker balloon	C. J. Dodd & ptnrs
	G-BHCM	Cessna F.172H	The English Connection Ltd/Panshanger
	G-BHCP	Cessna F.152	Fletcher Bros (Car Hire) Ltd/Doncaster
	G-BHCT	PA-23 Aztec 250	Colt Transport Ltd/Goodwood
	G-BHCW	PA-22 Tri-Pacer 150	B. Brooks
	G-BHCX	Cessna F.152	Graythorne Developments Ltd/Barton
	G-BHCZ	PA-38-112 Tomahawk	Sandwell Scaffold Co Ltd/Manchester
	G-BHDA	Shultz balloon	G. F. Fitzjohn
	G-BHDB	Maule M5-235 Lunar Rocket	Cleanacres Ltd/Staverton
	G-BHDD	V.668 Varsity T.1 (WL626)	G. & F. W. Vale/E. Midlands
	G-BHDE	SOCATA TB.10 Tobago	D. J. M. Wilson
	G-BHDH	Douglas DC-10-30	British Caledonian Airways *Sir Walter Scott*/Gatwick
	G-BHDI	Douglas DC-10-30	British Caledonian Airways *Robert The Bruce*/Gatwick
	G-BHDJ	Douglas DC-10-30	British Caledonian Airways *James S. McDonnell*/Gatwick
	G-BHDK	Boeing B-29A-BN (461748) ★	Imperial War Museum/Duxford
	G-BHDM	Cessna F.152 II	Tayside Aviation Ltd/Dundee
	G-BHDO	Cessna F.182Q II	S. Richman & ptnrs/Plymouth
	G-BHDP	Cessna F.182Q II	Rimmer Aviation Ltd/Elstree
	G-BHDR	Cessna F.152 II	G. Capes
	G-BHDS	Cessna F.152 II	Tayside Aviation Ltd/Dundee
	G-BHDT	SOCATA TB.10 Tobago	W. R. C. Foyle/Luton
	G-BHDU	Cessna F.152 II	P. N. Voysey/Sandown
	G-BHDV	Cameron V-77 balloon	E. D. Price
	G-BHDW	Cessna F.152	Air South Flying Group/Shoreham

Reg.	Type	Owner or Operator	Notes
G-BHDX	Cessna F.172N	D. M. Slama & T. Parsons/Sandown	
G-BHDZ	Cessna F.172N	J. Bines	
G-BHEC	Cessna F.152	Wickwell (UK) Ltd	
G-BHED	Cessna FA.152	TG Aviation Ltd/Manston	
G-BHEG	Jodel D.150	P. R. Underhill	
G-BHEH	Cessna 310G	D. R. McKnight-Coplowe	
G-BHEK	CP.1315C-3 Super Emeraude	D. B. Winstanley/Barton	
G-BHEL	Jodel D.117	J. C. Jefferies	
G-BHEM	Bensen B.8M	P. D. Widdicombe	
G-BHEN	Cessna FA.152	Leicestershire Aero Club Ltd	
G-BHEO	Cessna FR.182RG	Cosworth Engineering Ltd/Coventry	
G-BHEP	Cessna 172 RG Cutlass	Memec Systems Ltd/Conington	
G-BHER	SOCATA TB.10 Tobago	W. R. M. Dury/Biggin Hill	
G-BHET	SOCATA TB.10 Tobago	Michael Jackson Motors Ltd/ E. Midlands	
G-BHEU	Thunder Ax7-65 balloon	M. H. R. Govett *Polo Moche*	
G-BHEV	PA-28R Cherokee Arrow 200	N. J. Taylor & ptnrs/Ipswich	
G-BHEW	Sopwith Triplane Replica (N5430)	The Hon Patrick Lindsay/O. Warden	
G-BHEX	Colt 56A balloon	A. S. Dear & ptnrs *Super Wasp*	
G-BHEY	Pterodactyl O.R.	High School of Hang Gliding Ltd	
G-BHEZ	Jodel D.150	E. J. Horsfall/Blackpool	
G-BHFA	Pterodactyl O.R.	High School of Hang Gliding Ltd	
G-BHFB	Pterodactyl O.R.	High School of Hang Gliding Ltd	
G-BHFC	Cessna F.152	T. G. Aviation Ltd/Manston	
G-BHFE	PA-44-180 Seminole	Stanbride Ltd/Exeter	
G-BHFF	Jodel D.112	A. J. Maxwell/Blackpool	
G-BHFG	SNCAN SV-4C (45)	The Hon Patrick Lindsay/Booker	
G-BHFH	PA-34-200T-2 Seneca	Hendefern Ltd	
G-BHFI	Cessna F.152	The BAE (Warton) Flying Group/ Blackpool	
G-BHFK	PA-28-151 Warrior	Ilkeston Car Sales Ltd	
G-BHFL	PA-28 Cherokee 180	R. & C. Lord/Coventry	
G-BHFM	Murphy S.200 balloon	M. Murphy	
G-BHFR	Eiri PIK-20E-1	G. Mackie	
G-BHFS	Robin DR.400/180	Flair (Soft Drinks) Ltd/Shoreham	
G-BHFZ	Saffery S.200 balloon	D. Morris	
G-BHGA	PA-31-310 Navajo	Heltor Ltd	
G-BHGC	PA-18-150 Super Cub	Herefordshire Gliding Club Ltd/ Shobdon	
G-BHGF	Cameron V-56 balloon	I. T. & H. Seddon *Biggles*	
G-BHGG	Cessna F.172N	Bryan Aviation Ltd	
G-BHGJ	Jodel D.120	Q. M. B. Oswell	
G-BHGK	Sikorsky S-76	Bond Helicopters Ltd/Bourn	
G-BHGM	Beech 76 Duchess	R. J. A. Brown/Guernsey	
G-BHGN	Evans VP-1	A. R. Cameron	
G-BHGO	PA-32 Cherokee Six 260	Oxford Aero Charter Ltd/Kidlington	
G-BHGP	SOCATA TB.10 Tobago	J. McLeary & ptnrs/Edinburgh	
G-BHGR	Robin DR.315	Headcorn Flying School Ltd	
G-BHGU	WMB.2 Windtracker balloon	I. D. Bamber & ptnrs	
G-BHGV	Kiers captive balloon	K. J. Faulkner	
G-BHGX	Colt 56B balloon	S. Doyle *Prospect*	
G-BHGY	PA-28R Cherokee Arrow 200	Inca Marketing Ltd/Southend	
G-BHHB	Cameron V-77 balloon	I. G. N. Franklin	
G-BHHE	Jodel DR.1051/M1	B. E. Lowe-Lauri/Biggin Hill	
G-BHHG	Cessna F.152	Northamptonshire School of Flying Ltd/ Sywell	
G-BHHH	Thunder Ax7-65 balloon	C. A. Hendley (Essex) Ltd	
G-BHHI	Cessna F.152	Hartmann Ltd	
G-BHHK	Cameron N-77 balloon	S. Bridge & ptnrs	
G-BHHN	Cameron V-77 balloon	Itchen Valley Balloon Group	
G-BHHR	Robin DR.400/180R	D. B. Meeks/Booker	
G-BHHX	Jodel D.112	C. F. Walter	
G-BHHY	G.164 Turbo AgCat D	Miller Aerial Spraying Ltd/Wickenby	
G-BHHZ	Rotorway Scorpion 133	P. A. Gunn & D. Willingham	
G-BHIA	Cessna F.152	W. H. Wilkins/Southampton	
G-BHIB	Cessna F.182Q	ISF Aviation Ltd/Leicester	
G-BHIC	Cessna F.182Q	General Building Services Ltd/Leeds	
G-BHID	SOCATA TB.10 Tobago	W. B. Pinckney & Sons Farming Co Ltd	
G-BHIH	Cessna F.172N	Watkiss Group Aviation Ltd/Biggin Hill	
G-BHII	Cameron V-77 balloon	Starcrete Ltd	
G-BHIJ	Eiri PIK-20E-1	R. W. Hall & ptnrs/Swanton Morley	
G-BHIK	Adam RA-14 Loisirs	L. Lewis	
G-BHIL	PA-28-161 Warrior II	SFT Aviation Ltd/Bournemouth	

Notes	Reg.	Type	Owner or Operator
	G-BHIM	Jodel D.112	G. G. Hughes
	G-BHIN	Cessna F.152	Doncaster Aero Club Ltd
	G-BHIR	PA-28R Cherokee Arrow 200	Cheshire Flying Services Ltd/ Manchester
	G-BHIS	Thunder Ax7-65 balloon	Hedgehoppers Balloon Group
	G-BHIT	SOCATA TB.9 Tampico	EAC Components Ltd
	G-BHIY	Cessna F.150K	W. H. Cole
	G-BHIZ	PA-31 Navajo	Oxford Aero Charter Ltd/Kidlington
	G-BHJA	Cessna A.152	Wycombe Air Centre/Booker
	G-BHJB	Cessna A.152	E. E. Fenning & Son
	G-BHJF	SOCATA TB.10 Tobago	D. G. Ledman/Leavesden
	G-BHJI	Mooney 20J	T. R. Bamber & B. Refson/Elstree
	G-BHJK	Maule M5-235C Lunar Rocket	G. A. & B. J. Finch
	G-BHJN	Fournier RF-4D	G. G. Milton/Sibson
	G-BHJO	PA-28-161 Warrior II	Nairn Flying Services Ltd/Inverness
	G-BHJP	Partenavia P-68C	Shirlstar Container Transport Ltd
	G-BHJR	Saffery S.200 balloon	R. S. Sweeting
	G-BHJS	Partenavia P-68B	Tewin Aviation/Panshanger
	G-BHJU	Robin DR.400/2+2	Harlow Transport Services Ltd/ Headcorn
	G-BHJW	Cessna F.152	Leicestershire Aero Club Ltd
	G-BHJY	EMB-110P1 Bandeirante	Euroair Transport Ltd/Gatwick
	G-BHJZ	EMB-110P2 Bandeirante	Jersey European Airways
	G-BHKA	Evans VP-1	M. L. Perry
	G-BHKB	Westland-Bell 47G-3B1 (Soloy)	Helinorth Ltd/Woodvale
	G-BHKE	Bensen B.8MV	V. C. Whitehead
	G-BHKH	Cameron O-65 balloon	D. G. Body
	G-BHKJ	Cessna 421C	United Nations/Pakistan
	G-BHKR	Colt 14A balloon	British Balloon Museum
	G-BHKT	Jodel D.112	R. Featherstone & ptnrs/Old Sarum
	G-BHKV	AA-5A Cheetah	Metronote Business Machines Ltd/ Biggin Hill
	G-BHKW	Westland-Bell 47G-3B1	Dollar Air Services Ltd/Coventry
	G-BHKX	Beech 76 Duchess	R. M. English & Son Ltd
	G-BHKY	Cessna 310R II	Airwork Ltd/Perth
	G-BHLC	Beech 200 Super King Air	Tal-Air Ltd/Stansted
	G-BHLE	Robin DR.400/180	L. H. Mayall
	G-BHLF	H.S.125 Srs 700B	The Marconi Co Ltd/Luton
	G-BHLH	Robin DR.400/180	Trinecare Ltd/Southend
	G-BHLJ	Saffery-Rigg S.200 balloon	I. A. Rigg
	G-BHLK	GA-7 Cougar	Autair International Ltd/Cranfield
	G-BHLM	Cessna 421C	Brush Electrical Co Ltd/E. Midlands
	G-BHLP	Cessna 441	Automobile Association/Coventry
	G-BHLT	D.H.82A Tiger Moth	R. L. Godwin/Enstone
	G-BHLU	Fournier RF-3	G. G. Milton/Felthorpe
	G-BHLV	CP.301A Emeraude	K. E. Armstrong
	G-BHLW	Cessna 120	I. G. & M. Glenn
	G-BHLX	AA-5B Tiger	Tiger Aviation Ltd (Jersey)
	G-BHLY	Sikorsky S-76A	Bristow Helicopters Ltd
	G-BHLZ	GY-30 Supercab	C. R. & M. C. Sims/Goodwood
	G-BHMA	SIPA 903	Fairwood Flying Club/Swansea
	G-BHMC	M.S.880B Rallye Club	The G-BHMC Group/Bodmin
	G-BHMD	Rand KR-2	W. D. Francis
	G-BHME	WMB.2 Windtracker balloon	I. R. Bell & ptnrs
	G-BHMF	Cessna FA.152	Merrett Aviation Ltd
	G-BHMG	Cessna FA.152	Birmingham Aero Centre Ltd
	G-BHMH	Cessna FA.152	Birmingham Aero Centre Ltd
	G-BHMJ	Avenger T.200-2112 balloon	R. Light *Lord Anthony 1*
	G-BHMK	Avenger T.200-2112 balloon	P. Kinder *Lord Anthony 2*
	G-BHML	Avenger T.200-2112 balloon	L. Caulfield *Lord Anthony 3*
	G-BHMM	Avenger T.200-2112 balloon	M. Murphy *Lord Anthony 4*
	G-BHMO	PA-20M Cerpa Special (Pacer)	J. D. Campbell
	G-BHMR	Stinson 108-3	J. R. Rowell/Sandown
	G-BHMT	Evans VP-1	P. E. J. Sturgeon
	G-BHMU	Colt 21A balloon	J. R. Parkington & Co Ltd
	G-BHMW	F.27 Friendship Mk 200	Air UK/Norwich
	G-BHMX	F.27 Friendship Mk 200	Air UK/Norwich
	G-BHMY	F.27 Friendship Mk 200	Air UK/Norwich
	G-BHMZ	F.27 Friendship Mk 200	Air UK/Norwich
	G-BHNA	Cessna F.152	W. E. B. Wordsword/Stapleford
	G-BHNC	Cameron O-65 balloon	D. & C. Bareford
	G-BHND	Cameron N-65 balloon	Hunter & Sons (Wells) Ltd
	G-BHNE	Boeing 727-2J4	Dan-Air Services Ltd/Gatwick
	G-BHNF	Boeing 727-2J4	Dan-Air Services Ltd/Gatwick

Reg.	Type	Owner or Operator	Notes
G-BHNG	PA-23 Aztec 250	Southern Wings Ltd/Shoreham	
G-BHNI	Cessna 404 Titan	Donington Aviation Ltd/E. Midlands	
G-BHNK	Jodel D.120A	F. G. Miskelly	
G-BHNL	Jodel D.112	J. A. Harding	
G-BHNM	PA-44-180 Seminole	Cearte Tiles Ltd	
G-BHNN	PA-32R-301 Saratoga SP	H. Young Transport Ltd/Southampton	
G-BHNO	PA-28-181 Archer II	Davison Plant Hire Co/Compton Abbas	
G-BHNP	Eiri PIK-20E-1	M. Astley/Husbands Bosworth	
G-BHNR	Cameron N-77 balloon	Bath University Hot-Air Balloon Club	
G-BHNT	Cessna F.172N	I. J. Boyd & D. J. McCooke	
G-BHNU	Cessna F.172N	B. Swindell (Haulage) Ltd/Barton	
G-BHNV	Westland-Bell 47G-3B1	Leyline Helicopters Ltd	
G-BHNX	Jodel D.117	A. Scott & O. M. Hammond	
G-BHNY	Cessna 425	Sinclair Research Ltd/Cambridge	
G-BHOA	Robin DR.400/160	M. F. Bunn/Headcorn	
G-BHOF	Sikorsky S-61N	Bristow Helicopters Ltd	
G-BHOG	Sikorsky S-61N	Bristow Helicopters Ltd	
G-BHOH	Sikorsky S-61N	Bristow Helicopters Ltd	
G-BHOI	Westland-Bell 47G-3B1	Helicopter Hire Ltd/Southend	
G-BHOL	Jodel DR.1050	B. D. Deubelbeiss	
G-BHOM	PA-18 Super Cub 95	W. J. C. Scrope	
G-BHOO	Thunder Ax7-65 balloon	D. Livesey & J. M. Purves *Scraps*	
G-BHOP	Thunder Ax3 balloon	R. G. Griffin & R. Blackwell	
G-BHOR	PA-28-161 Warrior II.	D. P. Stringfield	
G-BHOT	Cameron V-65 balloon	Dante Balloon Group	
G-BHOU	Cameron V-65 balloon	F. W. Barnes	
G-BHOW	Beech 95-58P Baron	Anglo-African Machinery Ltd/ Birmingham	
G-BHOZ	SOCATA TB.9 Tampico	GT Aviation	
G-BHPJ	Eagle Microlite	G. Breen/Enstone	
G-BHPK	Piper J-3C-65 Cub (479865)	H. W. Sage/Thruxton	
G-BHPL	C.A.S.A. 1.131 Jungmann	M. G. Jeffries	
G-BHPM	PA-18 Super Cub 95	P. Morgans	
G-BHPN	Colt 14 balloon	Colt Balloons Ltd	
G-BHPO	Colt 14A balloon	C. Boxall	
G-BHPS	Jodel D.120A	C. J. Francis/Swansea	
G-BHPT	Piper J-3C-65 Cub	Airmiles Ltd	
G-BHPV	Cessna U.206G	R. M. Dobie	
G-BHPX	Cessna 152	Air South Flying Group/Shoreham	
G-BHPY	Cessna 152	Christopher Lunn & Co	
G-BHPZ	Cessna 172N	O'Brian Properties Ltd/Redhill	
G-BHRA	R. Commander 114A	M. I. Edwards/Feltwell	
G-BHRB	Cessna F.152	Light Planes (Lancashire) Ltd/Barton	
G-BHRC	PA-28-161 Warrior II	Sherwood Flying Club Ltd/Tollerton	
G-BHRD	D.H.C.1 Chipmunk 22 (WP977)	ISF Aviation Ltd/Wellesbourne	
G-BHRE	Persephone S.200 balloon	Cupro-Sapphire Ltd	
G-BHRF	Airborne Industries AB400 gas balloon	Balloon Stable Ltd	
G-BHRH	Cessna FA.150K	Merlin Flying Club Ltd/Hucknall	
G-BHRI	Saffery S.200 balloon	Cupro-Sapphire Ltd *Can-Can*	
G-BHRM	Cessna F.152	Angelsword Ltd	
G-BHRN	Cessna F.152	Channel Islands Aero Holdings Ltd/ Jersey	
G-BHRO	R. Commander 112A	John Raymond Transport Ltd/Cardiff	
G-BHRP	PA-44-180 Seminole	A. J. Hows	
G-BHRR	CP.301A Emeraude	T. W. Offen/Biggin Hill	
G-BHRS	ICA IS-28M2	British Aerospace Aircraft Group/ Woodford	
G-BHRU	Saffery S.1000 balloon	Cupro-Sapphire Ltd *Petunia*	
G-BHRV	Mooney M.20J	Tecnovil Equipamentos Industriales	
G-BHRW	Jodel DR.221	J. T. M. Ball/Redhill	
G-BHRY	Colt 56A balloon	Hot Air Balloon Co Ltd	
G-BHSA	Cessna 152	Skyviews & General Ltd/Sherburn	
G-BHSB	Cessna 172N	W. R. Craddock & Son Ltd/Sturgate	
G-BHSD	Scheibe SF.25E Super Falke	Lasham Gliding Soc Ltd	
G-BHSE	R. Commander 114	604 Sqn Flying Group Ltd	
G-BHSF	AA-5A Cheetah	D.S. Plant Hire Ltd (G-BHAS)	
G-BHSG	AB-206A JetRanger	Pryme Helicopters Ltd/Edinburgh	
G-BHSI	Jodel D.9	J. A. Rees	
G-BHSL	C.A.S.A. 1.131 Jungmann	Cotswold Flying Group/Badminton	
G-BHSM	AB-206B JetRanger 2	Dollar Air Services Ltd/Coventry	
G-BHSN	Cameron N-56 balloon	Ballooning Endeavours Ltd	
G-BHSP	Thunder Ax7-77Z balloon	Chicago Instruments Ltd	
G-BHSS	Pitts S-1C Special	J. Elsdon-Davies/Sandown	

Notes	Reg.	Type	Owner or Operator
	G-BHST	Hughes 369D	Abbey Hill Vehicle Services
	G-BHSU	H.S.125 Srs 700B	Shell Aircraft Ltd/Heathrow
	G-BHSV	H.S.125 Srs 700B	Shell Aircraft Ltd/Heathrow
	G-BHSW	H.S.125 Srs 700B	Shell Aircraft Ltd/Heathrow
	G-BHSY	Jodel DR.1050	S. R. Orwin & T. R. Allebone
	G-BHTA	PA-28-236 Dakota	Stenloss Ltd/Sywell
	G-BHTC	Jodel DR.1050/M1	T. A. Carpenter/Popham
	G-BHTD	Cessna T.188C AgHusky	Dallah-ADS Ltd/Egypt
	G-BHTG	Thunder Ax6-56 balloon	F. R. & Mrs S. H. MacDonald
	G-BHTH	T-6G Texan (2807)	Keenair Services Ltd/Liverpool
	G-BHTI	SA.102.5 Cavalier	R. Cochrane
	G-BHTM	Cameron 80 Can SS balloon	BP Oil Ltd
	G-BHTP	PA-31T-500 Cheyenne I	Ugland (UK) Ltd/Stansted
	G-BHTR	Bell 206B JetRanger 3	J. S. Bloor Ltd/Tollerton
	G-BHTT	Cessna 500 Citation	Lucas Industries Ltd/Birmingham
	G-BHTV	Cessna 310R	Air Atlantique Ltd/Jersey
	G-BHUB	Douglas C-47 (315509) ★	Imperial War Museum/Duxford
	G-BHUE	Jodel DR.1050	G. A. Mason/Andrewsfield
	G-BHUG	Cessna 172N	Cloudshire Ltd/Wellesbourne
	G-BHUH	Cremer PC.14 balloon	P. A. Cremer
	G-BHUI	Cessna 152	J. MacDonald/Bournemouth
	G-BHUJ	Cessna 172N	Three Counties Aero Club Ltd/ Blackbushe
	G-BHUL	Beech E90 King Air	Cega Aviation Ltd (G-BBKM)/Stansted
	G-BHUM	D.H.82A Tiger Moth	S. G. Towers
	G-BHUN	PZL-104 Wilga 35	W. Radwanski/Lasham
	G-BHUO	Evans VP-2	R. A. Povall
	G-BHUP	Cessna F.152	Light Planes (Lancs) Ltd/Barton
	G-BHUR	Thunder Ax3 balloon	B. F. G. Ribbons
	G-BHUU	PA-25 Pawnee 235	Farmwork Services (Eastern) Ltd
	G-BHUV	PA-25 Pawnee 235	Farmwork Services (Eastern) Ltd
	G-BHVB	PA-28-161 Warrior II	Norwich Air Training Ltd
	G-BHVC	Cessna 172RG Cutlass	Ian Willis Publicity Ltd/Panshanger
	G-BHVE	Saffery S.330 balloon	P. M. Randles
	G-BHVF	Jodel D.150A	C. A. Parker/Sywell
	G-BHVH	Boeing 737-2T5	Orion Airways Ltd/E. Midlands
	G-BHVI	Boeing 737-2T5	Orion Airways Ltd/E. Midlands
	G-BHVM	Cessna 152	B. W. Wells & Burbage Farms Ltd
	G-BHVN	Cessna 152	Three Counties Aero Club Ltd/ Blackbushe
	G-BHVP	Cessna 182Q	Air Tows/Lasham
	G-BHVR	Cessna 172N	Air Tows/Blackbushe
	G-BHVS	Enstrom F-28A-UK	Southern Air Ltd/Shoreham
	G-BHVT	Boeing 727-212	Dan-Air Services Ltd/Gatwick
	G-BHVV	Piper J-3C-65 Cub	A. E. Molton
	G-BHVY	AA-5B Tiger	R. Barren
	G-BHVZ	Cessna 180	R. Moore/Blackpool
	G-BHWA	Cessna F.152	Wickenby Aviation Ltd
	G-BHWB	Cessna F.152	Wickenby Aviation Ltd
	G-BHWE	Boeing 737-204ADV	Britannia Airways Ltd *Sir Sidney Camm*/ Luton
	G-BHWF	Boeing 737-204ADV	Britannia Airways Ltd *Lord Brabazon of Tara*/Luton
	G-BHWG	Mahatma S.200SR balloon	H. W. Gandy *Spectrum*
	G-BHWH	Weedhopper JC-24A	G. A. Clephane
	G-BHWK	M.S.880B Rallye Club	T. M. W. Webster & ptnrs/Defford
	G-BHWN	WMB.3 Windtracker 200 balloon	C. J. Dodd & G. J. Luckett
	G-BHWO	WMB.4 Windtracker II balloon	C. J. Dodd
	G-BHWR	AA-5A Cheetah	Alexander Aviation
	G-BHWS	Cessna F.152	Stapleford Flying Club
	G-BHWW	Cessna U.206G	Aerotime Ltd/Glenrothes
	G-BHWY	PA-28R-200 Cherokee Arrow	Rack Delta Ltd
	G-BHWZ	PA-28-181 Archer II	Symtec Computer Service Ltd
	G-BHXD	Jodel D.120	S. E. Richards
	G-BHXE	Thunder Ax3 balloon	C. Benning
	G-BHXJ	Nord 1203/2 Norecrin (103)	R. E. Coates/Popham
	G-BHXK	PA-28 Cherokee 140	I. R. F. Hammond
	G-BHXL	Evans VP-2	T. W. Woolley
	G-BHXN	Van's RV.3	P. R. Hing
	G-BHXO	Colt 14A balloon	Colt Balloons Ltd/Sweden
	G-BHXR	Thunder Ax7-65 balloon	Thunder Balloons Ltd/Brazil
	G-BHXS	Jodel D.120	S. Billington
	G-BHXT	Thunder Ax6-56Z balloon	Ocean Traffic Services Ltd

Reg.	Type	Owner or Operator	Notes
G-BHXU	AB-206B JetRanger 3	Castle Air Charters Ltd	
G-BHXV	AB-206B JetRanger 3	G. Greenall	
G-BHXX	PA-23 Aztec 250	Andwell Mill Trout Farm Ltd	
G-BHXY	Piper J-3C-65 Cub (44-79609)	Manchester Air Shows Ltd	
G-BHYA	Cessna R.182RG II	MLP Aviation Ltd/Elstree	
G-BHYB	Sikorsky S-76A	British Airways Helicopters Ltd/Beccles	
G-BHYC	Cessna 172RG Cutlass	TDS Circuits (Blackburn) Ltd/Blackpool	
G-BHYD	Cessna R.172K XP	Sylmar Aviation Services Ltd	
G-BHYE	PA-34-200T-2 Seneca	C.S.E. Aviation Ltd/Kidlington	
G-BHYF	PA-34-200T-2 Seneca	C.S.E. Aviation Ltd/Kidlington	
G-BHYG	PA-34-200T-2 Seneca	C.S.E. Aviation Ltd/Kidlington	
G-BHYI	Stampe SV-4A	R. J. Knights	
G-BHYN	Evans VP-2	A. B. Cameron	
G-BHYO	Cameron N-77 balloon	C. Sisson	
G-BHYP	Cessna F.172M	J. Burgess & ptnrs/Blackpool	
G-BHYR	Cessna F.172M	Alumvale Ltd/Stapleford	
G-BHYV	Evans VP-1	L. Chiappi	
G-BHYW	AB-206B JetRanger	Gleneagles Helicopter Services (Scotland) Ltd/Perth	
G-BHYX	Cessna 152	Stanton Aircraft Management Ltd	
G-BHZA	Piper J-3C-65 Cub	R. G. Warwick	
G-BHZE	PA-28-181 Archer II	E. O. Smith & Co Ltd/Tollerton	
G-BHZF	Evans VP-2	D. Silsbury/Dunkeswell	
G-BHZG	Monnet Sonerai II	R. A. Gardiner & B. Chapman/Prestwick	
G-BHZH	Cessna F.152	Havelet Leasing Ltd	
G-BHZI	Thunder Ax3 balloon	Thunder Balloons Ltd/W. Germany	
G-BHZJ	Hughes Stratosphere 150 balloon	P. J. Hughes	
G-BHZK	AA-5B Tiger	Achandunie Farming Co	
G-BHZM	Jodel DR.1050	G. H. Wylde/Manchester	
G-BHZN	AA-5B Tiger	Peacock Salt Ltd/Carlisle	
G-BHZO	AA-5A Cheetah	Peacock Salt Ltd/Glasgow	
G-BHZU	Piper J-3C-65 Cub	J. K. Tomkinson	
G-BHZV	Jodel D.120A	J. G. Munro/Glasgow	
G-BHZX	Thunder Ax7-65A balloon	Thermark (Plastic Processing) Ltd	
G-BHZY	Monnet Sonerai II	C. A. Keech	
G-BIAA	SOCATA TB.9 Tampico	Mona Aviation Ltd	
G-BIAB	SOCATA TB.9 Tampico	M. V. Male	
G-BIAC	M.S.894E Rallye Minerva	M. R. Carlton/Biggin Hill	
G-BIAH	Jodel D.112	D. Mitchell	
G-BIAI	WMB.2 Windtracker balloon	I. Chadwick	
G-BIAK	SOCATA TB.10 Tobago	Trent Combustion Components Ltd/ Tollerton	
G-BIAL	Rango NA.8 balloon	A. M. Lindsay	
G-BIAO	Evans VP-2	J. Stephenson	
G-BIAP	PA-16 Clipper	I. M. Callier & P. J. Bish/White Waltham	
G-BIAR	Rigg Skyliner II balloon	I. A. Rigg	
G-BIAT	Sopwith Pup Replica	G. A. Black	
G-BIAU	Sopwith Pup Replica (N6452)	Fleet Air Arm Museum/Yeovilton	
G-BIAV	Sikorsky S-76A	British Airways Helicopters Ltd/ Aberdeen	
G-BIAW	Sikorsky S-76A	British Airways Helicopters Ltd/ Aberdeen	
G-BIAX	Taylor JT.2 Titch	G. F. Rowley	
G-BIAY	AA-5 Traveler	Medi-Cine Productions Ltd/ Elstree	
G-BIBA	SOCATA TB.9 Tampico	Compak Board Ltd/Ipswich	
G-BIBB	Mooney M.20C	Gloucestershire Flying Club/Staverton	
G-BIBC	Cessna 310R	Airwork Ltd/Perth	
G-BIBD	Rotec Rally 2B	A. Clarke/Sturgate	
G-BIBE	EMB-110P1 Bandeirante	Jersey European Airways	
G-BIBF	Smith A12 Sport balloon	T. J. Smith	
G-BIBG	Sikorsky S-76A	British Caledonian Helicopters Ltd/ Aberdeen	
G-BIBJ	Enstrom F-280C-UK Shark	W. W. Kendrick & Sons Ltd/ Halfpenny Green	
G-BIBK	Taylor JT.2 Titch	T. C. Horner	
G-BIBN	Cessna FA.150K	P. H. Lewis	
G-BIBO	Cameron V-65 balloon	Southern Balloon Group	
G-BIBP	AA-5A Cheetah	Peacock Salt Ltd/Carlisle	
G-BIBS	Cameron P-20 balloon	Cameron Balloons Ltd	
G-BIBT	AA-5B Tiger	Fergusons (Blyth) Ltd	
G-BIBU	Morris Ax7-77 balloon	K. Morris	

Notes	Reg.	Type	Owner or Operator
	G-BIBV	WMB.3 Windtracker balloon	P. B. Street
	G-BIBW	Cessna F.172N	Deltair Ltd/Chester
	G-BIBX	WMB.2 Windtracker balloon	I. A. Rigg
	G-BIBY	Beech F33A Bonanza	Carl Peterson Ltd/Bournemouth
	G-BIBZ	Thunder Ax3 balloon	F. W. Barnes
	G-BICB	Rotec Rally 2B	J. D. Lye & A. P. Jones
	G-BICC	Vulture Tx3 balloon	C. P. Clitheroe
	G-BICD	Auster 5	J. A. S. Baldry & ptnrs
	G-BICE	AT-6C Harvard IIA (CE)	C. M. L. Edwards
	G-BICF	GA-7 Cougar	London School of Flying/Elstree
	G-BICG	Cessna F.152	R. M. Clarke/Coventry
	G-BICI	Cameron R-833 balloon	Ballooning Endeavours Ltd
	G-BICJ	Monnet Sonerai II	J. R. S. Heaton
	G-BICM	Colt 56A balloon	T. A. R. & S. Turner
	G-BICN	F.8L Falco	R. J. Barber
	G-BICO	Neal Mitefly balloon	T. J. Neale
	G-BICP	Robin DR.360	Bravo India Flying Group/Barton
	G-BICR	Jodel D.120A	S. W. C. Hall & ptnrs/Redhill
	G-BICS	Robin R.2100A	Propex (UK) Ltd
	G-BICT	Evans VP-1	A. S. Coombe & D. L. Tribe
	G-BICU	Cameron V-56 balloon	I. S. Clarke
	G-BICW	PA-28-161 Warrior II	Fastraven Ltd/Panshanger
	G-BICX	Maule M5-235C Lunar Rocket	Stanton Aircraft Management Ltd/ Headcorn
	G-BICY	PA-23 Apache 160	A. M. Lynn/Sibson
	G-BIDA	SOCATA Rallye Club 100ST	R. M. W. Mogg/Compton Abbas
	G-BIDB	BAe 167 Strikemaster	British Aerospace
	G-BIDD	Evans VP-1	J. E. Wedgbury
	G-BIDE	CP.301A Emeraude	D. Elliott
	G-BIDF	Cessna F.172P	Horizon Flying Club/Ipswich
	G-BIDG	Jodel D.150A	D. R. Gray/Barton
	G-BIDH	Cessna 152	Birmingham Aerocentre Ltd
	G-BIDI	PA-28R-201 Arrow III	M. J. Webb/Birmingham
	G-BIDJ	PA-18-150 Super Cub	Marchington Gliding Club
	G-BIDK	PA-18-150 Super Cub	Scottish Gliding Union Ltd
	G-BIDM	Cessna F.172H	J. F. Packaging/Ingoldmells
	G-BIDO	CP.301A Emeraude	J. Middlebrook
	G-BIDP	PA-28-181 Archer II	Staverton Flying Services Ltd
	G-BIDT	Cameron A375 balloon	Ballooning Endeavours Ltd
	G-BIDU	Cameron V-77 balloon	E. Eleazor
	G-BIDV	Colt 14A balloon	International Distillers & Vintners (House Trade) Ltd
	G-BIDW	Sopwith 1½ Strutter replica (A8226) ★	RAF Museum
	G-BIDX	Jodel D.112	H. N. Nuttall & R. P. Walley
	G-BIDY	WMB.2 Windtracker balloon	D. M. Campion
	G-BIDZ	Colt 21A balloon	Hot Air Balloon Co Ltd/S. Africa
	G-BIEC	AB-206A JetRanger 2	Autair Helicopters Ltd/Panshanger
	G-BIED	Beech F90 King Air	United Biscuits (Foods) Ltd/Denham
	G-BIEF	Cameron V-77 balloon	D. S. Bush
	G-BIEH	Sikorsky S-76A	Bond Helicopters Ltd/Bourn
	G-BIEJ	Sikorsky S-76A	Bristow Helicopters Ltd
	G-BIEK	WMB.4 Windtracker balloon	P. B. Street
	G-BIEL	WMB.4 Windtracker balloon	A. T. Walden
	G-BIEM	D.H.C.-6 Twin Otter 310	Loganair Ltd/Glasgow
	G-BIEN	Jodel D.120A	J. C. Mansell & R. V. Smith
	G-BIEO	Jodel D.112	R. Scott/Barton
	G-BIER	Rutan Long-Ez	V. Mossor
	G-BIES	Maule M5-235C Lunar Rocket	William Proctor Farms
	G-BIET	Cameron O-77 balloon	G. M. Westley
	G-BIEV	AA-5A Cheetah	A. J. Hows
	G-BIEW	Cessna U.206G	G. D. Atkinson/Guernsey
	G-BIEX	Andreasson BA-4B	H. P. Burrill/Sherburn
	G-BIEY	PA-28-151 Warrior	J. A. Pothecary/Shoreham
	G-BIEZ	Beech F90 King Air	Bass PLC
	G-BIFA	Cessna 310R-II	Land & Estates Consultants Ltd/ Cranfield
	G-BIFB	PA-28 Cherokee 150	C. J. Reed/Elstree
	G-BIFC	Colt 14A balloon	Colt Balloons Ltd
	G-BIFD	R. Commander 114	Foxgrove Construction Ltd/Fairoaks
	G-BIFE	Cessna A.185F	Conguess Aviation Ltd
	G-BIFN	Bensen B.8M	K. Willows
	G-BIFO	Evans VP-1	P. Raggett/Filton
	G-BIFP	Colt 56C balloon	J. Philp

Reg.	Type	Owner or Operator	Notes
G-BIFT	Cessna F.150L	Phoenix Aviation (Bedford) Ltd/ Cranfield	
G-BIFU	Short Skyhawk balloon	D. K. Short	
G-BIFV	Jodel D.150	J. H. Kirkham/Barton	
G-BIFW	Scruggs BL.2 Wunda balloon	D. Morris	
G-BIFY	Cessna F.150L	Phoenix Aviation (Bedford) Ltd/ Cranfield	
G-BIFZ	Partenavia P.68C	Abbey Hill Vehicle Services	
G-BIGC	Cameron O-42 balloon	C. M. Moroney	
G-BIGD	Cameron V-77 balloon	D. L. Clark	
G-BIGE	Champion Cloudseeker balloon	A. Foster	
G-BIGF	Thunder Ax7-77 balloon	M. D. Stever & C. A. Allen	
G-BIGG	Saffery S.200 balloon	R. S. Sweeting	
G-BIGH	Piper L-4H Cub	W. McNally	
G-BIGI	Mooney M.20J	Melinco Marketing (Jersey) Ltd	
G-BIGJ	Cessna F.172M	Essex Aero Services Ltd/Southend	
G-BIGK	Taylorcraft BC-12D	M. A. C. Stephenson	
G-BIGL	Cameron O-65 balloon	A. H. K. Olpin *Scorpio*	
G-BIGM	Avenger T.200-2112 balloon	M. Murphy	
G-BIGN	Attic Srs 1 balloon	G. Nettleship	
G-BIGP	Bensen B.8M	R. H. S. Cooper	
G-BIGR	Avenger T.200-2112 balloon	R. Light	
G-BIGU	Bensen B.8M	J. R. Martin	
G-BIGX	Bensen B.8M	J. R. Martin	
G-BIGY	Cameron V-65 balloon	Dante Balloon Group	
G-BIGZ	Scheibe SF.25B Falke	K. Ballington	
G-BIHB	Scruggs BL.2 Wunda balloon	D. Morris	
G-BIHC	Scruggs BL.2 Wunda balloon	P. D. Kiddell	
G-BIHD	Robin DR.400/160	G. R. Pope & ptnrs/Biggin Hill	
G-BIHE	Cessna FA.152	Inverness Flying Services Ltd	
G-BIHF	SE-5A Replica (F943)	K. J. Garrett/Booker	
G-BIHG	PA-28 Cherokee 140	T. Parmenter/Clacton	
G-BIHH	Sikorsky S-61N	British Caledonian Helicopters Ltd/ Aberdeen	
G-BIHI	Cessna 172M	J. H. A. Rogers	
G-BIHN	Skyship 500 airship	Airship Industries Ltd/Cardington	
G-BIHO	D.H.C.-6 Twin Otter 310	Brymon Aviation Ltd/Plymouth	
G-BIHP	Van Den Bemden gas balloon	J. J. Harris	
G-BIHR	WMB.2 Windtracker balloon	R. S. Sweeting	
G-BIHT	PA-17 Vagabond	G. D. Thomson/Wellesbourne	
G-BIHU	Saffery S.200 balloon	B. L. King	
G-BIHW	Aeronca A65TAC (2-7767)	J. Milton/Shobdon	
G-BIHX	Bensen B.8M	C. C. Irvine	
G-BIHY	Isaacs Fury	D. E. Olivant/Ronaldsway	
G-BIIA	Fournier RF-3	M. K. Field/Brize Norton	
G-BIIB	Cessna F.172M	S. L. Hawkins/Biggin Hill	
G-BIIC	Scruggs BL.2 Wunda balloon	S. J. Hodder & D. Cockerill	
G-BIID	PA-18 Super Cub 95	L. Dickson & M. Winter/Aberdeen	
G-BIIE	Cessna F.172P	Shoreham Flight Simulation Ltd/ Bournemouth	
G-BIIF	Fournier RF-4D	A. P. Walsh (G-BVET)	
G-BIIG	Thunder Ax-6-56Z balloon	The Larter Group Ltd	
G-BIIH	Scruggs BL.2T Turbo balloon	B. M. Scott	
G-BIIJ	Cessna F.152	Leicestershire Aero Club Ltd	
G-BIIK	M.S.883 Rallye 115	H. Russell & ptnrs	
G-BIIL	Thunder Ax6-56 balloon	G. W. Reader	
G-BIIM	Scruggs BL.2A Wunda balloon	K. D. Head	
G-BIIT	PA-28-161 Warrior II	Tayside Aviation Ltd/Dundee	
G-BIIV	PA-28-181 Archer II	Stratton Motor Co Ltd	
G-BIIW	Rango NA.10 balloon	Rango Kite Co	
G-BIIX	Rango NA.12 balloon	Rango Kite Co	
G-BIIZ	Great Lakes 2T-1A Sport Trainer	Hon P. Lindsay/Booker	
G-BIJA	Scruggs BL.2A Wunda balloon	P. L. E. Bennett	
G-BIJB	PA-18-150 Super Cub	Essex Gliding Club/North Weald	
G-BIJC	AB-206A JetRanger	Veritair Ltd/Cardiff	
G-BIJD	Bo 208C Junior	D. J. Dulborough/Redhill	
G-BIJE	Piper L-4A Cub	J. H. T. Davies & ptnrs/Breighton	
G-BIJS	Luton LA-4A Minor	I. J. Smith	
G-BIJT	AA-5A Cheetah	G. W. Plowman & Son Ltd	
G-BIJU	CP.301A Emeraude	C. J. Norman & M. Howard (G-BHTX)/ Fairoaks	
G-BIJV	Cessna F.152	Falcon Flying Services/Biggin Hill	
G-BIJW	Cessna F.152	G. K. Mitchell Ltd	

Notes	Reg.	Type	Owner or Operator
	G-BIJX	Cessna F.152	Civil Service Flying Club Ltd/ Biggin Hill
	G-BIJZ	Skyventurer Mk 1 balloon	R. Sweeting
	G-BIKA	Boeing 757-236	British Airways *Dover Castle*/ Heathrow
	G-BIKB	Boeing 757-236	British Airways *Windsor Castle*/ Heathrow
	G-BIKC	Boeing 757-236	British Airways *Edinburgh Castle*/ Heathrow
	G-BIKD	Boeing 757-236	British Airways *Caernarvon Castle*/ Heathrow
	G-BIKE	PA-28R Cherokee Arrow 200	R. V. Webb Ltd/Elstree
	G-BIKF	Boeing 757-236	British Airways *Carrikfergus Castle*/ Heathrow
	G-BIKG	Boeing 757-236	British Airways *Stirling Castle*/ Heathrow
	G-BIKH	Boeing 757-236	British Airways *Richmond Castle*/ Heathrow
	G-BIKI	Boeing 757-236	British Airways *Tintagel Castle*/ Heathrow
	G-BIKJ	Boeing 757-236	British Airways *Conway Castle*/ Heathrow
	G-BIKK	Boeing 757-236	British Airways *Eilean Donan Castle*/ Heathrow
	G-BIKL	Boeing 757-236	British Airways *Nottingham Castle*/ Heathrow
	G-BIKM	Boeing 757-236	British Airways *Glamis Castle*/ Heathrow
	G-BIKN	Boeing 757-236	British Airways *Bodiam Castle*/ Heathrow
	G-BIKO	Boeing 757-236	British Airways *Enniskillen Castle*/ Heathrow
	G-BIKP	Boeing 757-236	British Airways *Corfe Castle*/ Heathrow
	G-BIKR	Boeing 757-236	British Airways/Heathrow
	G-BIKS	Boeing 757-236	British Airways/Heathrow
	G-BIKT	Boeing 757-236	British Airways*Carisbrooke Castle*/ Heathrow
	G-BIKU	Boeing 757-236	British Airways *Inverrary Castle*/ Heathrow
	G-BIKV	Boeing 757-236	British Airways *Raglan Castle*/ Heathrow
	G-BIKW	Boeing 757-236	British Airways *Colchester Castle*/ Heathrow
	G-BIKX	Boeing 757-236	British Airways *Warwick Castle*/ Heathrow
	G-BIKY	Boeing 757-236	British Airways *Leeds Castle*/ Heathrow
	G-BIKZ	Boeing 757-236	British Airways *Kenilworth Castle*/ Heathrow
	G-BILA	Daletol DM.165L Viking	R. Lamplough
	G-BILB	WMB.2 Windtracker balloon	B. L. King
	G-BILE	Scruggs BL.2B balloon	P. D. Ridout
	G-BILF	Practavia Sprite 125	G. Harfield
	G-BILG	Scruggs BL.2B balloon	P. D. Ridout
	G-BILI	Piper J-3C-65 Cub	A. Dodd/Cranwell
	G-BILJ	Cessna FA.152	Shoreham Flight Simulation Ltd/ Bournemouth
	G-BILK	Cessna FA.152	A. Blair/Biggin Hill
	G-BILL	PA-25 Pawnee 235	Farmair (Kent) Ltd/Headcorn
	G-BILP	Cessna 152	Skyviews & General Ltd
	G-BILR	Cessna 152	Skyviews & General Ltd
	G-BILS	Cessna 152	Skyviews & General Ltd
	G-BILT	Cessna F.172P	Kanestar Ltd
	G-BILU	Cessna 172RG	Melrose Pigs Ltd
	G-BILX	Colt 31A balloon	Hot Air Balloon Co Ltd
	G-BILZ	Taylor JT.1 Monoplane	G. Beaumont
	G-BIMK	Tiger T.200 Srs 1 balloon	M. K. Baron
	G-BIML	Turner Super T.40A	R. T. Callow
	G-BIMM	PA-18 Super Cub 135	D. S. & I. M. Morgan
	G-BIMN	Steen Skybolt	C. R. Williamson
	G-BIMO	Stampe SV-4C	R. K. G. Hannington/Middle Wallop
	G-BIMT	Cessna FA.152	Staverton Flying Services Ltd

Reg.	Type	Owner or Operator	Notes
G-BIMU	Sikorsky S-61N	British Caledonian Helicopters Ltd/ Aberdeen	
G-BIMX	Rutan Vari-Eze	A. S. Knowles	
G-BIMZ	Beech 76 Duchess	Barrein Engineers Ltd/Lulsgate	
G-BINA	Saffery S.9 balloon	A. P. Bashford	
G-BINB	WMB.2A Windtracker balloon	S. R. Woolfries	
G-BINC	Tour de Calais balloon	Cupro Sapphire Ltd	
G-BIND	M.S.894E Rallye 235	G. Archer/Biggin Hill	
G-BINE	Scruggs BL.2A Wunda balloon	M. Gilbey	
G-BINF	Saffery S.200 balloon	T. Lewis	
G-BING	Cessna F.172P	J. E. M. Patrick/Humberside	
G-BINH	D.H.82A Tiger Moth	Arrow Air Services (Engineering) Ltd/ Felthorpe	
G-BINI	Scruggs BL.2C balloon	S. R. Woolfries	
G-BINJ	Rango NA.12 balloon	M. R. Haslam	
G-BINL	Scruggs BL.2B balloon	P. D. Ridout	
G-BINM	Scruggs BL.2B balloon	P. D. Ridout	
G-BINN	Unicorn UE.1A balloon	Unicorn Group	
G-BINO	Evans VP-1	J. I. Visser	
G-BINR	Unicorn UE.1A balloon	Unicorn Group	
G-BINS	Unicorn UE.2A balloon	Unicorn Group	
G-BINT	Unicorn UE.1A balloon	Unicorn Group	
G-BINU	Saffery S.200 balloon	T. Lewis	
G-BINV	Saffery S.200 balloon	R. S. Harris	
G-BINW	Scruggs BL.2B balloon	P. G. Macklin	
G-BINX	Scruggs BL.2B balloon	P. D. Ridout	
G-BINY	Oriental balloon	J. L. Morton	
G-BINZ	Rango NA.8 balloon	T. J. Sweeting & M. O. Davies	
G-BIOA	Hughes 369D	Weetabix Ltd/Sywell	
G-BIOB	Cessna F.172P	Hunting Surveys & Consultants Ltd	
G-BIOC	Cessna F.150L	Seawing Flying Club/Southend	
G-BIOF	Short SD3-30	Short Bros PLC (*leased in Nigeria*)	
G-BIOI	Jodel DR.1051-M	R. Cochrane	
G-BIOJ	R. Commander 112TCA	N. J. Orr/Denham	
G-BIOK	Cessna F.152	Hartmann Ltd/Booker	
G-BIOM	Cessna F.152	Gordon King (Aviation) Ltd/ Biggin Hill	
G-BION	Cameron V-77 balloon	Elliott's Pharmacy Ltd	
G-BIOO	Unicorn UE.2B balloon	Unicorn Group	
G-BIOP	Scruggs BL.2D balloon	J. P. S. Donnellan	
G-BIOR	M.S.880B Rallye Club	C. R. Galloway/Kidlington	
G-BIOS	Scruggs BL.2B balloon	D. Eaves	
G-BIOT	Bensen B.8M	M. W. & J. A. Joynes	
G-BIOU	Jodel D.117A	M. S. Printing & Graphics Machinery Ltd/Booker	
G-BIOW	Slingsby T.67A	Specialist Flying Training Ltd/Carlisle	
G-BIOX	Potter Crompton PRO.1 balloon	G. M. Potter	
G-BIOY	PAC-14 Special Shape balloon	P. A. Cremer	
G-BIPA	AA-5B Tiger	J. Campbell/Barrow	
G-BIPB	Weedhopper JC-24B	E. H. Moroney	
G-BIPC	PAC-14 Hefferlump balloon	P. A. Cremer	
G-BIPF	Scruggs BL.2C balloon	D. Morris	
G-BIPG	Global Mini balloon	P. Globe	
G-BIPH	Scruggs BL.2B balloon	C. M. Dewsnap	
G-BIPI	Everett Blackbird Mk 1	R. J. Everett (Engineering) Ltd	
G-BIPJ	PA-36-375 Brave	G. B. Pearce/Shoreham	
G-BIPK	Saffery S.200 balloon	P. J. Kelsey	
G-BIPL	AA-5A Cheetah	Denham School of Flying Ltd	
G-BIPM	Flamboyant Ax7-65 balloon	Pepsi Cola International Ltd/S. Africa	
G-BIPN	Fournier RF-3	M. R. Shelton	
G-BIPO	Mudry/CAARP CAP.20LS-200	BIPO Aviation Ltd/Booker	
G-BIPR	Sikorsky S-76A	Bristow Helicopters Ltd/Aberdeen	
G-BIPS	SOCATA Rallye 100ST	Lighter Than Air Ltd	
G-BIPT	Jodel D.112	C. R. Davies	
G-BIPU	AA-5B Tiger	Aero Group 78/Netherthorpe	
G-BIPV	AA-5B Tiger	I. D. Longfellow/Southampton	
G-BIPW	Avenger T.200-2112 balloon	B. L. King	
G-BIPX	Saffery S.9 balloon	J. R. Havers	
G-BIPY	Bensen B.8	A. J. Wood	
G-BIPZ	McCandless Mk 4-4	B. McIntyre	
G-BIRA	SOCATA TB.9 Tampico	Goldangel Ltd/Swansea	
G-BIRB	M.S.880B Rallye 100T	E. Smith/Newtownards	
G-BIRD	Pitts S-1C Special	B. K. Lecomber/Booker	

Notes	Reg.	Type	Owner or Operator
	G-BIRE	Colt 56 Bottle balloon	Hot Air Balloon Co Ltd
	G-BIRG	M.S.880B Rallye Club	Air Touring Services Ltd/Biggin Hill
	G-BIRH	PA-18 Super Cub 135	I. R. F. Hammond/Lee-on-Solent
	G-BIRI	C.A.S.A. 1.131E Jungmann	L. B. Jefferies
	G-BIRK	Avenger T.200-2112 balloon	D. Harland
	G-BIRL	Avenger T.200-2112 balloon	R. Light
	G-BIRM	Avenger T.200-2112 balloon	P. Higgins
	G-BIRN	Short SD3-30	Short Bros PLC/Sydenham
	G-BIRO	Cessna 172P	M. C. Grant
	G-BIRP	Arena Mk 17 Skyship balloon	A. S. Viel
	G-BIRS	Cessna 182P	D. P. Cranston & Bob Crowe Aircraft Sales Ltd (G-BBBS)/Cranfield
	G-BIRT	Robin R.1180TD	W. D'A. Hall/Booker
	G-BIRU	H.S.125 Srs 700B	Goodman Air Taxis/Heathrow
	G-BIRV	Bensen B.8MV	R. Hart
	G-BIRW	M.S.505 Criquet (F+IS)	Museum of Flight/E. Fortune
	G-BIRX	Scruggs RS.500 balloon	J. H. Searle
	G-BIRY	Cameron V-77 balloon	J. J. Winter
	G-BIRZ	Zenair CH.250	B. A. Arnall & M. Hanley/Biggin Hill
	G-BISA	Hase IIIT balloon	M. A. Hase
	G-BISB	Cessna F.152 II	Sheffield Aero Club Ltd/Netherthorpe
	G-BISC	Robinson R-22	Hamilton Bland (Consultants) Ltd
	G-BISF	Robinson R-22	Compuster Ltd
	G-BISG	FRED Srs 3	R. A. Coombe
	G-BISH	Cameron O-42 balloon	Zebedee Balloon Service
	G-BISI	Robinson R-22	Sloane Helicopters Ltd/Luton
	G-BISJ	Cessna 340A	Castle Aviation/Leeds
	G-BISK	R. Commander 112B	P. A. Warner
	G-BISL	Scruggs BL.2B balloon	P. D. Ridout
	G-BISM	Scruggs BL.2B balloon	P. D. Ridout
	G-BISN	Boeing Vertol 234LR Chinook	British Airways Helicopters Ltd/ Aberdeen
	G-BISP	Boeing Vertol 234LR Chinook	British Airways Helicopters Ltd/ Aberdeen
	G-BISR	Boeing Vertol 234LR Chinook	British Airways Helicopters Ltd/ Aberdeen
	G-BISS	Scruggs BL.2C balloon	P. D. Ridout
	G-BIST	Scruggs BL.2C balloon	P. D. Ridout
	G-BISU	B.170 Freighter 31M	Atlantic Air Transport/Stansted
	G-BISV	Cameron O-65 balloon	Hylyne Rabbits Ltd
	G-BISW	Cameron O-65 balloon	Hylyne Rabbits Ltd
	G-BISX	Colt 56A balloon	Long John International Ltd
	G-BISY	Scruggs BL-2C balloon	P. T. Witty
	G-BISZ	Sikorsky S-76A	Bristow Helicopters Ltd
	G-BITA	PA-18-150 Super Cub	P. M. D. Wiggins
	G-BITE	SOCATA TB.10 Tobago	M. A. Smith & R. J. Bristow/Fairoaks
	G-BITF	Cessna F.152	Bristol & Wessex Aeroplane Club/ Bristol
	G-BITG	Cessna F.152	Bristol & Wessex Aeroplane Club/ Bristol
	G-BITH	Cessna F.152	Bristol & Wessex Aeroplane Club/ Bristol
	G-BITI	Scruggs RS.5000 balloon	A. E. Smith
	G-BITK	FRED Srs 2	B. J. Miles
	G-BITL	Horncastle LL-901 balloon	M. J. Worsdell
	G-BITM	Cessna F.172P	D. G. Crabtree
	G-BITN	Short Albatross balloon	D. K. Short
	G-BITO	Jodel D.112D	A. Dunbar/Barton
	G-BITR	Sikorsky S-76A	Bristow Helicopters Ltd
	G-BITS	Drayton B-56 balloon	M. J. Betts
	G-BITT	Bo 208C Junior	E. T. Green/Lydd
	G-BITV	Short SD3-30	Fairflight Ltd/Biggin Hill
	G-BITW	Short SD3-30	Fairflight Ltd (G-EASI)/Biggin Hill
	G-BITX	Short SD3-30	Guernsey Airlines Ltd
	G-BITY	FD.31T balloon	A. J. Bell
	G-BITZ	Cremer Sandoe PACDS.14 balloon	P. A. Cremer & C. D. Sandoe
	G-BIUG	BN-2A Islander	Pilatus BN Ltd/Bembridge
	G-BIUH	BN-2A Islander	Pilatus BN Ltd/Bembridge
	G-BIUL	Cameron 60 SS balloon	Engineering Appliances Ltd
	G-BIUM	Cessna F.152	Sheffield Aero Club Ltd/ Netherthorpe
	G-BIUN	Cessna F.152	Sheffield Aero Club Ltd/ Netherthorpe

Reg.	Type	Owner or Operator	Notes
G-BIUP	SNCAN NC.854C	Questair Ltd	
G-BIUR	Boeing 727-155C	Dan-Air Services Ltd/Gatwick	
G-BIUT	Scruggs BL.2C balloon	N. J. Ball	
G-BIUU	PA-23 Aztec 250	Kingsmetal Ltd/Lydd	
G-BIUV	H.S.748 Srs 2A	Dan-Air Services Ltd (G-AYYH)/ Gatwick	
G-BIUW	PA-28-161 Warrior II	Staeng Ltd/Bodmin	
G-BIUX	PA-28-161 Warrior II	C.S.E. Aviation Ltd/Kidlington	
G-BIUY	PA-28-181 Archer II	Mega Yield Ltd	
G-BIUZ	Slingsby T.67B	Slingsby Aviation Ltd/ Kirkbymoorside	
G-BIVA	Robin R.2112	Cotswold Aero Club Ltd/Staverton	
G-BIVB	Jodel D.112	R. J. Lewis/Bodmin	
G-BIVC	Jodel D.112	The Harrier Flying Group/Bourn	
G-BIVF	CP.301C-3 Emeraude	J. Casker	
G-BIVI	Cremer PAC.500 airship	P. A. Cremer	
G-BIVJ	Cessna F.152	Reedtrend Ltd/Birmingham	
G-BIVK	Bensen B.8	J. G. Toy	
G-BIVL	Bensen B.8	T. E. Davies	
G-BIVO	G.164D Ag-Cat	Miller Aerial Spraying Ltd/ Wickenby	
G-BIVR	Featherlight Mk 1 balloon	A. P. Newman & N. P. Kemp	
G-BIVS	Featherlight Mk 2 balloon	J. M. J. Roberts & S. R. Rushton	
G-BIVT	Saffery S.80 balloon	L. F. Guyot	
G-BIVU	AA-5A Cheetah	Royan Consultants International Ltd/Tees-side	
G-BIVV	AA-5A Cheetah	W. Dass/Shobdon	
G-BIVW	Z.326 Trener Master	G. C. Masterson	
G-BIVX	Saffery S.80 balloon	P. T. Witty	
G-BIVY	Cessna 172N	R. J. Scott/Blackbushe	
G-BIVZ	D.31A Turbulent	Tiger Club Ltd/Redhill	
G-BIWA	Stevendon Skyreacher balloon	S. D. Barnes	
G-BIWB	Scruggs RS.5000 balloon	P. D. Ridout	
G-BIWC	Scruggs RS.5000 balloon	P. D. Ridout	
G-BIWD	Scruggs RS.5000 balloon	D. Eaves	
G-BIWE	Scruggs BL.2D balloon	M. D. Saunders	
G-BIWF	Warren balloon	P. D. Ridout	
G-BIWG	Zelenski Mk 2 balloon	P. D. Ridout	
G-BIWH	Cremer Super Fliteliner balloon	G. Lowther	
G-BIWI	Cremer WS.1 balloon	P. A. Cremer	
G-BIWJ	Unicorn UE.1A balloon	B. L. King	
G-BIWK	Cameron V-65 balloon	I. R. Williams & R. G. Bickerdale	
G-BIWL	PA-32-301 Saratoga	D. Hammant/Southampton	
G-BIWN	Jodel D.112	C. R. Coates	
G-BIWO	Scruggs RS.5000 balloon	D. Morris	
G-BIWP	Mooney M.20J	Tropair Cooling Ltd/Biggin Hill	
G-BIWR	Mooney M.20F	C. W. Yarnton & J. D. Heykoop/ Redhill	
G-BIWS	Cessna 182R	Anglian Double Glazing Ltd/ Norwich	
G-BIWU	Cameron V-65 balloon	J. T. Whicker & J. W. Unwin	
G-BIWV	Cremer PAC-550T balloon	P. A. Rutherford	
G-BIWW	AA-5 Traveler	B&K Aviation/Cranfield	
G-BIWX	AT-16 Harvard IV (FT239)	A. E. Hutton/White Waltham	
G-BIWY	Westland WG.30	British Airways Helicopters Ltd/ Beccles	
G-BIXA	SOCATA TB.9 Tampico	McClean & Gibson (Engineers) Ltd	
G-BIXB	SOCATA TB.9 Tampico	Ospreystar Ltd/Stapleford	
G-BIXH	Cessna F.152	Cambridge Aero Club Ltd	
G-BIXI	Cessna 172RG Cutlass	J. F. P. Lewis/Sandown	
G-BIXJ	Saffery S.40 balloon	T. M. Pates	
G-BIXK	Rand KR.2	R. G. Cousins	
G-BIXL	P-51D Mustang (472216)	R. Lamplough/Duxford	
G-BIXN	Boeing A.75N1 Stearman	I. L. Craig-Wood & ptnrs	
G-BIXR	Cameron A-140 balloon	Skysales Ltd	
G-BIXS	Avenger T.200-2112 balloon	M. Stuart	
G-BIXT	Cessna 182R	W. Lipka/Leavesden	
G-BIXU	AA-5B Tiger	Peacock Salt Ltd/Glasgow	
G-BIXV	Bell 212	Bristow Helicopters Ltd	
G-BIXW	Colt 56B balloon	J. R. Birkenhead	
G-BIXX	Pearson Srs 2 balloon	D. Pearson	
G-BIXY	Piper J-3C-90 Cub	A. J. E. Ditheridge/Ipswich	
G-BIXZ	Grob G-109	K. E. White/Booker	
G-BIYI	Cameron V-65 balloon	Sarnia Balloon Group	

Notes	Reg.	Type	Owner or Operator
	G-BIYJ	PA-19 Super Cub 95	S. Russell
	G-BIYK	Isaacs Fury	R. S. Martin
	G-BIYM	PA-32R-301 Saratoga SP	E. O. Liebert/Jersey
	G-BIYN	Pitts S-1S Special	R. P. Lewis
	G-BIYO	PA-31-310 Turbo Navajo	Northern Executive Aviation Ltd/ Manchester
	G-BIYP	PA-20 Pacer 135	R. A. Lloyd-Hubbard & R. J. Whitcombe/Goodwood
	G-BIYR	PA-18 Super Cub 135	Delta Foxtrot Flying Group/ Exeter
	G-BIYT	Colt 17A balloon	A. F. Selby
	G-BIYU	Fokker S.11.1 Instructor (E-15)	H. R. Smallwood/Denham
	G-BIYV	Cremer 14.700-15 balloon	G. Lowther & ptnrs
	G-BIYW	Jodel D.112	W. J. Tanswell
	G-BIYX	PA-28 Cherokee 140	Telepoint Ltd/Manchester
	G-BIYY	PA-19 Super Cub 95	A. E. & W. J. Taylor/Ingoldmells
	G-BIZB	AB-206 JetRanger 3	Martin Butler Associates Ltd/Fairoaks
	G-BIZE	SOCATA TB.9 Tampico	M. J. Reid/Lydd
	G-BIZF	Cessna F.172P	C. M. Vlieland-Boddy/Bourn
	G-BIZG	Cessna F.152	Aero Group 78/Netherthorpe
	G-BIZI	Robin DR.400/120	Headcorn Flying School Ltd
	G-BIZJ	Nord 3202	Keenair Services Ltd/Liverpool
	G-BIZK	Nord 3202	Keenair Services Ltd/Liverpool
	G-BIZL	Nord 3202	Keenair Services Ltd/Liverpool
	G-BIZM	Nord 3202	Keenair Services Ltd/Liverpool
	G-BIZN	Slingsby T.67A	Specialist Flying Training Ltd/ Carlisle
	G-BIZO	PA-28R Cherokee Arrow 200	Penny (Mechanical Services) Ltd
	G-BIZR	SOCATA TB.9 Tampico	Martin Ltd/Biggin Hill
	G-BIZT	Bensen B.80D	J. Ferguson
	G-BIZU	Thunder Ax6-56Z balloon	S. L. Leigh
	G-BIZV	PA-19 Super Cub 95	J. T. Heaton/Thruxton
	G-BIZW	Champion 7GCBC Citabria	G. Read & Son
	G-BIZY	Jodel D.112	C. R. A. Wood
	G-BIZZ	Cessna 500 Citation	Vickers Ltd/S. Marston
	G-BJAA	Unicorn UE.1A balloon	K. H. Turner
	G-BJAD	FRED Srs 2	C. Allison
	G-BJAE	Starck AS.80 Lavadoux	D. J. & S. A. E. Phillips/Coventry
	G-BJAF	Piper J-3C-65 Cub	P. J. Cottle
	G-BJAG	PA-28-181 Archer II	H. Hunter
	G-BJAH	Unicorn UE.1A balloon	A. D. Hutchings
	G-BJAJ	AA-5B Tiger	Batrade Ltd/Biggin Hill
	G-BJAK	Mooney M.20C	D. B. Jay/Stapleford
	G-BJAL	C.A.S.A. 1.131E Jungmann	Buccaneer Aviation Ltd/Booker
	G-BJAN	SA.102-5 Cavalier	J. Powlesland
	G-BJAO	Bensen B.8M	G. L. Stockdale
	G-BJAP	D.H.82A Tiger Moth	J. Pothecary
	G-BJAR	Unicorn UE.3A balloon	Unicorn Group
	G-BJAS	Rango NA.9 balloon	A. Lindsay
	G-BJAU	PZL-104 Wilga 35	Anglo Polish Sailplanes Ltd/Booker
	G-BJAV	GY-80 Horizon 160	R. Pickett/Leicester
	G-BJAW	Cameron V-65 balloon	G. W. McCarthy
	G-BJAX	Pilatus P2-05 (J-108)	The Old Flying Machine Co
	G-BJAY	Piper J-3C-65 Cub	K. L. Clarke/Ingoldmells
	G-BJAZ	Thunder Ax7-77 balloon	R. C. Weyda
	G-BJBA	Cessna 152	I. Graham/Wellesbourne
	G-BJBB	Cessna 152	I. Graham/Wellesbourne
	G-BJBI	Cessna 414A	Fosters Shopfitters (Southern) Ltd
	G-BJBJ	Boeing 737-2T5	Orion Airways Ltd/E. Midlands
	G-BJBK	PA-19 Super Cub 95	J. D. Campbell/White Waltham
	G-BJBL	Unicorn UE.1A balloon	Unicorn Group
	G-BJBM	Monnet Sonerai II	J. Pickerell & ptnrs/Southend
	G-BJBN	Ball JB.980 balloon	J. D. Ball
	G-BJBO	Jodel DR.250/160	T. P. Bowen
	G-BJBP	Beech A200 Super King Air	All Charter Ltd (G-HLUB)/Bournemouth
	G-BJBR	Robinson R-22	Stenoak Fencing & Construction Co Ltd
	G-BJBS	Robinson R-22	Cosworth Engineering Ltd
	G-BJBV	PA-28-161 Warrior II	C.S.E. Aviation Ltd/Kidlington
	G-BJBW	PA-28-161 Warrior II	C.S.E. Aviation Ltd/Kidlington
	G-BJBX	PA-28-161 Warrior II	C.S.E. Aviation Ltd/Kidlington
	G-BJBY	PA-28-161 Warrior II	C.S.E. Aviation Ltd/Kidlington
	G-BJBZ	Rotorway 133 Executive	Rotorway (UK) Ltd
	G-BJCA	PA-28-161 Warrior II	Birmingham Aerocentre Ltd

Reg.	Type	Owner or Operator	Notes
G-BJCC	Unicorn UE.1A balloon	R. J. Pooley	
G-BJCD	Bede BD-5BH	Brockmoor-Bede Aircraft (UK) Ltd	
G-BJCE	Cessna F.172P	Sulair Ltd	
G-BJCF	CP.1310-C3 Super Emeraude	M. W. Wooldridge & P. Palmer/ Felthorpe	
G-BJCH	Ocset 1 balloon	B.H.M.E.D. Balloon Group	
G-BJCI	PA-18-150 Super Cub	The Borders (Milfield) Aero-Tour Club Ltd	
G-BJCJ	PA-28-181 Archer II	Simmons Electronics Ltd/Elstree	
G-BJCL	Morane Saulnier M.S.230 1049)	B. J. S. Grey/Booker	
G-BJCM	FRED Srs 2	J. C. Miller	
G-BJCP	Unicorn UE.2B balloon	Unicorn Group	
G-BJCR	Partenavia P.68C	Nullifire Ltd/Coventry	
G-BJCS	Meagher Mk 2 balloon	S. A. Fowler	
G-BJCU	Boeing 737-204ADV	Britannia Airways Ltd/Luton	
G-BJCV	Boeing 737-204ADV	Britannia Airways Ltd *Viscount Trenchard*/Luton	
G-BJCW	PA-32R-301 Saratoga SP	Viscount Chelsea/Kidlington	
G-BJCY	Slingsby T.67A	Specialist Flying Training Ltd/Carlisle	
G-BJDE	Cessna F.172M	S. Lynn	
G-BJDF	M.S.880B Rallye 100T	W. R. Savin & ptnrs	
G-BJDG	SOCATA TB.10 Tobago	K. G. Meadows	
G-BJDI	Cessna FR.182RG	Spoils Kitchen Reject Shops Ltd	
G-BJDJ	H.S.125 Srs 700B	Consolidated Contractors (UK) Services Ltd/Heathrow	
G-BJDK	European E.14 balloon	Aeroprint Tours	
G-BJDL	Rango NA.9 balloon	D. Lawrence	
G-BJDM	SA.102-5 Cavalier	J. D. McCracken	
G-BJDO	AA-5A Cheetah	Border Transport/Southampton	
G-BJDP	Cremer Cloudcruiser balloon	P. J. Petitt & M. J. Harper	
G-BJDS	British Bulldog balloon	A. J. Cremer	
G-BJDT	SOCATA TB.9 Tampico	Tampico Group/Old Sarum	
G-BJDU	Scruggs BL.2B-2 balloon	C. D. Ibell	
G-BJDV	Kingram balloon	T. J. King & S. Ingram	
G-BJDW	Cessna F.172M	Suffolk Aero Club Ltd/Ipswich	
G-BJDX	Scruggs BL.2D-2 balloon	A. R. Maple	
G-BJDZ	Unicorn UE.1A balloon	A. P. & K. E. Chown	
G-BJEI	PA-19 Super Cub 95	H. J. Cox & S. J. Bushell/Bicester	
G-BJEL	Nord NC.854	J. P. Taylor/Lulsgate	
G-BJEM	Cube balloon	A. J. Cremer	
G-BJEN	Scruggs RS.5000 balloon	N. J. Richardson	
G-BJEO	PA-34-220T Seneca III	D. W. Clark Land Drainage Ltd (G-TOMF)	
G-BJES	Scruggs RS.5000 balloon	J. E. Christopher	
G-BJET	Cessna 425	Gatwick Air Taxis Ltd	
G-BJEU	Scruggs BL.2D-2 balloon	G. G. Kneller	
G-BJEV	Aeronca 11AC Chief	M. A. Musselwhite	
G-BJEW	Cremer balloon	C. D. Sandoe	
G-BJEX	Bo 208C Junior	G. D. H. Crawford/Thruxton	
G-BJEY	BHMED Srs 1 balloon	D. R. Meades & J. S. Edwards	
G-BJFB	Mk 1A balloon	Aeroprint Tours	
G-BJFC	European E.8 balloon	P. D. Ridout	
G-BJFD	BHMED Srs 1 balloon	D. G. Dance & I. R. Bell	
G-BJFE	PA-19 Super Cub 95 (L-18C)	C. C. Lovell	
G-BJFH	Boeing 737-2S3	Air Europe Ltd *Sandie*/Gatwick	
G-BJFI	Bell 47G-2A1	Helicopter Supplies & Engineering Ltd/ Bournemouth	
G-BJFK	Short SD3-30	Jersey European Airways Ltd	
G-BJFL	Sikorsky S-76A	Bristow Helicopters Ltd	
G-BJFM	Jodel D.120	M. L. Smith & ptnrs/Popham	
G-BJFN	Mk IV balloon	Windsor Balloon Group	
G-BJFO	Mk II balloon	Windsor Balloon Group	
G-BJFP	Mk III balloon	Windsor Balloon Group	
G-BJFR	Mk IV balloon	Windsor Balloon Group	
G-BJFS	Mk IV balloon	Windsor Balloon Group	
G-BJFT	Mk IV balloon	Windsor Balloon Group	
G-BJFU	Mk IV balloon	Windsor Balloon Group	
G-BJFV	Mk V balloon	Windsor Balloon Group	
G-BJFW	Mk V balloon	Windsor Balloon Group	
G-BJFX	Mk V balloon	Windsor Balloon Group	
G-BJFY	Mk I balloon	Windsor Balloon Group	
G-BJFZ	Mk II balloon	Windsor Balloon Group	
G-BJGA	Mk IV balloon	Windsor Balloon Group	

Notes	Reg.	Type	Owner or Operator
	G-BJGB	Mk I balloon	Windsor Balloon Group
	G-BJGC	Mk IV balloon	Windsor Balloon Group
	G-BJGD	Mk IV balloon	Windsor Balloon Group
	G-BJGE	Thunder Ax3 balloon	C. E. Weston-Baker/W. Germany
	G-BJGF	Mk 1 balloon	D. & D. Eaves
	G-BJGG	Mk 2 balloon	D. & D. Eaves
	G-BJGH	Slingsby T.67A	Specialist Flying Training Ltd/Carlisle
	G-BJGK	Cameron V-77 balloon	A. Simpson & R. Bailey
	G-BJGL	Cremer balloon	G. Lowther
	G-BJGM	Unicorn UE.1A balloon	D. Eaves & P. D. Ridout
	G-BJGN	Scruggs RS.5000 balloon	K. H. Turner
	G-BJGO	Cessna 172N	Golf Oscar Ltd/Birmingham
	G-BJGS	Cremer balloon	C. A. Larkins
	G-BJGT	Mooney M.20K	Air Bristol
	G-BJGW	M.H.1521M Broussard (92)	G. A. Warner/Duxford
	G-BJGX	Sikorsky S-76A	Bristow Helicopters Ltd
	G-BJGY	Cessna F.172P	Derek Crouch PLC
	G-BJHA	Cremer balloon	G. Cope
	G-BJHB	Mooney M.20J	TI Services Ltd/Biggin Hill
	G-BJHC	Swan 1 balloon	C. A. Swan
	G-BJHD	Mk 3B balloon	S. Meagher
	G-BJHE	Osprey 1B balloon	R. B. Symonds & J. M. Hopkins
	G-BJHG	Cremer balloon	P. A. Cremer & H. J. A. Green
	G-BJHJ	Osprey 1C balloon	D. Eaves
	G-BJHK	EAA Acro Sport	J. H. Kimber
	G-BJHL	Osprey 1C balloon	E. Bartlett
	G-BJHM	Osprey 1B balloon	W. P. Fulford
	G-BJHN	Osprey 1B balloon	J. E. Christopher
	G-BJHO	Osprey 1C balloon	G. G. Kneller
	G-BJHP	Osprey 1C balloon	N. J. Richardson
	G-BJHR	Osprey 1B balloon	J. E. Christopher
	G-BJHS	S.25 Sunderland V	Sunderland Ltd/Chatham
	G-BJHT	Thunder Ax7-65 balloon	A. H. & L. Symonds
	G-BJHU	Osprey 1C balloon	G. G. Kneller
	G-BJHV	Voisin Replica	M. P. Sayer/O. Warden
	G-BJHW	Osprey 1C balloon	N. J. Richardson
	G-BJHX	Osprey 1C balloon	A. B. Gulliford
	G-BJHY	Osprey 1C balloon	T. J. King & S. Ingram
	G-BJHZ	Osprey 1C balloon	M. Christopher
	G-BJIA	Allport balloon	D. J. Allport
	G-BJIB	D.31 Turbulent	N. H. Lemon
	G-BJIC	Dodo 1A balloon	P. D. Ridout
	G-BJID	Osprey 1B balloon	P. D. Ridout
	G-BJIE	Sphinx balloon	P. T. Witty
	G-BJIF	Bensen B.8M	H. Redwin
	G-BJIG	Slingsby T.67A	Specialist Flying Training Ltd/Carlisle
	G-BJII	Sphinx balloon	I. French
	G-BJIJ	Osprey 1B balloon	R. Hownsell
	G-BJIR	Cessna 550 Citation II	Supreme Securities Ltd
	G-BJIS	Mk 1 balloon	P. Paine
	G-BJIU	Bell 212	Bristow Helicopters Ltd
	G-BJIV	PA-18-150 Super Cub	Yorkshire Gliding Club (Pty) Ltd
	G-BJIW	T-1 balloon	S. Holland & G. Watmore
	G-BJIX	T-1 balloon	S. Holland & G. Watmore
	G-BJIY	Cessna T337D	Air Charter (Scotland) Ltd/Glasgow
	G-BJJE	Dodo Mk 3 balloon	D. Eaves
	G-BJJF	Dodo Mk 4 balloon	D. Eaves
	G-BJJG	Dodo Mk 5 balloon	D. Eaves
	G-BJJI	SAS balloon	R. Hounsell & M. R. Rooke
	G-BJJJ	Bitterne balloon	R. Hounsell & M. R. Rooke
	G-BJJK	Bitterne balloon	R. Hounsell & M. R. Rooke
	G-BJJL	SAS balloon	M. R. Rooke
	G-BJJN	Cessna F.172M	Ospreystar Ltd/Stapleford
	G-BJJO	Bell 212	Bristow Helicopters Ltd
	G-BJJP	Bell 212	Bristow Helicopters Ltd
	G-BJJS	Sphinx balloon	C. N. Childs
	G-BJJT	Mabey balloon	M. W. Mabey
	G-BJJU	Sphinx balloon	T. M. Bates
	G-BJJW	Mk B balloon	S. Meagher
	G-BJJX	Mk B balloon	S. Meagher
	G-BJJY	Mk B balloon	S. Meagher
	G-BJJZ	Unicorn UE.1A balloon	R. Woodley
	G-BJKA	SA.365C Dauphin 2	Bond Helicopters Ltd/Bourn
	G-BJKB	SA.365C Dauphin 2	Bond Helicopters Ltd/Bourn

Reg.	Type	Owner or Operator	Notes
G-BJKC	Mk B balloon	S. Meagher	
G-BJKD	Mk B balloon	S. Meagher	
G-BJKE	Mk A balloon	D. Addison	
G-BJKF	SOCATA TB.9 Tampico	K. Bollmann/Shoreham	
G-BJKG	Mk A balloon	D. Addison	
G-BJKH	Mk A balloon	D. Addison	
G-BJKI	Mk A balloon	D. Addison	
G-BJKJ	Mk A balloon	D. Addison	
G-BJKK	Mk A balloon	D. Addison	
G-BJKL	Mk A balloon	D. Addison	
G-BJKM	Mk II balloon	S. Meagher	
G-BJKN	Mk 1 balloon	D. Addison	
G-BJKO	Mk 1 balloon	D. Addison	
G-BJKP	Mk 7 balloon	D. Addison	
G-BJKR	Mk 1 balloon	D. Addison	
G-BJKS	Mk 1 balloon	D. Addison	
G-BJKT	Mk B balloon	S. Meagher	
G-BJKU	Osprey 1B balloon	S. A. Dalmas & P. G. Tarr	
G-BJKV	Opsrey 1F balloon	B. Diggle	
G-BJKW	Wills Aera II	J. K. S. Wills	
G-BJKX	Cessna F.152	Eglinton Flying Club	
G-BJKY	Cessna F.152	Westair Flying Services Ltd/Blackpool	
G-BJKZ	Osprey 1F balloon	M. J. N. Kirby	
G-BJLA	Osprey 1B balloon	D. Lawrence	
G-BJLB	Nord NC.854S	M. J. Barnby/Cardiff	
G-BJLC	Monnet Sonerai IIL	J. P. Whitham	
G-BJLD	Eagle 8 Mk 2 balloon	R. M. Richards	
G-BJLE	Osprey 1B balloon	I. Chadwick	
G-BJLF	Unicorn UE.1C balloon	I. Chadwick	
G-BJLG	Unicorn UE.1B balloon	I. Chadwick	
G-BJLH	PA-19 Super Cub 95 (K-33)	M. F. Newman/Ipswich	
G-BJLJ	Cameron D-50 balloon	Cameron Balloons Ltd	
G-BJLK	Short SD3-30	Fairflight Ltd/Biggin Hill	
G-BJLN	Featherlight Mk 3 balloon	A. P. Newman & T. J. Sweeting	
G-BJLO	PA-31-310 Navajo	Linco (Poultry Machinery) Ltd/ Biggin Hill	
G-BJLP	Featherlight Mk 3 balloon	N. P. Kemp & M. O. Davies	
G-BJLR	Featherlight Mk 3 balloon	M. O. Davies & S. R. Roberts	
G-BJLT	Featherlight Mk 3 balloon	J. M. J. Roberts & C. C. Marshall	
G-BJLU	Featherlight Mk 3 balloon	T. J. Sweeting & N. P. Kemp	
G-BJLV	Sphinx balloon	L. F. Guyot	
G-BJLW	Gleave CJ-I balloon	C. J. Gleave	
G-BJLX	Cremer balloon	P. W. May	
G-BJLY	Cremer balloon	P. Cannon	
G-BJLZ	Cremer balloon	S. K. McLean	
G-BJMA	Colt 21A balloon	Colt Balloons Ltd	
G-BJMB	Osprey 1B balloon	S. Meagher	
G-BJMG	European E.26C balloon	D. Eaves & A. P. Chown	
G-BJMH	Osprey Mk 3A balloon	D. Eaves	
G-BJMI	European E.84 balloon	D. Eaves	
G-BJMJ	Bensen B.8M	P. R. Snowdon	
G-BJMK	Cremer balloon	B. J. Larkins	
G-BJML	Cessna 120	D. F. Lawlar/Panshanger	
G-BJMO	Taylor JT.1 Monoplane	R. C. Mark	
G-BJMP	Brugger Colibri M.B.2	F. Skinner	
G-BJMR	Cessna 310R	A-One Transport (Leeds) Ltd/Sherburn	
G-BJMT	Osprey Mk 1E balloon	M. J. Sheather	
G-BJMU	European E.157 balloon	A. C. Mitchell	
G-BJMV	BAC One-Eleven 531FS	Dan-Air Services Ltd/Gatwick	
G-BJMW	Thunder Ax8-105 balloon	G. M. Westley	
G-BJMX	Jarre JR.3 balloon	P. D. Ridout	
G-BJMZ	European EA.8A balloon	P. D. Ridout	
G-BJNA	Arena Mk 117P balloon	P. D. Ridout	
G-BJNB	WAR F4U Corsair	A. V. Francis	
G-BJNC	Osprey Mk 1E balloon	G. Whitehead	
G-BJND	Osprey Mk 1E balloon	A. Billington & D. Whitmore	
G-BJNE	Osprey Mk 1E balloon	D. R. Sheldon	
G-BJNF	Cessna F.152	Exeter Flying Club Ltd	
G-BJNG	Slingsby T.67A	Specialist Flying Training Ltd/ Carlisle	
G-BJNH	Osprey Mk 1E balloon	D. A. Kirk	
G-BJNI	Osprey Mk 1C balloon	M. J. Sheather	
G-BJNL	Evans VP-2	K. Morris	
G-BJNN	PA-38-112 Tomahawk	Apollo Leasing Ltd/Edinburgh	

Notes	Reg.	Type	Owner or Operator
	G-BJNO	AA-5B Tiger	Kaal Electrics Ltd/Elstree
	G-BJNP	Rango NA.32 balloon	N. H. Ponsford
	G-BJNX	Cameron O-65 balloon	B. J. Petteford
	G-BJNY	Aeronca 11CC Super Chief	P. I. & D. M. Morgans
	G-BJNZ	PA-23 Aztec 250	Distance No Object Ltd (G-FANZ)/ Stansted
	G-BJOA	PA-28-181 Archer II	Channel Islands Aero Holdings (Jersey) Ltd
	G-BJOB	Jodel D.140C	T. W. M. Beck & M. J. Smith
	G-BJOD	Hollman HA-2M Sportster	H. J. Goddard
	G-BJOE	Jodel D.120A	Jodair Flying Group/Fenland
	G-BJOG	BN-2T Turbo Islander	Pilatus BN Ltd/Bembridge
	G-BJOI	Isaacs Special	J. O. Isaacs
	G-BJOP	BN-2B Islander	Loganair Ltd/Glasgow
	G-BJOT	Jodel D.117	R. L. A. Davies/Birmingham
	G-BJOV	Cessna F.150K	Ingham Aviation Ltd/Bristol
	G-BJOZ	Scheibe SF.25B Falke	P. W. Hextall
	G-BJPA	Osprey Mk 3A balloon	N. D. Brabham
	G-BJPB	Osprey Mk 4A balloon	C. B. Rundle
	G-BJPC	Cremer 1 gyroplane	P. A. Cremer
	G-BJPD	Osprey Mk 4D balloon	E. L. Fuller
	G-BJPE	Osprey Mk 1E balloon	M. A. Hase
	G-BJPI	Bede BD-5G	M. D. McQueen
	G-BJPJ	Osprey Mk 3A	K. R. Bundy
	G-BJPK	Osprey Mk 1B balloon	G. M. Hocquard
	G-BJPL	Osprey Mk 4A balloon	M. Vincent
	G-BJPM	Bursell PW.1 balloon	I. M. Holdsworth
	G-BJPN	JK Mk 1 balloon	A. Kaye & J. Corcoran
	G-BJPO	B&C balloon	S. Browne & J. Cheetham
	G-BJPU	Osprey Mk 4B balloon	P. Globe
	G-BJPV	Haigh balloon	M. J. Haigh
	G-BJPW	Osprey Mk 1C balloon	P. J. Cooper & M. Draper
	G-BJPX	Phoenix balloon	Cupro Sapphire Ltd
	G-BJPY	Cremer balloon	P. A. Cremer & P. V. M. Green
	G-BJPZ	Osprey Mk 1C balloon	C. E. Newman
	G-BJRA	Osprey Mk 4B balloon	E. Osborn
	G-BJRB	European E.254 balloon	D. Eaves
	G-BJRC	European E.84R balloon	D. Eaves
	G-BJRD	European E.84R balloon	D. Eaves
	G-BJRF	Saffery S.80 balloon	C. F. Chipping
	G-BJRG	Osprey Mk 4B balloon	A. de Gruchy
	G-BJRH	Rango NA.36 balloon	N. H. Ponsford
	G-BJRI	Osprey Mk 4D balloon	G. G. Kneller
	G-BJRJ	Osprey Mk 4D balloon	G. G. Kneller
	G-BJRK	Osprey Mk 1E balloon	G. G. Kneller
	G-BJRL	Osprey Mk 4B balloon	G. G. Kneller
	G-BJRO	Osprey Mk 4D balloon	M. Christopher
	G-BJRP	Cremer balloon	M. Williams
	G-BJRR	Cremer balloon	M. Wallbank
	G-BJRS	Cremer balloon	P. Wallbank
	G-BJRT	BAC One-Eleven 528	British Caledonian Airways Ltd/ Gatwick
	G-BJRU	BAC One-Eleven 528	British Caledonian Airways Ltd *City of Edinburgh*/Gatwick
	G-BJRV	Cremer balloon	M. D. Williams
	G-BJRW	Cessna U.206G	A. I. Walgate & Son Ltd
	G-BJRX	RMB Mk 1 balloon	R. J. MacNeil
	G-BJRY	PA-28-151 Warrior	Eastern Counties Aero Club Ltd/ Southend
	G-BJRZ	Partenavia P.68C	W. P. J. Davison
	G-BJSA	BN-2A Islander	Harvest Air Ltd/Southend
	G-BJSC	Osprey Mk 4D balloon	N. J. Richardson
	G-BJSD	Osprey Mk 4D balloon	N. J. Richardson
	G-BJSE	Osprey Mk 1E balloon	J. E. Christopher
	G-BJSF	Osprey Mk 4B balloon	N. J. Richardson
	G-BJSG	V.S.361 Spitfire LF.IXE (ML417)	B. J. S. Grey/Booker
	G-BJSH	Sindlinger Hurricane 5/8 scale replica	A. F. Winstanley
	G-BJSI	Osprey Mk 1E balloon	N. J. Richardson
	G-BJSJ	Osprey Mk 1E balloon	M. Christopher
	G-BJSK	Osprey Mk 4B balloon	N. J. Richardson
	G-BJSL	Flamboyant Ax7-65 balloon	Pepsi Cola International Ltd
	G-BJSM	Bursell Mk 1 balloon	M. C. Bursell
	G-BJSP	Guido 1A Srs 61 balloon	G. A. Newsome

Reg.	Type	Owner or Operator	Notes
G-BJSR	Osprey Mk 4B balloon	C. F. Chipping	
G-BJSS	Allport balloon	D. J. Allport	
G-BJST	CCF Harvard 4	V. Norman & M. Lawrence	
G-BJSU	Bensen B.8M	J. D. Newlyn	
G-BJSV	PA-28-161 Warrior II	A. F. Aviation Ltd/Stansted	
G-BJSW	Thunder Ax7-65 balloon	Sandcliffe Garage Ltd	
G-BJSX	Unicorn UE-1C balloon	N. J. Richardson	
G-BJSY	Beech E90 King Air	Allcharter Ltd/Bournemouth	
G-BJSZ	Piper J-3C-65 Cub	H. Gilbert	
G-BJTA	Osprey Mk 4B balloon	C. F. Chipping	
G-BJTB	Cessna A.150M	Leisure Lease Aviation/Southend	
G-BJTD	Colt AS-90 airship	Bulk-More Dairy Shops Ltd/Canada	
G-BJTF	Skyrider Mk 1 balloon	D. A. Kirk	
G-BJTG	Osprey Mk 4B balloon	M. Millen	
G-BJTH	Kestrel AC Mk 1 balloon	G. Whitehead	
G-BJTI	Woodie K2400J-2 balloon	M. J. Woodward	
G-BJTJ	Osprey Mk 4B balloon	G. Hocquard	
G-BJTK	Taylor JT.1 Monoplane	E. N. Simmons (G-BEUM)	
G-BJTN	Osprey Mk 4B balloon	M. Vincent	
G-BJTO	Piper L-4H Cub	K. R. Nunn	
G-BJTP	PA-19 Super Cub 95	J. T. Parkins/Sywell	
G-BJTS	Osprey Mk 4B balloon	G. Hocquard	
G-BJTT	Sphinx SP.2 balloon	N. J. Godfrey	
G-BJTU	Cremer Cracker balloon	D. R. Green	
G-BJTV	M.S.880B Rallye Club	J. M. Kirk	
G-BJTW	European E.107 balloon	C. J. Brealey	
G-BJTX	PA-31-325 Turbo Navajo	Truvelo Manufacturers Ltd	
G-BJTY	Osprey Mk 4B balloon	A. E. de Gruchy	
G-BJTZ	Osprey Mk 4A balloon	M. J. Sheather	
G-BJUA	Sphinx SP.12 balloon	T. M. Pates	
G-BJUB	BVS Special 01 balloon	P. G. Wild	
G-BJUC	Robinson R-22	Jones & Brooks Ltd	
G-BJUD	Robin DR.400/180R	Southern Sailplanes Ltd	
G-BJUE	Osprey Mk 4B balloon	M. Vincent	
G-BJUG	SOCATA TB.9 Tampico	G. N. Taylor/Denham	
G-BJUI	Osprey Mk 4B balloon	B. A. de Gruchy	
G-BJUK	Short SD3-30	Jersey European Airways (G-OCAS)	
G-BJUN	Unicorn UE.1C balloon	R. J. Rallysport	
G-BJUP	Osprey Mk 4B balloon	W. J. Pill	
G-BJUR	PA-38-112 Tomahawk	Truman Aviation Ltd/Tollerton	
G-BJUS	PA-38-112 Tomahawk	Panshanger School of Flying	
G-BJUU	Osprey Mk 4B balloon	M. Vincent	
G-BJUV	Cameron V-20 balloon	Cameron Balloons Ltd	
G-BJUW	Osprey Mk 4B balloon	C. F. Chipping	
G-BJUX	Bursell balloon	I. M. Holdsworth	
G-BJUY	Colt Ax-77 balloon	Colt Balloons Ltd	
G-BJUZ	BAT Mk II balloon	A. R. Thompson	
G-BJVA	BAT Mk I balloon	B. L. Thompson	
G-BJVB	Cremcorn Ax1.4 balloon	P. A. Cremer & I. Chadwick	
G-BJVC	Evans VP-2	R. G. Fenn/Leicester	
G-BJVF	Thunder Ax3 balloon	A. G. R. Calder & F. J. Spite	
G-BJVG	Thunder Ax8-105 balloon	Thunder Balloons Ltd	
G-BJVH	Cessna F.182Q	A. R. G. Brooker Engineering Ltd	
G-BJVI	Osprey Mk 4D balloon	S. M. Colville	
G-BJVJ	Cessna F.152	Cambridge Aero Club Ltd	
G-BJVK	Grob G-109	B. Kimberley/Enstone	
G-BJVL	Saffery Hermes balloon	Cupro Sapphire Ltd	
G-BJVM	Cessna 172M	Angelsword Ltd/Wellesbourne	
G-BJVO	Cameron D-50 airship	Cameron Balloons Ltd	
G-BJVS	CP.1315C-3 Super Emeraude	Aerofel & Super Emeraude Group/ Norwich	
G-BJVT	Cessna F.152	Cambridge Aero Club Ltd	
G-BJVU	Thunder Ax6-56 balloon	G. V. Beckwith	
G-BJVV	Robin R.1180	Medway Flying Group Ltd/Rochester	
G-BJVX	Sikorsky S-76A	Bristow Helicopters Ltd	
G-BJVZ	Sikorsky S-76A	Bristow Helicopters Ltd	
G-BJWC	Saro Skeeter AOP.12 ★	J. E. Wilkie	
G-BJWD	Zenith CH.300	D. Winton	
G-BJWF	Ayres S2R-R3S Thrush Commander	Shoreham Flight Simulation Ltd/ Bournemouth	
G-BJWH	Cessna F.152	Hartmann Ltd	
G-BJWI	Cessna F.172P	Shoreham Flight Simulation Ltd/ Bournemouth	
G-BJWJ	Cameron V-65 balloon	R. G. Turnbull & S. G. Forse	

Notes	Reg.	Type	Owner or Operator
	G-BJWL	BN-2A-8 Islander	Harvest Aviation Ltd (G-BBMC)/ Southend
	G-BJWM	BN-2A-26 Islander	Harvest Air Ltd (G-BCAE)/Southend
	G-BJWN	BN-2A-8 Islander	Harvest Air Ltd (G-BALO)/Southend
	G-BJWO	BN-2A-8 Islander	Harvest Air Ltd (G-BAXC)/Southend
	G-BJWP	BN-2A-26 Islander	Harvest Air Ltd (G-BCEJ)/Southend
	G-BJWR	D.H.82A Tiger Moth	D. R. Whitby & ptnrs
	G-BJWT	Wittman W.10 Tailwind	J. F. Bakewell & R. A. Shelley
	G-BJWV	Colt 17A balloon	Lighter-Than-Air Ltd
	G-BJWW	Cessna F.172N	Westair Flying Services Ltd/Blackpool
	G-BJWX	PA-19 Super Cub 95	D. E. Lamb/Wickenby
	G-BJWY	Sikorsky S-55 Whirlwind 21	J. E. Wilkie
	G-BJWZ	PA-19 Super Cub 95	G. V. Harfield/Thruxton
	G-BJXA	Slingsby T.67A	Specialist Flying Training Ltd/Carlisle
	G-BJXB	Slingsby T.67A	Light Planes (Lancs) Ltd/Barton
	G-BJXD	Colt 17A balloon	Hot Air Balloon Co Ltd
	G-BJXJ	Boeing 737-219	Dan-Air Services Ltd/Gatwick
	G-BJXK	Fournier RF-5	P. Storey & ptnrs
	G-BJXL	Boeing 737-2T4	Dan-Air Services Ltd/Gatwick
	G-BJXN	Boeing 747-230B	British Caledonian Airways *Mungo Park — The Scottish Explorer*/Gatwick
	G-BJXO	Cessna 441	Hatfield Executive Aviation Ltd
	G-BJXP	Colt 56B balloon	Lighter-Than-Air Ltd
	G-BJXR	Auster AOP.9 (XR267)	Cotswold Aircraft Restoration Group
	G-BJXU	Thunder Ax7-77 balloon	Perdix Ltd
	G-BJXX	PA-23 Aztec 250E	Creative Conferences Aviation Ltd
	G-BJXZ	Cessna 172N	J. R. Kettle/Halfpenny Green
	G-BJYC	Cessna 425	Carters Aviation Ltd
	G-BJYD	Cessna F.152 II	Cleveland Flying School Ltd/ Tees-side
	G-BJYF	Colt 56A balloon	Hot Air Balloon Co Ltd
	G-BJYG	PA-28-161 Warrior II	Channel Aviation Ltd/Guernsey
	G-BJYK	Jodel D.120A	D. R. Emmett/Bourn
	G-BJYL	BAC One-Eleven 515FB	Dan-Air Services Ltd (G-AZPE)/Gatwick
	G-BJYM	BAC One-Eleven 531FS	Dan-Air Services Ltd/Gatwick
	G-BJYN	PA-38-112 Tomahawk	Panshanger School of Flying Ltd (G-BJTE)
	G-BJYO	PA-38-112 Tomahawk	Panshanger School of Flying Ltd
	G-BJYZ	BN-2B Islander	Pilatus BN Ltd/Bembridge
	G-BJZA	Cameron N-65 balloon	A. D. Pinner
	G-BJZB	Evans VP-2	A. Graham
	G-BJZC	Thunder Ax7-65Z balloon	Greenpeace (UK) Ltd/S. Africa
	G-BJZD	Douglas DC-10-10	Cal Air (G-GFAL)/Gatwick
	G-BJZE	Douglas DC-10-10	Cal Air (G-GSKY)/Gatwick
	G-BJZF	D.H.82A Tiger Moth	C. A. Parker/Sywell
	G-BJZH	Colt 77B balloon	Colt Balloons Ltd
	G-BJZK	Cessna T.303	Standard Aviation Ltd
	G-BJZL	Cameron V-65 balloon	S. L. G. Williams
	G-BJZM	Slingsby T.67A	Specialist Flying Training Ltd/Carlisle
	G-BJZN	Slingsby T-67A	Light Planes (Lancs) Ltd/Barton
	G-BJZR	Colt 42A balloon	C. F. Sisson
	G-BJZT	Cessna FA.152	Hartmann Ltd/Booker
	G-BJZU	Cessna FA.152	Hartmann Ltd/Booker
	G-BJZX	Grob G.109	Oxfordshire Sport Flying Ltd/Enstone
	G-BJZY	Bensen B.8MV	D. E. & M. A. Cooke
	G-BJZZ	Hispano HA.1112 (14)	R. Mulqueen
	G-BKAA	H.S.125 Srs 700B	Abbey Investment Co Ltd
	G-BKAB	ICA Brasov IS-28M2	R. B. Woodhouse
	G-BKAC	Cessna F.150L	Andrewsfield Flying Club Ltd (G-BAIO)
	G-BKAE	Jodel D.120	J. S. Lewer
	G-BKAF	FRED Srs 2	L. G. Millen/Headcorn
	G-BKAG	Boeing 727-217	Dan-Air Services Ltd/Gatwick
	G-BKAK	Beech C90 King Air	Navigations Ltd/Luton
	G-BKAM	Slingsby T.67M Firefly	Slingsby Aviation Ltd/Kirkbymoorside
	G-BKAN	Cessna 340A	Northair Aviation Ltd/Leeds
	G-BKAO	Jodel D.112	E. Carter & G. Higgins
	G-BKAR	PA-38-112 Tomahawk	C.S.E. Aviation Ltd/Kidlington
	G-BKAS	PA-38-112 Tomahawk	C.S.E. Aviation Ltd/Kidlington
	G-BKAT	Pitts S-1C Special	I. M. G. Senior & J. G. Harper
	G-BKAY	R. Commander 114	Costello Gears Ltd/Biggin Hill
	G-BKAZ	Cessna 152	Skyviews & General Ltd
	G-BKBB	Hawker Fury replica	The Hon P. Lindsay/Booker

Reg.	Type	Owner or Operator	Notes
G-BKBC	D.H.C.-6 Twin Otter 310	Jersey European Airways/Jersey	
G-BKBD	Thunder Ax3 balloon	D. Clark	
G-BKBE	AA-5A Cheetah	G. W. Plowman & Sons Ltd/Elstree	
G-BKBF	M.S.894A Rallye Minerva 220	Callow Aviation/Staverton	
G-BKBI	Quickie Q.2	R. H. Gibbs	
G-BKBK	Stampe SV-4A	B. M. O'Brien/Redhill	
G-BKBM	H.S.125 Srs 600B	R. Hitchin & Co Ltd (G-BCCL)/Luton	
G-BKBN	SOCATA TB.10 Tobago	Cross Bros Ltd	
G-BKBO	Colt 17A balloon	Bridges Van Hire Ltd/Tollerton	
G-BKBP	Bellanca 7GCBC Scout	L. B. Jefferies	
G-BKBR	Cameron Chateau 84 balloon	Forbes Europe Ltd/France	
G-BKBS	Bensen B.8MV	C. R. Dawe	
G-BKBV	SOCATA TB.10 Tobago	J. Bett/Prestwick	
G-BKBW	SOCATA TB.10 Tobago	P. Murphy/Fairoaks	
G-BKBY	Bell 206B JetRanger 3	Real Time Control Ltd	
G-BKCB	PA-28R Cherokee Arrow 200	P. G. Kitchingham & A. Dugdale	
G-BKCC	PA-28 Cherokee 180	Classic Aeroplane Ltd/Staverton	
G-BKCD	H.S.125 Srs 600B	McAlpine Aviation Ltd (G-BDOA)/Luton	
G-BKCE	Cessna F.172P-II	Fleetdelta Ltd/Norwich	
G-BKCF	Rutan Long Ez	I. C. Fallows	
G-BKCH	Thompson Cassutt	S. C. Thompson/Redhill	
G-BKCI	Brugger M.B.2 Colibri	E. R. Newall	
G-BKCJ	Oldfield Baby Great Lakes	S. V. Roberts/Sleap	
G-BKCK	CCF Harvard IV	E. T. & T. C. Webster/Bahrain	
G-BKCL	PA-30 Twin Comanche 160	Jubilee Airways Ltd (G-AXSP)/ Conington	
G-BKCM	Bell 206B JetRanger 3	S. W. Electricity Board/Lulsgate	
G-BKCN	Currie Wot	S. E. Tomlinson	
G-BKCR	SOCATA TB.9 Tampico	Automated Data Systems Ltd	
G-BKCS	Cessna T.207	G. H. Nolan	
G-BKCT	Cameron V-77 balloon	Quality Products General Engineering (Wickwat) Ltd	
G-BKCU	Sequoia F.8L Falco	J. J. Anziani & D. F. Simpson	
G-BKCV	EAA Acro Sport II	M. J. Clark	
G-BKCW	Jodel D.120A	T. Rayner & P. McIntosh/Dundee	
G-BKCX	Mudry CAARP CAP.10	D. E. Starkey & M. J. Heudebourck/ Denham	
G-BKCY	PA-38-112 Tomahawk II	Norwich Air Training Ltd	
G-BKCZ	Jodel D.120A	P. Penn-Sayers Model Services Ltd/ Shoreham	
G-BKDA	AB-206B JetRanger	Dollar Air Services Ltd/Coventry	
G-BKDC	Monnet Sonerai II	J. Boobyer	
G-BKDD	Bell 206B JetRanger	Dollar Air Services Ltd/Coventry	
G-BKDE	Kendrick I Motorglider	J. K. Rushton	
G-BKDF	Kendrick II Motorglider	J. K. Rushton	
G-BKDG	PA-19 Super Cub 95	M. Fleetwood	
G-BKDH	Robin DR.400/120	W. R. C. Foyle/Luton	
G-BKDI	Robin DR.400/120	Wickwell Solo Leasing	
G-BKDJ	Robin DR.400/120	W. R. C. Foyle/Luton	
G-BKDK	Thunder Ax7-77Z balloon	Thunder Balloons Ltd	
G-BKDN	Short SD3-30	Fairflight Ltd/Biggin Hill	
G-BKDO	Short SD3-30	Fairflight Ltd/Biggin Hill	
G-BKDP	FRED Srs 3	M. Whittaker	
G-BKDR	Pitts S.1S Special	T. R. G. Barnby/Redhill	
G-BKDT	S.E.5A replica (F943)	J. H. Tetley & W. A. Sneesby/Sherburn	
G-BKDW	K.1260/3 Stu balloon	P. C. Carlton	
G-BKDX	Jodel DR.1050	F. A. L. Castleden & ptnrs	
G-BKDY	Jodel D.120A	J. J. Pratt & A. Lumley/Sturgate	
G-BKEA	BN-2B Islander	Pilatus BN Ltd/Bembridge	
G-BKEF	BN-2B Islander	Pilatus BN Ltd/Bembridge	
G-BKEG	BN-2B Islander	Pilatus BN Ltd/Bembridge	
G-BKEH	BN-2B Islander	Bitteswell Ltd	
G-BKEK	PA-32 Cherokee Six 300	Cruspane Ltd/Stapleford	
G-BKEM	SOCATA TB.9 Tampico	D. V. D. Reed/Dunkeswell	
G-BKEN	SOCATA TB.10 Tobago	Berglen Products Ltd/Elstree	
G-BKEP	Cessna F.172M	Reedtrend Ltd	
G-BKER	S.E.5A replica (F5447)	N. K. Geddes	
G-BKES	Cameron SS bottle balloon	Lighter-Than-Air Ltd	
G-BKET	PA-19 Super Cub 95	M. Ingamells & P. Southerington	
G-BKEU	Taylor JT.1 Monoplane	R. J. Whybrow & J. M. Springham	
G-BKEV	Cessna F.172M	Guernsey Flight Training Ltd	
G-BKEW	Bell 206B JetRanger 3	N. R. Foster	
G-BKEX	Rich Prototype glider	D. B. Rich	
G-BKEY	FRED Srs 3	G. S. Taylor	

Notes	Reg.	Type	Owner or Operator
	G-BKEZ	PA-19 Super Cub 95	A. N. G. Gardiner
	G-BKFA	Monnet Sonerai IIL	R. F. Bridge
	G-BKFC	Cessna F.152-II	D. W. Walton/Sywell
	G-BKFG	Thunder Ax3 balloon	H. V. Wallis/Canada
	G-BKFI	Evans VP-1	F. A. R. de Lavergne
	G-BKFK	Isaacs Fury II	G. C. Jones
	G-BKFL	Aerosport Scamp	I. D. Daniels
	G-BKFM	Rutan Quickie	R. I. Davidson & P. J. Cheyney
	G-BKFN	Bell 214ST	British Caledonian Helicopters Ltd
	G-BKFP	Bell 214ST	British Caledonian Helicopters Ltd
	G-BKFR	CP.301C Emeraude	I. N. Jennison/Barton
	G-BKFT	Cessna F.152-II	Hartmann Ltd/Booker
	G-BKFV	Rand KR.2	F. H. French/Swansea
	G-BKFW	P.56 Provost T.1	M. Howson
	G-BKFX	Colt 17A balloon	Colt Balloons Ltd
	G-BKFY	Beech C90 King Air	Medop Aviation Ltd/Staverton
	G-BKFZ	PA-28R Cherokee Arrow 200	P. H. de Havilland
	G-BKGA	M.S.892E Rallye 150GT	Harwoods of Essex Ltd
	G-BKGB	Jodel D.120	R. W. Greenwood
	G-BKGC	Maule M.6	Stol-Air Ltd/Sibson
	G-BKGD	Westland WG.30 Srs 100	British Airways Helicopters Ltd (G-BKBJ)/Beccles
	G-BKGE	Evans VP-2	D. Johnstone & L. Ward
	G-BKGF	Saxon II	Saxon Aircraft Co
	G-BKGH	Bell 205A-1	Rotair Ltd/Panshanger
	G-BKGK	PA-31T3 T1040	Vickers Shipbuilding & Engineering Ltd/ Barrow
	G-BKGL	Beech 18 (164)	G. A. Warner/Duxford
	G-BKGN	Cessna U.206-II	Shorrock Security Systems Ltd/ Blackpool
	G-BKGO	Piper J-3C-65 Cub	J. A. S. & I. K. Baldry
	G-BKGR	Cameron O-65 balloon	S. R. Bridge
	G-BKGT	SOCATA Rallye 110ST	Cambridge Discount Heating & Plumbing Ltd
	G-BKGW	Cessna F.152-II	Leicestershire Aero Club Ltd
	G-BKGX	Isaacs Fury	I. L. McMahon
	G-BKGZ	Bensen B.8	C. F. Simpson
	G-BKHA	WS.55 Whirlwind HAR.10 (XJ763)	D. Wilson
	G-BKHB	WS.55 Whirlwind HAR.10 (XJ407)	R. Windley
	G-BKHC	WS.55 Whirlwind HAR.10 (XP328)	R. Windley
	G-BKHD	Baby Great Lakes	P. J. Tanulak
	G-BKHE	Boeing 737-204	Britannia Airways Ltd *Sir Francis Chichester*/Luton
	G-BKHF	Boeing 737-204	Britannia Airways Ltd *Sir Alliot Verdon Roe*/Luton
	G-BKHG	Piper J-3C-65 Cub	K. G. Wakefield
	G-BKHH	Thunder Ax10-160Z balloon	R. Carr/France
	G-BKHI	SA Jetstream 3102	Peregrine Air Services Ltd/Edinburgh
	G-BKHJ	Cessna 182P	Augur Films Ltd/Shipdham
	G-BKHL	Thunder Ax9-140 balloon	R. Carr/France
	G-BKHM	Ben Air Sparrowhawk VL12/35	Ben Air Ltd
	G-BKHO	Boeing 737-2T4	Orion Airways Ltd/E. Midlands
	G-BKHP	P.56 Provost T.1 (WW397)	M. J. Crymble/Lyneham
	G-BKHR	Luton LA-4 Minor	R. J. Parkhouse
	G-BKHT	BAe 146-100	Dan-Air Services Ltd/Gatwick
	G-BKHV	Taylor JT.2 Titch	P. D. Holt
	G-BKHW	Stoddart-Hamilton Glassair SH.2RG	N. Clayton
	G-BKHX	Bensen B.8M	D. H. Greenwood
	G-BKHY	Taylor JT.1 Monoplane	J. Hall/Southampton
	G-BKHZ	Cessna F.172P	Birmingham Aerocentre Ltd
	G-BKIA	SOCATA TB.10 Tobago	Briseheart Finance Ltd/Redhill
	G-BKIB	SOCATA TB.9 Tampico	Bobbington Aviation Ltd/ Halfpenny Green
	G-BKIC	Cameron V-77 balloon	C. A. Butler
	G-BKIF	Fournier RF-6B	G. G. Milton
	G-BKIH	AS.355F1 Twin Squirrel	McAlpine Helicopters Ltd/Hayes
	G-BKII	Cessna F.172N	M. S. Knight/Goodwood
	G-BKIJ	Cessna F.172M	WTFA Ltd/Stapleford
	G-BKIK	Cameron DG-10 airship	Cameron Balloons Ltd
	G-BKIM	Unicorn UE.5A balloon	I. Chadwick & K. H. Turner

Reg.	Type	Owner or Operator	Notes
G-BKIN	Alon A.2A Aircoupe	Bob Crowe Aircraft Sales Ltd/Cranfield	
G-BKIP	Beech C90-1 King Air	Reckitt & Colman Products Ltd/ Norwich	
G-BKIR	Jodel D.117	R. Shaw & D. M. Hardaker	
G-BKIS	SOCATA TB.10 Tobago	E. Bilney Garage Ltd/Shipdham	
G-BKIT	SOCATA TB.9 Tampico	Martin Ltd/Biggin Hill	
G-BKIU	Colt 17A balloon	Robert Pooley Ltd	
G-BKIV	Colt 21A balloon	Colt Balloons Ltd	
G-BKIX	Cameron V-31 balloon	P. G. Dunnington	
G-BKIY	Thunder Ax3 balloon	A. Hornak	
G-BKIZ	Cameron V-31 balloon	A. P. Greathead	
G-BKJB	PA-18 Super Cub 135	Cormack (Aircraft Services) Ltd/ Glasgow	
G-BKJD	Bell 214ST	British Caledonian Helicopters Ltd/ Aberdeen	
G-BKJE	Cessna 172N	The G-BKJE Group	
G-BKJF	M.S.880B Rallye 100T	D. A. Smart & E. Jones	
G-BKJG	BN-2B Islander	Pilatus BN Ltd/Bembridge	
G-BKJH	BN-2B Islander	Pilatus BN Ltd/Bembridge	
G-BKJI	BN-2B Islander	Pilatus BN Ltd/Bembridge	
G-BKJK	BN-2B Islander	Pilatus BN Ltd/Bembridge	
G-BKJL	BN-2B Islander	Pilatus BN Ltd/Bembridge	
G-BKJM	BN-2B Islander	Pilatus BN Ltd/Bembridge	
G-BKJR	Hughes 269C	Thirsk Aero Services Ltd	
G-BKJS	Jodel D.120A	S. Walmsley/Blackpool	
G-BKJT	Cameron O-65 balloon	K. A. Ward	
G-BKJU	Sikorsky S-76A	Bristow Helicopters Ltd	
G-BKJW	PA-23 Aztec 2150	Alan Williams Entertainments Ltd	
G-BKJZ	G.159 Gulfstream 1	Rolls-Royce Ltd/Filton	
G-BKKB	Cessna A.188B	Northair Aviation Ltd/Leeds	
G-BKKI	Westland WG.30 Srs 100	British Airways Helicopters/Beccles	
G-BKKK	—	—	
G-BKKL	McCullogh J.2	R. J. Everett	
G-BKKM	Aeronca 7AC Champion	M. McChesney	
G-BKKN	Cessna 182R	Marvagraphic Ltd/Panshanger	
G-BKKO	Cessna 182R	B. & G. Jebson Ltd/Leeds	
G-BKKP	Cessna 182R	ISF Aviation Ltd/Leicester	
G-BKKR	Rand KR-2	D. R. Trouse	
G-BKKS	Mercury Dart Srs 1	B. A. Mills	
G-BKKY	BAe Jetstream 3102	Peregrine Air Services Ltd/Inverness	
G-BKKZ	Pitts S-1D Special	G. C. Masterson	
G-BKLB	S2R Thrush Commander	Ag-Air	
G-BKLC	Cameron V-56 balloon	B. J. & L. A. Workman	
G-BKLJ	Westland Scout AH.1 ★	J. E. Wilkie	
G-BKLM	Thunder Ax9-140 balloon	Balloon & Airship Co Ltd	
G-BKLO	Cessna F.172M	Reedtrend Ltd	
G-BKLP	Cessna F.172N	Reedtrend Ltd/Liverpool	
G-BKLR	—	—	
G-BKLS	SA.341G Gazelle	Helicopter Services Ltd	
G-BKLT	SA.341G Gazelle	Helicopter Services Ltd	
G-BKLU	SA.341G Gazelle	Helicopter Services Ltd	
G-BKLV	SA.341G Gazelle	Helicopter Services Ltd	
G-BKLW	SA.341G Gazelle	Helicopter Services Ltd	
G-BKLX	Colt 105A balloon	Colt Balloons Ltd	
G-BKLZ	Vinten-Wallis WA-116MC	W. Vinten Ltd	
G-BKMA	Mooney M.20J Srs 201	Clement Garage Ltd/Stapleford	
G-BKMB	Mooney M.20J Srs 201	R. Matthews	
G-BKMD	SC.7 Skyvan Srs 3	London Parachuting Ltd (G-BAHK)/ Pampisford	
G-BKME	SC.7 Skyvan Srs 3	Flightspares PLC/Southend	
G-BKMF	SC.7 Skyvan Srs 3	Flightspares PLC/Southend	
G-BKMG	Handley Page 0/400 replica	M. G. King	
G-BKMH	Flamboyant Ax7-65 balloon	Pepsi-Cola International Ltd/S. Africa	
G-BKMI	V.S.359 Spitfire HF VIII	Fighter Wing Display Ltd/Duxford	
G-BKMK	PA-38-112 Tomahawk	Cormack (Aircraft Services) Ltd/ Glasgow	
G-BKML	Cessna 210H	Bryant Bros Ltd/Southampton	
G-BKMM	Cessna 180K	Rationorth Ltd/Cardiff	
G-BKMN	BAe 146-100	Dan-Air Services Ltd (G-ODAN)/ Gatwick	
G-BKMP	—	—	
G-BKMR	Thunder Ax3 balloon	B. F. G. Ribbons	
G-BKMT	PA-32R-301 Saratoga SP	Hillary Investments Ltd	
G-BKMU	Short SD3-30	Guernsey Airlines Ltd	

Notes	Reg.	Type	Owner or Operator
	G-BKMW	Short SD3-30 Sherpa	Short Bros Ltd/Sydenham
	G-BKMX	Short SD3-60	Loganair Ltd/Glasgow
	G-BKNA	Cessna 421	Star Paper Ltd/Blackpool
	G-BKNB	Cameron V-42 balloon	S. A. Burnett
	G-BKND	Colt 56A balloon	Hot Air Balloon Co Ltd
	G-BKNE	PA-28-161 Warrior II	J. R. Coughlan/Andrewsfield
	G-BKNF	Westland Bell 47G-3B1	Rotair Ltd/Panshanger
	G-BKNH	Boeing 737-2E7	Dan-Air Services Ltd/Gatwick
	G-BKNI	GY-80 Horizon 160D	A. Hartigan & ptnrs/Fenland
	G-BKNJ	Grob G.109	Oxfordshire Sport Flying Ltd/Enstone
	G-BKNK	Rutan Vari-Eze	P. & S. H. Sutcliffe/Manchester
	G-BKNL	Cameron D-96 airship	Cameron Balloons Ltd
	G-BKNN	Cameron Minar E Pakistan balloon	Forbes Europe Ltd/France
	G-BKNO	Monnet Sonerai IIL	S. Tattersfield & K. Bailey
	G-BKNX	SA.102.5	G. D. Horn
	G-BKNY	Bensen B.8M-P-VW	D. A. C. MacCormack
	G-BKNZ	CP.301A Emeraude	R. Evernden/Barton
	G-BKOA	M.S.893E Rallye 180GT	Cheshire Flying Services Ltd/ Manchester
	G-BKOP	Barnes 65 balloon	Robert Pooley Ltd
	G-BKOR	Barnes 77 balloon	Robert Pooley Ltd
	G-BKOS	P.56 Provost T.51	J. G. Cassidy/Woodvale
	G-BKOT	Wassmer WA.81 Piranha	B. D. Denbelbeiss
	G-BKOU	P.84 Jet Provost T.3 (XN637)	A. Topen/Cranfield
	G-BKOV	Jodel DR.220A	E. H. Ellis & G. C. Winter
	G-BKOW	Cameron 77A balloon	Hot Air Ballon Co Ltd
	G-BKOX	—	—
	G-BKOY	Barnes 105 balloon	Robert Pooley Ltd
	G-BKOZ	Barnes 77 balloon	Robert Pooley Ltd
	G-BKPA	Hoffman H-36 Dimona	Airmark Aviation Ltd/Booker
	G-BKPB	Aerosport Scamp	R. Scroby
	G-BKPC	Cessna A.185F	Black Knights Parachute Centre
	G-BKPD	Viking Dragonfly	P. E. J. Sturgeon
	G-BKPE	Jodel DR.250/160	H. Best-Devereux/Panshanger
	G-BKPG	Luscombe Rattler Strike	Luscombe Aircraft Ltd/Lympne
	G-BKPH	Luscombe Valiant	Luscombe Aircraft Ltd/Lympne
	G-BKPK	John McHugh Gyrocopter	J. C. McHugh
	G-BKPM	Schempp-Hirth HS.5 Nimbus 2	J. L. Rolls
	G-BKPN	Cameron N-77 balloon	Flamboyant Promotions Ltd
	G-BKPS	AA-5B Tiger	Eyewitness Ltd/Southampton
	G-BKPT	M.H.1521M Broussard	Wessex Aviation & Transport Ltd
	G-BKPU	M.H.1521M Broussard	M. G. Pickering/Thruxton
	G-BKPV	Stevex 250.1	A. F. Stevens
	G-BKPW	Boeing 767-204	Britannia Airways Ltd *Sir Winston Churchill*/Luton
	G-BKPX	Jodel D.120A	P. A. Davey/Cranwell
	G-BKPY	Saab 91B/2 Safir (56321)★	Newark Air Museum Ltd
	G-BKPZ	Pitts S-1 Special	P. G. Kynsey & J. Harper
	G-BKRA	AT-6G Harvard (51-15227)	Pulsegrove Ltd/Shoreham
	G-BKRB	Cessna 172N	Saunders Caravans Ltd/Staverton
	G-BKRD	Cessna 320E	Thackwell Motorsports Ltd/Fairoaks
	G-BKRE	—	—
	G-BKRF	PA-18 Super Cub 95	K. M. Bishop
	G-BKRG	Beechcraft C-45G	Aces High Ltd/North Weald
	G-BKRH	Brugger MB.2 Colibri	M. R. Benwell
	G-BKRI	Cameron V-77 balloon	J. R. Lowe & R. J. Fuller
	G-BKRJ	Colt 105A balloon	Owners Abroad Group PLC
	G-BKRK	SNCAN Stampe SV-4C	J. M. Alexander & ptnrs/Aberdeen
	G-BKRL	Designability Leopard	Chichester-Miles Consultants Ltd
	G-BKRM	Boeing 757-236	Air Europe Ltd/British Airways
	G-BKRN	Beechcraft D.18S	Scottish Aircraft Collection/Perth
	G-BKRP	S2R Thrush Commander	D. W. Craig
	G-BKRR	Cameron N-56 balloon	S. L. G. Williams
	G-BKRS	Cameron V-56 balloon	M. Z. & L. A. Rawson
	G-BKRT	PA-34-220T-3 Seneca	Paucristar Ltd
	G-BKRU	Ensign Crossley Racer	M. Crossley
	G-BKRV	Hovey Beta Bird	A. V. Francis
	G-BKRW	Cameron 0-160 balloon	Bondbaste Ltd
	G-BKRX	Cameron 0-160 balloon	Bondbaste Ltd
	G-BKRY	Thunder Ax6-56Z balloon	Thunder Balloons Ltd
	G-BKRZ	Dragon 77 balloon	Anglia Balloon School Ltd
	G-BKSB	Cessna T.310Q	P. S. King
	G-BKSC	Saro Skeeter AOP.12 (XN351)	R. A. L. Falconer

Reg.	Type	Owner or Operator	Notes
G-BKSD	Colt 56A balloon	M. J. & G. C. Casson	
G-BKSE	Quickie Quickie	C. G. Taylor & ptnrs	
G-BKSG	Hoffman H-36 Dimona	B. J. Wilson & F. C. Y. Cheung/H. Kong	
G-BKSH	Colt 21A balloon	Greenham Trading Ltd	
G-BKSJ	Cameron N-108 balloon	Cameron Balloons Ltd	
G-BKSK	Quickie Q-2	M. J. Sullivan/Elstree	
G-BKSO	Cessna 421C	Anglian Double Glazing Co Ltd/Norwich	
G-BKSP	Schleicher ASK.14	M. R. Shelton	
G-BKSR	Cessna 550 Citation II	Osiwell Ltd/Leavesden	
G-BKSS	Jodel D.150	D. H. Wilson-Spratt	
G-BKST	Rutan Vari-Eze	R. Towle	
G-BKSU	Short SD3-30	Fairflight Ltd	
G-BKSX	SNCAN Stampe SV-4C	R. J. Partridge/Denham	
G-BKSZ	Cessna P.210N	Clark Mast Ltd/Sandown	
G-BKTA	PA-18 Super Cub 95	J. I. Evans/Southend	
G-BKTB	Mooney M.20E	A. T. Bruce	
G-BKTC	Pitts S-2E	E. B. Bray	
G-BKTF	H.S.125 Srs 800A	British Aerospace PLC/Chester	
G-BKTG	Enstrom F-280 Shark	Trowell Plant Sales Ltd/Tollerton	
G-BKTH	CCF Hawker Sea Hurricane IB (Z7015)	Shuttleworth Trust/Duxford	
G-BKTJ	Cessna 404 Titan	Donington Aviation Ltd/E. Midlands	
G-BKTM	PZL SZD-45A Ogar	Repclif Aviation Ltd/Liverpool	
G-BKTN	BAe Jetstream 3102	McAlpine Aviation Ltd/Luton	
G-BKTO	Beech 58P Baron	Gold Key Trust Ltd	
G-BKTR	Cameron V-77 balloon	G. F. & D. D. Bouten	
G-BKTS	Cameron 0-65 balloon	C. H. Pearce & Sons (Contractors) Ltd	
G-BKTT	Cessna F.152	Stapleford Flying Club Ltd	
G-BKTU	Colt 56A balloon	E. Ten Houten	
G-BKTV	Cessna F.152	London Flight Centre Ltd/Stapleford	
G-BKTW	Cessna 404 Titan II	Hawk Aviation Ltd (G-WTVE)/ E. Midlands	
G-BKTY	SOCATA TB.10 Tobago	R. J. Alford/Biggin Hill	
G-BKTZ	Slingsby T.67M Firefly	Slingsby Aviation Ltd (G-SFTV)/ Kirkbymoorside	
G-BKUB	Bell 206A JetRanger	Rotair Ltd/Panshanger	
G-BKUC	Mudry/CAARP CAP.10B	D. M. Britten	
G-BKUE	SOCATA TB.9 Tampico	Martin Ltd/Biggin Hill	
G-BKUI	D.31 Turbulent	A. Onoufriou/Breighton	
G-BKUJ	Thunder Ax6-56 balloon	J. M. Albury	
G-BKUM	AS.350B Ecureuil	T. W. Walker Ltd/Tees-side	
G-BKUN	Cessna 404 Titan	Berrard Ltd/Blackbushe	
G-BKUP	—	—	
G-BKUR	CP.301A Emeraude	P. Gilmour/Perth	
G-BKUS	Bensen B.80	I. J. Lawson	
G-BKUT	M.S.880B Rallye Club	J. J. Hustwitt/Bodmin	
G-BKUU	Thunder Ax7-77-1 balloon	City of London Balloon Group	
G-BKUV	—	—	
G-BKUX	Beech C90 King Air	Marchwiel Aviation Ltd/ Halfpenny Green	
G-BKUY	BAe Jetstream 3102	McAlpine Aviation Ltd/Luton	
G-BKUZ	Zenair CH.250	K. Morris	
G-BKVA	SOCATA Rallye 180T	R. Evans	
G-BKVB	SOCATA Rallye 110ST	Martin Ltd/Biggin Hill	
G-BKVC	SOCATA TB.9 Tampico	Martin Ltd/Biggin Hill	
G-BKVE	Rutan Vari-Eze	H. R. Rowley (G-EZLT)	
G-BKVF	FRED Srs 3	N. E. Johnson	
G-BKVG	Scheibe SF.25E Super Falke	Westland Flying Club Ltd/Yeovil	
G-BKVI	—	—	
G-BKVJ	Colt 21A balloon	Colt Balloons Ltd	
G-BKVK	Auster AOP.9	R. J. Starling & C. W. Monsell/Norwich	
G-BKVL	Robin DR.400/160	The Cotswold Aero Club Ltd/Staverton	
G-BKVM	PA-18 Super Cub 150	W. R. C. Foyle/Luton	
G-BKVN	PA-23 Aztec 250F	VG Instruments Ltd/Headcorn	
G-BKVO	Pietenpol Aircamper	M. J. Honeychurch	
G-BKVP	Pitts S-21D Special	P. J. Leggo	
G-BKVR	PA-28 Cherokee 140	Page Aviation Ltd/Stapleford	
G-BKVS	Bensen B.8M	V. Scott	
G-BKVT	PA-23 Aztec 250E	T. S. Grimshaw Ltd (G-HARV)/ Cardiff	
G-BKVV	Beech 95-B55 Baron	L. Mc. G. Tulloch	
G-BKVW	Airtour 56 balloon	Airtour Balloon Co Ltd	
G-BKVX	Airtour 56 balloon	Airtour Balloon Co Ltd	
G-BKVY	Airtour 31 balloon	Airtour Balloon Co Ltd	

Notes	Reg.	Type	Owner or Operator
	G-BKVZ	Boeing 767-204	Britannia Airways Ltd *Earl Mountbatten of Burma*/Luton
	G-BKWA	Cessna 404 Titan	Hawk Aviation Ltd (G-BELV)/ E. Midlands
	G-BKWB	EMB-110P2 Bandeirante	Euroair Ltd (G-CHEV)/Gatwick
	G-BKWD	Taylor JT.2 Titch	E. Shouler
	G-BKWE	Colt 17A balloon	Hot-Air Balloon Co Ltd
	G-BKWF	Jodel DR.1051	L. L. Vickers & D. A. C. Clissett
	G-BKWG	PZL-104 Wilga	Anglo-Polish Sailplanes Ltd/Booker
	G-BKWH	Cessna F.172P	W. H. & J. Rogers Group Ltd/Cranfield
	G-BKWI	Pitts S-2A	R. A. Seeley/Denham
	G-BKWO	—	—
	G-BKWR	Cameron V-65 balloon	April & Gilbert Games Photographers
	G-BKWT	Airbus A.310-203	British Caledonian Airways Ltd *John Logie Baird*/Gatwick
	G-BKWU	Airbus A.310-203	British Caledonian Airways Ltd *Robert Watson-Watt*/Gatwick
	G-BKWW	Cameron 0-77 balloon	A. M. Marten
	G-BKWY	Cessna F.152	Cambridge Aero Club
	G-BKWZ	—	—
	G-BKXA	Robin R.2100	G. J. Anderson & ptnrs
	G-BKXB	Steen Skybolt	P. W. Scott
	G-BKXC	Cameron V-77 balloon	P. Sarretti
	G-BKXD	SA.365N Dauphin 2	Bond Helicopters Ltd/Bourn
	G-BKXE	SA.365N Dauphin 2	Bond Helicopters Ltd/Bourn
	G-BKXF	PA-28R Cherokee Arrow 200	P. L. Brunton
	G-BKXG	Cessna T.303	Lampson Group Ltd
	G-BKXH	Robinson R-22	Fastflight Ltd/Denham
	G-BKXI	Cessna T.303	Crane Investments Ltd/Guernsey
	G-BKXK	SA.365N Dauphin 2	The Marconi Co Ltd
	G-BKXL	Cameron Bottle 70 balloon	Cameron Balloons Ltd
	G-BKXM	Colt 17A balloon	R. G. Turnbull
	G-BKXN	ICA IS-28M2A	British Aerospace PLC
	G-BKXO	Rutan Long-Ez	P. J. Wareham
	G-BKXP	Auster AOP.6	R. Skingley
	G-BKXR	D.31 Turbulent	S. B. Churchill
	G-BKXS	Colt 56A balloon	Hot-Air Balloon Co Ltd
	G-BKXT	Cameron D-50 airship	Cameron Balloons Ltd
	G-BKXU	Cameron Dairy Queen Cone balloon	Cameron Balloons Ltd
	G-BKXW	TB-25J Mitchell ★	Aces High Ltd/Duxford
	G-BKXX	Cameron V-65 balloon	A. J. Legg & C. H. Harbord
	G-BKXY	Westland WG.30 Srs 100-60	Westland Helicopters Ltd/Yeovil
	G-BKYA	Boeing 737-236	British Airways *River Derwent*/ Heathrow
	G-BKYB	Boeing 737-236	British Airways *River Stour*/Heathrow
	G-BKYC	Boeing 737-236	British Airways *River Wye*/Heathrow
	G-BKYD	Boeing 737-236	British Airways *River Conway*/ Heathrow
	G-BKYE	Boeing 737-236	British Airways *River Lagan*/Heathrow
	G-BKYF	Boeing 737-236	British Airways *River Spey*/Heathrow
	G-BKYG	Boeing 737-236	British Airways *River Exe*/Heathrow
	G-BKYH	Boeing 737-236	British Airways *River Dart*/Heathrow
	G-BKYI	Boeing 737-236	British Airways *River Waveney*/ Heathrow
	G-BKYJ	Boeing 737-236	British Airways *River Neath*/Heathrow
	G-BKYK	Boeing 737-236	British Airways *River Foyle*/Heathrow
	G-BKYL	Boeing 737-236	British Airways *River Ayr*/Heathrow
	G-BKYM	Boeing 737-236	British Airways *River Cam*/Heathrow
	G-BKYN	Boeing 737-236	British Airways *River Isis*/Heathrow
	G-BKYO	Boeing 737-236	British Airways *River Kennet*/ Heathrow
	G-BKYP	Boeing 737-236	British Airways *River Ystwyth*/ Heathrow
	G-BKYR	—	British Airways/Heathrow
	G-BKYS	—	British Airways/Heathrow
	G-BKYT	—	British Airways/Heathrow
	G-BKYU	—	British Airways/Heathrow
	G-BKYV	—	British Airways/Heathrow
	G-BKYW	—	British Airways/Heathrow
	G-BKYX	—	British Airways/Heathrow
	G-BKYY	—	British Airways/Heathrow
	G-BKYZ	—	British Airways/Heathrow
	G-BKZA	Cameron N-77 balloon	University of Bath Students Union

Reg.	Type	Owner or Operator	Notes
G-BKZB	Cameron V-77 balloon	A. J. Montgomery	
G-BKZC	Cessna A.152	Montaguis Ltd/Kuwait	
G-BKZD	Cessna A.152	Montaguis Ltd/Kuwait	
G-BKZE	AS.332L Super Puma	British Airways Helicopters/Aberdeen	
G-BKZF	Cameron V-56 balloon	G. M. Hobster	
G-BKZG	AS.332L Super Puma	British Airways Helicopters/Aberdeen	
G-BKZH	AS.332L Super Puma	British Airways Helicopters/Aberdeen	
G-BKZI	Bell 206B JetRanger	J. M. V. Butterfield	
G-BKZJ	Bensen B.8V	S. H. Kirkby	
G-BKZK	Robinson R-22A	Helicopters (W. Midlands) Ltd	
G-BKZL	Colt AS-42 airship	Colt Balloons Ltd	
G-BKZM	Isaacs Fury II (K2060)	R. J. Smyth	
G-BKZT	FRED Srs 2	A. E. Morris	
G-BKZU	Colt 105A balloon	Colt Balloons Ltd	
G-BKZV	Bede BD-4A	—/Headcorn	
G-BKZW	Beech C90 King Air	National Airways/Elstree	
G-BKZY	Cameron N-77 balloon	W. Counties Automobile Co Ltd	
G-BLAA	Fournier RF-5	A. D. Wren/Southend	
G-BLAC	Cessna FA.152	Lancashire Aero Club/Barton	
G-BLAD	Thunder Ax7-77-1 balloon	Balloon & Airship Co Ltd	
G-BLAF	Stolp V-Star SA.900	J. E. Malloy	
G-BLAG	Pitts S-1D Special	S. A. W. Becker	
G-BLAH	Thunder Ax7-77-1 balloon	T. Donnelly	
G-BLAI	Monnet Sonerai IIL	T. Simpson	
G-BLAJ	Pazmany PL.4A	J. D. LePine	
G-BLAM	Jodel DR.360	B. F. Baldock	
G-BLAN	SA.341G Gazelle	Specialist Flying Training Ltd/Carlisle	
G-BLAO	SA.341G Gazelle	Specialist Flying Training Ltd/Carlisle	
G-BLAR	—	—	
G-BLAS	V.S.361 Spitfire F.IX	Aero Vintage Ltd	
G-BLAT	Jodel D.150	R. Tyler	
G-BLAU	Bell 47G-4	Bridge Helicopters Ltd	
G-BLAW	PA-28-181 Archer II	Lion Systems Developments Ltd	
G-BLAX	Cessna FA.152	Shoreham Flight Simulation Ltd/ Bournemouth	
G-BLAY	Robin HR.100/200B	B. A. Mills	
G-BLAZ	Cessna 421B	R. McMillan Aviation Ltd/Biggin Hill	
	The G-BLBA-BZ batch has been reserved for British Airways.		
G-BLCA	Bell 206B JetRanger	RMH Stainless Ltd	
G-BLCC	Thunder Ax7-77Z balloon	P. Hassell Ltd	
G-BLCF	EAA Acrosport 2	M. J. Watkins & ptnrs	
G-BLCG	SOCATA TB.10 Tobago	P.A.C.K. Enterprise (Sussex) Ltd (G-BHES)	
G-BLCH	Colt 56D balloon	A. D. McCutcheon	
G-BLCI	EAA Acrosport	P. A. Falter/Biggin Hill	
G-BLCK	V.S.361 Spitfire F.IX	Aero Vintage Ltd	
G-BLCM	SOCATA TB.9 Tampico	Repclif Aviation Ltd/Liverpool	
G-BLCT	Jodel DR.220 2+2	H. W. Jemmett	
G-BLCU	Scheibe SF.25B Falke	B. Lumb & ptnrs	
G-BLCV	Hoffman H-36 Dimona	P. E. Villiers	
G-BLCW	Evans VP-1	K. D. Pearce	
G-BLCX	Glaser-Dirks DG.400	B. A. Eastwell	
G-BLCY	Thunder Ax7-65Z balloon	Thunder Balloons Ltd	
G-BLCZ	Cessna 441	Northair Aviation Ltd/Leeds	
G-BLDA	SOCATA Rallye 110ST	Martin Ltd/Biggin Hill	
G-BLDB	Taylor JT.1 Monoplane	C. J. Bush	
G-BLDC	K&S Jungster 1	C. A. Laycock	
G-BLDD	WAG-Aero Cuby AcroTrainer	C. A. Laycock	
G-BLDE	Boeing 737-2E7	Dan-Air Services Ltd/Gatwick	
G-BLDF	Bell 47G-5	Helicopter Farming Ltd	
G-BLDG	PA-25 Pawnee 260C	L. G. & M. Appelbeck	
G-BLDH	BAC One-Eleven 475EZ	McAlpine Aviation Ltd/Luton	
G-BLDJ	PA-28-161 Warrior II	SFT Aviation Ltd/Bournemouth	
G-BLDK	Robinson R-22	William Towns Ltd	
G-BLDL	Cameron Truck 56 balloon	Cameron Balloons Ltd	
G-BLDM	Hiller UH-12E	G. & S. G. Neal (Helicopters) Ltd	
G-BLDN	Rand KR-2	R. Y. Kendal	
G-BLDO	BAe Jetstream 3102	McAlpine Aviation Ltd/Luton	
G-BLDP	Slingsby T.67M Firefly	Cavendish Aviation Ltd/Netherthorpe	
G-BLDT	BN-2B Islander	Pilatus BN Ltd/Bembridge	
G-BLDU	BN-2B Islander	Pilatus BN Ltd/Bembridge	

Notes	Reg.	Type	Owner or Operator
	G-BLDV	BN-2B Islander	Pilatus BN Ltd/Bembridge
	G-BLDW	BN-2B Islander	Pilatus BN Ltd/Bembridge
	G-BLDX	BN-2B Islander	Pilatus BN Ltd/Bembridge
	G-BLDY	Bell 212	Bristow Helicopters Ltd
	G-BLDZ	Cameron N-77 balloon	Flamboyant Promotions Ltd
	G-BLEB	Colt 69A balloon	I. R. M. Jacobs
	G-BLEC	BN-2B-27 Islander	LEC Refrigeration PLC (G-BJBG)
	G-BLEI	BN-2B-26 Islander	Transavia Ltd/Luton
	G-BLEJ	PA-28-161 Warrior II	Express Aviation Services Ltd/ Biggin Hill
	G-BLEL	Ax7-77-245 balloon	T. S. Price
	G-BLEP	Cameron V-65 balloon	D. Chapman
	G-BLER	Slingsby T.67M Firefly	I. C. Fallows
	G-BLES	SA.750 Acroduster Too	W. G. Hosie & ptnrs
	G-BLET	Thunder Ax7-77-1 balloon	Colt Balloons Ltd
	G-BLEU	Bensen B.80	B. D. S. Preston
	G-BLEV	AS.355F Twin Squirrel	Chivas Bros Ltd/Glasgow
	G-BLEW	Cessna F.182Q	Interair Aviation Ltd/Bournemouth
	G-BLEY	SA.365N Dauphin 2	Bond Helicopters Ltd/Bourn
	G-BLEZ	SA.365N Dauphin 2	Bond Helicopters Ltd/Bourn
	G-BLFC	Edgley EA-7 Optica	Edgley Aircraft Ltd/Old Sarum
	G-BLFE	Cameron Sphinx SS balloon	Forbes Europe Inc
	G-BLFF	Cessna F.172M	Air Advertising UK Ltd
	G-BLFI	PA-28-181 Archer II	Leavesden Flight Centre Ltd
	G-BLFJ	F.27 Friendship Mk 100	Air UK Ltd (G-OMAN/G-SPUD)/Norwich
	G-BLFL	Douglas C-47B	Aces High Ltd/North Weald
	G-BLFT	P.56 Provost T.1	B. W. H. Parkhouse
	G-BLFV	Cessna 182R	Goddard Kay Rogers & Associates Ltd/ Booker
	G-BLFW	AA-5 Traveler	M. A. C. Stephenson & P. Wilkinson
	G-BLFY	Cameron V-77 balloon	A. N. F. Pertwee
	G-BLFZ	PA-31-310C Turbo Navajo	City Flight Ltd/Biggin Hill
	G-BLGB	Short SD3-60	Loganair Ltd/Glasgow
	G-BLGG	Short SD3-30	Short Bros Ltd/Sydenham
	G-BLGH	Robin DR.300/180R	Booker Gliding Club Ltd
	G-BLGI	—	—
	G-BLGK	Cameron Propane Bottle balloon	Cameron Balloons Ltd
	G-BLGL	—	—
	G-BLGM	Cessna 425	Northair Aviation Ltd/Leeds
	G-BLGN	Skyhawk Gyroplane	S. M. Hawkins
	G-BLGO	Bensen B.8M	R. J. Bent
	G-BLGP	—	—
	G-BLGR	Bell 47G-4A	Land Air Ltd
	G-BLGS	SOCATA Rallye 180T	Lasham Gliding Society Ltd
	G-BLGT	PA-18 Super Cub 95	T. A. Reed
	G-BLGV	Bell 206B JetRanger	Helicrops Ltd
	G-BLGW	F.27 Friendship Mk 200	Air UK Ltd/Norwich
	G-BLGX	Thunder Ax7-65 balloon	Harper & Co (Glasgow) Ltd
	G-BLGY	Grob G.109B	T. I. Pale-Harris & K.N.C. (One) Ltd
	G-BLHA	Thunder Ax10-160 balloon	Thunder Balloons Ltd
	G-BLHB	Thunder Ax10-160 balloon	Thunder Balloons Ltd
	G-BLHC	BAe Jetstream 3102	McAlpine Aviation Ltd/Luton
	G-BLHD	BAC One-Eleven 492GM	McAlpine Aviation Ltd/Luton
	G-BLHE	Pitts S-1E Special	W. R. Penaluna
	G-BLHF	Nott/Cameron ULD.2 balloon	J. R. P. Nott
	G-BLHG	Hoffman H-36 Dimona	C. H. Dobson
	G-BLHH	Jodel DR.315	G. G. Milton
	G-BLHI	—	—
	G-BLHJ	Cessna F.172P	P. P. D. Howard-Johnston/Edinburgh
	G-BLHK	Colt 105A balloon	Hale Hot-Air Balloon Club
	G-BLHM	PA-18 Super Cub 95	J. S. Simmonds
	G-BLHN	Robin HR.100/285	H. M. Bouquiere/Biggin Hill
	G-BLHO	AA-5A Cheetah	Triple Airways/Southampton
	G-BLHR	GA-7 Cougar	G. A. F. Tilley
	G-BLHS	Bellanca 7ECA Citabria	C. H. A. Bott
	G-BLHT	Varga 2150A Kachina	G. G. L. Thomas/Swansea
	G-BLHU	—	—
	G-BLHV	—	—
	G-BLHW	Varga 2150A Kachina	Willoughby Farms Ltd
	G-BLHX	—	—
	G-BLHY	—	—
	G-BLHZ	Varga 2150A Kachina	MLP Aviation Ltd/Elstree
	G-BLID	D.H.112 Venom FB.50	Aces High Ltd/Duxford
	G-BLIE	D.H.112 Venom FB.50	Air Charter (Scotland) Ltd/Glasgow

Reg.	Type	Owner or Operator	Notes
G-BLIF	D.H.112 Venom FB.50	Aces High Ltd/Duxford	
G-BLIG	Cameron V-65 balloon	W. Davison	
G-BLIH	PA-18 Super Cub 135	I. R. F. Hammond	
G-BLIK	Wallis WA-116/F/S	K. H. Wallis	
G-BLIO	Cameron R-42 gas balloon	Cameron Balloons Ltd	
G-BLIP	Cameron N-77 balloon	Systems 80 Group Ltd	
G-BLIR	Cessna 441	Northair Aviation Ltd/Leeds	
G-BLIT	Thorp T-18 CW	A. J. Waller	
G-BLIV	Cameron 0-105 balloon	A. M. Thompson	
G-BLIW	P.56 Provost T.51	Pulsegrove Ltd/Shoreham	
G-BLIX	Saro Skeeter Mk 12	A. P. Nowicki	
G-BLIY	M.S.892A Rallye Commodore	L. Everex & Sons Ltd	
G-BLIZ	PA-46-301P Malibu	Malibu Flying Services Ltd	
G-BLJD	Glaser-Dirks DG.400	P. A. Hearne & ptnrs	
G-BLJE	A-Bell 206B JetRanger	Window Boxes Ltd & Veritair Ltd	
G-BLJF	Cameron 0-65 balloon	D. Fowler	
G-BLJG	Cameron N-105 balloon	New DFS Furniture Ltd	
G-BLJH	Cameron N-77 balloon	A. J. Clarke & J. M. Hallam	
G-BLJI	Colt 105A balloon	Colt Balloons Ltd	
G-BLJJ	Cessna 305 Bird Dog	P. Dawe	
G-BLJK	Evans VP-2	R. R. Pierce	
G-BLJM	Beech 95-B55 Baron	Advanced Marketing Management Ltd/ Elstree	
G-BLJN	Nott/Cameron ULD balloon	J. R. P. Nott	
G-BLJO	Cessna F.152	M. J. Endacott	
G-BLJP	Cessna F.150L	F. & S. E. Horridge/Lasham	
G-BLJW	Glaser-Dirks DG.400	B. J. Willson	
G-BLJX	Bensen B.8M	R. Snow	
G-BLJY	Sequoia F.8L Falco	K. Morris	
G-BLJZ	PA-31P Navajo	Avionics Research Ltd	
G-BLKA	D.H.112 Venom FB-54 (WR149)	A. Topen/Cranfield	
G-BLKB	Boeing 737-3T5	Orion Airways Ltd/E. Midlands	
G-BLKC	Boeing 737-3T5	Orion Airways Ltd/E. Midlands	
G-BLKD	Boeing 737-3T5	Orion Airways Ltd/E. Midlands	
G-BLKE	Bgeing 737-3T5	Orion Airways Ltd/E. Midlands	
G-BLKF	Thunder Ax10-160 balloon	Thunder Balloons Ltd	
G-BLKG	Thunder Ax10-160 balloon	Thunder Balloons Ltd	
G-BLKH	Thunder Ax10-160 balloon	Thunder Balloons Ltd	
G-BLKI	Thunder Ax10-160 balloon	Thunder Balloons Ltd	
G-BLKJ	Thunder Ax7-65 balloon	D. T. Watkins	
G-BLKK	Evans VP-1	R. W. Burrows	
G-BLKL	D.31 Turbulent	R. G. Halliam	
G-BLKM	Jodel DR.1051	P. Earnshaw	
G-BLKP	BAe Jetstream 3102	British Aerospace Ltd/Prestwick	
G-BLKR	Westland WG.30 Srs 100	Westland Helicopters Ltd/Yeovil	
G-BLKS	—	—	
G-BLKT	—	—	
G-BLKU	Colt 56 SS balloon	Hot-Air Balloon Co Ltd	
G-BLKV	Boeing 767-204	Britannia Airways Ltd/Luton	
G-BLKW	Boeing 767-204	Britannia Airways Ltd/Luton	
G-BLKX	PA-38-112 Tomahawk	Panshanger School of Flying Ltd	
G-BLKY	Beech 95-58	Kebbell Holdings Ltd/Leavesden	
G-BLKZ	Pilatus P2-05	Fairoaks Aviation Services Ltd/ Blackbushe	
G-BLLA	Bensen B.8M	K. T. Donaghey	
G-BLLB	Bensen B.8M	D. H. Moss	
G-BLLC	Beech 200 Super King Air	British Airways (G-LKOW)/Heathrow	
G-BLLD	Cameron 0-77 balloon	J. P. Edge	
G-BLLE	Cameron 60 Burger King SS balloon	Burger King UK Ltd	
G-BLLF	Westland WG.30 Srs 100-60	Westland Helicopters Ltd/Yeovil	
G-BLLG	—	—	
G-BLLH	Jodel DR.220A 2+2	A. L. Burton & J. Sumner	
G-BLLM	PA-23 Aztec 250E	C. & M. Thomas (G-BBNM)/Swansea	
G-BLLN	PA-18 Super Cub 95	W. H. Pelly	
G-BLLO	PA-18 Super Cub 95	D. G. & M. G. Marketts	
G-BLLP	Slingsby T.67B	Slingsby Aviation Ltd/Kirkbymoorside	
G-BLLR	Slingsby T.67B	Slingsby Aviation Ltd/Kirkbymoorside	
G-BLLS	Slingsby T.67B	Slingsby Aviation Ltd/Kirkbymoorside	
G-BLLT	AA-5B Tiger	Alpha Welding & Engineering Ltd	
G-BLLU	Cessna 421C	J. Rowe/Manchester	
G-BLLV	Slingsby T.67B	Fleet Delta Ltd/Norwich	
G-BLLW	Colt 56B balloon	J. C. Stupples	
G-BLLX	SA.315B Lama	Dollar Air Services Ltd/Coventry	

Notes	Reg.	Type	Owner or Operator
	G-BLLY	Cessna 340A	David Sutton Motorsport Ltd/Leeds
	G-BLLZ	Rutan Long-Ez	G. E. Relf & ptnrs
	G-BLMA	Zlin 326 Trener Master	G. C. Masterson
	G-BLMC	Avro 698 Vulcan B.2 ★	Aeropark/E. Midlands
	G-BLMD	Robinson R-22	Sloane Helicopters Ltd/Luton
	G-BLME	Robinson R-22	Kindell Motors
	G-BLMF	—	—
	G-BLMG	Grob G.109B	Bowker Air Services Ltd/Rush Green
	G-BLMI	PA-18 Super Cub 95	J. D. Atkinson & G. V. Horfield
	G-BLML	F.27 Friendship Mk 200	Air UK Ltd/Norwich
	G-BLMN	Rutan Long-Ez	N. J. & R. A. Farrington
	G-BLMO	Cameron 60 Demestica Bottle SS balloon	Cameron Balloons Ltd
	G-BLMP	PA-17 Vagabond	M. Austin
	G-BLMR	PA-18 Super Cub 150	Cormack (Aircraft Services) Ltd/ Glasgow
	G-BLMS	PA-19 Super Cub 90	Cormack (Aircraft Services) Ltd/ Glasgow
	G-BLMT	PA-18 Super Cub 135	Cormack (Aircraft Services) Ltd/ Glasgow
	G-BLMU	Isaacs Fury II	P. R. Skeels
	G-BLMV	—	—
	G-BLMW	Nipper T.66 RA45/3	S. L. Millar
	G-BLMX	Cessna FR.172H	A. J. Fuller & ptnrs
	G-BLMY	Grob G.109B	Soaring (Oxford) Ltd/Enstone
	G-BLMZ	Colt 105A balloon	M. J. Hutchins
	G-BLNA	Beech B90 King Air	Navigations Ltd (G-BHGT/G-AWWK)/ Elstree
	G-BLNB	V.802 Viscount	British Air Ferries (G-AOHV)/Southend
	G-BLNC	BN-2B Islander	Pilatus BN Ltd/Bembridge
	G-BLND	BN-2B Islander	Pilatus BN Ltd/Bembridge
	G-BLNF	BN-2B Islander	Pilatus BN Ltd/Bembridge
	G-BLNG	BN-2B Islander	Pilatus BN Ltd/Bembridge
	G-BLNH	BN-2B Islander	Pilatus BN Ltd/Bembridge
	G-BLNI	BN-2B Islander	Pilatus BN Ltd/Bembridge
	G-BLNJ	BN-2B Islander	Pilatus BN Ltd/Bembridge
	G-BLNK	BN-2B Islander	Pilatus BN Ltd/Bembridge
	G-BLNL	BN-2B Islander	Pilatus BN Ltd/Bembridge
	G-BLNM	BN-2B Islander	Pilatus BN Ltd/Bembridge
	G-BLNN	PA-38-112 Tomahawk	Nalson Aviation Ltd/Biggin Hill
	G-BLNO	FRED Srs 3	L. W. Smith
	G-BLNP	—	—
	G-BLNS	BN-2B Islander	Pilatus BN Ltd/Bembridge
	G-BLNT	BN-2B Islander	Pilatus BN Ltd/Bembridge
	G-BLNU	BN-2B Islander	Pilatus BN Ltd/Bembridge
	G-BLNV	BN-2B Islander	Pilatus BN Ltd/Bembridge
	G-BLNW	BN-2B Islander	Pilatus BN Ltd/Bembridge
	G-BLNX	BN-2B Islander	Pilatus BN Ltd/Bembridge
	G-BLNY	BN-2B Islander	Pilatus BN Ltd/Bembridge
	G-BLNZ	BN-2B Islander	Pilatus BN Ltd/Bembridge
	G-BLOA	V.806 Viscount	British Air Ferries Ltd (G-AOYJ)/ Southend
	G-BLOB	Colt Ax4-31A balloon	Jacques W. Soukup Ltd
	G-BLOC	Rand KR-2	F. Woodhouse
	G-BLOD	Colt Ax7-77A balloon	Thunder & Colt Ltd
	G-BLOE	PA-31-350 Navajo Chieftain	Gill Aviation Ltd (G-NITE)/Newcastle
	G-BLOF	—	—
	G-BLOG	Cameron 0-77 balloon	British Airtours Ltd
	G-BLOJ	Thunder Ax7-77 Srs 1 balloon	J. W. Cato
	G-BLOK	Colt Ax7-77A balloon	Thunder & Colt Ltd
	G-BLOL	SNCAN Stampe SV-4A	Skysport Engineering
	G-BLOM	—	—
	G-BLON	—	—
	G-BLOO	Sopwith Dove Replica	C. O'Brien & T. Moore
	G-BLOP	Cessna 404	Crosbyglow Ltd (G-OEMA)
	G-BLOR	PA-30 Twin Comanche 160	C. R. Cox/Fairoaks
	G-BLOS	Cessna 185A (floatplane)	Duocane Ltd
	G-BLOT	Colt Ax6-56B balloon	Thunder & Colt Ltd
	G-BLOU	Rand KR-2	D. Cole
	G-BLOV	Colt Ax5-42 Srs 1 balloon	Thunder & Colt Ltd
	G-BLOW	—	—
	G-BLOX	—	—
	G-BLOY	—	—
	G-BLOZ	Cameron N-105 balloon	Cameron Balloons Ltd

Reg.	Type	Owner or Operator	Notes
G-BLPA	Piper J-3C-65 Cub	G. A. Card	
G-BLPB	Turner TSW Hot Two Wot	J. R. Woolford & K. M. Thomas	
G-BLPD	—	—	
G-BLPE	PA-18 Super Cub 95	A. Haig-Thomas	
G-BLPF	Cessna FR.172G	E. J. McMillan/Perth	
G-BLPG	J/1N Alpha	L. A. & P. Groves (G-AZIH)	
G-BLPH	Cessna FRA.150L	R. A. Rigelsford & V. A. Jennings	
G-BLPI	Slingsby T.67B	Slingsby Aviation Ltd/Kirkbymoorside	
G-BLPJ	—	—	
G-BLPK	Cameron V-65 balloon	A. J. & C. P. Nicholls	
G-BLPL	A-Bell 206B JetRanger	Bristow Helicopters Ltd	
G-BLPN	—	—	
G-BLPO	Rotorcraft	R. Jefferson	
G-BLPP	Cameron V-77 balloon	L. P. Purfield	
G-BLPR	Westland WG.30 Srs 100/60	Westland Helicopters Ltd/Yeovil	
G-BLPV	Short SD3-60	Air UK Ltd/Norwich	
G-BLPY	Short SD3-60	Air UK Ltd/Norwich	
G-BLPZ	D.H.104 Devon C.2	Air Leasing Ltd	
G-BLRB	D.H.104 Devon C.2 (VP962)	V. S. E. Norman	
G-BLRC	PA.18 Super Cub 135	R. A. L. Hubbard	
G-BLRD	MBB Bo.209 Monsun 150FF	G. N. Richardson	
G-BLRE	Slingsby T.67C	Slingsby Aviation Ltd/Kirkbymoorside	
G-BLRF	Slingsby T.67N	Slingsby Aviation Ltd/Kirkbymoorside	
G-BLRG	Slingsby T.67D	Slingsby Aviation Ltd/Kirkbymoorside	
G-BLRH	Rutan Long Ez	G. L. Thompson	
G-BLRJ	Jodel DR.1051	M. P. Hallam	
G-BLRK	PA-42-720 Cheyenne IIIA	McAlpine Aviation Ltd/Luton	
G-BLRL	CP.301C-1 Emeraude	R. A. Abrahams	
G-BLRM	Glaser Dirks DG-400	R. L. McLean & J. N. Ellis	
G-BLRN	D.H.104 Devon C.2	C. W. Simpson	
G-BLRO	—	—	
G-BLRP	FMA 1A58-A Pucara	Grampian Helicopters International Ltd	
G-BLRS	—	—	
G-BLRT	Short SD3-60	Fairflight Ltd/Biggin Hill	
G-BLRV	SA.365N Dauphin II	McAlpine Helicopters Ltd/Hayes	
G-BLRW	Cameron 77 Elephant balloon	Cameron Balloons Ltd	
G-BLRX	SOCATA TB.9 Tampico	Air Touring Services Ltd/Biggin Hill	
G-BLRZ	SOCATA TB.9 Tampico	Martin Ltd/Biggin Hill	
G-BLSA	PA-42-720 Cheyenne IIIA	McAlpine Aviation Ltd/Luton	
G-BLSC	Consolidated PBY-5A Catalina (JV928)	J. N. Watts & J. P. W. Wilson	
G-BLSD	D.H.112 Venom FB.54	Aces High Ltd/Duxford	
G-BLSF	AA-5A Cheetah	J. P. E. Walsh	
G-BLSG	—	—	
G-BLSH	Cameron V-77 balloon	Property Six	
G-BLSI	Colt AS-56 airship	Colt Balloons Ltd	
G-BLSJ	Thunder Ax8-90 balloon	Thunder Balloons Ltd	
G-BLSK	Colt 77A balloon	HR & H. Marketing Research International Ltd	
G-BLSM	H.S.125 Srs 700B	Dravidian Air Services Ltd/Heathrow	
G-BLSN	Colt AS-42 airship	Colt Balloons Ltd	
G-BLSO	Colt AS-42 airship	K. L. C. M. Busemeyer	
G-BLSR	—	—	
G-BLSS	Cessna F.150J	Seawing Flying Club Ltd/Southend	
G-BLST	Cessna 421C	Cecil Aviation Ltd/Cambridge	
G-BLSU	Cameron A-210 balloon	Skysales Ltd	
G-BLSV	—	—	
G-BLSX	Cameron O-105 balloon	B. J. Petteford	
G-BLSY	Bell 222A	Air Hanson Sales Ltd/Weybridge	
G-BLTA	Colt 77A balloon	Thunder & Colt Balloons Ltd	
G-BLTB	PA-42-720 Cheyenne IIIA	McAlpine Aviation Ltd/Luton	
G-BLTC	D.31 Turbulent	G. P. Smith & A. W. Burton	
G-BLTE	Cessna F.182G	Northair Aviation Ltd/Leeds	
G-BLTF	Robinson R-22A	Forest Dale Hotels Ltd	
G-BLTG	WAR Sea Fury	P. R. Pykett	
G-BLTH	Cessna 404	Banline Ltd (G-BKVH/G-WTVA)	
G-BLTI	Cessna 402B	Simpen Ltd (G-BCBI)/E. Midlands	
G-BLTJ	—	—	
G-BLTK	R. Commander 112TC	B. Rogalewski	
G-BLTL	—	—	
G-BLTM	Robin HR.200/100	P. D. Wheatland	
G-BLTN	Thunder Ax7-65 balloon	J. A. Liddle	
G-BLTO	Short SD3-60	Air UK Ltd/Fairflight Ltd	
G-BLTP	H.S.125 Srs 700B	Dravidian Air Services Ltd/Heathrow	

Notes	Reg.	Type	Owner or Operator
	G-BLTR	Scheibe SF.25B Falke	V. Mallon
	G-BLTS	Rutan Long Ez	R. W. Cutler
	G-BLTT	Slingsby T.67B	Slingsby Aviation PLC/Kirkbymoorside
	G-BLTU	Slingsby T.67B	Slingsby Aviation PLC/Kirkbymoorside
	G-BLTV	Slingsby T.67B	Slingsby Aviation PLC/Kirkbymoorside
	G-BLTW	Slingsby T.67B	Slingsby Aviation PLC/Kirkbymoorside
	G-BLTX	—	—
	G-BLTY	Westland WG.30 Srs 100-60	Westland Helicopters Ltd/Yeovil
	G-BLTZ	SOCATA TB.10 Tobago	Martin Hill/Biggin Hill
	G-BLUA	Robinson R-22	P. J. Millward/Booker
	G-BLUB	—	—
	G-BLUE	Colting Ax7-77A balloon	M. R. & C. Cumpston
	G-BLUF	Thunder Ax10-180 balloon	Thunder & Colt Ltd
	G-BLUG	Thunder Ax10-180 balloon	Thunder & Colt Ltd
	G-BLUH	Thunder Ax10-180 balloon	Thunder & Colt Ltd
	G-BLUI	Thunder Ax7-65 balloon	S. M. Marshall
	G-BLUJ	Cameron V-56 balloon	J. N. W. West
	G-BLUK	Bond Sky Dancer	J. Owen
	G-BLUL	Jodel DR.1051/M1	J. Owen
	G-BLUM	SA.365N Dauphin 2	Bond Helicopters Ltd
	G-BLUN	SA.365N Dauphin 2	Bond Helicopters Ltd
	G-BLUO	SA.365N Dauphin 2	Bond Helicopters Ltd
	G-BLUP	SA.365N Dauphin 2	Bond Helicopters Ltd
	G-BLUS	L.1011 TriStar 500	British Airways *Laggan Bay*/Heathrow
	G-BLUT	L.1011 TriStar 500	British Airways *Dunnet Bay*/Heathrow
	G-BLUV	Grob G.109B	Go-Grob Ltd
	G-BLUW	H.S.125 Srs 600B	Field Aviation Trading Ltd/Heathrow
	G-BLUX	Slingsby T.67C	Slingsby Aviation Ltd/Kirkbymoorside
	G-BLUY	Colt 69A balloon	The Balloon Goes Up Ltd
	G-BLUZ	D.H.82B Queen Bee	B. Bayes
	G-BLVA	—	—
	G-BLVB	—	—
	G-BLVC	—	—
	G-BLVE	Boeing 747-2B4B	British Airways *City of Lincoln*/ Heathrow
	G-BLVF	Boeing 747-2B4B	British Airways *City of Lancaster*/ Heathrow
	G-BLVG	EMB-110P1 Bandeirante	Euroair Transport Ltd (G-RLAY)/Gatwick
	G-BLVH	Boeing 757-236	Air Europe Ltd/Gatwick
	G-BLVI	Slingsby T.67M	Slingsby Aviation Ltd/Kirkbymoorside
	G-BLVJ	Colt AS-56 airship	Colt Balloons Ltd
	G-BLVK	CAARP CAP-10B	BAC Aviation Ltd/Southend
	G-BLVL	PA-28-161 Warrior II	C.S.E. Aviation Ltd/Kidlington
	G-BLVM	—	—
	G-BLVN	Cameron N-77 balloon	B. Hodge
	G-BLVO	—	—
	G-BLVP	—	—
	G-BLVR	—	—
	G-BLVS	Cessna 150M	W. Lancashire Aero Club Ltd/Woodvale
	G-BLVT	Cessna FR.172J	London Flight Centre (Stansted) Ltd
	G-BLVU	Pitts S-2A	P. C. Henry/Denham
	G-BLVV	Bell 206B JetRanger	Bristow Helicopters Ltd
	G-BLVW	Cessna F.172H	Crop Aviation (UK) Ltd
	G-BLVX	Colt 105A balloon	Colt Balloons Ltd
	G-BLVY	Colt 21A balloon	Colt Balloons Ltd
	G-BLVZ	R. Commander 114	Botsford & Willard Ltd/Panshanger
	G-BLWB	Thunder Ax6-56 balloon	G. G. Bacon
	G-BLWC	SA.365N Dauphin 2	McAlpine Helicopters Ltd/Hayes
	G-BLWD	PA-34-200T Seneca	C.S.E. Aviation Ltd/Kidlington
	G-BLWE	Colt 90A balloon	Thunder & Colt Ltd
	G-BLWF	Robin HR.100/210	Hill Leigh Group Ltd
	G-BLWG	Varga 2150A Kachina	MLP Aviation Ltd/Elstree
	G-BLWH	Fournier RF-6B-100	R. J. Fray
	G-BLWL	Colt 31A balloon	Hot Air Balloon Co Ltd
	G-BLWM	Bristol M.1C replica	D. M. Cashmore
	G-BLWO	Cameron N-77 balloon	Cameron Balloons Ltd
	G-BLWP	PA-38-112 Tomahawk	A. Dodd/Redhill
	G-BLWR	—	—
	G-BLWS	ICA IS-28M2A	Classic Aeroplane Ltd
	G-BLWT	Evans VP-1	G. B. O'Neill
	G-BLWU	Bell 206B JetRanger	P. J. Millward/Booker
	G-BLWV	Cessna F.152	Belverhurst Ltd/Doncaster
	G-BLWW	Taylor Mini Imp Model C	W. E. Wilks
	G-BLWX	Cameron N-56 balloon	Skipton Building Soc

Reg.	Type	Owner or Operator	Notes
G-BLWY	Robin 2161D	A. D. Russell	
G-BLWZ	—	—	
G-BLXA	SOCATA TB.20 Trinidad	Street Construction (Wigan) Ltd	
G-BLXB	Colt 240A balloon	Colt Balloons Ltd	
G-BLXC	Colt 240A balloon	Colt Balloons Ltd	
G-BLXD	Slingsby T.67B	Slingsby Aviation Ltd/Kirkbymoorside	
G-BLXE	Slingsby T.67B	Slingsby Aviation Ltd/Kirkbymoorside	
G-BLXF	Cameron V-77 balloon	D. I. Gray-Fisk	
G-BLXG	Colt 21A balloon	Balloon & Airship Co Ltd	
G-BLXH	Fournier RF-3	E. J. Horsfall	
G-BLXI	CP.1310-C3 Super Emeraude	J. L. Phelps	
G-BLXJ	SA.315B Lama	Autair Ltd/Cranfield	
G-BLXK	Agusta-Bell 205	Autair Helicopters Ltd/Cranfield	
G-BLXL	Colt AS-105 airship	Thunder & Colt Ltd	
G-BLXM	—	—	
G-BLXO	Jodel D.150	P. R. Powell	
G-BLXP	PA-28R Cherokee Arrow 200	London Flight Centre (Stansted) Ltd	
G-BLXR	AS.332L Super Puma	Bristow Helicopters Ltd	
G-BLXS	AS.332L Super Puma	Bristow Helicopters Ltd	
G-BLXT	Eberhardt SE-5A	P. Lindsay/Booker	
G-BLXU	—	—	
G-BLXX	PA-23 Aztec 250	Hockstar Ltd (G-PIED)	
G-BLXY	Cameron V-65 balloon	Gone With The Wind Ltd	
G-BLYB	Beech B200 Super King Air	Alfred McAlpine Aviation Ltd	
G-BLYC	PA-38-112 Tomahawk	Automated Data Systems Ltd	
G-BLYD	SOCATA TB.20 Trinidad	R. J. Crocker	
G-BLYE	SOCATA TB.10 Tobago	Air Touring Services Ltd/Biggin Hill	
G-BLYJ	Cameron V-77 balloon	E. E. Clark & J. A. Lomas	
G-BLYK	PA-34-220T Seneca 2	G. E. Walker/Jersey	
G-BLYM	—	—	
G-BLYP	Robin 3000/120	Air Touring Services Ltd/Biggin Hill	
G-BLYR	—	—	
G-BLYS	Robin DR.400/180	Soaring (Oxford) Ltd/Enstone	
G-BLYT	—	—	
G-BLYU	—	—	
G-BLYV	—	—	
G-BLYW	P-51D-25NA Mustang	C. J. Warrilow	
G-BLYX	—	—	
G-BLYY	PA-28-181 Archer II	MLP Aviation/Elstree	
G-BLYZ	Edgley EA.7 Optica	Edgley Aircraft Ltd/Old Sarum	
G-BLZA	Scheibe SF.25B Falke	P. Downes	
G-BLZB	Cameron N-65 balloon	D. Bareford	
G-BLZC	Flamboyant Ax7-65 balloon	T. A. Adams	
G-BLZD	Robin R.1180T	Moto Baldet (Northampton) Ltd/Sywell	
G-BLZE	Cessna F.152	Flairhire Ltd (G-CSSC)/Redhill	
G-BLZF	Thunder Ax7-77 balloon	J. A. Snowball & ptnrs	
G-BLZH	Cessna F.152	Shoreham Flight Simulation Ltd/ Bournemouth	
G-BLZI	—	—	
G-BLZK	—	—	
G-BLZL	—	—	
G-BLZM	Rutan Long Ez	B. C. Barton	
G-BLZN	Bell 206B JetRanger	J. P. Millward	
G-BLZO	Cessna F.172P	Rogers Aviation Ltd/Cranfield	
G-BLZP	Cessna F.152	W. H. & J. Rogers Group Ltd/Cranfield	
G-BLZR	Cameron A-140 balloon	Clipper Worldwide Trading Ltd	
G-BLZS	Cameron O-77 balloon	Cameron Balloons Ltd	
G-BLZT	Short SD3-60	Air UK Ltd/Norwich	
G-BLZX	—	—	
G-BLZY	—	—	
G-BLZZ	CAARP CAP.21	D. M. Britten	
G-BMAA	Douglas DC-9-15	British Midland Airways Ltd *Dovedale*/ (G-BFIH)/E. Midlands	
G-BMAB	Douglas DC-9-15	British Midland Airways Ltd *Ulster*/ E. Midlands	
G-BMAC	Douglas DC-9-15	British Midland Airways Ltd/E. Midlands	
G-BMAD	Cameron V-77 balloon	F. J. J. Fielder	
G-BMAE	F-27 Friendship Mk 200	British Midland Airways Ltd/E. Midlands	
G-BMAF	Cessna 180F	N. F. Hemming (G-BDVR)/Staverton	
G-BMAG	Douglas DC-9-15	British Midland Airways Ltd/E. Midlands	
G-BMAH	Douglas DC-9-14	British Midland Airways Ltd/E. Midlands	
G-BMAI	Douglas DC-9-14	British Midland Airways Ltd/E. Midlands	

Notes	Reg.	Type	Owner or Operator
	G-BMAJ	Short SD3-60	British Midland Airways Ltd (G-BKPO)/ E. Midlands
	G-BMAK	Douglas DC-9-30	British Midland Airways Ltd/E. Midlands
	G-BMAL	Sikorsky S-76A	Bond Helicopters Ltd/Bourn
	G-BMAM	Douglas DC-9-30	British Midland Airways Ltd/E. Midlands
	G-BMAO	Taylor JT.1 Monoplane	V. A. Wordsworth
	G-BMAP	F-27 Friendship Mk 200	British Midland Airways Ltd/E. Midlands
	G-BMAR	Short SD3-60	British Midland Airways Ltd (G-BLCR)/ E. Midlands
	G-BMAT	V.813 Viscount	British Midland Airways Ltd (G-AZLT)/ E. Midlands
	G-BMAU	F.27 Friendship Mk 200	British Midland Airways Ltd/E. Midlands
	G-BMAV	AS.350B Ecureuil	Timothy Laing Aviation
	G-BMAW	F-27 Friendship Mk 200	British Midland Airways Ltd/E. Midlands
	G-BMAX	FRED Srs 2	P. Cawkwell
	G-BMAY	PA-18 Super Cub 135	G. V. Harfield/Popham
	G-BMBB	Cessna F.150L	Falcon Flying Services/Biggin Hill
	G-BMBC	PA-31-350 Navajo Chieftain	Air Charter (Scotland) Ltd/Glasgow
	G-BMBD	—	—
	G-BMBE	PA-46-310P Malibu	Artix Ltd
	G-BMBF	—	—
	G-BMBG	Colt 21A balloon	Thunder & Colt Ltd
	G-BMBI	PA-31-350 Navajo Chieftain	Computaplane Ltd/Glasgow
	G-BMBJ	Schempp-Hirth Janus CM	Oxfordshire Sportflying Ltd/Enstone
	G-BMBN	Slingsby T.67M-200	Slingsby Aviation PLC/Kirkbymoorside
	G-BMBO	—	—
	G-BMBP	Colt Whisky Bottle balloon	Thunder & Colt Ltd
	G-BMBR	Issoire Motor Iris D77-M	G. R. Horner
	G-BMBS	Colt 105A balloon	A. P. Hardiman & H. G. Davies
	G-BMBT	Thunder Ax8-90 balloon	Capital Balloon Club Ltd
	G-BMBU	Robinson R-22	J. P. Millward
	G-BMBW	Bensen B.80	M. Vahdat
	G-BMBX	Robinson R-22	Sloane Helicopters Ltd/Sywell
	G-BMBY	Beech A36 Bonanza	Shadow Photographic Ltd
	G-BMBZ	Scheibe SF.25E Falke	Cairngorm Gliding Club
	G-BMCB	Partenavia P.68B	Air Kilroe Ltd/Manchester
	G-BMCC	Thunder Ax7-77 balloon	H. N. Harben Ltd
	G-BMCD	Cameron V-65 balloon	M. C. Drye
	G-BMCE	Bensen B.8M	J. Lee
	G-BMCF	—	—
	G-BMCG	Grob G.109B	Soaring (Oxford) Ltd/Enstone
	G-BMCH	AB-206 JetRanger	Autair Ltd/Luton
	G-BMCI	Cessna F.172H	P. P. D. Howard-Johnston/Edinburgh
	G-BMCJ	PA-31-350 Navajo Chieftain	Chelsea Land (Finance) Ltd
	G-BMCK	Cameron O-77 balloon	D. L. Smith
	G-BMCM	Grob G.109B	Sonardyne Ltd/Blackbushe
	G-BMCN	Cessna F.152	Lincoln Aero Club Ltd/Sturgate
	G-BMCO	Colomban MC.15 Cri-Cri	G. P. Clarke
	G-BMCS	PA-22 Tri-Pacer 135	W. A. P. Darbishire
	G-BMCT	Cameron D-50 airship	Cameron Balloons Ltd
	G-BMCU	AS.350B Ecureuil	McAlpine Helicopters Ltd/Hayes
	G-BMCV	Cessna F.152	Leicester Aero Club Ltd
	G-BMCW	AS.332L Super Puma	Bristow Helicopters Ltd
	G-BMCX	AS.332L Super Puma	Bristow Helicopters Ltd
	G-BMCY	AS.355F-1 Twin Squirrel	McAlpine Helicopters Ltd/Hayes
	G-BMCZ	Colt 69A balloon	Thunder & Colt Ltd
	G-BMDA	Colt 17A balloon	Thunder & Colt Ltd
	G-BMDB	SE-5A replica	D. Biggs
	G-BMDC	PA-32-301 Saratoga	MLP Aviation Ltd/Elstree
	G-BMDD	Slingsby T.29	D. I. H. Johnstone & L. Ward
	G-BMDE	Pientenpol Aircamper	D. Silsbury & B. P. Irish
	G-BMDF	Boeing 737-2E7	Dan-Air Services Ltd/Gatwick
	G-BMDG	Cameron O-105 balloon	Buddy Bombard Balloons Ltd
	G-BMDH	Cameron O-105 balloon	Buddy Bombard Balloons Ltd
	G-BMDI	Thunder Ax8-105Z balloon	Buddy Bombard Balloons Ltd
	G-BMDJ	Price Ax7-77S balloon	T. S. Price
	G-BMDK	PA-34-220T Seneca 3	Skyline Helicopters Ltd/Booker
	G-BMDL	Cessna 402C	Skyline Helicopters Ltd/Booker
	G-BMDM	Cessna 340A	Skyline Helicopters Ltd/Booker
	G-BMDO	ARV Super 2	H. L. Wensley
	G-BMDP	Partenavia P.64B Oscar 200	D. Foey
	G-BMDR	—	—
	G-BMDS	Jodel D.120	D. Stansfield & M. Smith
	G-BMDT	BN-2A Islander	Pilatus BN Ltd/Bembridge

Reg.	*Type*	*Owner or Operator*	*Notes*
G-BMDU	Bell 214ST	British Caledonian Helicopters Ltd	
G-BMDV	Bell 47G-5	Trent Air Services Ltd/Cranfield	
G-BMDW	Dangerous Sports Club/Colt Hoppalong 1 balloon	D. A. C. Kirke	
G-BMDX	Grumman HU-16B Albatross	Daedalus Aviation Ltd	
G-BMDY	GA-7 Cougar	Eastern Air Taxis/Elstree	
G-BMDZ	Cessna 310Q	Computaplane Ltd/Glasgow	
G-BMEA	PA-19 Super Cub 95	I. M. Callier	
G-BMEB	Rotorway Scorpion 145	I. M. Bartlett	
G-BMED	Edgley EA.7 Optica	Edgley Aircraft Ltd/Old Sarum	
G-BMEE	Cameron O-105 balloon	A. G. R. Calder	
G-BMEF	Beech C90 King Air	Navigations Ltd/Elstree	
G-BMEG	SOCATA TB.10 Tobago	Air Touring Services Ltd/Biggin Hill	
G-BMEH	Jodel Super Special Mascaret	E. J. Horsfall/Blackpool	
G-BMEI	Partenavia P.68C	Northair Aviation Ltd/Leeds	
G-BMEJ	—	—	
G-BMEK	Mooney M.20K	Overseas Flight Sales Ltd	
G-BMEM	Fournier RF-4D	A. M. Witt	
G-BMET	Taylor JT.1 Monoplane	M. K. A. Blyth	
G-BMEU	Isaacs Fury II	A. W. Austin	
G-BMEV	—	—	
G-BMEW	Lockheed 18-56 Lodestar	Aces High Ltd/N. Weald	
G-BMEX	Cessna A.150K	Air Sale PLC/Booker	
G-BMEY	PA-32-301 Saratoga	Express Aviation Services Ltd	
G-BMEZ	Cameron D-50 airship	Cameron Balloons Ltd	
G-BMFA	—	—	
G-BMFB	Douglas AD-4W Skyraider	Coys of Kensington (Petrol Sales) Ltd	
G-BMFC	Douglas AD-4W Skyraider	Coys of Kensington (Petrol Sales) Ltd	
G-BMFD	PA-23 Aztec 250	Bomford & Evershed Ltd/Coventry	
G-BMFE	Cameron Truck balloon	Cameron Balloons Ltd	
G-BMFF	—	—	
G-BMFG	Dornier Do.27A-4	Onderstar Aviation Ltd/Booker	
G-BMFH	Dornier Do.27A-4	Onderstar Aviation Ltd/Booker	
G-BMFI	PZL SZD-45A Ogar	Anglo Polish Sailplanes Ltd/Booker	
G-BMFJ	Thunder Ax7-77 balloon	Thunder & Colt Ltd	
G-BMFK	PA-28-236 Dakota	C.S.E. Aviation Ltd/Kidlington	
G-BMFL	Rand KR-2	E. W. B. Comber & M. F. Leusby	
G-BMFN	Quickie 200	W. Blair-Hickman	
G-BMFO	—	—	
G-BMFP	PA-28-161 Warrior II	Fairprime Ltd/Blackbushe	
G-BMFR	—	—	
G-BMFS	—	—	
G-BMFT	H.S.748 Srs 2A	Euroair Transport Ltd/Gatwick	
G-BMFU	Cameron N-90 balloon	Cameron Balloons Ltd	
G-BMFV	Cameron N-56 balloon	Cameron Balloons Ltd	
G-BMFW	Hughes 369E	Ebor Finance Co Ltd	
G-BMFX	—	—	
G-BMFY	Grob G.109B	P. J. Shearer	
G-BMFZ	Cessna F.152	Cornwall Flying Club Ltd	
G-BMGA	—	—	
G-BMGB	PA-28 Cherokee Arrow 200	G-Air Ltd/Goodwood	
G-BMGC	Fairey Swordfish Mk II	Strathallan Aircraft Collection	
G-BMGD	Colt 17A balloon	Thunder & Colt Ltd	
G-BMGF	Cessna 310R	Computaplane Ltd/Glasgow	
G-BMGG	Cessna 152	A. S. Bamrah	
G-BMGH	PA-31-325 Navajo	Skyline Helicopters Ltd	
G-BMGI	Beech 58 Baron	Skyline Helicopters Ltd	
G-BMGJ	—	—	
G-BMGK	—	—	
G-BMGL	—	—	
G-BMGM	—	—	
G-BMGN	—	—	
G-BMGO	—	—	
G-BMGP	Hughes 269C	March Helicopters Ltd	
G-BMGR	Grob G.109B	J. E. T. Woodger	
G-BMGS	—	—	
G-BMGT	Cessna 310R	Overson Services Ltd	
G-BMGU	Cessna 210M	E. McGinnity	
G-BMGV	Robinson R-22	Sloane Helicopters Ltd/Sywell	
G-BMGW	—	—	
G-BMGY	Lake LA-4-200 Buccaneer	M. A. Ashmole (G-BWKS/G-BDDI)	
G-BMGZ	AS.332L Super Puma	Bristow Helicopters Ltd	
G-BMHA	Rutan Long Ez	S. F. Elvins	
G-BMHB	—	—	

Notes	Reg.	Type	Owner or Operator
	G-BMHC	Cessna U206F	D. R. Bromiley
	G-BMHD	—	—
	G-BMHE	Cessna T310R	Computaplane Ltd/Glasgow
	G-BMHF	Mooney M.20J	Computaplane Ltd/Glasgow
	G-BMHI	Cessna F.152	Belverhurst Ltd
	G-BMHJ	—	—
	G-BMHK	Cameron V-77 balloon	B. J. Workman
	G-BMHL	Wittman W.8 Tailwind	T. G. Hoult
	G-BMHM	PA-28R-201T Arrow	G-Air Ltd/Goodwood
	G-BMHN	Robinson R-22A	A. & P. M. Ford
	G-BMHO	Colt 105A balloon	Thunder & Colt Ltd
	G-BMHP	PA-34-200T Seneca	BI Aviation Services Ltd
	G-BMHR	Grob G.109B	HRN Aviation Ltd
	G-BMHS	—	—
	G-BMHT	PA-28RT-201T Turbo Arrow	Xylored Ltd
	G-BMHU	Viking Dragonfly	H. A. Bancroft-Wilson
	G-BMHX	Short SD3-60	Short Bros PLC/Sydenham
	G-BMHY	Short SD3-60	Short Bros PLC/Sydenham
	G-BMHZ	PA-28RT-201T Turbo Arrow	Xylored Ltd
	G-BMIA	Thunder Ax8-90 balloon	Thunder & Colt Ltd
	G-BMIB	Bell 206B JetRanger	J. P. Millward
	G-BMID	Jodel D.120	A. W. Cooke
	G-BMIE	Thunder Ax8-105 balloon	Thunder & Colt Ltd
	G-BMIF	AS.350B Ecureuil	McAlpine Helicopters Ltd
	G-BMIG	—	—
	G-BMIH	H.S.125 srs 700B	Jaguar Aviation Ltd
	G-BMII	—	—
	G-BMIJ	Thunder Ax7-65 balloon	D. Cole
	G-BMIK	—	—
	G-BMIL	—	—
	G-BMIM	Rutan Long Ez	R. M. Smith
	G-BMIO	Glassair RG	A. H. Carrington
	G-BMIP	Jodel D.112	M. T. Kinch
	G-BMIR	Westland Wasp HAS.1	R. Windley
	G-BMIS	—	—
	G-BMIU	Enstrom F-28A	Executive Helicopters Ltd
	G-BMIV	PA-28R-201T Turbo Arrow	Xylored Ltd
	G-BMIW	PA-28-181 Archer II	Xylored Ltd
	G-BMIX	SOCATA TB.20 Trinidad	Air Touring Services Ltd/Biggin Hill
	G-BMIY	Oldfield Baby Great Lakes	J. B. Scott (G-NOME)
	G-BMJA	—	—
	G-BMJB	—	—
	G-BMJC	—	—
	G-BMJD	Cessna 152	Aberdeen Aviation Ltd
	G-BMJF	—	—
	G-BMJG	—	—
	G-BMJI	Cameron 75 Watch balloon	Cameron Balloons Ltd
	G-BMJJ	Cameron 75 Watch balloon	Cameron Balloons Ltd
	G-BMJK	Cameron 75 Watch balloon	Cameron Balloons Ltd
	G-BMJL	R. Commander 114	Computaplane Ltd
	G-BMJM	Evans VP-1	J. A. Mawby
	G-BMJN	Cameron 0-65 balloon	V. J. Bishton
	G-BMJO	PA-34-220T Seneca	C.S.E. Aviation Ltd/Kidlington
	G-BMJP	Colt AS-56 airship	Thunder & Colt Ltd
	G-BMJR	Cessna T.337H	John Roberts Services Ltd (G-NOVA)
	G-BMJS	Thunder Ax7-77 balloon	Anglia Balloon School Ltd
	G-BMJT	Beech 76 Duchess	Mike Osborne Insurance Services Ltd
	G-BMJU	H.S.748c srs 2B	British Aerospace PLC
	G-BMJV	Hughes 369D	Bristol Estates Ltd
	G-BMJW	AT-6D Harvard IIA	J. Woods
	G-BMJX	—	—
	G-BMJY	Yakolev C-18M	Coys of Kensington (Petrol Sales) Ltd
	G-BMJZ	Cameron N-90 balloon	Windsor Pharmaceuticals Ltd
	G-BMKA	Robin 3000/120	Air Touring Services Ltd/Biggin Hill
	G-BMKB	PA-18 Super Cub 135	G. Cormack
	G-BMKC	Piper J-3C-65 Cub	J. J. Anziani
	G-BMKD	—	—
	G-BMKE	—	—
	G-BMKF	—	—
	G-BMKG	—	—
	G-BMKH	—	—
	G-BMKI	—	—
	G-BMKJ	Cameron V-77 balloon	R. C. Thursby
	G-BMKK	PA-28R Cherokee Arrow 200	J. D. Poole

Reg.	Type	Owner or Operator	Notes
G-BMKL	PA-28-181 Archer II	RJS Aviation Ltd	
G-BMKM	—	—	
G-BMKN	—	—	
G-BMKO	PA-28-181 Archer II	CP Aviation Ltd	
G-BMKP	Cameron V-77 balloon	Jacques W. Soukup Enterprises Ltd	
G-BMKR	PA-28-161 Warrior II	E. J. Cannings (G-BGKR)/Kidlington	
G-BMKS	Aerosport Scamp	J. N. Hamlen	
G-BMKT	SA.315B Lama	Dollar Air Services Ltd	
G-BMKU	Colt Apple balloon	Thunder & Colt Ltd	
G-BMKV	Thunder Ax7-77 balloon	A. Hornak	
G-BMKW	Cameron V-77 balloon	H. J. H. Lonsdale	
G-BMKX	—	—	
G-BMKY	—	—	
G-BMKZ	—	—	
G-BMLA	Bell UH-1H	Grampian Helicopters International Ltd	
G-BMLB	Jodel D.120A	W. O. Brown	
G-BMLC	—	—	
G-BMLD	—	—	
G-BMLE	—	—	
G-BMLF	—	—	
G-BMLG	—	—	
G-BMLI	Beech F33A Bonanza	Bermuda Aviation Associates (UK) Ltd	
G-BMLN	Fuji FA-200-160	Air Associates Ltd	
G-BMLT	Pietenpol Aircamper	R. A. Hawke	
G-BMNF	Beech 200 Super King Air	Bernard Matthews PLC	
G-BMOL	PA-23 Aztec 250	LDL Enterprises (G-BBSR)/Elstree	
G-BMON	Boeing 737-2K9	Monarch Airlines Ltd/Luton	
G-BMSB	V.S.509 Spitfire IX	M. S. Bayliss (G-ASOZ)	
G-BMSC	Evans VP-2	G. J. Taylor	
G-BMSF	PA-38-112 Tomahawk	M.S.F. Aviation Ltd/Manchester	
G-BMTC	AS.355F-1 Twin Squirrel	The Marley Tile Co Ltd (G-BKUK)	
G-BMTD	Short SD3-30	Brencham Leasing Ltd (G-BLTD)	
G-BMUD	Cessna 182P	Ingham Aviation Ltd	
G-BMVV	Rutan Vari-Viggen	G. B. Roberts	
G-BMYU	Jodel D.120	G. Davies	

Out-of-Sequence Registrations

Notes	Reg.	Type	Owner or Operator
	G-BNAA	V.806 Viscount	British Air Ferries (G-AOYH)/Southend
	G-BNAB	GA-7 Cougar	Noortman & Brod Ltd (G-BGYP)/Elstree
	G-BNBY	Beech 95-B55A Baron	E. L. Klinge (G-AXXR)/Biggin Hill
	G-BNDX	H.S.125 Srs 600B	Genavco Air Ltd (G-BAYT)/Heathrow
	G-BNFL	WHE Airbuggy	Roger Savage (Photography) (G-AXXN)
	G-BNHP	Saffrey S.330 balloon	N. H. Ponsford *Alpha II*
	G-BNJF	PA-32RT-300 Turbo Lance II	Mike Mansfield Enterprises Ltd
	G-BNOC	EMB-110P1 Bandeirante	Connectair Ltd/Biggin Hill
	G-BNPD	PA-23 Aztec 250	Lion Air Lease/Glasgow
	G-BNSH	Sikorsky S-76A	Bond Helicopters Ltd/Bourn
	G-BOAA	Concorde 102	British Airways (G-N94AA)/Heathrow
	G-BOAB	Concorde 102	British Airways (G-N94AB)/Heathrow
	G-BOAC	Concorde 102	British Airways (G-N81AC)/Heathrow
	G-BOAD	Concorde 102	British Airways (G-N94AD)/Heathrow
	G-BOAE	Concorde 102	British Airways (G-N94AE)/Heathrow
	G-BOAF	Concorde 102	British Airways (G-N94AF/G-BFKX)/ Heathrow
	G-BOAG	Concorde 102	British Airways (G-BFKW)/Heathrow
	G-BOBI	Cessna 152	R. M. Seath (G-BHJD)/Sherburn
	G-BOBS	Quickie Q.2	R. Stevens
	G-BOBY	Monnet Sonerai II	R. G. Hallam/Sleap
	G-BOIS	PA-31-300 Navajo	Woods Management Services Ltd (G-AYNB)
	G-BOLT	R. Commander 114	Hooper & Jones Ltd/Kidlington
	G-BOMB	Cassutt Racer	R. W. L. Breckell
	G-BOND	Sikorsky S-76	Bond Helicopters Ltd/Bourn
	G-BONE	Pilatus P2-06 (U-142)	Aeromech Ltd
	G-BOOB	Cameron N-65 balloon	I. J. Sadler
	G-BOOK	Pitts S-1S Special	B. K. Lecomber/Denham
	G-BOOM	Hunter T.7	Hunter Promotions Ltd/Bournemouth
	G-BOOZ	Cameron N-77 balloon	J. A. F. Croft
	G-BOTL	Colt 42R balloon	Colt Balloons Ltd
	G-BOVA	PA-31-310 Turbo Navajo	Air Bristol (G-BECP)
	G-BPAH	Colt 69A balloon	International Distillers & Vintners Ltd
	G-BPAJ	D.H.82A Tiger Moth	P. A. Jackson (G-AOIX)
	G-BPAM	Jodel D.150A	A. J. Symes-Bullen
	G-BPAV	FRED Srs 2	P. A. Valentine
	G-BPBP	Brugger Colibri Mk II	M. F. Collett
	G-BPCH	Beech 300 Super King Air	Eagle Beechcraft Ltd/Leavesden
	G-BPEG	Currie Wot	D. M. Harrington
	G-BPFA	Knight Swallow GK-2	G. Knight & D. G. Pridham
	G-BPGW	Boeing 757-236	Air Europe Ltd *Anna-Marie*/Gatwick
	G-BPJH	PA-19 Super Cub 95	P. J. Heron
	G-BPMB	Maule M5-235C Lunar Rocket	P. M. Breton
	G-BPMN	Super Coot Model A	P. Napp
	G-BPOP	Aircraft Designs Sheriff	Sheriff Aerospace Ltd/Sandown
	G-BPPN	Cessna F.182Q	Hunt Business Consultancy Ltd/ Shoreham
	G-BPUF	Thunder Ax6-56Z balloon	Buf-Puf Balloon Group *Buf-Puf*
	G-BPYN	Piper J-3C-65 Cub	D. W. Stubbs & ptnrs/White Waltham
	G-BPZD	Nord NC.858S	T. G. Solomon/Shoreham
	G-BRAD	Beech 95-B55 Baron	C. Walker
	G-BRAF	V. S. Spitfire XVIII (SM969)	D. W. Arnold
	G-BRAG	Taylor JT.2 Titch	A. R. Greenfield
	G-BRAL	G.159 Gulfstream 1	Ford Motor Co Ltd/Stansted
	G-BREF	Cessna 421C	Refair Ltd/Jersey
	G-BREL	Cameron O-77 balloon	BICC Research & Engineering Ltd
	G-BREW	PA-31-350 Navajo Chieftain	Whitbread & Co Ltd/Biggin Hill
	G-BRFC	P.57 Sea Prince T.1 (WP321)	Rural Naval Air Service/Bourn
	G-BRGH	FRED Srs 2	F. G. Hallam
	G-BRGV	PA-31-350 Navajo Chieftain	Express Aviation Services Ltd/ Biggin Hill
	G-BRGW	GY-201 Minicab	R. G. White
	G-BRIK	Tipsy Nipper 3	C. W. R. Piper
	G-BRIT	Cessna 421C	Britannia Airways Ltd/Luton
	G-BRIX	PA-32-301 Saratoga SP	Taylor Maxwell & Co Ltd/Bristol
	G-BRJP	Boeing 737-2S3	Air Europe Ltd (*Louise*)/Gatwick
	G-BRJS	BAe 146-100	British Aerospace (G-OBAF/G-SCHH)
	G-BRJW	Bellanca 7GCBC Citabria	Comarket/Staverton

Reg.	Type	Owner or Operator	Notes
G-BRMA	WS-51 Dragonfly Mk 5 (WG719) ★	British Rotorcraft Museum	
G-BRMB	B.192 Belvedere Mk 1 (XG452) ★	British Rotorcraft Museum	
G-BRMC	Stampe SV-4B	A. Cullen/Andrewsfield	
G-BRMH	Bell 206B JetRanger 2	R. M. H. Stainless Ltd (G-BBUX)	
G-BROD	H.S. 125 srs 403B	Aravia (CI) Ltd	
G-BROM	ICA IS-28M2	Westlake & Co Ltd/Biggin Hill	
G-BRPM	Nipper T.66 Srs 3	R. P. Morris	
G-BRSL	Cameron N-56 ballogn	S. Budd	
G-BRUX	PA-44-180 Seminole	Hambrair Ltd/Tollerton	
G-BRWG	Maule M5-235C Lunar Rocket	D. H. Tonkin/Exeter	
G-BRWN	G.159 Gulfstream 1	Brown Air Services Ltd/Leeds	
G-BRYA	D.H.C. 7-110 Dash Seven	Brymon Aviation Ltd/Plymouth	
G-BRYB	D.H.C. 7-110 Dash Seven	Brymon Aviation Ltd/Plymouth	
G-BRYC	D.H.C. 7-110 Dash Seven	Brymon Aviation Ltd/Plymouth	
G-BRYL	Agusta A.109A	Castle Air Charters Ltd (G-ROPE/G-OAMH)	
G-BSAN	G.1159A Gulfstream 3	Shell Aviation Ltd/Heathrow	
G-BSBH	Short SD3-30	Short Bros Ltd/Sydenham	
G-BSDL	SOCATA TB.10 Tobago	Consort Aviation Ltd	
G-BSEL	Slingsby T-61G	RAFGSA/Bicester	
G-BSFC	PA-38-112 Tomahawk	Sherwood Flying Club Ltd/Tollerton	
G-BSFL	PA-23 Aztec 250	T. Kilroe & Sons Ltd/Manchester	
G-BSFT	PA-31-300 Navajo	SFT Aviation Ltd (G-AXYC)/ Bournemouth	
G-BSFZ	PA-25 Pawnee 235	A. Edwards (G-ASFZ)/Fairoaks	
G-BSHL	H.S.125 Srs 600B	S. H. Services Ltd (G-BBMD)/Luton	
G-BSHR	Cessna F.172N	H. Rothwell (G-BFGE)/Dundee	
G-BSIS	Pitts S-1S Special	R. A. Mills	
G-BSPC	Jodel D.140C	B. E. Cotton/Headcorn	
G-BSPE	Cessna F.172P	P. & M. Jones	
G-BSRS	PA-28R-201T Turbo Arrow III	Xylored Ltd	
G-BSSL	Beech B80 Queen Air	Parker & Heard Ltd (G-BFEP)/Biggin Hill	
G-BSST	Concorde 002 ★	Fleet Air Arm Museum	
G-BSUS	Taylor JT.1 Monoplane	R. Parker	
G-BSVP	PA-23 Aztec 250F	C. D. Weiswall/Elstree	
G-BTAL	Cessna F.152	TG Aviation Ltd	
G-BTAN	Thunder Ax7-65Z balloon	The BTAN Balloon Group	
G-BTCG	PA-23 Aztec 250	Eagle Tugs Ltd (G-AVRX)/Mombasa	
G-BTDK	Cessna 421B	Surplus Machinery Exports Ltd/ Manchester	
G-BTEA	Cameron N-105 balloon	Southern Balloon Group	
G-BTFC	Cessna F.152 II	Tayside Aviation Ltd/Dundee	
G-BTFH	Cessna 414A	C. Taylor & Co Ltd/Birmingham	
G-BTGS	Smyth Sidewinder	T. G. Soloman	
G-BTHL	PA-31-350 Navajo Chieftain	Air Charter (Scotland) Ltd/Glasgow	
G-BTHS	PA-23 Aztec 250F	Partlease Ltd/Stansted	
G-BTIE	SOCATA TB.10 Tobago	Rotaters Ltd/Manchester	
G-BTJM	Taylor JT.2 Titch	T. J. Miller/Dunkeswell	
G-BTJR	Short SD3-30	Short Bros PLC (G-BGNA)/Sydenham	
G-BTLE	PA-31-350 Navajo Chieftain	Andwell Mill Trout Farm Ltd	
G-BTOM	PA-38-112 Tomahawk	C. R. Timber Ltd	
G-BTOW	SOCATA Rallye 180T	Aerospecial Ltd/Biggin Hill	
G-BTSC	Evans VP-2	B. P. Irish	
G-BTUG	SOCATA Rallye 180T	Lasham Gliding Soc Ltd	
G-BTWW	AB-206B JetRanger	Dollar Air Services Ltd/Coventry	
G-BUCC	C.A.S.A. 1.131 Jungmann	E. J. McEntee/White Waltham	
G-BUCK	C.A.S.A. 1.131E Jungmann (BU+CK)	E. J. McEntee/White Waltham	
G-BUDG	Cessna 421C	A. F. Budge Ltd (G-BKWX)/E. Midlands	
G-BUDY	Colt 17A balloon	Bondbaste Ltd	
G-BUFF	Jodel D.112	D. J. Buffham/Fenland	
G-BUMP	PA-28-181 Archer II	Cordscan Ltd	
G-BURD	Cessna F.172N	RJS Aviation Ltd/Halfpenny Green	
G-BURT	PA-28-161 Warrior II	A. T. Howarth/Biggin Hill	
G-BUSY	Thunder Ax6-56A balloon	B. R. & Mrs M. Boyle *Busy Bodies*	
G-BUTL	PA-24 Comanche 250	D. Buttle (G-ARLB)/Blackbushe	
G-BUZZ	AB-206B JetRanger 2	Adifer Ltd	
G-BVMM	Robin HR.200/100	M. G. Owen	
G-BVMZ	Robin HR.100/210	Chiltern Handbags (London) Ltd	
G-BVPI	Evans VP-1	N. L. E. & R. A. Dupee/Dunkeswell	
G-BVPM	Evans VP-2	P. Marigold	
G-BWEC	Cassutt-Colson Variant	C. E. Bellhouse	

Notes	Reg.	Type	Owner or Operator
	G-BWFC	Boeing Vertol 234LR Chinook	British Airways Helicopters Ltd/ Aberdeen
	G-BWFJ	Evans VP-1	W. F. Jones
	G-BWHO	Thunder Ax6-56Z balloon	Thunder Balloons Ltd (G-BJZP)
	G-BWIG	G.17S replica	K. Wigglesworth
	G-BWJB	Thunder Ax8-105 balloon	Justerini & Brooks Ltd *Whiskey J. & B.*
	G-BWKK	Auster AOP.9 (XP279)	C. B. Godkin
	G-BWMB	Jodel D.119	Tony Dyer Television
	G-BWMP	Gulfstream 695A	R. B. Tyler (Plant) Ltd
	G-BWRB	D.H.C.-6 Twin Otter 310	Brymon Aviation Ltd/Plymouth
	G-BWSI	K&S SA.102.5 Cavalier	B. W. Shaw
	G-BWWW	BAe Jetstream 3102	The Distillers Co PLC/Bournemouth
	G-BXAZ	PA-31 Turbo Navajo	Michian Aviation Ltd (G-AXAZ)
	G-BXNW	SNCAN SV-4C	Telepoint Ltd/Barton
	G-BXPU	H.S.125 Srs 3B/RA	McAlpine Aviation Ltd (G-AXPU/ G-IBIS)/Luton
	G-BXVI	V.S.361 Spitfire F.XVI	D. Arnold
	G-BXYZ	R. Turbo Commander 690C	British Airports Authority/Gatwick
	G-BYLL	F.8L Falco	N. J. Langrick
	G-BYRD	Mooney M.20K	Birds Garage Ltd/Denham
	G-BYSE	AB-206B JetRanger 2	Bewise Ltd (G-BFND)
	G-BZAC	Sikorsky S-76A	British Airways Helicopters Ltd/ Aberdeen
	G-BZBH	Thunder Ax6-65 balloon	R. S. Whittaker & P. E. Sadler
	G-BZBY	Colt 56 Buzby balloon	British Telecom
	G-BZKK	Cameron V-56 balloon	P. J. Green & C. Bosley *Gemini II*
	G-BZZZ	Enstrom F-28C-UK	Star Groundwork Ltd (G-BBBZ)
	G-CALL	PA-23 Aztec 250F	Woodgate Aviation Ltd/Ronaldsway
	G-CBEA	BAe Jetstream 3102-01	Birmingham Executive Airways Ltd
	G-CBIA	BAC One-Eleven 416	British Island Airways (G-AWXJ) *Island Ensign*/Gatwick
	G-CBIL	Cessna 182K	N. Law/Manchester
	G-CCAA	H.S.125 Srs 700B	Civil Aviation Authority (G-DBBI)/ Stansted
	G-CCAR	Cameron N-77 balloon	Colt Car Co Ltd *Colt*
	G-CCCC	Cessna 172H	D. A. Pearce & D. M. Power
	G-CCIX	V.S.361 Spitfire LF.IX	Charles Church (Spitfires) Ltd (G-BIXP)
	G-CCOZ	Monnet Sonerai II	P. R. Cozens
	G-CCUB	Piper J-3C-65 Cub	Cormack (Aircraft Services) Ltd
	G-CDAH	Taylor Super Coot A	D. A. Hood
	G-CDAN	V.S.361 Spitfire LF.XVIC	B. J. S. Grey/Duxford
	G-CDGA	Taylor JT.1 Monoplane	D. G. Anderson (*Stored*)/Prestwick
	G-CDGL	Saffery S.330 balloon	C. J. Dodd & G. J. Luckett *Penny*
	G-CEGA	PA-34-200T-2 Seneca	Cega Aviation Ltd/Goodwood
	G-CEGB	AS.355F1 Twin Squirrel	Central Electricity Generating Board (G-BLJL)
	G-CELL	PA-32R-301 Saratoga SP	CEL Electronics Ltd
	G-CETA	Cessna E.310Q	CETA Video Ltd (G-BBIM)/Fairoaks
	G-CETC	Aeronca 15AC Sedan	G. Churchill/Finmere
	G-CEZY	Thunder Ax9-140 balloon	R. Carr/France
	G-CFBI	Colt 56A balloon	Lighter-Than-Air Ltd
	G-CFLY	Cessna 172F	Cee-Fly
	G-CGHM	PA-28 Cherokee 140	CGH Managements Ltd/Elstree
	G-CHIK	Cessna F.152	Wickwell (UK) Ltd (G-BHAZ)
	G-CHIP	PA-28-181 Archer II	Cross Lane Properties/Sherburn
	G-CHOP	Westland Bell 47G-3B1	Time Choppers Ltd/Chilbolton
	G-CHTT	Varga 2150A Kachina	C.H.T. Trace
	G-CINE	Bell 206L-1 LongRanger	PLM Helicopters Ltd
	G-CITY	PA-31-350 Navajo Chieftain	Woodgate Aviation Ltd/Ronaldsway
	G-CJBC	PA-28 Cherokee 180	J. B. Cave/Halfpenny Green
	G-CJCB	Bell 206L LongRanger	J. C. Bamford (Excavators) Ltd (G-LIII)/ E. Midlands
	G-CJCI	Pilatus P2-06	C. Church/Blackbushe
	G-CJHI	Bell 206B JetRanger	Tudorbury Air Services Ltd (G-BBFB)
	G-CJIM	Taylor JT.1 Monoplane	J. Crawford
	G-CLEA	PA-28-161 Warrior II	Roger Head Motors/Staverton
	G-CLEM	Bo 208A2 Junior	G. Clements (G-ASWE)/Netherthorpe
	G-CLIK	PA-18 Super Cub 95	N. J. R. Empson/Ipswich
	G-CLUB	Cessna FRA.150M	B.L.A. Ltd/Birmingham
	G-CLUX	Cessna F.172N	D. Bardsley/Manchester
	G-CNET	AS.355F-1 Twin Squirrel	McAlpine Helicopters Ltd (G-MCAH)/ Hayes
	G-CNIS	Partenavia P.68B	P. Davison (G-BJOF/G-PAUL)

Reg.	Type	Owner or Operator	Notes
G-COAL	Bell 206B JetRanger 3	Brencham Helicopter Management Ltd	
G-COCO	Cessna F.172M	Capel Aviation/Ipswich	
G-COIN	Bell 206B JetRanger 2	P. Woodward	
G-COLD	Cessna T.337D	Coldspec Ltd/Bournemouth	
G-COLL	Enstrom F-280C	Decoy Engineering Projects Ltd	
G-COLN	AS.350B Ecureuil	C. M. Meade (G-BHIV)	
G-COMB	PA-30 Twin Comanche 160	J. A. Ranscombe (G-AVBL)	
G-COMM	PA-23 Aztec 250	Commair Aviation Ltd (G-AZMG)/ E. Midlands	
G-CONI	L.749A Constellation (N7777G) ★	Science Museum/Wroughton	
G-COOL	Cameron O-31 balloon	Swire Bros *Sprite*/Hong Kong	
G-COOP	Cameron N-31 balloon	Balloon Stable Ltd *Co-op*	
G-COPE	Enstrom F-280C-UK-2 Shark	A. Cope	
G-COPS	Piper J-3C-65 Cub	W. T. Sproat	
G-COPY	AA-5A Cheetah	D. G. Lewerdon (G-BIEU)	
G-CORR	AS.355F-1 Twin Squirrel	Gorran Foods Ltd	
G-COTT	Cameron 60 SS balloon	Nottingham Building Soc	
G-COZY	Rutan Cozy	J. F. Mackay	
G-CPFC	Cessna F.152	B. W. Wells & Burbage Farms Ltd	
G-CPPC	PA-23 Aztec 250	L.&M. Food Group Ltd	
G-CPTS	AB-206B JetRanger 2	A. R. B. Aspinall	
G-CRAN	Robin R.1180T	Slea Aviation Ltd/Cranwell	
G-CRIC	Colomban Cri-Cri MC-15	A. J. Maxwell	
G-CRIL	R. Commander 112B	K. Cochrane	
G-CRIS	Taylor JT.1 Monoplane	C. J. Bragg	
G-CRTI	PA-28RT-201 Arrow IV	Decoy Engineering Projects Ltd	
G-CRZY	Thunder Ax8-105 balloon	R. Carr/France	
G-CSBM	Cessna F.150M	Coventry (Civil) Aviation Ltd	
G-CSFC	Cessna 150L	Knack Air Ltd	
G-CSFT	PA-23 Aztec 250	SFT Aviation Ltd (G-AYKU)/ Bournemouth	
G-CSNA	Cessna 421C	Nouvair Ltd	
G-CSZB	V.807B Viscount	British Air Ferries *Viscount Scotland* (G-AOXU)/Southend	
G-CTIX	V.S.509 Spitfire T.IX	Charles Church (Spitfires) Ltd	
G-CTKL	CCF Harvard IV	C. T. K. Lane	
G-CTRN	Enstrom F-28C-UK	W. E. Taylor & Son Ltd	
G-CTRX	H.P.137 Jetstream 200	Centrax Ltd (G-BCWW/G-AXUN)/Exeter	
G-CTSI	Enstrom F-280C Shark	Lemlyne Ltd (G-BKIO)	
G-CUBB	PA-18-150 Super Cub	Booker Gliding Club Ltd	
G-CUBI	PA-18-135 Super Cub	Hambletons Gliding Club Ltd	
G-CUBJ	PA-18 Super Cub 150	A. K. Leasing (Jersey) Ltd	
G-CUKL	Beech 200 Super King Air	Conoco (UK) Ltd (G-CNSI/G-OSKA)/ Inverness	
G-CWOT	Currie Wot	D. A. Lord	
G-CYII	H.S.125 Srs 600B	Yeates of Leicester Ltd (G-BART)	
G-CYMA	GA-7 Cougar	CYMA Petroleum Ltd (G-BKOM)/Elstree	
G-DAAH	PA-28R-201T Turbo Arrow IV	A. A. Hunter	
G-DACA	P.57 Sea Prince T.1	Atlantic & Caribbean Aviation Ltd/ Staverton	
G-DAFS	Cessna 404 Titan	Dept. of Agriculture & Fisheries for Scotland (G-BHNH)/Edinburgh	
G-DAJW	K & S Jungster 1	A. J. Walters	
G-DAKS	Dakota 3 (KG874)	Aces High Ltd/North Weald	
G-DAND	SOCATA TB.10 Tobago	Whitemoor Engineering Co Ltd	
G-DANN	Stampe SV-4B	D. R. Scott-Songhurst/White Waltham	
G-DART	Rollason Beta B2	M. G. Ollis	
G-DASI	Short SD3-60	Air UK Ltd (G-BKKW)/Norwich	
G-DATA	EMB-110P2 Bandeirante	Euroair Ltd (G-BGNK)/Gatwick	
G-DAVE	Jodel D.112	D. A. Porter/Sturgate	
G-DAVY	Evans VP-2	D. Morris	
G-DBAL	H.S.125 Srs 3B	Falcon Jet Centre Ltd (G-BSAA)	
G-DCAN	PA-38-112 Tomahawk	Airways Aero Associations Ltd/Booker	
G-DCCC	H.S.125 Srs 800B	British Aerospace PLC/Chester	
G-DCIO	Douglas DC-10-30	British Caledonian Airways *Flora McDonald — The Scottish Heroine*	
G-DCKK	Cessna F.172N	Bulldog Aviation Ltd/Andrewsfield	
G-DDCD	D.H.104 Dove 8	C. Daniel (G-ARUM)/Biggin Hill	
G-DDDV	Boeing 737-2S3	Air Europe Ltd *Peggy*/Gatwick	
G-DEBS	Colt AA-150 gas balloon	Hot-Air Balloon Co Ltd	
G-DELI	Thunder Ax7-77 balloon	C. Delius	
G-DENS	CP.301S Smaragd	D. Russel	

Notes	Reg.	Type	Owner or Operator
	G-DEVN	D.H.104 Devon C.2	T. P. Luscombe & P. Gaston/Lympne
	G-DEVS	PA.28 Cherokee 180	H. C. Devonish (G-BGVJ)/Southend
	G-DEXP	ARV Super 2	ARV Aviation Ltd
	G-DFIN	SA.365N Dauphin 2	McAlpine Helicopters Ltd/Hayes
	G-DFLY	PA-38-112 Tomahawk	Airways Aero Associations Ltd/Booker
	G-DFTS	Cessna FA.152	Hartmann Ltd/Booker
	G-DFUB	Boeing 737-2K9	Monarch Airlines Ltd/Luton
	G-DGDG	Glaser-Dirks DG-400/17	I. L. McKelvie & ptnrs
	G-DGDP	Boeing 737-2T7	Monarch Airlines Ltd/Luton
	G-DHTM	D.H.82A Tiger Moth replica	C. R. Hardiman
	G-DICK	Thunder Ax6-56Z balloon	Bandag Tyre Co
	G-DINA	AA-5B Tiger	Simon Deverall Print Ltd/ Compton Abbas
	G-DIPS	Taylor JT.1 Monoplane	B. J. Halls
	G-DISC	Cessna U.206A	I. A. Louttit (G-BGWR)/Exeter
	G-DJBE	Cessna 550 Citation II	Fisons PLC/E. Midlands
	G-DJHB	Beech A23-19 Musketeer	Wayfree Ltd (G-AZZE)/Andrewsfield
	G-DJHH	Cessna 550 Citation II	York Aviation Ltd/Humberside
	G-DJIM	MHCA-I	J. Crawford
	G-DJMJ	H.S.125 Srs 1B	McAlpine Aviation Ltd (G-AWUF)/Luton
	G-DKDP	Grob G.109B	Diss Aviation Ltd
	G-DLRA	BN-2T Islander	Pilatus BN Ltd (G-BJYU)/Bembridge
	G-DMAF	Bell 222A	Genavco Air Ltd (G-BLSZ)
	G-DMCH	Hiller UH-12E	D. McK. Carnegie & ptnrs
	G-DMCS	PA.28R Cherokee Arrow 200-2	Adeptus Drawing Office & Studio Supplies Ltd (G-CPAC)/Manston
	G-DODD	Cessna F.172P-II	Johtyne Ltd
	G-DOGS	Cessna R.182RG	Usoland Ltd/Elstree
	G-DOOK	Beech 60 Duke	I. K. Stewart (G-AXEN)
	G-DORE	Partenavia P.68C	Nullifire Ltd/Coventry
	G-DOVE	Cessna 182Q	S. G. Lawrence
	G-DRAY	Taylor JT.1 Monoplane	L. J. Dray
	G-DSHW	Boeing 737-300	Monarch Airlines Ltd/Luton
	G-DTOO	PA-38-112 Tomahawk	Airways Aero Associations Ltd/Booker
	G-DUET	Wood Duet	C. Wood
	G-DUNN	Zenair CH.250	A. Dunn
	G-DUVL	Cessna F.172N	Duval Studios Ltd/Denham
	G-DVON	D.H.104 Devon C.2 (VP955)	C. L. Thatcher/Staverton
	G-DWHH	Boeing 737-2T7	Monarch Airlines Ltd/Luton
	G-DWMI	Bell 206L-1 LongRanger	Glenwood Helicopters Ltd/Fairoaks
	G-DYOU	PA-38-112 Tomahawk	Airways Aero Associations Ltd/Booker
	G-EAGL	Cessna 421C	Associated Dairies Ltd/Leeds
	G-EBJI	Hawker Cygnet Replica	A. V. Francis
	G-ECBH	Cessna F.150K	Air Fenland Ltd
	G-ECGC	Cessna F.172N-II	Leicestershire Aero Club Ltd
	G-ECHO	Enstrom F-280C-UK-2 Shark	ALP Electrical (Maidenhead) Ltd (G-LONS/G-BDIB)/White Waltham
	G-ECMA	PA-31-325 Navajo	Elliot Bros (London) Ltd/Rochester
	G-ECOX	Grega GN.1 Air Camper	H. C. Cox
	G-EDDY	PA-28RT-201 Arrow IV	Supaglide Ltd/Stapleford
	G-EDEN	SOCATA TB.10 Tobago	N. I. Mandell & J. D. Wittich/ Elstree
	G-EDHE	PA-24 Comanche 180	Hughes Engineers (Devon) Ltd (G-ASFH)/Exeter
	G-EDIF	Evans VP-2	R. Simpson
	G-EDNA	PA-38-112 Tomahawk	Sandwell Scaffold Co Ltd/Manchester
	G-EEEE	Slingsby T.31 Motor Glider	R. F. Selby
	G-EENY	GA-7 Cougar	Charles Henry Leasing/Elstree
	G-EESE	Cessna U.206G	D. Penny
	G-EEUP	SNCAN SV-4C	Meridian Drilling Co Ltd
	G-EEZE	Rutan Vari-Eze	A. J. Nurse
	G-EGEE	Cessna 310Q	C. & D. Extended Warranties Ltd (G-AZVY)/Elstree
	G-EGGS	Robin DR.400/180	R. Foot/Bickmarsh
	G-EGLE	Christen Eagle II	Airmore Aviation Ltd/Elstree
	G-EHAP	Sportavia-Pützer RF.7	M. J. Revill/Exeter
	G-EHMM	Robin DR.400/180R	Booker Gliding Club Ltd
	G-EIIR	Cameron N-77 balloon	Major C. J. T. Davey *Silver Jubilee*
	G-EIWT	Cessna FR.182RG	P. P. D. Howard-Johnston/Edinburgh
	G-EJCB	Agusta A.109A Mk2	J. C. Bamford Excavators Ltd
	G-EJGO	Zlin Z.226HE Trener	N. J. Radford
	G-EKOE	Robin DR.400/180R	London Gliding Club (Pty) Ltd/ Dunstable

Reg.	Type	Owner or Operator	Notes
G-ELEC	Westland WG.30 Srs 200	Westland Helicopters Ltd (G-BKNV)/ Yeovil	
G-EMAK	PA-28R-201 Cherokee Arrow II	H. Clayton & J. O. Carlisle	
G-EMKM	Jodel D.120A	Cawdor Flying Group/Inverness	
G-EMMA	Cessna F.182Q	Watkiss Group Aviation	
G-EMMS	PA-38-112 Tomahawk	Surrey & Kent Flying Club Ltd/ Biggin Hill	
G-EMMY	Rutan Vari-Eze	M. J. Tooze	
G-ENCE	Partenavia P.68B	P. A. de Courcy-Swoffer (G-OROY/G-BFSU)	
G-ENIE	Nipper T.66 Srs 3	C. N. Harrison/Denham	
G-ENII	Cessna F.172M	G. Weale/Shoreham	
G-ENOA	Cessna F.172F	M. Konstantinovic (G-ASZW)/ Stapleford	
G-ENSI	Beech F33A Bonanza	F. B. Gibbons & Sons Ltd	
G-EOFF	Taylor JT.2 Titch	G. H. Wylde	
G-EORG	PA-38-112 Tomahawk	Airways Aero Association/Booker	
G-EPDI	Cameron N-77 balloon	R. Moss & Pegasus Aviation Ltd	
G-ERIC	R. Commander 112TC	Beech Group/Cardiff	
G-ERMS	Thunder AS33 Airship	Thunder Balloons Ltd	
G-ERRY	AA-5B Tiger	T. R. Bamber (G-BFMJ)/Denham	
G-ERTY	D.H.82A Tiger Moth	E. R. Thomas (G-ANDC)	
G-ESSX	PA-28-161 Warrior II	S. Harcourt (G-BHYY)/Shoreham	
G-EURA	Agusta-Bell 47J-2	E. W. Schnedlitz (G-ASNV)	
G-EVAN	Taylor JT.2 Titch	E. Evans	
G-EVNS	Cessna 441	Northair Aviation Ltd/Leeds	
G-EWBJ	SOCATA TB.10 Tobago	Crocker Air Services/Biggin Hill	
G-EWIZ	Pitts S-2E Special	J. E. Davies	
G-EXEC	PA-34-200 Seneca	Roys Cooked Meats Ltd/Swansea	
G-EXEX	Cessna 404	Hubbard Air Ltd/Norwich	
G-EXIT	SOCATA Rallye 180GT	G-Exit Ltd/Rochester	
G-EZEE	Rutan Vari-Eze	M. G. E. Hutton	
G-EZOS	Rutan Vari-Eze	O. Smith/Tees-side	
G-FAGT	Cessna 337G	Acorn Computers Ltd/Cambridge	
G-FAIR	SOCATA TB.10 Tobago	Sally Marine Ltd/Guernsey	
G-FALC	Aeromere F.8L Falco	P. W. Hunter (G-AROT)/Elstree	
G-FALK	Sequoia F.8L Falco 4	I. Chancellor	
G-FALL	Cessna 182L	D. M. Penny	
G-FANG	AA-5A Cheetah	Reedtrend Ltd/Biggin Hill	
G-FANL	Cessna FR.172K XP-II	J. Woodhouse & Co/Staverton	
G-FARM	SOCATA Rallye 235GT	M. J. Jardine-Paterson	
G-FARR	Jodel D.150	G. H. Farr	
G-FAYE	Cessna F.150M	Cheshire Air Training School Ltd/ Liverpool	
G-FBDC	Cessna 340A	P. L. & M. J. E. Builder (G-BFJS)	
G-FBWH	PA-28R Cherokee Arrow 180	Servicecentre Systems (Cambs) Ltd	
G-FCAS	PA-23 Aztec 250	R. A. Berry Ltd	
G-FCHJ	Cessna 340A	Tilling Associates Ltd (G-BJLS)	
G-FDGM	Beech B60 Duke	Fisher & Donaldson (G-BFEZ)	
G-FELT	Cameron N-77 balloon	Allan Industries Ltd	
G-FERY	Cessna 550 Citation II	European Ferries Ltd (G-DJBI)/Gatwick	
G-FFEN	Cessna F.150M	Suffolk Aero Club Ltd/Ipswich	
G-FFLY	Slingsby T.67M Firefly	Slingsby Aviation Ltd/ Kirkbymoorside	
G-FHAS	Scheibe SF.25E Super Falke	Fourth Harrow Aviation/Lasham	
G-FIRE	V.S.379 Spitfire XIVc	Classic Air Displays Ltd/Elstree	
G-FISH	Cessna 310R-II	Boston Deep Sea Fisheries Ltd/ Humberside	
G-FIST	Fieseler Fi.156C Storch	Spoils Kitchen Reject Shops Ltd	
G-FIZZ	PA-28-161 Warrior II	N. C. L. & J. E. Wright	
G-FJKI	Cessna 404 Titan	Spartan Aviation Ltd (G-VWGB)	
G-FLCH	AB-206B JetRanger 3	Fletchair (G-BGGX)/Leeds	
G-FLCO	Sequoia F.8L Falco	J. B. Mowforth	
G-FLEA	SOCATA TB.10 Tobago	Fleair Trading Co/Biggin Hill	
G-FLIC	Cessna FA.152	Stanton Aircraft Management Ltd (G-BILV)/Biggin Hill	
G-FLIK	Pitts S.1S Special	R. P. Millinship	
G-FLIP	Cessna FA.152	Brailsford Aviation Ltd/Netherthorpe	
G-FLIX	Cessna E.310P	Stuart Hurrion & Co Solicitors (G-AZFL)	
G-FLPI	R. Commander 112A	Tuscany Ltd/Leicester	
G-FLYI	PA-34-200 Seneca	Airmore Aviation Ltd (G-BHVO)/Elstree	
G-FMUS	Robinson R-22	F. M. Usher-Smith (G-BJBT)	

Notes	Reg.	Type	Owner or Operator
	G-FOAM	M.S.892A Rallye Commodore	McAully Flying Group Ltd (G-AVPL)/ Little Snoring
	G-FOCK	WAR Focke-Wulf Fw.190-A	P. R. Underhill
	G-FOOD	Beech B200 Super King Air	Airmore Aviation Ltd/Elstree
	G-FORC	SNCAN Stampe SV-4A	J. A. Sykes
	G-FORD	SNCAN SV-4B	P. Meeson
	G-FORT	Boeing B-17G-VE	Fairoaks Aviation Services Ltd
	G-FOTO	PA-23 Aztec 250	Davis Gibson Advertising Ltd (G-BJDH/ G-BDXV)/Booker
	G-FOUX	AA-5A Cheetah	Baryn Finance Ltd/Elstree
	G-FRAG	PA-32-300 Cherokee Six	R. Goodwin & Co Ltd/Southend
	G-FRED	FRED Srs 2	R. Cox
	G-FREE	Pitts S-2A Special	Pegasus Flying Group
	G-FRJB	Britten Sheriff SA-1 ★	Aeropark/E. Midlands
	G-FROZ	Cessna 421C	European Air Charter Ltd
	G-FRST	PA-44T Turbo Seminole 180	Frost & Frost/Kidlington
	G-FSCL	AB-206B JetRanger	Thirsk Aero Services Ltd
	G-FSDA	AB-206B JetRanger 2	Southern Air Ltd (G-AWJW)/Shoreham
	G-FSDG	AB-206B JetRanger	Flair (Soft Drinks) Ltd (G-ROOT/G-JETR)
	G-FSPL	PA-32R-300 Lance	Klingair Ltd/Conington
	G-FTAX	Cessna 421C	Arbour Finance Ltd
	G-FUEL	Robin DR.400/180	R. Darch/Compton Abbas
	G-FUJI	Fuji FA200-180	K. J. Farrance & G. Willson/Luton
	G-FULL	PA-28R-200-2 Cherokee Arrow	Fuller Aviation (G-HWAY/G-JULI)
	G-FUND	Thunder Ax7-65Z balloon	Schroder Life Assurance Ltd
	G-FUZZ	PA-19 Super Cub 95	G. W. Cline
	G-FVEE	Monnet Sonerai I	D. R. Sparke
	G-FWRP	Cessna 421C	Vange Scaffolding & Engineering Co Ltd
	G-FXIV	V.S.379 Spitfire FR.XIV (MV370)	R. Lamplough
	G-FZZZ	Colt 56A balloon	Hot-Air Balloon Co Ltd
	G-GABD	GA-7 Cougar	Turnhouse Flying Club Ltd
	G-GACA	P.57 Sea Prince T.1	Atlantic & Caribbean Aviation Ltd/ Staverton
	G-GAEL	H.S.125 Srs 800B	British Aerospace PLC/Hatfield
	G-GALE	PA-34-200T-2 Seneca	Gale Construction Co Ltd/Norwich
	G-GAMA	Beech B58 Baron	Gama Aviation Ltd (G-BBSD)/Fairoaks
	G-GAME	Cessna T.303	Lindsay Advertising Ltd
	G-GANE	Sequoia F.8L Falco	S. J. Gane
	G-GANJ	Fournier RF-6B-100	Soaring Equipment Ltd/Coventry
	G-GASA	Hughes 369HS	Flair (Soft Drinks) Ltd (G-TATI)/ Shoreham
	G-GASB	Hughes 369HS	Nelson Helicopter Services Ltd/ Shoreham
	G-GASC	Hughes 369HS	Southern Air Ltd (G-WELD/G-FROG)/ Shoreham
	G-GAYL	Learjet 35A	AA Travel Services Ltd (G-ZING)/ Coventry
	G-GAZE	Robinson R-22A	Property Associates Ltd
	G-GBAO	Robin R.1180TD	J. Kay-Movat
	G-GBCA	Agusta A.109A Mk II	British Car Auctions (Aviation) Ltd
	G-GBLP	Cessna F.172M	B. L. Pratt
	G-GBLR	Cessna F.150L	B. L. Pratt/Luton
	G-GBSL	Beech 76 Duchess	George Barlow & Sons Ltd (G-BGVG)
	G-GCAA	PA-28R Cherokee Arrow 200	Southern Air Ltd/Shoreham
	G-GCAB	PA-30 Twin Comanche 180	Southern Air Ltd/Shoreham
	G-GCAT	PA-28 Cherokee 140B	G. S. B. Large (G-BFRH)/Denham
	G-GCKI	Mooney M.20K	Imperial Group Ltd/Tollerton
	G-GDAM	PA-18 Super Cub 135	G. D. A. Martin/Booker
	G-GEAR	Cessna FR.182Q	Ashcombe Distributors
	G-GEEP	Robin R.1180T	Organic Concentrates Ltd/Booker
	G-GEES	Cameron N-77 balloon	Mark Jarvis Ltd *Mark Jarvis*
	G-GEEZ	Cameron N-77 balloon	Charnwood Forest Turf Accountants Ltd
	G-GEIL	H.S.125 Srs 800B	Heron Management Ltd
	G-GENE	Cessna 501 Citation	Ambrian Aircraft Trading Ltd
	G-GEOF	Pereira Osprey 2	G. Crossley
	G-GEUP	Cameron N-77 balloon	P. Clark
	G-GFLY	Cessna F.150L	Knack Air Ltd
	G-GGGG	Thunder Ap7-77A balloon	T. A. Gilmour
	G-GHNC	AA-5A Cheetah	Chamberlain Leasing/Andrewsfield
	G-GHRW	PA-28RT-201 Arrow IV	Distance No Object Ltd (G-ONAB/ G-BHAK)/Stansted

Reg.	Type	Owner or Operator	Notes
G-GIGI	M.S.893A Rallye Commodore	Holding & Barnes Ltd (G-AYVX)	
G-GINA	AS.350B Ecureuil	Endeavour Aviation Ltd	
G-GKNB	Beech 200 Super King Air	GKN Group Services Ltd/Birmingham	
G-GLEN	Bell 212	Autair International Ltd/Panshanger	
G-GLOS	H.P.137 Jetstream 200	British Aerospace (G-BCGU/G-AXRI)	
G-GLUE	Cameron N-65 balloon	M. F. Glue	
G-GMSI	SOCATA TB.9 Tampico	A. R. Gray/Prestwick	
G-GNAT	H.S. Gnat T.1 (XS101)	Ruanil Investments Ltd/Cranfield	
G-GOGO	Hughes 369D	A. W. Alloys Ltd	
G-GOLD	Thunder Ax6-56A balloon	John Terry & Sons Ltd	
G-GOLF	SOCATA TB.10 Tobago	K. Piggott/Fairoaks	
G-GOMM	PA-32R-300 Lance	Embermere Ltd/Blackbushe	
G-GONE	D.H.112 Venom FB.50	Glylynn Ltd/Bournemouth	
G-GOOS	Cessna F.182Q	Roger Clark (Air Transport) Ltd	
G-GOSS	Jodel DR.221	M. I. Goss	
G-GOZO	Cessna R.182	Transmatic Fyllan Ltd (G-BJZO)/ Cranfield	
G-GRAY	Cessna 172N	BLS Aviation Ltd/Elstree	
G-GREG	Jodel DR.220 2+2	G. Long	
G-GRIF	R. Commander 112TCA	Z. I. Bilbeisi (G-BHXC)	
G-GROB	Grob G.109	G-GROB Ltd	
G-GROW	Cameron N-77 balloon	Derbyshire Building Society	
G-GTPL	Mooney M.20K	Advance Air Charter Ltd (G-BHOS)/ Manchester	
G-GUNN	Cessna F.172H	J. G. Gunn (G-AWGC)	
G-GWHH	AS.355F Twin Squirrel	Wimpey Homes Holdings Ltd (G-BKUL)	
G-GWIL	AS.350B Ecureuil	Talan Ltd	
G-GWYN	Cessna F.172M	C. Bosher	
G-GYRO	Bensen B.8	N. A. Pitcher & A. L. Howell	
G-HADI	G.1159 Gulfstream 2	Arab Express Ltd/Heathrow	
G-HAEC	Commonwealth Mustang 22	R. G. Hanna	
G-HALL	PA-22 Tri-Pacer 160	F P. Hall (G-ARAH)	
G-HALP	SOCATA TB.10 Tobago	D. Halpera (G-BITD)/Elstree	
G-HAMA	Beech 200 Super King Air	Gama Aviation Ltd	
G-HANK	Cessna FR.172H	J. H. & R. Hankinson (G-AYTH)	
G-HANS	Robin DR.400 2+2	Headcorn Flying School Ltd	
G-HAPR	B.171 Sycamore HR.14 (XG547) ★	British Rotorcraft Museum	
G-HASL	AA-5A Cheetah	D.B.G. Ltd (G-BGSL)/Biggin Hill	
G-HAWK	H.S.1182 Hawk	British Aerospace/Dunsfold	
G-HBCA	Agusta A.109A-11	British Car Auctions (Aviation) Ltd	
G-HBUS	Bell 206L-1 LongRanger	Toleman Delivery Service Ltd	
G-HCHU	Cessna TU.206G	Barons (UK) Ltd (G-BKKJ)	
G-HDBA	H.S.748 Srs 2B	British Airways/Glasgow	
G-HDBB	H.S.748 Srs 2B	British Airways/Glasgow	
G-HEAD	Colt 56 balloon	Colt Balloons Ltd	
G-HELI	Saro Skeeter Mk 12 (XM556) ★	British Rotorcraft Museum	
G-HELX	Cameron N-31 balloon	Hot-Air Balloon Co Ltd	
G-HELY	Agusta 109A	Castle Air Charters Ltd	
G-HENS	Cameron N-65 balloon	Horrells Dairies Ltd	
G-HEWI	Piper J-3C-65 Cub	Parker Airways Ltd (G-BLEN)/Denham	
G-HEWS	Hughes 369D	Apex Tubulars Ltd	
G-HEYY	Cameron N-65 balloon	Hot-Air Balloon Co Ltd	
G-HFCI	Cessna F.150L	Horizon Flying Club Ltd/Ipswich	
G-HFCT	Cessna F.152	P. Ratcliffe	
G-HGPC	BN-2A-27 Islander	Halfpenny Green Parachute Centre Ltd (G-FANS)	
G-HHOI	H.S.125 Srs 700B	Trust House Forte Airport Services Ltd (G-BHTJ)/Heathrow	
G-HIFI	PA-28R-201 Arrow III	Partipak Ltd (G-BFTB)/White Waltham	
G-HIGG	Beech 200 Super King Air	Navigations Ltd (G-BLKN)/Elstree	
G-HIGS	Cessna 404 Titan	Hubbardair Ltd (G-ODAS)/Norwich	
G-HIHO	Boeing 747-123	Highland Express Ltd	
G-HILR	Hiller UH-12E	G. & S. G. Neal (Helicopters) Ltd	
G-HIRE	GA-7 Cougar	London Aerial Tours Ltd (G-BGSZ)/ Biggin Hill	
G-HIVE	Cessna F.150M	M. P. Lynn (G-BCXT)/Sibson	
G-HLFT	SC.5 Belfast 2	HeavyLift Cargo Airlines Ltd/Stansted	
G-HLIX	Cameron 80 Oil Can balloon	Hot-Air Balloon Co Ltd	
G-HOFM	Cameron N-56 balloon	Hot-Air Balloon Co Ltd	
G-HOLT	Taylor JT.1 Monoplane	K. D. Holt	
G-HOME	Colt 77A balloon	Anglia Balloon School *Tardis*	
G-HOOK	Hughes 369D	Auto Alloys (Foundries) Ltd	
G-HOPE	Beech F33A Bonanza	Eurohaul Ltd/Southampton	

Notes	Reg.	Type	Owner or Operator
	G-HORN	Cameron V-77 balloon	Travel Gas (Midlands) Ltd
	G-HOST	Cameron N-77 balloon	A. J. Clarke & J. M. Hallam
	G-HOTS	Thunder Colt AS-80 airship	Island Airship Ltd
	G-HOUL	FRED Srs 2	D. M. M. Richardson
	G-HOUS	Colt 31A balloon	Anglia Balloons Ltd
	G-HOVA	Enstrom F-280C-UK Shark	Nelson Helicopter Services Ltd (G-BEYR)/Shoreham
	G-HPVC	Partenavia P.68B	Airtime (Hampshire) Ltd
	G-HRLM	Brugger MB.2 Colibri	R. A. Harris
	G-HRZN	Colt 77A balloon	D. Gaze
	G-HSDW	Bell 206B JetRanger	Wisfield Shoe Co Ltd
	G-HSKY	Hughes 369HM	Skyline Helicopters Ltd (G-VNPP/ G-BDKL)
	G-HSON	Cessna 441	Paul Sykes Group Ltd/Leeds
	G-HUBB	Partenavia P.68B	Hubbardair Ltd/Norwich
	G-HUEY	Bell UH-1H	RAF Benevolent Fund/Odiham
	G-HUFF	Cessna 182P	J. R. W. Keates & Z. Husein-Ellis/ Shoreham
	G-HUGE	Boeing 747-2D3B	British Caledonian Airways Ltd/Gatwick
	G-HUKT	PA-28-181 Archer II	P. Meyrick
	G-HULL	Cessna F.150M	Oldment Ltd/Grindale
	G-HUMP	Beech 95-B55 Baron	J. H. Humphreys (G-BAMI)/Guernsey
	G-HUNT	Hunter F.51	Berowell Management Ltd/ Bournemouth
	G-HUNY	Cessna F.150G	T. J. Lynn (G-AVGL)
	G-HURI	CCF Hawker Hurricane IIB	B. J. S. Grey
	G-HWBK	Agusta A.109A	Camlet Helicopters Ltd
	G-HYGA	H.S.125 Srs 800B	York Aviation Ltd
	G-IBCA	Beech 200 Super King Air	British Car Auctions (Aviation) Ltd (G-BMCA)
	G-IBFW	PA-28R-201 Arrow III	B. Walker & Co (Dursley) Ltd & J. & C. Ward (Holdings) Ltd/ Staverton
	G-IBLL	R. Commander 690D	Imperial Brewing & Leisure Ltd
	G-ICRU	Bell 206A JetRanger	Specialist Flying Training Ltd/ Carlisle
	G-ICUB	Piper J-3C-65 Cub	G. Cormack/Glasgow
	G-IDDY	D.H.C.I Super Chipmunk	N. A. Brendish (G-BBMS)/Southend
	G-IDEA	AA.5A Cheetah	Autohover Ltd (G-BGNO)
	G-IDJB	Cessna 150L	Leisure Lease Aviation/Southend
	G-IDWR	Hughes 369HS	Ryburn Air Ltd (G-AXEJ)
	G-IESH	D.H.82A Tiger Moth	I. E. S. Huddleston (G-ANPE)
	G-IFIT	PA-31-350 Navajo Chieftain	Noortman & Brod Ltd (G-NABI/ G-MARG)/Elstree
	G-IFLI	AA-5A Cheetah	Archpoint Ltd/Elstree
	G-IFTA	PA-31-350 Navajo Chieftain	Interflight (Air Charters) Ltd (G-BAVM)
	G-IGAR	PA-31-310C Turbo Navajo	J. H. Jackson (Estate Agents) Ltd
	G-IGON	PA-31-310 Turbo Navajo	Air Charter (Scotland) Ltd/Glasgow
	G-IKIS	Cessna 210M	Bob Crowe Aircraft Sales Ltd/Cranfield
	G-ILFC	Boeing 737-2U4	Dan-Air Services Ltd (G-BOSL)/Gatwick
	G-ILLY	PA-28-181 Archer II	A. G. & K. M. Spiers
	G-ILSE	Corby CJ-1 Starlet	S. Stride
	G-IMBE	PA-31-310 Navajo	Ambrian Aviation Ltd (G-BXYB/ G-AXYB)
	G-IMLH	Bell 206A JetRanger 3	Sabaru (UK) Ltd
	G-INDC	Cessna T.303	Biograft Medical Group Ltd (G-BKFH)
	G-INMO	PA-31-310 Turbo Navajo	Sabaru (UK) Ltd/Coventry
	G-INNY	SE-5A Replica (F5459)	R. M. Ordish/Old Sarum
	G-INOW	Monnet Moni	M. C. Searle & M. J. Sullivan/Elstree
	G-IOMA	F.27 Friendship Mk 100	Loganair Ltd/Glasgow
	G-IOOI	Robin DR.400/160	Moto Baldet (Northampton) Ltd/ Sywell
	G-IOOO	Gulfstream Commander 1000	Falcon Jet Centre Ltd/Heathrow
	G-IOSI	Jodel DR.1051	R. G. E. Simpson & A. M. Alexander
	G-IPEC	SIAI-Marchetti S.205-18F	G. E. Taylor (G-AVEG)
	G-IPPM	SA.102-5 Cavalier	I. D. Perry & P. S. Murfitt
	G-IPRA	Beech A200 Super King Air	J. H. Ritblat (G-BGRD)/Stansted
	G-IPSY	Rutan Vari-Eze	R. A. Fairclough/Biggin Hill
	G-IRLS	Cessna FR.172J	Starvillas Ltd/Luton
	G-ISIS	D.H.82A Tiger Moth	D. R. & M. Wood (G-AODR)
	G-ISLE	Short SD3-60	Manx Airlines Ltd (G-BLEG)/ Ronaldsway
	G-ITTU	PA-23 Aztec 250	Grasmere Hotels Ltd (G-BCSW)

Reg.	Type	Owner or Operator	Notes
G-IVAN	Rutan Vari-Eze	I. Shaw	
G-IWPL	Cessna F.172M	Reedy Supplies Ltd/Exeter	
G-JADE	Beech 95-58 Baron	Liaison & Consultant Services Ltd	
G-JAFC	Cameron N-77 balloon	J. A. F. Croft	
G-JAJV	Partenavia P.68C	Matthew Royce Ltd	
G-JAKE	D.H.C.I Chipmunk 22	J. M. W. Henstock (G-BBMY)/ Netherthorpe	
G-JAKY	PA-31-325 Navajo	Air Charter Europe/Glasgow	
G-JANE	Cessna 340A	Malcolm Air Ltd/Birmingham	
G-JANS	Cessna FR.172J	I. G. Aizlewood/Luton	
G-JASM	Robinson R-22A	J. L. Lawrence & ptnrs	
G-JASP	PA-23 Aztec 250	Landsurcon (Air Survey) Ltd/Staverton	
G-JAZZ	AA-5A Cheetah	Biggin Hill School of Flying	
G-JBUS	FRED Srs 2	R. V. Joyce	
G-JCUB	PA-18 Super Cub 135	Piper Cub Consortium Ltd/Jersey	
G-JDEE	SOCATA TB.20 Trinidad	John Dee Transport Ltd (G-BKLA)	
G-JDHI	Enstrom F-28C-UK	Valiant Press Ltd (G-BCOT)	
G-JDIX	Mooney M.20B	J. E. Dixon (G-ARTB)	
G-JEFF	PA-38-112 Tomahawk	Channel Aviation Ltd/Guernsey	
G-JELY	PA-18A Super Cub 150	W. R. M. C. Foyle	
G-JENA	Mooney M.20K	P. Leverkuehn/Biggin Hill	
G-JENN	AA-5B Tiger	Bambair/Denham	
G-JENS	SOCATA Rallye 100ST	B. H. Burnet (G-BDEG)/Dunkeswell	
G-JENY	Baby Great Lakes	J. M. C. Pothecary	
G-JETA	Cessna 550 Citation II	IDS Aircraft Ltd/Heathrow	
G-JETB	Cessna 550 Citation II	IDS Aircraft Ltd/Heathrow	
G-JETC	Cessna 550 Citation II	IDS Aircraft Ltd/Heathrow	
G-JETD	Cessna 550 Citation II	IDS Aircraft Ltd/Heathrow	
G-JETE	Cessna 500 Citation	IDS Aircraft Ltd (G-BCKM)/Heathrow	
G-JETH	Hawker Sea Hawk FGA.6	Brencham Historic Aircraft Ltd/ Bournemouth	
G-JETM	Gloster Meteor T.7	Brencham Historic Aircraft Ltd/ Bournemouth	
G-JETP	Hunting Jet Provost T.54A	Hunter Promotions Ltd/Bournemouth	
G-JETS	A.61 Terrier 2	J. E. Tootell (G-ASOM)	
G-JFWI	Cessna F.172N	J. F. Wallis/Goodwood	
G-JGCL	Cessna 414A	Johnson Group Management Services Ltd/Blackpool	
G-JGFF	AB-206B JetRanger 3	S.W. Electricity Board/Lulsgate	
G-JILL	R. Commander 112TCA	Hanover Aviation/Elstree	
G-JIMS	Cessna 340A-II	Granpack Ltd (G-PETE)/Leavesden	
G-JIMY	PA-28 Cherokee 140	Decoy Engineering Projects Ltd (G-AYUG)/Coventry	
G-JJCB	H.S.125 Srs 800B	J. C. Bamford Ltd	
G-JJSG	Learjet 35A	Smurfit Ltd	
G-JLBI	Bell 206L-1 Long Ranger	Alton Towers Ltd	
G-JLCO	AS.355F-1 Twin Squirrel	John Laing Construction Ltd	
G-JLTB	Varga 2150A Kachina	Acorn Ltd/Elstree	
G-JMCC	Beech 95-58 Baron	Ibis Enterprises Ltd/Jersey	
G-JMFW	Taylor JT.1 Monoplane	G. J. M. F. Winder	
G-JMVB	AB-206B JetRanger 3	J. M. V. Butterfield (G-OIML)	
G-JMWT	SOCATA TB.10 Tobago	Halton Communications Ltd/Liverpool	
G-JOAN	AA-5B Tiger	Oldment Ltd (G-BFML)/Netherthorpe	
G-JOES	Cessna 421B	Sipson Coachworks Ltd (G-BLOH/ G-NAIR/G-KACT)	
G-JOEY	BN-2A Mk III-2 Trislander	Aurigny Air Services (G-BDGG)/ Guernsey	
G-JOKE	AB-206B JetRanger 3	Jokeman Ltd (G-CSKY/G-TALY)	
G-JOLY	Cessna 120	T. E. G. Burkett/White Waltham	
G-JONE	Cessna 172M	Glibbery Electronics Ltd/Stapleford	
G-JONI	Cessna FA.152	Luton Flight Training Ltd (G-BFTU)	
G-JONS	PA-31-350 Navajo Chieftain	Topflight Aviation Ltd/Fairoaks	
G-JORR	AS.350B Ecureuil	C.E. Aviation Ltd (G-BJMY)/Staverton	
G-JOSE	Cessna U.206G	Safari Skylink Enterprises Ltd	
G-JRBI	AS.350B Ecureuil	Berkeley Leisure Group Ltd (G-BKJY)	
G-JRCM	Hawker Fury Mk I Replica	J. R. C. Morgan	
G-JRMM	R. Turbo Commander 690B	R. A. Berry Ltd/Bournemouth	
G-JSSD	SA. Jetstream 3001	British Aerospace (G-AXJZ)/Prestwick	
G-JTCA	PA-23 Aztec 250	J. D. Tighe (G-BBCU)/Sturgate	
G-JTIE	Cessna 421C	Eastern Air Executive (G-RBBE)/ Sturgate	
G-JUDI	AT-6D Harvard III (FX301)	A. Haig-Thomas	
G-JUDY	AA-5A Cheetah	BLS Aviation Ltd/Elstree	

Notes	Reg.	Type	Owner or Operator
	G-JULY	AA-5A Cheetah	W. H. Wilkins Ltd (G-BHTZ)
	G-JURG	R. Commander 114A	Jurgair Ltd
	G-JVMR	Partenavia P.68B	Sonardyne Ltd (G-JCTI/G-OJOE)/ Blackbushe
	G-JWIV	Jodel DR.1051	J. W. West
	G-KADY	Rutan Long Ez	M. W. Caddy
	G-KAFC	Cessna 152	B. W. Wells
	G-KAIR	PA-28-181 Archer II	Academy Lithoplates Ltd/Aldergrove
	G-KARI	Fuji FA.200-160	C. J. Zetter (G-BBRE)
	G-KATE	Westland WG.30 Srs 100	Helicopter Hire Ltd/Southend
	G-KATH	Cessna P.210N	Avionics Research Ltd
	G-KATS	PA-28 Cherokee 140	ALP Electrical (Maidenhead) Ltd/ White Waltham
	G-KBPI	PA-28-161 Warrior II	K. B. Page (Aviation) Ltd (G-BFSZ)/ Shoreham
	G-KCIG	Sportavia RF-5B	K. A. Nicholls
	G-KDFF	Scheibe SF.25E Super Falke	Deeside Super Falke Group
	G-KDIX	Jodel D.9 Bebe	K. Barlow
	G-KEEN	Stolp SA.300 Starduster Too	F. Holmes/Andrewsfield
	G-KEMC	Grob G.109B	Eye-Fly Ltd
	G-KENY	Enstrom F-280C-UK-2	A. J. & B. Pond/Booker
	G-KERC	Nord NC.854S	Kirk Aviation
	G-KERR	Cessna FR.172K-XP	A. G. Chrismas Ltd/Shoreham
	G-KERY	PA-28 Cherokee 180	Kerrytype Ltd (G-ATWO)/Goodwood
	G-KEYS	PA-23 Aztec 250	Ferguson Aviation/Newtownards
	G-KFIT	Beech F90 King Air	Kwik Fit Euro Ltd (G-BHUS)/Edinburgh
	G-KHRE	M.S.893E Rallye 150SV	Kenlyn Enterprises Ltd/Shoreham
	G-KIAM	Grob G.109B	D. T. Hulme
	G-KIDS	PA-34-220T-3 Seneca	Holding & Barnes Ltd
	G-KILT	AA-5A Cheetah	London School of Flying (G-BJFA)/ Elstree
	G-KINE	AA-5A Cheetah	Audio Kinectic (UK) Ltd/Elstree
	G-KING	PA-38-112 Tomahawk	Gordon King (Aviation) Ltd/Biggin Hill
	G-KIRK	Piper J-3C-65 Cub	M. Kirk
	G-KISS	Rand KR-2	A. J. Barthwick
	G-KLAY	Enstrom F-280C Shark	Apollo Manufacturing (Derby) Ltd (G-BGZD)
	G-KOOL	D.H.104 Devon C.2 ★	E. Surrey Technical College/nr Redhill
	G-KRIS	Maule M5-235C Lunar Rocket	Lord Howard de Walden
	G-KSBF	Hughes 369D	Ken Stokes (Business Forms) Ltd (G-BMJH)
	G-KUKU	Pfalzkuku (BS676)	A. D. Lawrence
	G-KUTU	Quickie Q2	J. Parkinson & ptnrs
	G-KWAX	Cessna 182E Skylane	J. Harper
	G-KWIK	Partenavia P.68B	Birchwood Aviation Ltd
	G-KYAK	Yakolev C-11 (00)	R. Lamplough/Duxford
	G-LADE	PA-32 Cherokee Six 300E	Appleby Glade Ltd/Kidlington
	G-LAKI	Jodel DR.1050	V. Panteli
	G-LANA	SOCATA TB.10 Tobago	Pektron Ltd
	G-LANC	Avro 683 Lancaster X	D. Arnold
	G-LANE	Cessna F.172N	Michael Newman Aviation/Denham
	G-LARK	Helton Lark 95	J. Fox
	G-LASH	Monnet Sonerai II	A. Lawson
	G-LASS	Rutan Vari-Eze	G. Lewis/Liverpool
	G-LATC	EMB-110P1 Bandeirante	Euroair Transport Ltd/Biggin Hill
	G-LAZE	Jodel DR.1050	N. B. Holmes
	G-LCOK	Colt 69A balloon	Hot-Air Balloon Co Ltd (G-BLWI)
	G-LDYS	Colt 56A balloon	A. Green
	G-LEAM	PA-28-236 Dakota	Gleam Aviation (G-BHLS)/Fairoaks
	G-LEAN	Cessna FR.182	J. G. Hogg (G-BGAP)
	G-LEAR	Learjet 35A	Northern Executive Aviation Ltd/ Manchester
	G-LEAU	Cameron N-31 balloon	Balloon Stable Ltd
	G-LEEM	PA-28R-200-2 Cherokee Arrow	B. J. Mounce (G-BJXW)
	G-LEGS	Short SD3-60	Manx Airlines Ltd (G-BLEF)/Ronaldsway
	G-LENS	Thunder Ax7-77Z balloon	P. W. A. Browitt
	G-LEON	PA-31-350 Navajo Chieftain	Chauffair Ltd/Blackbushe
	G-LEXI	Cameron N-77 balloon	R. H. Welch
	G-LEZE	Rutan Long Ez	K. G. M. Loyal & ptnrs
	G-LFCA	Cessna F.152	Midland Aircraft Leasing Ltd
	G-LFIX	V.S.509 Spitfire LF.IX (ML407)	Island Trading Ltd

Reg.	Type	Owner or Operator	Notes
G-LIBS	Hughes 369HS	Satinclass Ltd	
G-LIDD	D.H.104 Dove 8A	Acme Jewellery Ltd (G-ARSN)/Coventry	
G-LIDE	PA-31-350 Navajo Chieftain	Oxford Aero Charter Ltd/Kidlington	
G-LIFE	Thunder Ax6-56Z balloon	Schroder Life Assurance Ltd	
G-LIMA	R. Commander 114	Cargo Care Ltd/Ronaldsway	
G-LING	Thunder Ax7-65 balloon	Bridges Van Hire Ltd	
G-LINK	Sikorsky S-61N	British Caledonian Airways/Gatwick	
G-LINT	Pitts S.1S Special	P. L. Moss/Elstree	
G-LIOA	Lockheed 10A Electra ★ (NC5171N)	Science Museum/Wroughton	
G-LION	PA-18-135 Super Cub	Holding & Barnes Ltd	
G-LITE	R. Commander 112A	Rhoburt Ltd/Manchester	
G-LIVE	Grob G.109B	C. Baldwin	
G-LLAI	Colt 21A balloon	Lighter-Than-Air Ltd	
G-LOAG	Cameron N-77 balloon	Matthew Gloag & Son Ltd	
G-LOCH	Piper J-3C-90 Cub	J. M. Greenland	
G-LOND	V.806 Viscount	London European Airways Ltd (G-AOYI)/Luton	
G-LONG	Bell 206L LongRanger	Air Hanson Ltd/Brooklands	
G-LOOP	Pitts S-1C Special	K. P. Miller/Seething	
G-LORI	H.S.125 Srs 403B	Re-Enforce Trading Co Ltd (G-AYOJ)	
G-LORY	Thunder Ax4-31Z balloon	A. J. Moore	
G-LOSM	Gloster Meteor NF.11 (WM167)	Brencham Historic Aircraft Ltd/ Bournemouth	
G-LOTI	Bleriot XI (replica)	M. L. Beach	
G-LOVX	Cessna 441 Conquest	Lovaux Ltd (G-BLCJ)	
G-LOWE	Monnet Sonerai II	P. Archer	
G-LOYD	SA.341G Gazelle Srs 1	Appollo Manufacturing (Derby) Ltd (G-SFTC)	
G-LRII	Bell 206L LongRanger	Castle Air Charters Ltd	
G-LSFI	AA-5A Cheetah	G. W. Plowman & Son Ltd (G-BGSK)/ Elstree	
G-LSMI	Cessna F.152	Hartmann Ltd/Booker	
G-LUAR	SOCATA TB.10 Tobago	L. da Costa Saiago	
G-LUBE	Cameron N-77 balloon	Century Oils Ltd	
G-LUCA	Thunder Ax7-77Z balloon	Thunder & Colt Ltd	
G-LUCK	Cessna F.150M	M. Carrigan/Humberside	
G-LUKE	Rutan Long-Ez	S. G. Busby	
G-LULU	Grob G.109	B. A. & M. L. M. Langevad	
G-LUNA	PA-32RT-300T Turbo Lance II	W. Surrey Engineering Ltd	
G-LUSC	Luscombe 8E Silvaire	M. Fowler	
G-LYDE	Eiri PIK-20E	J. F. McAulay	
G-LYNN	PA-32RT-300 Lance II	Bearing Supplies Ltd (G-BGNY)/ Norwich	
G-LYNX	Westland WG.13 Lynx	Westland Helicopters Ltd/Yeovil	
G-MAAG	PA-30 Twin Comanche 160	M. A. Grayburn (G-ASOB)/Southend	
G-MABI	Cessna F.150L	Anglian Flight Training Ltd (G-BGOJ)/ Norwich	
G-MACH	SIAI-Marchetti SF.260	Cheyne Motors Ltd/Popham	
G-MACK	PA-28R Cherokee Arrow 200	Grumman Travel (Surrey) Ltd	
G-MAFF	BN-2T Islander	Ministry of Agriculture, Food & Fisheries (G-BJEO)	
G-MAGG	Pitts S-1SE Special	R. J. Pickin	
G-MAGS	Cessna 340A	Goldstar Publications Ltd/Biggin Hill	
G-MAGY	AS.350B Ecureuil	Quantel Ltd (G-BIYC)	
G-MALA	PA-28-181 Archer II	H. Burtwhistle & Son	
G-MALC	AA-5 Traveler	Air Coventry Ltd (G-BCPM)/Coventry	
G-MALI	BN-2A-26 Islander	Malinair Ltd (G-DIVE/G-BEXA)/Glasgow	
G-MALK	Cessna F.172N	R. R. & M. Mackay/Liverpool	
G-MALN	BA-2A-26 Islander	Pilatus BN Ltd (G-BLYO)/Bembridge	
G-MALS	Mooney M.20K-231	M. A. Lenihan/Netherthorpe	
G-MANT	Cessna 210L	Bob Crowe Aircraft Sales Ltd (G-MAXY)	
G-MANX	FRED Srs 2	P. Williamson	
G-MARC	AS.350B Ecureuil	Denis Ferranti Hoverknights Ltd (G-BKHU)	
G-MARY	Cassutt Special 1	J. Chadwick/Redhill	
G-MATP	BAe 748ATP	British Aerospace PLC/Woodford	
G-MATT	Robin R.2160	Sierra Flying Group (G-BKRC)/ Newcastle	
G-MAWL	Maule M4-210C Rocket	D. Group	
G-MAXI	PA-34-200T-2 Seneca	C. W. Middlemass	

Notes	Reg.	Type	Owner or Operator
	G-MAYO	PA-28-161 Warrior II	Creedair & Jermyk Engineering/ Fairoaks
	G-MCAR	PA-32 Cherokee Six 300D	Miller Aerial Spraying Ltd (G-LADA/G-AYWK)/Wickenby
	G-MCDS	Cessna 210N	Merseyside Car Delivery (G-BHNB)/ Liverpool
	G-MCOX	Fuji FA.200-180AO	W. Surrey Engineering (Shepperton) Ltd
	G-MDAS	PA-31-310 Navajo	Crosbyglow Ltd (G-BCJZ)
	G-MEBC	Cessna 310-1	Sarpedon Ltd (G-ROGA/G-ASVV)
	G-MELD	AA-5A Cheetah	Judgeshire Ltd (G-BHCB)/Biggin Hill
	G-MELT	Cessna F.172H	Alvair Aviation (Sales) Ltd (G-AWTI)/ Coventry
	G-MERI	PA-28-181 Archer II	Peacock Salt Ltd/Glasgow
	G-META	Bell 222	The Metropolitan Police/Lippitts Hill
	G-METB	Bell 222	The Metropolitan Police/Lippitts Hill
	G-METC	Bell 222	The Metropolitan Police (G-JAMC)/ Lippitts Hill
	G-METP	Short SD3-30	Brencham Leasing Ltd (G-METO/ G-BKIE)/Bournemouth
	G-MFMF	Bell 206B JetRanger 3	Micro Focus PLC (G-BJNJ)
	G-MFMM	Scheibe SF-25C Falke	S. Telfer-Evans & ptnrs
	G-MHBD	Cameron 0-105 balloon	M. H. B. Dawson Ltd
	G-MICK	Cessna F.172N	S. Grant & ptnrs
	G-MIKE	Hornet Gyroplane	M. H. J. Goldring
	G-MILK	SOCATA TB.10 Tobago	G. Whincup
	G-MINI	Currie Wot	D. Collinson
	G-MINT	Pitts S-1S Special	T. G. Anderson/Tollerton
	G-MIOO	Miles GM.100 Student 2	M. Woodley (G-APLK)/N. Weald
	G-MISS	Taylor JT.2 Titch	A. Brennan
	G-MIST	Cessna T.210K	Allzones Travel Ltd (G-AYGM)/ Biggin Hill
	G-MITS	Cameron N-77 balloon	Colt Car Co Ltd
	G-MKAY	Cessna 172N	Limbros Demolition Ltd
	G-MKEE	EAA Acro Sport	G. M. McKee
	G-MKIV	Bristol Blenheim IV	G. A. Warner/Duxford
	G-MKIX	V.S.361 Spitfire F.IX (NH238)	D. W. Arnold/Blackbushe
	G-MLAS	Cessna 182E	Mark Luton Aviation Services
	G-MLBU	PA-46-310P Malibu	Cheshire Scaffold Co Ltd
	G-MLBY	Cessna 340A II	Saint Piran Services Ltd
	G-MLCS	Cessna 414A	Mountleigh Air Services (G-MGHI/ G-BHKK)/Leeds
	G-MLGL	Colt 21A balloon	Colt Balloons Ltd
	G-MOBL	EMB-110P2 Bandeirante	Tal-Air Ltd/Stansted
	G-MOGG	Cessna F.172N	J. G. James (G-BHDY)
	G-MOLY	PA-23 Apache 160	A. H. Hunt & ptnrs (G-APFV)/ St Just
	G-MONA	M.S.880B Rallye Club	K. Lockett (G-AWJK)
	G-MONB	Boeing 757-2T7	Monarch Airlines Ltd/Luton
	G-MONC	Boeing 757-2T7	Monarch Airlines Ltd/Luton
	G-MOND	Boeing 757-2T7	Monarch Airlines Ltd/Luton
	G-MONE	Boeing 757-2T7	Monarch Airlines Ltd/Luton
	G-MONF	Boeing 737-300	Monarch Airlines Ltd/Luton
	G-MONG	Boeing 737-300	Monarch Airlines Ltd/Luton
	G-MONI	Monnet Moni	R. P. Williams
	G-MONO	Taylor JT.1 Monoplane	A. Doughty
	G-MOTH	D.H.82A Tiger Moth (K2567)	M. C. Russell/Duxford
	G-MOVE	Aerostar 601P	Red Dragon Travel Ltd/Cardiff
	G-MOXY	Cessna 441	Brown Aviation Services Ltd (G-BHLN)/ Leeds
	G-MOZY	D.H.98 (replica)	J. Beck & G. L. Kemp
	G-MPWI	Robin HR.100/210	The AT Group/Bristol
	G-MRFB	H.S.125 Srs 3B	Food Brokers (Holdings) Ltd (G-AZVS)/Gatwick
	G-MRTY	Cameron N-77 balloon	R. A. & P. G. Vale
	G-MSDS	Cessna 404	Donington Aviation Ltd/E. Midlands
	G-MSFY	H.S.125 Srs 700B	Mohamed Said Fakhry/Heathrow
	G-MULL	Douglas DC-10-30	British Caledonian Airways Ltd/ Gatwick
	G-MUSO	Rutan Long-Ez	M. Moran
	G-NACI	Norman NAC.1 Srs 100	Norman Aeroplane Co Ltd/Sandown
	G-NASH	AA-5A Cheetah	Sky Rambler Ltd/Southampton
	G-NATT	R. Commander 114A	Northgleam Ltd

Reg.	Type	Owner or Operator	Notes
G-NAVY	D.H.104 Sea Devon C.20 (XJ348)	J. S. Flavell & K. Fehrenbach (G-AMXX)/Shoreham	
G-NBSI	Cameron N-77 balloon	Nottingham Building Soc	
G-NCUB	Piper J-3C-65 Cub	N. Thomson (G-BGXV)/Norwich	
G-NDGC	Grob G.109	Soaring Southwest	
G-NDNI	NDN-1 Firecracker	Norman Marsh Aircraft Ltd/Goodwood	
G-NEAL	PA-32 Cherokee Six 260	C. Goodliffe Neal & Co Ltd (G-BFPY)	
G-NEIL	Thunder Ax3 balloon	Islington Motors (Trowbridge) Ltd	
G-NELL	R. Commander 112A	M. J. Scott/E. Midlands	
G-NEUS	Brugger MB.2 Colibri	G. E. Smeaton	
G-NEWR	PA-31-350 Navajo Chieftain	Eastern Air Executive Ltd/Sturgate	
G-NEWS	Bell 206B JetRanger 3	Peter Press Ltd	
G-NGBI	AA-5B Tiger	Filemart Ltd (G-JAKK/G-BHWI)/ Biggin Hill	
G-NHRH	PA-28 Cherokee 140	H. Dodd	
G-NHVH	Maule M5-235C Lunar Rocket	Commercial Go-Karts Ltd/Exeter	
G-NIAL	AS.350B Ecureuil	Timothy Laing Aviation	
G-NICK	PA-19 Super Cub 95	J. G. O'Donnell & I. Woolacott	
G-NIII	BAC One-Eleven 408EF	Bacone Ltd (G-BIII/G-AWKJ)/Gatwick	
G-NIKY	PA-31-350 Navajo Chieftain	Air Charter Europe (G-BPAR)/Glasgow	
G-NILE	Colt 77A balloon	Zebedee Balloon Services	
G-NISR	R. Commander 690A	Z. I. Bilbeisi	
G-NITA	PA-28 Cherokee 180	The Genita Group (G-AVVG)/Redhill	
G-NIUK	Douglas DC-10-30	British Caledonian Airways Ltd/Gatwick	
G-NJAG	Cessna 207	G. H. Nolan Ltd	
G-NNAC	PA-18 Super Cub 135	P. A. Wilde	
G-NOBY	Rand KR-2	N. P. Rieser	
G-NODE	AA-5B Tiger	Curd & Green Ltd/Elstree	
G-NORD	Nord NC.854	R. G. E. Simpson & A. M. Alexander/ Panshanger	
G-NRDC	NDN-6 Fieldmaster	NDN Aeroculture Ltd/Sandown	
G-NROA	Boeing 727-217	Dan-Air Services Ltd (G-BKNG)/Gatwick	
G-NUNN	PA-24 Comanche 250	W. H. G. Nunn (G-AWKW)/Elstree	
G-NUTS	Cameron 35SS balloon	The Balloon Stable Ltd	
G-NUTZ	AS.355F-1 Twin Squirrel	Powersense Ltd (G-BLRI)	
G-NWPB	Thunder Ax7-77Z balloon	Lighter-Than-Air Ltd	
G-NWPR	Cameron N-77 balloon	Post Office N.W. Postal Board	
G-OABG	Hughes 369E	A. B. Gee of Ripley	
G-OABI	Cessna 421C	Mont Arthur Finance Ltd	
G-OADE	Cessna F.177RG	Vehicle Fleet Management Ltd (G-AZKH)/Coventry	
G-OAIM	Hughes 369HS	J. E. Clarke (G-BDFP)/Bournemouth	
G-OAKL	Beech 200 Super King Air	T. Kilroe & Sons (G-BJZG)/Manchester	
G-OAKS	Cessna 421C	Barratt Developments Ltd/Newcastle	
G-OAMY	Cessna 152	Birmingham Aviation Ltd	
G-OAPA	Pilatus PC-6/B2-H2 Turbo Porter	Army Parachute Association/ Netheravon	
G-OARV	ARV.1 Prototype	ARV Aviation Ltd/Sandown	
G-OATS	PA-38-112 Tomahawk	Truman Aviation Ltd/Tollerton	
G-OAUS	Sikorsky S-76A	Ashton Upthorpe Stud & Farms Ltd	
G-OBAC	AS.350B Ecureuil	BAC Aviation Ltd (G-EORR/G-FERG/ G-BGCW)/Southend	
G-OBAT	Cessna F.152	J. J. Baumhardt Associates Ltd	
G-OBCA	Cessna 421C	British Car Auctions Ltd	
G-OBEA	BAe Jetstream 3102-01	Birmingham Executive Airways Ltd	
G-OBED	PA-34-200T Seneca	G-Air Ltd	
G-OBEY	PA-23 Aztec 250	Creaton Aviation Services (G-BAAJ)	
G-OBMS	Cessna F.172N	BMS Electrical Services Ltd/ Birmingham	
G-OBMW	AA-5 Traveler	Fretcourt Ltd (G-BDPV)	
G-OBSV	Partenavia P.68B Observer	Northair Aviation Ltd/Leeds	
G-OBUD	Colt 69A balloon	Hot-Air Balloon Co Ltd	
G-OCAL	Partenavia P.68B	Grosvenor Aviation Services Ltd (G-BGMY)/Manchester	
G-OCAP	Bell 206B JetRanger	Powersense Ltd	
G-OCAT	Eiri PIK-20E	N. G. G. Hackett	
G-OCCC	H.S.125 Srs 800B	Consolidated Contractors (UK) Ltd	
G-OCDS	Aviamilano F.8L Falco II	D. I. Simpson & Computer Diskdrive Services ltd (G-VEGL)	
G-OCND	Cameron 0-77 balloon	J. P. Hatton & ptnrs	
G-OCPC	Cessna FA.152	Hampshire Aeroplane Co Ltd/ St Just	
G-OCUB	Piper J-3C-90 Cub	P. T. Middlebrook	
G-ODAY	Cameron N-56 balloon	C. O. Day (Estate Agents)	

Notes	Reg.	Type	Owner or Operator
	G-ODEL	Falconar F-II-3	A. Brinkley & R. H. Ford
	G-ODER	Cameron O-77 balloon	M. White
	G-ODON	AA-5B Tiger	Moynihan Motor Engineering Ltd/ Elstree
	G-ODSF	AA-5A Cheetah	Holmes Rentals (G-BEUW)
	G-OESX	PA-23 Aztec 250	J. J. Baumhardt Associates Ltd (G-BAJX)
	G-OEZE	Rutan Vari-Eze	S. Stride & ptnrs
	G-OFBL	Beech C90 King Air	Food Brokers Ltd (G-MEDI)/Shoreham
	G-OFCM	Cessna F172L	F. C. M Aviation Ltd (G-AZUN)/Guernsey
	G-OFHS	Hughes 369E	A. & P. M. Ford
	G-OFLY	Cessna 210L	A. P. Mothew/Stapleford
	G-OFOR	Thunder Ax3 balloon	T. Donnelly
	G-OFRB	Everett gyroplane	F. R. Blennerhassett
	G-OFRH	Cessna 421C	Flight Refuelling Ltd (G-NORX)/ Bournemouth
	G-OFRL	Cessna 414A	Flight Refuelling Ltd/Bournemouth
	G-OFUN	Valentin Taifun 17E	J. A. Sangster/Booker
	G-OGAS	Westland WG.30 Srs 100	Bristol Helicopters Ltd (G-BKNW)
	G-OGDN	Beech A200 Super King Air	A. Ogden & Sons Ltd/Leeds
	G-OGET	PA-39 Twin Comanche 160 C/R	Videochord Ltd (G-AYXY)
	G-OHCA	SC.5 Belfast (XR363)	HeavyLift Cargo Airlines Ltd/Southend
	G-OHTL	Sikorsky S-76A	Air Hanson Ltd/Brooklands
	G-OHUB	Cessna 404 Titan	Hubbardair Ltd/Norwich
	G-OIAN	M.S.880B Rallye Club	Ian Richard Transport Services Ltd
	G-OIAS	PA-31-350 Navajo Chieftain	Air Charter (Scotland) Ltd/Glasgow
	G-OICI	Quickie Q.2	Quickie Aircraft (Europe) Ltd (G-OGKN)
	G-OIFR	Cessna 172RG	J. J. Baumhardt Associates (G-BHJG)
	G-OILS	Cessna T.210L	Machine Music Ltd (G-BCZP)/ Blackbushe
	G-OING	AA-5A Cheetah	Abraxas Aviation Ltd (G-BFPD)/Denham
	G-OINK	Piper J-3C-65 Cub	A. R. Harding (G-BILD/G-KERK)
	G-OIOO	PA-23 Aztec 250	A. A. Kelly (G-AVLV)
	G-OITD	Cessna 310F	ITD Aviation Ltd (G-AROK)
	G-OJCB	AB-206B JetRanger 2	Air Hanson Ltd/Weybridge
	G-OJCT	Partenavia P.68C	Save the Children Fund (G-BHOV)
	G-OJCW	PA-32RT-300 Lance II	World Carrier (London) Ltd
	G-OJEE	Bede BD-4	G. Hodges
	G-OJFR	Bell 206B JetRanger	Daniel Martin Plant Ltd
	G-OJON	Taylor JT.2 Titch	J. H. Fell
	G-OJRS	Cessna F.152	Stanton Aircraft Management Ltd (G-BFFD)
	G-OJVC	J/1N Alpha	R. W. J. Holland (G-AHCL)/Sywell
	G-OJVH	Cessna F.150H	Yorkshire Light Aircraft Ltd (G-AWJZ)/ Leeds
	G-OKAY	Pitts S-1E Special	Aerial & Aerobatic Service/Booker
	G-OLDN	Bell 206L LongRanger	Autoklenz (UK) Ltd (G-TBCA/G-BFAL)
	G-OLDS	Colt AS-105 airship	Hot-Air Balloon Co Ltd
	G-OLDY	Luton LA-5 Major	M. P. & A. P. Sargent
	G-OLEE	Cessna F.152	Birmingham Aviation Ltd
	G-OLFC	PA-38-112 Tomahawk	Leavesden Flight Centre (G-BGZG)
	G-OLFT	R. Commander 114	W. R. M. C. Foyle (G-WJMN)/Luton
	G-OLIN	PA-30 Twin Comanche 160	Skyhawk Ltd (G-AWMB)/Stapleford
	G-OLLI	Cameron O-31 SS balloon	N. A. Robertson
	G-OLLY	PA-31-350 Navajo Chieftain	Robertson Foods Ltd (G-BCES)/Bristol
	G-OLMA	Partenavia P.68B	Landell Mills Associates Ltd (G-BGBT)
	G-OLVR	FRED Srs 2	A. R. Oliver
	G-OMAC	Cessna FR.172E	Skymedia Ltd
	G-OMAD	Cessna 210	Mantime Magic Services Ltd (G-BMDN)
	G-OMAV	AS.355F-1 Twin Squirrel	Massellaz Helicopters Ltd/Hayes
	G-OMCL	Cessna 550 Citation II	Quantel Ltd
	G-OMED	AA-5B Tiger	Caslon Ltd (G-BERL)/Elstree
	G-OMET	Beech C90 King Air	G. L. Group PLC (G-COTE/G-BBKN)
	G-OMHC	PA-28RT-201 Arrow IV	M. H. Cundley/Redhill
	G-OMNI	PA-28R Cherokee Arrow 200D	A. Somerville (G-BAWA)/Blackbushe
	G-ONOR	Cessna 425	Norcross Transport PLC (G-BKSA)/ Blackbushe
	G-ONPN	H.S.125 Srs 1B	Avonmore International Ltd (G-BAXG)
	G-ONTA	Hughes 369D	Explosive Developments
	G-ONZO	Cameron N-77 balloon	J. A. Kershaw
	G-OODE	SNCAN SV-4B	V. S. E. Norman (G-AZNN)
	G-OODI	Pitts S-1D Special	R. N. Goode (G-BBBU)/White Waltham

Reg.	Type	Owner or Operator	Notes
G-OODO	Stephens Akro	R. N. Goode/White Waltham	
G-OODS	Extra EA.230	R. N. Goode	
G-OOFY	Rollason Beta	G. Staples	
G-OOLY	Everett Gyroplane	N. A. Brandish/Southend	
G-OOSE	Rutan Vari-Eze	J. A. Towers	
G-OPAT	Beech 76 Duchess	Ray Holt (Land Drainage) Ltd/(G-BHAO)	
G-OPBN	BN-2T Islander	Pilatus BN Ltd (G-BJOH)/Bembridge	
G-OPIK	Eiri PIK-20E	R. L. McLean & J. N. Ellis	
G-OPJT	Enstrom F-280C Shark	Southern Air Ltd (G-BKCO)/Shoreham	
G-OPOP	Enstrom F-280C-UK-2 Shark	J. C. Elmer (G-OFED)/Shoreham	
G-OPPL	AA-5A Cheetah	London School of Flying Ltd (G-BGNN)/ Elstree	
G-OPSF	PA-38-112 Tomahawk	Panshanger School of Flying (G-BGZI)	
G-OPUP	B.121 Pup 2	G. C. Rhodes (G-AXEU)	
G-ORAV	Cessna 337D	R. J. Everett Engineering Ltd (G-AXGJ)/ Ipswich	
G-ORAY	Cessna F.182Q II	C. Robinson (G-BHDN)/Blackpool	
G-ORED	BN-2T Islander	Pilatus BN Ltd (G-BJYW)/Bembridge	
G-ORFC	Jurca MJ.5 Sirocco	RFC Flying Group/Popham	
G-ORJW	Laverda F.8L Falco IV	R. J. Willies	
G-OROG	Cessna 401	Porchland Ashphalt Ltd (G-ZEUS/ G-ODJM/G-BSIX/G-CAFE/G-AWXM)	
G-OSAL	Cessna 421C	Air Swift Ltd/Fairoaks	
G-OSCC	PA-32 Cherokee Six 300	Silkstone Construction Co Ltd (G-BGFD)	
G-OSDI	Beech 95-58 Baron	Systems Designers Aviation Ltd (G-BHFY)	
G-OSEA	BN-2B-26 Islander	South East Air Ltd (G-BKOL)	
G-OSKY	Cessna 172M	Integrated Hydraulics Ltd	
G-OSND	Cessna FRA.150M	J. J. Baumhardt Associates Ltd (G-BDOU)/Southend	
G-OSST	Colt 77A balloon	British Airways PLC	
G-OSSY	PA-28-181 Archer II	Bryan Goss Motorcycles Ltd/ Bournemouth	
G-OTAD	Cessna 421C	Chaseside Holdings Ltd (G-BEVL)	
G-OTMM	Cessna F.150F	D. J. Hockings (G-ATMM/Biggin Hill	
G-OTOW	Cessna 175BX	Sywell Air Services (G-AROC)	
G-OTRG	Cessna TR.182RG	G. F. Holdings (Contractors) Ltd/ Manchester	
G-OTSB	BN-2A Mk III-2 Trislander	Aurigny Air Services Ltd (G-BDTO)	
G-OTTA	Colt 1.5 MCB balloon	Colt Balloons Ltd	
G-OTUG	PA-18 Super Cub 150	Holding & Barnes Ltd	
G-OTUX	PA-28R-201T Turbo Arrow III	M. A. M. Quadrini/Newcastle	
G-OTVS	BN-2T Islander	TVS Television Ltd (G-BPBN/G-BCMY)/ Headcorn	
G-OULD	Gould Mk I balloon	C. A. Gould	
G-OVFR	Cessna F.172N	Sunningdale Aviation Services Ltd	
G-OVIP	G.1159 Gulfstream 2	H. Goodman (G-AYMI)	
G-OVMC	Cessna F.152 II	Staverton Flying Services Ltd	
G-OWAC	Cessna F.152	Stanton Aircraft Management Ltd (G-BHEB)	
G-OWAK	Cessna F.152	Stanton Aircraft Management Ltd (G-BHEA)	
G-OWEN	K & S Jungster	R. C. Owen	
G-OWER	PA-31-310 Turbo Navajo	Achela International Ltd (G-FOIL)	
G-OWIN	BN-2A-8 Islander	UK Parachute Services Ltd (G-AYXE)/ Ipswich	
G-OWJM	AB-206B JetRanger 3	J. M. Gow (G-BHXV)	
G-OZOI	Cessna R.182	R. Holt (G-ROBK)	
G-PACE	Robin R.1180T	Millicron Instruments Ltd/Coventry	
G-PACY	Rutan Vari-Viggen	E. Pace	
G-PADY	R. Commander 114	M. J. Brown/Coventry	
G-PAGE	Cessna F.150L	E. Shipley/Clacton	
G-PALS	Enstrom F-280C-UK-2 Shark	R. J. White	
G-PAMI	AS.355F-1 Twin Squirrel	Lynton Aviation Ltd (G-BUSA)	
G-PARA	Cessna 207	Paraski/Swansea	
G-PARI	Cessna 172RG Cutlass	Fairline Boats Ltd	
G-PARK	Lake LA-4-200 Buccaneer	Leisure Sport Ltd (G-BBGK)/ Headcorn	
G-PARS	Evans VP-2	A. Parsfield	
G-PART	Partenavia P.68B	Highway Windscreens Ltd/Stapleford	
G-PATT	Cessna 404 Titan	Casair Aviation Ltd (G-BHGL)/ Tees-side	
G-PATY	Colt Flying Sausage balloon	Colt Balloons Ltd	

Notes	Reg.	Type	Owner or Operator
	G-PAWL	PA-28 Cherokee 140	P. Lodge (G-AWEU)/Liverpool
	G-PAWS	AA-5A Cheetah	Reedtrend Ltd/Biggin Hill
	G-PAXX	PA-20 Pacer 135	D. W. & M. R. Grace
	G-PCUB	PA-18 Super Cub 135 (L-21B) (54-2474)	M. J. Wilson/Redhill
	G-PDCC	AS.350B Ecureuil	Devon & Cornwall Police Authority (G-PORR)
	G-PDHJ	Cessna T.182R	Northair Aviation Ltd
	G-PDOC	PA-44-180 Seminole	M. White (G-PVAF)
	G-PDON	WMB.2 Windtracker balloon	P. Donnellan
	G-PEAT	Cessna 421B	Forest Aviation Ltd (G-BBIJ)/ Manchester
	G-PEET	Cessna 401A	J. R. Fuller/Biggin Hill
	G-PENN	AA-5B Tiger	Compair
	G-PENY	Sopwith LC-IT Triplane	J. S. Penny
	G-PERR	Cameron 60 bottle balloon	The Balloon Stable Ltd
	G-PETR	PA-28 Cherokee 140	P. B. Donoghue (G-BCJL)
	G-PFAA	EAA Model P biplane	P. E. Barker
	G-PFAB	Colomban MC.15 Cri-Cri	P. Fabish
	G-PFAC	FRED Srs 2	Basit Flying Group
	G-PFAD	Wittman W.8 Tailwind	M. R. Stamp
	G-PFAE	Taylor JT.1 Monoplane	G. Johnson
	G-PFAF	FRED Srs 2	P. A. Smith
	G-PFAG	Evans VP-1	N. S. Giles-Townsend
	G-PFAH	Evans VP-1	J. A. Scott
	G-PFAI	Clutton EC.2 Easy Too	G. W. Cartledge
	G-PFAL	FRED Srs 2	J. McCullough
	G-PFAM	FRED Srs 2	W. C. Rigby
	G-PFAN	Avro 558 (replica)	N. P. Harrison
	G-PFAO	Evans VP-1	P. W. Price
	G-PFAP	Currie Wot/SE-5A (C1904)	P. G. Abbey
	G-PFAR	Isaacs Fury II (K2059)	C. J. Repik
	G-PFAS	GY-20 Minicab	J. Sproston & F. W. Speed
	G-PFAT	Monnet Sonerai II	H. B. Carter
	G-PFAU	Rand KR-2	D. E. Peace
	G-PFAV	D.31 Turbulent	B. A. Luckins
	G-PFAW	Evans VP-1	R. F. Shingler
	G-PFAX	FRED Srs 2	A. J. Dunston
	G-PFAY	EAA Biplane	A. K. Lang & A. L. Young
	G-PFAZ	Evans VP-1	B. Kylo
	G-PHIL	Hornet Gyroplane	A. J. Philpotts
	G-PICS	Cessna 182F	Astral Aerial Surveys Ltd (G-ASHO)
	G-PIES	Thunder Ax7-77Z balloon	Pork Farms Ltd
	G-PIGN	Bolmet Paloma Mk 1	T. P. Metson & J. A. Bollen
	G-PINT	Cameron 65 SS balloon	Charles Wells Ltd
	G-PIPE	Cameron N-56 SS balloon	Carreras Rothmans Ltd
	G-PITS	Pitts S-2AE	D. Rolfe
	G-PLAN	Cessna F.150L	Phoenix Aviation (Bedford) Ltd/ Cranfield
	G-PLAS	GA-7 Cougar	Birchwood Aviation Ltd (G-BGHL)/ Elstree
	G-PLAY	Robin R.2100A	Cotswold Aero Club Ltd/Staverton
	G-PLEV	Cessna 340	KJ Bill Aviation Ltd/Halfpenny Green
	G-PLIV	Pazmany PL.4	B. P. North
	G-PLOW	Hughes 269B	March Helicopters Ltd (G-AVUM)/ Sywell
	G-PLUM	Bell 206L LongRanger	PLM Helicopters Ltd/Inverness
	G-PLUS	PA-34-200T-2 Seneca	C. G. Strasser/Jersey
	G-PMAM	Cameron V-65 balloon	P. A. Meecham
	G-PMCN	Monnet Sonerai II	P. J. McNamee
	G-PMNL	Extra EA.230	Aerobatic Displays Ltd
	G-POLE	Rutan Long-Ez	A. M. Dutton
	G-POLO	PA-31-350 Navajo Chieftain	Grosvenor Aviation Services Ltd/ Manchester
	G-POLY	Cameron N-77 balloon	Empty Wallets Balloon Group
	G-PONY	Colt 31A balloon	International Motors Ltd
	G-POOH	Piper J-3C-65 Cub	P. & H. Robinson
	G-POPE	Eiri PIK-20E-1	C. J. Hadley
	G-PORK	AA-5B Tiger	P. H. Johnson (G-BFHS)
	G-POST	EMB-110P1 Bandeirante	Euroair Ltd/Gatwick
	G-POWA	PA-24 Comanche 400	G. F. Miller/Cardiff
	G-POWL	Cessna 182R	J. & B. Powell (Printers) Ltd
	G-PPLI	Pazmany PL.1	G. Anderson
	G-PRAG	Brugger MB.2 Colibri	R. J. Hodder & ptnrs

Reg.	Type	Owner or Operator	Notes
G-PRMC	H.S.125 Srs 700B	RMC Group Services Ltd (G-BFSP)/ Biggin Hill	
G-PROP	AA-5A Cheetah	Urban & City Properties Ltd (G-BHKU)/ Biggin Hill	
G-PROV	Hunting Jet Provost T.54A	Berowell Management Ltd/ Bournemouth	
G-PRXI	V.S.365 Spitfire PR.XI (PL983)	R. Fraissinet	
G-PSID	P-51D Mustang	Fairoaks Aviation Services Ltd	
G-PSVS	Beech 58 Baron	Astra Aviation Ltd/Guernsey	
G-PTER	Beech C90 King Air	Moseley Group (PSV) Ltd (G-BIEE)	
G-PTWB	Cessna T.303 Crusader	Northair Aviation Ltd/Leeds	
G-PTWO	Pilatus P2-05 (RF+16)	P. S. John	
G-PUBS	Colt 56 SS balloon	Flamboyant Promotions Ltd	
G-PUFF	Thunder Ax7-77A balloon	Intervarsity Balloon Club *Puffin II*	
G-PULL	PA-18 Super Cub 150	G. R. Janney/Lympne	
G-PUMA	AS.332L Super Puma	Bond Helicopters Ltd/Bourn	
G-PUMB	AS.332L Super Puma	Bond Helicopters Ltd/Bourn	
G-PUMD	AS.332L Super Puma	Bond Helicopters Ltd/Bourn	
G-PUME	AS.332L Super Puma	Bond Helicopters Ltd/Bourn	
G-PUMG	AS.332L Super Puma	Bond Helicopters Ltd/Bourn	
G-PUMH	AS.332L Super Puma	Bond Helicopters Ltd/Bourn	
G-PUMI	AS.332L Super Puma	Bond Helicopters Ltd/Bourn	
G-PURR	AA-5A Cheetah	W. H. Wilkins Ltd (G-BJDN)/Biggin Hill	
G-PUSH	Rutan Long-Ez	E. G. Peterson	
G-PVAM	Port Victoria 7 Grain Kitten	A. J. Manning	
G-PYRO	Cameron N-65 balloon	J. Stone	
G-RACA	P.57 Sea Prince T.1	Atlantic & Caribbean Aviation Ltd/ Staverton	
G-RACL	H.S.125 Srs 700B	Racal Avionics Ltd/Biggin Hill	
G-RADE	Cessna 210L	R. J. Herbert (G-CENT)	
G-RAEM	Rutan LongEz	G. F. H. Singleton	
G-RAFC	Robin R.2112	RAF Cranwell Flying Club	
G-RAFE	Thunder Ax7-77 balloon	A. J. W. Rose	
G-RAFF	Learjet 35A	Graff Aviation Ltd/Heathrow	
G-RAFT	Rutan Long-Ez	D. G. Foreman	
G-RAFW	Mooney M.20E	Warnell Motors Ltd (G-ATHW)	
G-RAIN	Maule M5-235C Lunar Rocket	J. S. Mehew	
G-RALI	Hughes 369HS	David Richards Autosport Ltd (G-BLKO)	
G-RALY	Robinson R-22	Malcolm Wilson (Motorsport) Ltd	
G-RAMS	PA-32R-301 Saratoga SP	Peacock & Archer Ltd/Manchester	
G-RAND	Rand KR-2	R. L. Wharmby	
G-RANY	Cessna 421C	McCarthy & Stone (Developments) Ltd (G-BHLA)	
G-RAPA	BN-2T Islander	R. G. Card	
G-RARE	Thunder Ax5-42 SS balloon	International Distillers & Vintners Ltd	
G-RASC	Evans VP-2	R. A. Codling	
G-RATE	AA-5A Cheetah	Hadley Green Garage Ltd (G-BIFF)/ Elstree	
G-RATS	Alexander/Todd Steen Skybolt	P. S. Dodd (G-RHFI)/Ipswich	
G-RAYS	Zenair CH.250	R. E. Delves	
G-RBIN	Robin DR.400/2+2	Headcorn Flying School Ltd	
G-RBOS	Colt AS-105 airship	Royal Bank of Scotland	
G-RCPW	AA-5A Cheetah	Paintwell (St Albans) Ltd (G-BERM)/ Elstree	
G-RDCI	R. Commander 112A	R. D. S. Cook (G-BFWG)/Shoreham	
G-RDON	WMB.2 Windtracker balloon	P. J. Donnellan (G-BICH)	
G-REAT	GA-7 Cougar	Hadley Green Garage Ltd/Elstree	
G-REEK	AA-5A Cheetah	Telemachus Ltd/Kidlington	
G-REEN	Cessna 340	Ernest Green International Ltd (G-AZYR)	
G-REES	Jodel D.140C	M. D. S. Hood/Redhill	
G-REID	Rotorway Scorpion 133	J. Reid (G-BGAW)	
G-REIS	PA-28R-201T Turbo Arrow III	H. Reis (Hard Chrome) Ltd/ Halfpenny Green	
G-RENO	SOCATA TB.10 Tobago	International Motors Ltd	
G-REST	Beech P35 Bonanza	C. R. Taylor (G-ASFJ)	
G-RETA	C.A.S.A. 1.131 Jungmann	Maronco Ltd/Leeds	
G-REXP	Beech 65-70 Queen Air	Parker & Heard Ltd (G-AYPC)/Biggin Hill	
G-REXS	PA-28-181 Archer II	Channel Islands Aero Holdings (Jersey) Ltd	
G-REXY	Beech 65-A80 Queen Air	Parker & Heard Ltd (G-AVNG)/Biggin Hill	
G-RHCN	Cessna FR.182RG	R. H. C. Neville	
G-RHHT	PA-32RT-300 Lance II	Hart Poultry Ltd	

Notes	Reg.	Type	Owner or Operator
	G-RICH	Cessna F.152	Stanton Aircraft Management Ltd
	G-RICK	Beech 95-B55 Baron	K. E. Lundquist (G-BAAG)/Elstree
	G-RIDE	Stephens Akro	R. Mitchell/Coventry
	G-RIGS	Aerostar 601P	Rigs Design Services Ltd/Fairoaks
	G-RILL	Cessna 421C	Maxwell Restaurants Ltd (G-BGZM)/ Elstree
	G-RILY	Monnet Sonerai II	K. D. Riley
	G-RIND	Cessna 335	ATA Grinding Processes/Leavesden
	G-RING	Cessna FR.182RG	A. Hopper
	G-RIST	Cessna 310R-II	Velcourt (East) Ltd & ptnrs (G-DATS)/ Staverton
	G-RJMI	AA-5A Cheetah	R. J. Mole
	G-RMAE	PA-31 Turbo Navajo	Logbirch Ltd (G-BAEG)
	G-RMAM	Musselwhite MAM.1	M. A. Musselwhite
	G-RMSS	Short SD3-60	Manx Airlines Ltd (G-BKKU)/ Ronaldsway
	G-RNAS	D.H.104 Sea Devon C.20 (XK896)	D. W. Hermiston-Hooper/Sandown
	G-RNCO	R. Commander 690C	Ranco Europe Ltd/Plymouth
	G-RNMO	Short SD3-30	Fairflight Ltd (G-BFZW)/Biggin Hill
	G-ROAN	Boeing E.75N-1 Stearman	Subtec Aviation/Shoreham
	G-ROAR	Cessna 401	Salon Productions Ltd (G-BZFL/ G-AWSF)/Biggin Hill
	G-ROBB	Grob G.109B	Grobb-Air
	G-ROBE	Grob G.109B	Corbett Farms Ltd/Shobdon
	G-ROBI	Grob G.109B	A. W. McGarrigle/Cardiff
	G-ROBN	Robin R.1180T	J. G. Beaumont
	G-ROBY	Colt 17A balloon	Lighter-Than-Air Ltd
	G-RODI	Isaacs Fury	J. R. C. Morgan
	G-RODS	A-Bell 206B JetRanger 2	Crook & Son (G-NOEL/G-BCWN)
	G-ROGR	Bell 206A JetRanger	P. Woodward (G-AXMM)
	G-ROLF	PA-28R-301 Saratoga SP	R. W. Burchardt
	G-ROLL	Pitts S-2A Special	RPM Aviation Ltd/Guernsey
	G-RONW	Fred Srs 2	P. J. D. Granow
	G-ROOF	Brantly B.2B	S. Lee (G-AXSR)
	G-ROOK	Cessna F.172P	Cejam Electronics Ltd/Biggin Hill
	G-ROOM	Short SD3-60	Short Bros Ltd (G-BSBL)/Sydenham
	G-ROPI	Hughes 369HS	Gt. Western Developments Ltd (G-ROMA/G-ONPP)
	G-RORO	Cessna 337B	Ronageny (Shipping) Ltd (G-AVIX)/ Blackpool
	G-ROSE	Evans VP-1	W. K. Rose
	G-ROSS	Practavia Pilot Sprite	F. M. T. Ross
	G-ROTA	Bensen B.8	D. Ellerton
	G-ROUP	Cessna F.172M	Stanton Aircraft Management Ltd (G-BDPH)/Biggin Hill
	G-ROUS	PA-34-200T-2 Seneca	C.S.E. Aviation Ltd/Kidlington
	G-ROVE	PA-18 Super Cub 135	Howard Avis Travel Ltd
	G-ROWL	AA-5B Tiger	C. P. Riddington/Biggin Hill
	G-ROWS	PA-28-151 Warrior	P. G. Lee
	G-ROYL	Taylor JT.1 Monoplane	R. L. Wharmby
	G-ROYS	D.H.C.I Chipmunk T.10	R. W. & S. Pullan
	G-ROZY	Cameron R.36 balloon	Jacques W. Soukup Ltd
	G-RPAH	Rutan Vari-Eze	B. Hanson
	G-RPEZ	Rutan Long-Ez	B. A. Fairston & D. Richardson
	G-RRRR	Privateer Motor Glider	R. F. Selby
	G-RRSG	Cameron V-77 balloon	J. N. Harley
	G-RRTM	Sikorsky S-70C	Rolls-Royce Ltd
	G-RTHL	Leivers Special	R. Leivers
	G-RUBB	AA-5B Tiger	Summerfield Group Ltd/Elstree
	G-RUDD	Cameron V-65 balloon	N. A. Apsey
	G-RUIA	Cessna F.172M	Delamere & Norley Finance Ltd/ Humberside
	G-RUMN	AA-1A Trainer	J. M. Horsburgh & A. Wilson
	G-RUNT	Cassutt IIIM	N. A. Brendish/Southend
	G-RUSH	Cessna 404	Kondair (G-BEMX)/Stansted
	G-RUSS	Cessna 172N	Leisure Lease/Southend
	G-RYAN	PA-28R-201T Turbo Arrow III	Lancing Service Station Ltd (G-BFMN)/ Shoreham
	G-SAAB	R. Commander 112TC	S. Richmen (G-BEFS)
	G-SAAS	Ayres S2R-T34 Thrush Commander	Shoreham Flight Simulation/ Bournemouth
	G-SABA	PA-28R-201T Turbo Arrow III	Barlow Tyrie Ltd (G-BFEN)

Reg.	Type	Owner or Operator	Notes
G-SACA	Cessna 152 II	Southern Air Ltd (G-HOSE)/Shoreham	
G-SACB	Cessna F.152 II	Southern Air Ltd (G-BFRB)/Shoreham	
G-SACD	Cessna F.172H	Southern Air Ltd (G-AVCD)/Shoreham	
G-SACE	Cessna F.150L	Southern Air Ltd (G-AZLK)/Shoreham	
G-SACF	Cessna 152 II	Southern Air Ltd (G-BHSZ)/Shoreham	
G-SAFE	Cameron N-77 balloon	Derbyshire Building Soc	
G-SAHI	Trago Mills SAH-1	Trago Mills Ltd/Bodmin	
G-SAIL	Boeing 707-323C	*Stored*	
G-SALA	PA-32-300 Cherokee Six	Golf-Sala Ltd/Elstree	
G-SALL	Cessna F.150L (Tailwheel)	Lubair (Transport Services) Ltd/ E. Midlands	
G-SALT	PA-23 Aztec 250	Peacock Salt Ltd (G-BGTH)/Glasgow	
G-SALU	Short SD3-60	Loganair Ltd (G-BKZR)/Glasgow	
G-SALV	Beech C90 King Air	Christian Salvensen Ltd (G-BIXM)/ Edinburgh	
G-SALY	Hawker Sea Fury FB.11	T. P. Luscombe & ptnrs/Lympne	
G-SAMG	Grob G.109B	RAFGSA/Bicester	
G-SAMS	M.S.880B Rallye Club	L. C. Salmon/Fenland	
G-SAMZ	Cessna 150D	N. E. Sames (G-ASSO)	
G-SARA	PA-28-181 Archer II	R. H. Ford/Elstree	
G-SARO	Saro Skeeter Mk 12	F. F. Chamberlain/Inverness	
G-SATO	PA-23 Aztec 250	Cleveland Flying School Ltd (G-BCXP)/ Tees-side	
G-SAVE	PA-31-350 Navajo Chieftain	Securicor Ltd/Birmingham	
G-SBRV	BRV Special	B. R. Vickers	
G-SCAH	Cameron V-77 balloon	S. C. A. Howarth	
G-SCAN	Vinten-Wallis WA-116/100	W. Vinten Ltd	
G-SCFO	Cameron O-77 balloon	D. V. Fowler	
G-SCHH	BAe 146-100	Dan-Air Services Ltd/Gatwick	
G-SCOT	PA-31-350 Navajo Chieftain	ATS Air Charter Ltd/Blackbushe	
G-SCUB	PA-18-135 Super Cub (542447)	N. D. Needham Farms	
G-SCUH	Boeing 737-3Q8	Dan-Air Services Ltd/Gatwick	
G-SEAH	Hawker Sea Hawk FB.3	Berowell Management Ltd/ Bournemouth	
G-SEAR	Pazmany PL.4	A. J. Sear	
G-SEBE	Learjet 35A	Siebe PLC (G-ZIPS/G-ZONE)	
G-SEBI	AS.350B Ecureuil	Walkfine Ltd (G-BMCU)	
G-SEED	Piper J-3C-65 Cub	J. H. Seed	
G-SEEK	Cessna T.210N	Abersoch Land & Sea Ltd	
G-SEJW	PA-28-161 Warrior II	Truman Aviation Ltd/Tollerton	
G-SELL	Robin DR.400/180	Sellair	
G-SEVA	SE-5A replica	I. D. Gregory	
G-SEWL	PA-28-151 Warrior	A. R. Sewell & Sons/Andrewsfield	
G-SEXY	AA-I Yankee	W. Davies (G-AYLM)/Cardiff	
G-SFHR	PA-23 Aztec 250	E. L. Becker & J. Harper (G-BHSO)	
G-SFTA	SA.341G Gazelle Srs 1	Specialist Flying Training Ltd/Carlisle	
G-SFTD	SA.341G Gazelle Srs 1	Specialist Flying Training Ltd/Carlisle	
G-SFTE	SA.341G Gazelle Srs 1	Specialist Flying Training Ltd/Carlisle	
G-SFTF	SA.341G Gazelle Srs 1	Specialist Flying Training Ltd/Carlisle	
G-SFTG	SA.341G Gazelle Srs 1	Specialist Flying Training Ltd/Carlisle	
G-SFTH	SA.341G Gazelle	Specialist Flying Training Ltd (G-BLAP)/Carlisle	
G-SFTR	NDN-1T Turbo Firecracker	Specialist Flying Training Ltd/Carlisle	
G-SFTS	NDN-1T Turbo Firecracker	Specialist Flying Training Ltd/Carlisle	
G-SFTT	NDN-1T Turbo Firecracker	Specialist Flying Training Ltd/Carlisle	
G-SFTX	Slingsby T.67M Firefly	Specialist Flying Training Ltd/Carlisle	
G-SFTZ	Slingsby T.67M Firefly	Specialist Flying Training Ltd/Carlisle	
G-SHAW	PA-30 Twin Comanche 160	Micro Metalsmiths Ltd	
G-SHEL	Cameron O-56 balloon	The Shell Company of Hong Kong Ltd	
G-SHIP	PA-23 Aztec 250 ★	Midland Air Museum/Coventry	
G-SHIV	GA-7 Cougar	G. L. Cailes/Southampton	
G-SHOE	Cessna 421C-II	Essair Ltd (G-BHGD)	
G-SHOT	Cameron V-77 balloon	Bucks Hot-Air Balloon Group	
G-SIBE	Beech 200 Super King Air	Siebe Gorman Holdings Ltd (G-MCEO/ G-BILY)	
G-SIGN	PA-39 Twin Comanche C/R	Liontravel Ltd/Elstree	
G-SILK	Aerostar 601P	R. Yorke QC/Biggin Hill	
G-SILV	Cessna 340A	Middle East Business Club/Guernsey	
G-SING	Beech B60 Duke	Tembo Records Ltd/Elstree	
G-SIPA	SIPA 903	M. Collins (G-BGBM)	
G-SITU	Partenavia P.68C	Insituform Holdings Ltd (G-NEWU/ G-BHJX)	
G-SJAB	PA-39 Twin Comanche 160 C/R	Foyle Flyers Ltd	

Notes	Reg.	Type	Owner or Operator
	G-SKAN	Cessna F.172M	Aircraft Rentals Humberside Ltd (G-BFKT)
	G-SKIM	AS.350B Ecureuil	Ernest George Aviation Ltd (G-BIVP)
	G-SKIP	Cameron N-77 balloon	Skipton Building Soc
	G-SKSA	Airship Industries SKS.500	Airship Industries Ltd/Cardington
	G-SKSB	Airship Industries SKS.500	Airship Industries Ltd/Cardington
	G-SKSC	Airship Industries SKS.600	Airship Industries Ltd/Cardington
	G-SKSD	Airship Industries SKS.600	Airship Industries Ltd/Cardington
	G-SKSF	Airship Industries SKS.600	Airship Industries Ltd/Cardington
	G-SKSG	Airship Industries SKS.600/03	Airship Industries Ltd/Cardington
	G-SKSH	Airship Industries SKS.500/06	Airship Industries Ltd/Cardington
	G-SKSJ	Airship Industries SKS.600/05	Airship Industries Ltd/Cardington
	G-SKYE	Cessna TU.206G	RAF Sport Parachute Association
	G-SKYH	Cessna 172N	Elgor Hire Purchase & Credit Ltd/ Southend
	G-SKYM	Cessna F.337E	Bencray Ltd (G-AYHW)/Blackpool
	G-SLEA	Mudry/CAARP CAP.10B	P. D. Southerington/Sturgate
	G-SLIK	Taylor JT.2 Titch	J. Jennings
	G-SLIM	Colt 56A balloon	Hot-Air Balloon Co Ltd
	G-SMHK	Cameron D-38 airship	San Miguel Brewery Ltd
	G-SMIG	Cameron 0-65 balloon	Hong Kong Balloon & Airship Club
	G-SMIT	Messerschmitt Bf.109G	Fairoaks Aviation Services Ltd/ Australia
	G-SMJJ	Cessna 414A	Gull Air Ltd/Guernsey
	G-SMRI	Westland-Bell 47G-3B1 (Soloy)	Helicrops Ltd (G-BHBV)
	G-SMUJ	Enstrom F-28C-UK	Rotorchamp Ltd (G-BHTF)
	G-SNAP	Cameron V-77 balloon	N. A. Apsey
	G-SNIP	Cessna F.172H	IWT Sheetmetal Ltd (G-AXSI)
	G-SNOW	Cameron V-77 balloon	M. J. Snow
	G-SOAR	Eiri PIK-20E	P. Rees
	G-SOFA	Cameron N-65 balloon	Northern Upholstery Ltd
	G-SOFY	PA-46-310P Malibu	Worldwide Freight Ltd
	G-SOLD	Robinson R-22A	Travel Management Ltd
	G-SOLO	Pitts S-2S Special	Fyat Ltd/Booker
	G-SONA	SOCATA TB.10 Tobago	J. Greenwood (G-BIBI)
	G-SORR	AS.350B Ecureuil	CE Aviation Ltd (G-BKMO)/Staverton
	G-SPEY	AB-206B JetRanger 3	Castle Air Charters Ltd (G-BIGO)
	G-SPIN	Pitts S-2A Special	R. N. Goode/White Waltham
	G-SPIT	V.S.379 Spitfire XIV (MV293)	B. J. S. Grey (G-BGHB)/Duxford
	G-SPOT	Partenavia P.68B Observer	Namaste (G-BCDK) (1)
	G-SRES	Beech 300 Super King Air	Vernair Transport Services Ltd Liverpool
	G-SSBS	Colting Ax77 balloon	R. J. Barr
	G-SSCH	BAe.146-100	British Aerospace Ltd (G-BIAF)/Hatfield
	G-SSSH	BAe.146-100	British Aerospace Ltd (G-BIAD)/Hatfield
	G-STAG	Cameron O-65 balloon	Holker Estates Ltd
	G-STAN	F.27 Friendship Mk.200	Air UK/Norwich
	G-STAT	Cessna U.206F	SMK Engineers Ltd
	G-STEF	Hughes 369HS	Barry Sheene Racing Ltd (G-BKTK)
	G-STEV	Jodel DR.221	S. W. Talbot/Long Marston
	G-STIO	ST.10 Diplomate	D. J. R. Gardhouse
	G-STMP	SNCAN Stampe SV-4A	W. Partridge
	G-STST	Bell 206B JetRanger 3	Petrochemical Supplies Ltd
	G-STWO	ARV Super 2	ARV Aviation Ltd
	G-SUPA	PA-18-135 Super Cub	Yorkshire Gliding Club (Pty) Ltd
	G-SUSI	Cameron V-77 balloon	H. S. & C. J. Dryden
	G-SUTT	Hughes 369E	Sutton Windows Ltd (G-OEPF/G-OMJH)
	G-SUZI	Beech 95-B55 Baron	M. Joy (G-BAXR)
	G-SUZY	Taylor JT.1 Monoplane	S. A. Kaniok/St. Just
	G-SVHA	Partenavia P.68B	D. Martin Couriers Ltd
	G-SWOT	Currie Super Wot	M. A. Kaye
	G-SWPR	Cameron N-56 balloon	Balloon Stable Ltd
	G-SYFW	Focke-Wulf Fw.190 replica	M. R. Parr
	G-TACA	P.57 Sea Prince T.1	Atlantic & Caribbean Aviation Ltd/ Staverton
	G-TACE	H.S.125 Srs 403B	Lynx Aviation Ltd (G-AYIZ)/Cranfield
	G-TACK	Grob G.109B	D. J. Tack/Spain
	G-TAFF	CASA 1.131 Jungmann	Custompac Ltd (G-BFNE)
	G-TAGL	Cessna T.182R	Mohammed Ali Hashemi (G-SAAM)/ Elstree
	G-TALI	AS.355F-1 Twin Squirrel	The Duke of Westminster
	G-TAMY	Cessna 421B	Abbergail Ltd/Luton
	G-TAPE	PA-23 Aztec 250	E. C. Murray (G-AWVW)

Reg.	Type	Owner or Operator	Notes
G-TARA	Christen Eagle II	F. E. P. Holmes	
G-TATT	GY-20 Minicab	L. Tattershall	
G-TAXI	PA-23 Aztec 250	Northern Executive Aviation Ltd/ Manchester	
G-TBIO	SOCATA TB.10 Tobago	Buchanan Electronics	
G-TBXX	SOCATA TB.20 Trinidad	H. Deville & ptnrs	
G-TBZO	SOCATA TB.20 Trinidad	M. Barnett/Elstree	
G-TCAR	Robin HR.100/210	Michael Jackson Motors Ltd	
G-TCAT	G.164D Turbo Ag-Cat	Miller Aerial Spraying Ltd/Wickenby	
G-TDAA	Cessna U.206G	Edward & Susan Dexter Ltd/Denham	
G-TEAC	AT-6C Harvard IIA (EX280)	E. C. English/Bourn	
G-TEAM	Cessna 414A	Imperial Brewing & Leisure Ltd/ (G-BHJT)	
G-TECH	R. Commander 114	P. A. Reed (G-BEDH)/Denham	
G-TEDS	SOCATA TB.10 Tobago	E. M. Fleet (G-BHCO)/Lydd	
G-TEES	Cessna F.152	Cleveland Flying School Ltd (G-BIUI)/ Tees-side	
G-TEFC	PA-28 Cherokee 140	Thames Estuary Flying Club/Southend	
G-TEFH	Cessna 500 Citation	T. B. T. (Transport) Ltd (G-BCII)/ E. Midlands	
G-TEMI	BN-2T Islander	Pilatus BN Ltd (G-BJYX)/Bembridge	
G-TESS	Quickie Q.2	D. Evans	
G-TFCI	Cessna FA.152	Tayside Aviation Ltd/Dundee	
G-TFUN	Valentin Taifun 17E	J. A. Sangster/Booker	
G-THAM	Cessna F.182Q	German Tourist Facilities Ltd/Luton	
G-THEA	Boeing E75 Stearman	L. M. Walton	
G-THOM	Thunder Ax6-56 balloon	T. H. Wilson	
G-THOR	Thunder Ax8-105 balloon	N. C. Faithful *Turncoat*	
G-THSL	PA-28R-201 Arrow II	G. Fearnley/Southend	
G-TIGA	D.H.82A Tiger Moth	Truman Aviation Ltd (G-AOEG)	
G-TIGB	AS.332L Super Puma	Bristow Helicopters Ltd (G-BJXC)	
G-TIGC	AS.332L Super Puma	Bristow Helicopters Ltd (G-BJYH)	
G-TIGE	AS.332L Super Puma	Bristow Helicopters Ltd (G-BJYJ)	
G-TIGF	AS.332L Super Puma	Bristow Helicopters Ltd	
G-TIGG	AS.332L Super Puma	Bristow Helicopters Ltd	
G-TIGH	AS.332L Super Puma	Bristow Helicopters Ltd	
G-TIGI	AS.332L Super Puma	Bristow Helicopters Ltd	
G-TIGJ	AS.332L Super Puma	Bristow Helicopters Ltd	
G-TIGK	AS.332L Super Puma	Bristow Helicopters Ltd	
G-TIGL	AS.332L Super Puma	Bristow Helicopters Ltd	
G-TIGM	AS.332L Super Puma	Bristow Helicopters Ltd	
G-TIGN	AS.332L Super Puma	Bristow Helicopters Ltd	
G-TIGO	AS.332L Super Puma	Bristow Helicopters Ltd	
G-TIGP	AS.332L Super Puma	Bristow Helicopters Ltd	
G-TIGR	AS.332L Super Puma	Bristow Helicopters Ltd	
G-TIGS	AS.332L Super Puma	Bristow Helicopters Ltd	
G-TIGT	AS.332L Super Puma	Bristow Helicopters Ltd	
G-TIGU	AS.332L Super Puma	Bristow Helicopters Ltd	
G-TIGV	AS.332L Super Puma	Bristow Helicopters Ltd	
G-TIGW	AS.332L Super Puma	Bristow Helicopters Ltd	
G-TIKI	Colt 105A balloon	Lighter-Than-Air Ltd (G-BKWV)	
G-TIMB	Rutan Vari-Eze	T. M. Bailey	
G-TIME	Aerostar 601P	J. J. Donn	
G-TIMJ	Rand KR-2	T. M. Justice	
G-TIMK	PA-28-181 Archer II	J. C. Lloyd	
G-TIMW	PA-28 Cherokee 140C	T. G. Watts (G-AXSH)/Fairoaks	
G-TINA	SOCATA TB.10 Tobago	A. Lister	
G-TIRE	Colt Flying Tyre balloon	Colt Balloons Ltd	
G-TISH	PA-31-310 Turbo Navajo	Northern Aviation Ltd (G-BFKJ)	
G-TJET	Lockheed T-33A-1-LO	Aces High Ltd/Duxford	
G-TKHM	AB-206B JetRanger 3	Red Rose Helicopters Ltd (G-MKAN/ G-DOUG)/Blackpool	
G-TLOL	Cessna 421C	Littlewoods Organisation Ltd/ Liverpool	
G-TMAS	H.S.125 Srs 600B	Clartacrest Ltd (G-MFEU)	
G-TMJH	Hughes 369E	Hughes of Beaconsfield	
G-TOBI	Cessna F.172K	John Bradley & Barry Ltd (G-AYVB)/ Sandown	
G-TOBY	Cessna 172B	J. A. Kelman (G-ARCM)/Elstree	
G-TOFF	AS.355F Twin Squirrel	Atlantic Computer Leasing PLC (G-BKJX)	
G-TOGA	PA-32-301 Saratoga	W. H. Carson/Elstree	

Notes	Reg.	Type	Owner or Operator
	G-TOMI	H.S.125 Srs 600B	Tom Walkinshaw Racing Ltd G-BBEP/G-BJOY)
	G-TOMS	PA-38-112 Tomahawk	Channel Aviation Ltd/Guernsey
	G-TONI	Cessna 421C	Rassler Aero Services/Booker
	G-TOUR	Robin R.2112	Finncharter Services
	G-TOWN	Piaggio FWP.149D	J. Townhill
	G-TOYS	Enstrom F-280C-UK-2	London Refurbishing Co Ltd (G-BISE)
	G-TPPH	AB-206B JetRanger 2	Winterbourne Construction Ltd (G-BCYP)
	G-TPTR	AB-206B JetRanger 3	Mann Aviation Ltd (G-LOCK)/Fairoaks
	G-TRAF	SA.365N Dauphin 2	Trans World Leasing Ltd (G-BLDR)
	G-TREV	Saffery S.330 balloon	T. W. Gurd
	G-TRIM	Monnet Moni	J. E. Bennell
	G-TRIP	PA-32R-301 Saratoga SP	A. P. J. Lavelle & I. Fabini (G-HOSK)
	G-TRIX	V.S.509 Spitfire T.IX	S. Atkins
	G-TRUK	Stoddard Hamilton Glassair RG	Archer Engineering (Leeds) Ltd
	G-TSAM	H.S.125 Srs 800B	BSM Holdings Ltd/Heathrow
	G-TSIX	AT-6C Harvard IIA	D. Taylor/E. Midlands
	G-TTAM	Taylor JT.2 Titch	A. J. Manning
	G-TTWO	Colt 56A balloon	P. N. Tilney
	G-TUBS	Beech 65-80 Queen Air	A. H. Bowers (G-ASKM)/Staverton
	G-TUBY	Cessna 310J	A. H. Bower (G-ASZZ)
	G-TUDR	Cameron V-77 balloon	Jacques W. Soukup Ltd
	G-TUGG	PA-18 Super Cub 150	Air Sale PLC/Booker
	G-TUKE	Robin DR.400/160	Tukair/Headcorn
	G-TURB	D.31 Turbulent	A. Ryan-Fecitt
	G-TVKE	Cessna 310R	Ewart & Co (Studio) Ltd (G-EURO)/ Elstree
	G-TVSI	Campbell Cricket	W. H. Beevers (G-AYHH)
	G-TWEL	PA-28-181 Archer II	T. W. Electrical Ltd/Sywell
	G-TWEY	Colt 69A balloon	British Telecom Thameswey
	G-TWIN	PA-44-180 Seminole	Leavesden Flight Centre Ltd
	G-TWOB	BN-2B-26 Islander	Pilatus BN Ltd (G-BKJJ)/Bembridge
	G-TYGA	AA-5B Tiger	Autohover Ltd (G-BHNZ)/Biggin Hill
	G-TYME	R. Commander 690B	Marlborough (London) Ltd
	G-TYRE	Cessna F.172M	Watts Aviation Ltd/Staverton
	G-UARD	Sequoia F.8L Falco	A. J. Baggarley
	G-UBHL	Beech B200 Super King Air	United Biscuits (UK) Ltd/Denham
	G-UERN	BN-2B-26 Islander	Air Sarnia Ltd (G-BHXI)/Guernsey
	G-UIDE	Jodel D.120	S. T. Gilbert/Popham
	G-UILD	Grob G.109B	Runnymede Consultants Ltd
	G-UKNO	Cessna U.206C	British Parachute Schools (G-BAMN)
	G-UPDN	Cameron V-65 balloon	R. J. O. Evans
	G-USAF	T-28C Trojan	M. B. Walker
	G-USAM	Cameron Uncle Sam balloon	Jacques W. Soukup Ltd
	G-USTO	Beech A24R Musketeer	E. Perrin Ltd (G-AYPA)/Biggin Hill
	G-USTY	FRED Srs 2	S. Styles
	G-VAGA	PA-15 Vagabond	Pyrochem Ltd/White Waltham
	G-VAJK	H.S.748 Srs 1	Euroair Ltd (G-BEKG)/Gatwick
	G-VANG	AB-206B JetRanger 3	Skyhook Lifting Ltd (G-BIZA)
	G-VARG	Varga 2150A Kachina	J. Hannibal/Halfpenny Green
	G-VAUN	Cessna 340	F. E. Peacock & Son (Thorney) Ltd
	G-VELT	Cessna 340A	Velcourt (East) Ltd/Staverton
	G-VENI	D.H.112 Venom FB.50	Source Premium & Promotional Consultants Ltd
	G-VEZE	Rutan Vari-Eze	P. J. Henderson
	G-VGIN	Boeing 747-243B	Virgin Atlantic Airways Ltd/ Gatwick
	G-VICK	PA-31 Turbo Navajo	Howard Richard & Co Ltd (G-AWED)/ Elstree
	G-VIDI	D.H.112 Venom FB.50	Source Premium & Promotional Consultants Ltd
	G-VIEW	Vinten-Wallis WA-116/100	W. Vinten Ltd
	G-VIKE	Bellanca 1730A Viking	Modular Business Computers Ltd/ Elstree
	G-VIRG	Boeing 747-287B	Virgin Atlantic Airways Ltd/Gatwick
	G-VIST	PA-30 Twin Comanche 160	B. P. Paine & ptnrs (G-AVHZ)/Elstree
	G-VITE	Robin R.1180T	Trans Global Aviation Supply Co Ltd/ Booker
	G-VIVA	Thunder Ax7-65 balloon	J. G. Spearing

Reg.	Type	Owner or Operator	Notes
G-VIXN	D.H.110 Sea Vixon FAW.2	Brencham Historic Aircraft Ltd/ Bournemouth	
G-VIZZ	Sportavia RS.180 Sportsman	Executive Air Sport Ltd/Exeter	
G-VJCT	Partenavia P.68C	JCT 600 Ltd/Leeds	
G-VMDE	Cessna P.210N	V. S. Evans & Horne & Sutton Ltd/ Cranfield	
G-VNOM	D.H.112 Venom FB.50	A. Topen/Cranfield	
G-VPTO	Evans VP-2	J. Cater	
G-VRES	Beech A200 Super King Air	Vernair Transport Services/Liverpool	
G-VSEL	Beech 200 Super King Air	Vickers Shipbuilding & Engineering Ltd (G-SONG/G-BKTI)/Barrow	
G-VSOP	Cameron SS balloon	J. R. Parkington & Co Ltd	
G-VTEN	Vinten-Wallis WA-117	W. Vinten Ltd	
G-VTII	D.H.115 Vampire T.11 (WZ507)	J. Turnbull & ptnrs/Cranfield	
G-VTOL	H.S. Harrier T52	British Aerospace/Dunsfold	
G-WAAC	Cameron N-56 balloon	Advertising Balloon Co	
G-WADE	Cessna F.172N	Wade Aviation (G-BHMI)	
G-WAGY	Cessna F.172N	J. B. Wagstaff/E. Midlands	
G-WARD	Taylor JT.1 Monoplane	G. & G. D. Ward	
G-WASP	Brantly B.2B	P. A. Taylor (G-ASXE)	
G-WAUS	BAe 146-200	British Aerospace PLC (G-WISC)/ Hatfield	
G-WAVE	Grob G.109B	M. L. Murdoch	
G-WCEI	M.S.894 Rallye 220GT	W. C. Evans (G-BAOC)	
G-WEBB	PA-23 Aztec 250	Brands Hatch Circuit Ltd (G-BJBU)	
G-WELI	Cameron N-77 balloon	London Life Assurance Ltd	
G-WELL	Beech E90 King Air	CEGA Aviation Ltd/Goodwood	
G-WEND	PA-28RT-201 Arrow IV	Trent Insulations Ltd/Birmingham	
G-WERY	SOCATA TB.20 Trinidad	G. J. Werry/Manston	
G-WEST	Agusta A.109A	Westland Helicopters Ltd/Yeovil	
G-WETI	Cameron N-31 balloon	Zebedee Balloon Service	
G-WGCS	PA-19 Super Cub 95	G. W. G. C. Sudlow	
G-WHIZ	Pitts S-1 Special	K. M. McLeod	
G-WHIZ	V.732 Viscount (fuselage only) ★	S. Wales Museum/Rhoose (G-ANRS)	
G-WICH	FRED Srs 2	R. H. Hearn	
G-WICK	Partenavia P.68B	Fortune Services Ltd (G-BGFZ)/ Manchester	
G-WILD	Pitts S-1T Special	B. K. Lecomber	
G-WILY	Rutan Long-Ez	W. S. Allen	
G-WINE	Thunder Ax7-77Z balloon	Thunder Balloons Ltd	
G-WITE	Cessna 414A	Essair Ltd (G-LOVO/G-KENT)	
G-WIXY	Mudry/CAARP CAP.10B	G. Tanner & P. O. Wicks Ltd/ Andrewsfield	
G-WIZZ	AB-206B JetRanger 2	Hubbard Reader Group Ltd/Ipswich	
G-WLAD	BAC One-Eleven 304AX	Airways International Cymru Ltd (G-ATPI)	
G-WMCC	BAe Jetstream 3102-01	Birmingham Executive Airways Ltd (G-TALL)	
G-WOLF	PA-28 Cherokee 140	E. Ford	
G-WOLL	G.164A Ag-Cat	Wickwell Holdings Ltd (G-AYTM)	
G-WOOD	Beech 95-B55 Baron	Bunyan Meyer & ptnrs (G-AYID)/ Fairoaks	
G-WOSP	Bell 206B JetRanger 3	Gleneagles Helicopter Services (Scotland) Ltd	
G-WOTG	BN-2T Islander	Secretary of State for Defence (G-BJYT)/Bicester	
G-WREN	Pitts S-2A Special	P. Meeson/Booker	
G-WRMN	Glaser-Dirks DG-400	W. R. McNair	
G-WROY	PA-32RT-300T Turbo Lance II	R. L. West (G-WRAY)/Norwich	
G-WSKY	Enstrom F-280C Shark	First Class Furniture Ltd (G-BEEK)/ Booker	
G-WSSL	PA-31-350 Navajo Chieftain	Foster Yeoman Ltd	
G-WTVB	Cessna 404 Titan	Casair Aviation Ltd/Tees-side	
G-WULF	WAR Focke-Wulf Fw.190 (O4)	A. C. Walker & ptnrs/Elstree	
G-WULL	AA-5A Cheetah	Canonbury Wine Ltd	
G-WWHL	Beech 200 Super King Air	Wharton Williams Ltd (G-BLAE)	
G-WWII	V.S. Spitfire 18 (SM832)	D. W. Arnold & ptnrs	
G-WWUK	Enstrom F-28A-UK	Wickwell (Holdings) Ltd (G-BFFN)	
G-WYNN	Rand KR-2	W. Thomas	
G-WYNT	Cameron N-56 balloon	Jacques W. Soukup Enterprises Ltd	
G-WYRL	Robinson R-22A	ITM Ltd	
G-WYTE	Bell 47G-2A-1	M. G. White/Southend	

Notes	Reg.	Type	Owner or Operator
	G-WZZZ	Colt AS-42 balloon	Hot-Air Balloon Co Ltd
	G-XCUB	PA-18-150 Super Cub	W. G. Fisher/Sandown
	G-XMAF	G.1159A Gulfstream 3	Fayair (Jersey) 1984 Ltd
	G-XTWO	EMB-121A Xingu II	Numerically Controlled Machine Tools Ltd (G-XING)
	G-YABU	Gulfstream Commander 695A	Celtic Leasing Services Ltd
	G-YBAA	Cessna FR.172J	J. Blackburn
	G-YIII	Cessna F.150L	Sherburn Aero Club Ltd
	G-YMRU	BAC One-Eleven 304AX	Airways International Cymru (G-ATPH)/ Cardiff
	G-YNOT	D.62B Condor	T. Littlefair (G-AYFH)
	G-YORK	Cessna F.172M	Sherburn Aero Club Ltd
	G-YPSY	Andreasson BA-4B	H. P. Burrill
	G-YROS	Bensen B.80-D	J. M. Montgomerie
	G-YTWO	Cessna F.172M	Sherburn Aero Club Ltd
	G-YULL	PA-28 Cherokee 180E	Lansdowne Chemical Co (G-BEAJ)/ Kidlington
	G-ZAZA	PA-18 Super Cub 95	M. C. Barraclough
	G-ZERO	AA-5B Tiger	Snowadem Ltd/Luton
	G-ZIPI	Robin DR.400/180	Stahl Engineering Co Ltd/Headcorn
	G-ZIPP	Cessna E.310Q	Bank Farm Ltd (G-BAYU)
	G-ZLIN	Z.526 Trener Master	V. S. E. Norman
	G-ZSOL	Zlin Z.50L	V. S. E. Norman/Kemble
	G-ZUMP	Cameron N-77 balloon	M. J. Allen *Gazump*
	G-ZZIM	Rutan Laser 200	J. G. M. Heathcote
	G-ZZZZ	Point Maker Mk. 1 balloon	M. J. Wakelin

Toy Balloons

Reg.	Type	Owner or Operator	Notes
G-FYAA	Osprey Mk 4D	C. Wilson	
G-FYAB	Osprey Mk 4B	M. R. Wilson	
G-FYAC	Portswood Mk XVI	J. D. Hall	
G-FYAD	Portswood Mk XVI	J. D. Hall	
G-FYAE	Portswood Mk XVI	J. D. Hall	
G-FYAF	Portswood Mk XVI	J. D. Hall	
G-FYAG	Portswood Mk XVI	J. D. Hall	
G-FYAH	Portswood Mk XVI	J. D. Hall	
G-FYAI	Portswood Mk XVI	J. D. Hall	
G-FYAJ	Kelsey	P. J. Kelsey	
G-FYAK	European E.21	J. E. Christopher	
G-FYAL	Osprey Mk 4E2	J. Goodman	
G-FYAM	Osprey Mk 4E2	P. Goodman	
G-FYAN	Williams	M. D. Williams	
G-FYAO	Williams	M. D. Williams	
G-FYAP	Williams Mk 2	G. E. Clarke	
G-FYAR	Williams Mk 2	S. T. Wallbank	
G-FYAS	Osprey Mk 4H2	K. B. Miles	
G-FYAT	Osprey Mk 4D	S. D. Templeman	
G-FYAU	Williams MK 2	P. Bowater	
G-FYAV	Osprey Mk 4E2	C. D. Egan & C. Stiles	
G-FYAW	Portswood Mk XVI	R. S. Joste	
G-FYAX	Osprey Mk 4B	S. A. Dalmas & P. G. Tarr	
G-FYAY	Osprey Mk 1E	M. K. Levenson	
G-FYAZ	Osprey Mk 4D2	M. A. Roblett	
G-FYBA	Portswood Mk XVI	C. R. Rundle	
G-FYBD	Osprey Mk 1E	M. Vincent	
G-FYBE	Osprey Mk 4D	M. Vincent	
G-FYBF	Osprey Mk V	M. Vincent	
G-FYBG	Osprey Mk 4G2	M. Vincent	
G-FYBH	Osprey Mk 4G	M. Vincent	
G-FYBI	Osprey Mk 4H	M. Vincent	
G-FYBK	Osprey Mk 4G2	A. G. Coe & S. R. Burgess	
G-FYBL	Osprey Mk 4D	P. A. Tilley	
G-FYBM	Osprey Mk 4G	P. C. Anderson	
G-FYBN	Osprey Mk 4G2	M. Ford	
G-FYBO	Osprey Mk 4B	D. Eaves	
G-FYBP	European E.84PW	D. Eaves	
G-FYBR	Osprey Mk 4G2	N. A. Partridge	
G-FYBT	Portswood Mk XVI	M. Hazelwood	
G-FYBU	Portswood Mk XVI	M. A. Roblett	
G-FYBV	Osprey Mk 4D2	D. I. Garrod	
G-FYBW	Osprey Mk 4D	N. I. McAllen	
G-FYBX	Portswood Mk XVI	I. Chadwick	
G-FYBY	Osprey Mk 4D	K. H. Turner	
G-FYBZ	Osprey Mk 1E	S. J. Showbridge	
G-FYCA	Osprey Mk 4D	R. G. Crewe	
G-FYCB	Osprey Mk 4B	I. R. Hemsley	
G-FYCC	Osprey Mk 4G2	A. Russell	
G-FYCD	BHMED	D. Meades	
G-FYCE	Portswood Mk XVI	R. S. Joste	
G-FYCF	Portswood Mk XVI	R. S. Joste	
G-FYCG	Portswood Mk XVI	R. S. Joste	
G-FYCH	Swan Mk 1	R. S. Joste	
G-FYCI	Portswood Mk XVI	R. S. Joste	
G-FYCJ	Osprey Mk 4H2	A. G. Coe & S. R. Burgess	
G-FYCK	Lovell Mk 1	G. P. Lovell	
G-FYCL	Osprey Mk 4G	P. J. Rogers	
G-FYCN	Osprey Mk 4D	C. F. Chipping	
G-FYCO	Osprey Mk 4B	C. F. Chipping	
G-FYCP	Osprey Mk 1E	C. F. Chipping	
G-FYCR	Osprey MK 4D	C. F. Chipping	
G-FYCS	Portswood Mk XVI	S. McDonald	
G-FYCT	Osprey Mk 4D	S. T. Wallbank	
G-FYCU	Osprey Mk 4D	G. M. Smith	
G-FYCV	Osprey Mk 4D	M. Thomson	
G-FYCW	Osprey Mk 4D	M. L. Partridge	
G-FYCX	Jefferson Mk IV	J. R. Sumner	
G-FYCZ	Osprey Mk 4D2	P. Middleton	
G-FYDA	Atom	H. C. Saffrey	

Notes	Reg.	Type	Owner or Operator
	G-FYDB	European E.84EL	D. Eaves
	G-FYDC	European EDH-1	D. Eaves & H. Goddard
	G-FYDD	Osprey Mk 4D	A. C. Mitchell
	G-FYDE	Osprey Mk 4D	P. F. Mitchell
	G-FYDF	Osprey Mk 4D	K. A. Jones
	G-FYDG	Osprey Mk 4D	M. D. Williams
	G-FYDH	Premier Voyage	H. C. Saffery
	G-FYDI	Williams Westwind Two	M. D. Williams
	G-FYDK	Williams Westwind Two	M. D. Williams
	G-FYDM	Williams Westwind Four	M. D. Williams
	G-FYDN	European 8C	P. D. Ridout
	G-FYDO	Osprey Mk 4D	N. L. Scallan
	G-FYDP	Williams Westwind Three	M. D. Williams
	G-FYDR	European 118	P. F. Mitchell
	G-FYDS	Osprey Mk 4D	N. L. Scallan
	G-FYDT	Viking Warrior Mk 1	D. G. Tomlin
	G-FYDU	Osprey Mk 4D	J. R. Moody
	G-FYDV	Osprey Mk 4D	A. J. Jackson
	G-FYDW	Osprey Mk 4B	R. A. Balfre
	G-FYDX	Osprey Mk 4B	G. T. Young
	G-FYDY	Osprey Mk 4B	P. S. Flanagan
	G-FYDZ	Portswood Mk XVI	S. M. Chance
	G-FYEA	Osprey Mk 4B	S. M. Chance
	G-FYEB	Rango Rega	N. H. Ponsford
	G-FYEC	Osprey Mk 4B	T. R. Spruce
	G-FYED	Osprey Mk 1C	T. P. Pusey
	G-FYEE	Osprey Mk 4B	T. R. Spruce
	G-FYEF	Portswood Mk XVI	T. R. Spruce
	G-FYEG	Osprey Mk 1C	P. E. Prime
	G-FYEI	Portswood Mk XVI	A. Russell
	G-FYEJ	Rango NA.24	N. H. Ponsford
	G-FYEK	Unicorn UE.1C	D. & D. Eaves
	G-FYEL	European E.84Z	D. Eaves
	G-FYEO	Eagle Mk.1	M. E. Scallon
	G-FYEP	Boing 746-200A	S. M. Colville & D. J. Hall
	G-FYER	Osprey Mk.4B	S. J. Menges
	G-FYES	Osprey Mk 2 SJM	S. J. Menges
	G-FYET	Markmite Mk 2	M. W. Mabey & M. Davies
	G-FYEU	Rango N.8	R. G. Scathdee
	G-FYEV	Osprey Mk.1C	M. E. Scallen
	G-FYEW	Saturn Mk 2A balloon	M. J. Sheather
	G-FYEY	Largess balloon	S. J. Menges
	G-FYEZ	Firefly Mk 1 balloon	M. E. & N. L. Scallan
	G-FYFA	European E.84LD balloon	D. Goddard & D. Eaves
	G-FYFB	Osprey Mk 1E	K. Marsh
	G-FYFC	European E.84NZ	R. MacPherson
	G-FYFD	Osprey Mk 2CM	M. Carp
	G-FYFE	Osprey Mk 2GB	G. Bone
	G-FYFF	Osprey Mk 2SW	S. Willis
	G-FYFG	European E.84DE	D. Eaves
	G-FYFH	European E.84DS	D. Eaves
	G-FYFI	European E.84DS	M. Stelling
	G-FYFJ	Westland 2	P. Feasey
	G-FYFK	Westland 2	D. Feasey
	G-FYFL	Osprey Mk 2CL	C. Kennedy

Microlights

Reg.	Type	Owner or Operator	Notes
G-MBAA	Hiway Skytrike Mk 2	Hiway Hang Gliders Ltd	
G-MBAB	Hovey Whing-Ding II	R. F. Morton	
G-MBAD	Weedhopper JC-24A	M. Stott	
G-MBAE	Lazair	H. A. Leek	
G-MBAF	R. J. Swift 3	C. G. Wrzesien	
G-MBAG	Skycraft Scout	B. D. Jones	
G-MBAH	Harker D. H.	D. Harker	
G-MBAI	Typhoon Tripacer 250	C. J. & K. Yarrow	
G-MBAJ	Chargus T.250	V. F. Potter	
G-MBAK	Eurowing Spirit	J. S. Potts	
G-MBAL	Hiway Demon	K. W. E. Brunnenkant	
G-MBAM	Skycraft Scout 2	J. E. Orbell	
G-MBAN	American Aerolights Eagle	R. W. Millward	
G-MBAP	Rotec Rally 2B	P. D. Lucas	
G-MBAR	Skycraft Scout	L. Chiappi	
G-MBAS	Typhoon Tripacer 250	T. J. Birkbeck	
G-MBAT	Hiway Skytrike	M. R. Gardiner	
G-MBAU	Hiway Skytrike	M. R. Gardiner	
G-MBAV	Weedhopper	L. F. Smith	
G-MBAW	Pterodactyl Ptraveller	J. C. K. Soardifield	
G-MBAX	Hiway Skytrike	D. Clarke	
G-MBAY	Skycraft Scout	G. M. Hayden	
G-MBAZ	Rotec Rally 2B	Western Skysports Ltd	
G-MBBA	Ultraflight Lazair	P. Roberts	
G-MBBB	Skycraft Scout 2	A. J. & B. Chalkley	
G-MBBC	Chargus T.250	R. R. G. Close-Smith	
G-MBBD	Pterodactyl Ptraveller	R. N. Greenshields	
G-MBBE	Striplin Skyranger	A. S. Coombes	
G-MBBF	Chargus Titan 38	Chargus Gliding Co	
G-MBBG	Weedhopper JC-24B	A. J. Plumbridge & G. E. Kershaw	
G-MBBH	Flexiform Sealander 160	J. A. Evans	
G-MBBI	Ultraflight Mirage	B. H. Trunkfield & A. A. Howard	
G-MBBJ	Hiway Demon Trike	E. B. Jones	
G-MBBL	Lightning Microlight	I. M. Grayland	
G-MBBM	Eipper Quicksilver MX	J. Brown	
G-MBBN	Eagle Microlight	S. Taylor & D. Williams	
G-MBBO	Rotec Rally 2B	P. T. Dawson	
G-MBBP	Chotia Weedhopper	G. L. Moon	
G-MBBR	Weedhopper JC-24B	J. G. Wallers	
G-MBBS	Chargus T.250	P. R. De Fraine	
G-MBBT	Ultrasports Tripacer 330	The Post Office	
G-MBBU	Southdown Savage	R. Venton-Walters	
G-MBBV	Rotec Rally 2B	Blois Aviation Ltd	
G-MBBW	Flexiform Hilander	R. J. Hamilton & W. J. Shaw	
G-MBBX	Chargus Skytrike	A. W. Buchan	
G-MBBY	Flexiform Sealander	J. E. Halsall	
G-MBBZ	Volmer Jensen VJ-24W	D. G. Cook	
G-MBCA	Chargus Cyclone T.250	E. M. Jelonek	
G-MBCB	Southdown Lightning	P. G. Huxham	
G-MBCD	La Mouette Atlas	M. G. Dean	
G-MBCE	American Aerolights Eagle	I. H. Lewis	
G-MBCF	Pterodactyl Pfledgling	T. C. N. Carroll	
G-MBCG	Ultrasports Tripacer T.250	A. G. Parkinson	
G-MBCI	Hiway Skytrike	J. R. Bridge	
G-MBCJ	Mainair Sports Tri-Flyer	J. R. North	
G-MBCK	Eipper Quicksilver MX	G. W. Rowbotham	
G-MBCL	Hiway Demon Triflyer	B. R. Underwood & P. J. Challis	
G-MBCM	Hiway Demon 175	G. M. R. & D. M. Walters	
G-MBCN	Hiway Super Scorpion	M. J. Hadland	
G-MBCO	Flexiform Sealander Buggy	P. G. Kavanagh	
G-MBCP	Mainair Tri-Flyer 250	J. B. Wincott	
G-MBCR	Ultraflight Mirage	B. N. Bower	
G-MBCS	American Aerolights Eagle	Pleasurecraft Ltd	
G-MBCT	American Aerolights Eagle	Pleasurecraft Ltd	
G-MBCU	American Aerolights Eagle	J. L. May	
G-MBCV	Hiway Skytrike	C. J. Greasley	
G-MBCW	Hiway Demon 175	C. Foster & S. B. Elwis	
G-MBCX	Airwave Nimrod 165	M. J. Ashley-Rogers	
G-MBCY	American Aerolights Eagle	R. D. Chiles	
G-MBCZ	Chargus Skytrike 160	R. M. Sheppard	

Notes	Reg.	Type	Owner or Operator
	G-MBDA	Rotec Rally 2B	Blois Aviation Ltd
	G-MBDB	Solar Wings Typhoon	D. J. Smith
	G-MBDC	Hornet Microlight	R. R. Wolfenden & G. Priestley
	G-MBDD	Skyhook Skytrike	D. Hancock
	G-MBDE	Flexiform Skytrike	R. W. Chatterton
	G-MBDF	Rotec Rally 2B	J. R. & B. T. Jordan
	G-MBDG	Eurowing Goldwing	N. W. Beadle & ptnrs
	G-MBDH	Hiway Demon Triflyer	A. T. Delaney
	G-MBDI	Flexiform Sealander	K. Bryan
	G-MBDJ	Flexiform Sealander Triflyer	J. W. F. Hargrave
	G-MBDK	Solar Wings Typhoon	A. O'Brien
	G-MBDL	Lone Ranger Microlight	Aero & Engineering Services Ltd
	G-MBDM	Southdown Sigma Trike	A. R. Prentice
	G-MBDN	Hornet Atlas	R. Burton
	G-MBDO	Flexiform Sealander Trike	K. Kerr
	G-MBDP	Flexiform Sealander Skytrike	D. Mackillop
	G-MBDR	U.A.S. Stormbuggy	P. J. D. Kerr
	G-MBDT	American Aerolights Eagle	I. D. Stokes
	G-MBDU	Chargus Titan 38	Property Associates Ltd
	G-MBDV	Pterodactyl Ptraveller	C. J. R. Miller
	G-MBDW	Ultrasports Tripacer Skytrike A	J. T. Meager
	G-MBDX	Electraflyer Eagle	Ardenco Ltd
	G-MBDY	Weedhopper 2	G. N. Mayes
	G-MBDZ	Eipper Quicksilver MX	M. Risdale
	G-MBEA	Hornet Nimrod	M. Holling
	G-MBEB	Hiway Skytrike 250 Mk II	M. A. H. Milne
	G-MBEC	Hiway Super Scorpion	C. S. Wates
	G-MBED	Chargus Titan 38	G. G. Foster
	G-MBEE	Hiway Super Scorpion Skytrike 160	P. H. Risdale & ptnrs
	G-MBEJ	Electraflyer Eagle	D. J. Royce & C. R. Gale
	G-MBEL	Electraflyer Eagle	J. R. Fairweather
	G-MBEN	Eipper Quicksilver MX	S. Cordova
	G-MBEO	Flexiform Sealander	H. W. Williams
	G-MBEP	American Aerolights Eagle	R. W. Lavender
	G-MBER	Skyhook Sailwings TR-1	Skyhook Sailwings Ltd
	G-MBES	Skyhook Sailwings TR-2	Skyhook Sailwings Ltd
	G-MBET	MEA Mistral Trainer	J. W. V. Edmunds
	G-MBEU	Hiway Demon T.250	R. C. Smith
	G-MBEV	Chargus Titan 38	A. K. Hatenboer
	G-MBEW	UAS Solar Buggy	A. G. Davies
	G-MBEZ	Pterodactyl Ptraveller II	P. A. Smith
	G-MBFA	Hiway Skytrike 250	P. S. Jones
	G-MBFD	Gemini Hummingbird	Micro Aviation Ltd
	G-MBFE	American Aerolights Eagle	P. W. Cole
	G-MBFF	Southern Aerosports Scorpion	H. Redwin
	G-MBFG	Skyhook Sabre	R. E. Rushton
	G-MBFH	Hiway Skytrike	P. Baldwin
	G-MBFI	Hiway Skytrike II	J. R. Brabbs
	G-MBFJ	Chargus Typhoon T.250	A. W. Knowles
	G-MBFK	Hiway Demon	D. W. Stamp
	G-MBFL	Hiway Demon	J. C. Houghton
	G-MBFM	Hiway Hang Glider	G. P. Kimmons & T. V. O. Mahony
	G-MBFN	Hiway Skytrike II	R. Williamson
	G-MBFO	Eipper Quicksilver MX	F. G. Shepherd
	G-MBFP	Southern Aerosports Scorpion	J. G. Beesley
	G-MBFR	American Aerolights Eagle	W. G. Bradley
	G-MBFS	American Aerolights Eagle	R. Fox
	G-MBFT	Southdown Sigma 12 Meter	D. P. Watts
	G-MBFU	Ultrasports Tripacer	T. H. J. Prowse
	G-MBFV	Comet Skytrike	R. Willis
	G-MBFX	Hiway Skytrike 250	J. W. Broadhead
	G-MBFY	Mirage II	J. P. Metcalf
	G-MBFZ	M. S. S. Goldwing	I. T. Barr
	G-MBGA	Solar Wings Typhoon	A. L. Rogers
	G-MBGB	American Aerolights Eagle	J. C. Miles
	G-MBGD	Pterodactyl 430C Replica	C. Wilkinson
	G-MBGE	Hiway Scorpion Trike	J. A. Rudd
	G-MBGF	Twamley Trike	T. B. Woolley
	G-MBGG	Chargus Titan 38	Solar Wings Ltd
	G-MBGH	Chargus T.250	A. G. Doubtfire
	G-MBGI	Chargus Titan 38	A. G. Doubtfire
	G-MBGJ	Hiway Skytrike Mk 2	B. C. Norris & J. R. Edwards
	G-MBGK	Electra Flyer Eagle	R. J. Osbourne

Reg.	Type	Owner or Operator	Notes
G-MBGL	Flexiform Sealander Skytrike	H. Field	
G-MBGM	Eipper Quicksilver MX	G. G. Johnson	
G-MBGN	Weedhopper Model A	D. Roberts	
G-MBGO	American Aerolights Eagle	J. E. Bennison	
G-MBGP	Solar Wings Typhoon Skytrike	M. F. R. Collett	
G-MBGR	Eurowing Goldwing	G. A. J. Salter	
G-MBGS	Rotec Rally 2B	P. C. Bell	
G-MBGT	American Aerolights Eagle	D. C. Lloyd	
G-MBGV	Skyhook Cutlass	D. M. Parsons	
G-MBGW	Hiway Skytrike	G. W. R. Cooke	
G-MBGX	Southdown Lightning	R. B. D. Baker	
G-MBGY	Hiway Demon Skytrike	W. Hopkins	
G-MBGZ	American Aerolights Eagle	Flying Machine (Circa 1910) Ltd	
G-MBHA	Trident Trike	P. Jackson	
G-MBHB	Cenrair Moto Delta G-11	Moto Baldet (Northampton) Ltd	
G-MBHC	Chargus Lightning T.250	R. E. Worth	
G-MBHD	Hiway Vulcan Trike	D. Kiddy	
G-MBHE	American Aerolights Eagle	D. K. W. Paterson	
G-MBHF	Pterodactyl Ptraveller	D. B. Girry	
G-MBHH	Flexiform Sealander Skytrike	G. G. & G. J. Norris	
G-MBHI	Ultrasports Tripacer 250	P. T. Anstey	
G-MBHJ	Hornet Skyhook Cutlass	J. O'Neill	
G-MBHK	Flexiform Skytrike	E. Barfoot	
G-MBHL	Skyhook Skytrike	C. R. Brewitt	
G-MBHM	Weedhopper	J. Hopkinson	
G-MBHN	Weedhopper	S. Hopkinson	
G-MBHO	Skyhook Super Sabre Trike	E. Smith	
G-MBHP	American Aerolights Eagle II	P. V. Trollope & H. Caldwell	
G-MBHR	Flexiform Skytrike	Y. P. Osbourne	
G-MBHT	Chargus T.250	S. F. Dawe	
G-MBHU	Flexiform Hilander Skytrike	R. Bridgstock	
G-MBHV	Pterodactyl Ptraveller	H. Partridge	
G-MBHW	American Aerolights Eagle	P. D. Lloyd-Davies	
G-MBHX	Pterodactyl Ptraveller	P. Samal	
G-MBHZ	Pterodactyl Ptraveller	T. Deeming	
G-MBIA	Flexiform Sealander Skytrike	I. P. Cook	
G-MBIB	Mainair Flexiform Sealander	A. D. Pearson	
G-MBIC	Maxair Hummer	R. Houseman	
G-MBID	American Aerolights Eagle	D. A. Campbell	
G-MBIE	Flexiform Striker	Flying Machine (Circa 1910) Ltd	
G-MBIF	American Aerolights Eagle	Flying Machine (Circa 1910) Ltd	
G-MBIG	American Aerolights Eagle	H. G. I. Goodheart	
G-MBIH	Flexiform Skytrike	M. Hurtley	
G-MBII	Hiway Skytrike	C. L. Salter	
G-MBIK	Wheeler Scout	D. K. McDonald	
G-MBIL	Southern Aerosports Scorpion 1	D. V. Collier	
G-MBIM	American Aerolights Sea Eagle	A. J. Sheardown	
G-MBIN	Wheeler Sea Scout	I. F. Kerr	
G-MBIO	American Aerolights Eagle Z Drive	B. J. C. Hill	
G-MBIP	Gemini Hummingbird	Micro Aviation Ltd	
G-MBIR	Gemini Hummingbird	Micro Aviation Ltd	
G-MBIS	American Aerolights Eagle	I. R. Bendall	
G-MBIT	Hiway Demon Skytrike	Kuernaland (UK) Ltd	
G-MBIU	Wills Microlight	M. E. Wills	
G-MBIV	Flexiform Skytrike	C. D. Weaver & ptnrs	
G-MBIW	Hiway Demon Tri-Flyer Skytrike	Computer Mart Ltd	
G-MBIX	Ultra Sports	S. A. V. Smith	
G-MBIY	Ultra Sports	J. C. Beatham	
G-MBIZ	Mainair Tri-Flyer	E. F. Clapham & ptnrs	
G-MBJA	Eurowing Goldwing	J. L. Gaunt	
G-MBJB	Hiway Skytrike Mk II	P. Cooper	
G-MBJC	American Aerolights Eagle	R. Jenkins	
G-MBJD	American Aerolights Eagle	R. W. F. Boarder	
G-MBJE	Airwave Nimrod	M. E. Glanvill	
G-MBJF	Hiway Skytrike Mk II	A. P. Clark	
G-MBJG	Airwave Nimrod	B. Tetley	
G-MBJH	Chargus Titan	IOW Microlight Club Training Centre	
G-MBJI	Southern Aerosports Scorpion	Robert Montgomery Ltd	
G-MBJJ	Mirage Mk II	H. Glover	
G-MBJK	American Aerolights Eagle	B. W. Olley	
G-MBJL	Airwave Nimrod	A. G. Lowe	
G-MBJM	Striplin Lone Ranger	C. K. Brown	
G-MBJN	Electraflyer Eagle	Manx Eagle Club	

Notes	Reg.	Type	Owner or Operator
	G-MBJO	Birdman Cherokee	T. A. Hinton
	G-MBJP	Hiway Skytrike	L. A. Seers
	G-MBJR	American Aerolights Eagle	M. P. Skelding
	G-MBJS	Mainair Tri-Flyer	T. W. Taylor
	G-MBJT	Hiway Skytrike II	R. A. Kennedy
	G-MBJU	American Aerolights Eagle	J. Basford
	G-MBJV	Rotec Rally 2B	C. J. G. Welch
	G-MBJW	Hiway Demon Mk II	M. J. Grace
	G-MBJX	Hiway Super Scorpion	D. G. Hughes
	G-MBJY	Rotec Rally 2B	C. R. V. Hitch
	G-MBJZ	Eurowing Catto CP.16	Neville Chamberlain Ltd
	G-MBKA	Mistral Trainer	Tricraft Ltd
	G-MBKB	Pterodactyl Ptraveller	W. H. Foddy
	G-MBKC	Southdown Lightning	R. D. Rayner
	G-MBKD	Chargus T.250	T. Knight
	G-MBKE	Eurowing Catto CP.16	R. S. Tuberville
	G-MBKF	Striplin Skyranger	P. R. Botterill
	G-MBKG	Batchelor-Hunt Skytrike	M. J. Batchelor & ptnrs
	G-MBKH	Southdown Skytrike	R. Turner
	G-MBKI	Solar Wings Typhoon	S. T. Jones
	G-MBKJ	Chargus TS.440 Titan 38	Westair Microlights
	G-MBKK	Pterodactyl Ascender	T. D. Baker
	G-MBKL	Hiway Demon Skytrike	D. C. Bedding
	G-MBKN	Chargus TS.440 Titan	Chargus Gliding Co Ltd
	G-MBKO	Chargus TS.440 Titan	Chargus Gliding Co Ltd
	G-MBKP	Hiway Skytrike 160	R. A. Davies
	G-MBKR	Hiway Skytrike	C. J. Macey
	G-MBKS	Hiway Skytrike 160	J. H. M. Houldridge
	G-MBKT	Mitchell Wing B.10	T. Beckett
	G-MBKU	Hiway Demon Skytrike	P. W. Twizell
	G-MBKV	Eurowing Goldwing	J. Bell
	G-MBKW	Pterodactyl Ptraveller	R. C. H. Russell
	G-MBKY	American Aerolights Eagle	B. Fussell
	G-MBKZ	Hiway Skytrike	S. I. Harding
	G-MBLA	Flexiform Skytrike	L. Tracey
	G-MBLB	Eipper Quicksilver MX	Southern Microlight Centre Ltd
	G-MBLD	Flexiform Striker	K. Akister
	G-MBLE	Hiway Demon Skytrike II	R. E. Harvey
	G-MBLF	Hiway Demon 195 Tri Pacer	A. P. Rostron
	G-MBLG	Chargus Titan T.38	P. R. F. Glenville
	G-MBLH	Flexwing Trike Hornet	A. Brown
	G-MBLJ	Eipper Quicksilver MX	Flylight South East
	G-MBLK	Southdown Puma	C. S. Hales
	G-MBLM	Hiway Skytrike	W. N. Natson
	G-MBLN	Pterodactyl Ptraveller	G. van der Gaag
	G-MBLO	Sealander Skytrike	A. R. Fawkes
	G-MBLP	Pterodactyl Ptraveller	R. N. Greenshields
	G-MBLR	Ultrasports Tripacer	N. Hyde
	G-MBLS	MEA Mistral	I. D. Stokes
	G-MBLT	Chargus TS.440 Titan	P. J. Harvey
	G-MBLU	Southdown Lightning L.195	R. J. Honey
	G-MBLV	Ultrasports Hybrid	M. A. Gosden
	G-MBLW	Hiway Scorpion	J. Davies
	G-MBLX	Eurowing Goldwing	W. B. Thomas
	G-MBLY	Flexiform Sealander Trike	J. Clithero
	G-MBLZ	Southern Aerosports Scorpion	J. P. Bennett-Snewin
	G-MBMA	Eipper Quicksilver MX	M. Maxwell
	G-MBMC	Waspair Tomcat	F. D. Buckle
	G-MBMD	Eurowing CP.16	S. Dorrance
	G-MBME	American Aerolights Eagle Z Drive	Perme Westcott Flying Club
	G-MBMF	Rotec Rally 2B	J. G. Woods
	G-MBMG	Rotec Rally 2B	J. R. Pyper
	G-MBMH	American Aerolights Eagle	M. S. Scott
	G-MBMI	Chargus T.440	G. Durbin
	G-MBMJ	Mainair Tri-Flyer	P. A. Gardner
	G-MBMK	Weedhopper Model B	P. W. Grange
	G-MBML	American Aerolights Zenoah Eagle	R. C. Jones
	G-MBMN	Skyhook Silhouette	A. D. F. Clifford
	G-MBMO	Hiway Skytrike 160	D. F. Chambers
	G-MBMP	Mitchell Wing B.10	J. Pavelin
	G-MBMR	Ultrasports Tripacer Typhoon	L. Mills
	G-MBMS	Hornet	R. L. Smith

Reg.	Type	Owner or Operator	Notes
G-MBMT	Mainair Tri-Flyer	T. R. Yeomans	
G-MBMU	Eurowing Goldwing	P. R. Wason	
G-MBMV	Chargus TS.440 Titan 38	C. Churchill	
G-MBMW	Solar Wings Typhoon	J. Cunliffe	
G-MBMY	Pterodactyl Fledge	G. Clarke	
G-MBMZ	Sealander Tripacer	T. D. Otho-Briggs	
G-MBNA	American Aerolights Eagle	N. D. Hall	
G-MBNC	Southdown Sailwings Puma	Southern Airsports Ltd	
G-MBND	Skyhook Sailwings SK TR.2	Eastern Microlight Aircraft Centre Ltd	
G-MBNF	American Aerolights Eagle	D. Read	
G-MBNG	Hiway Demon Skytrike	C. J. Clayson	
G-MBNH	Southern Airsports Scorpion	R. F. Thomas	
G-MBNJ	Eipper Quicksilver MX	C. Lamb	
G-MBNK	American Aerolights Eagle	R. Moss	
G-MBNL	Hiway Skytrike C.2	K. V. Shail & H. W. Preston	
G-MBNM	American Aerolights Eagle	D. W. J. Orchard	
G-MBNN	Southern Microlight Gazelle P.160N	N. A. Pitcher	
G-MBNP	Eurowing Catto CP.16	M. H. C. Bishop	
G-MBNR	Flexiform Skysails Striker	D. T. Kaberry	
G-MBNS	Chargus Titan 38	P. N. Lynch	
G-MBNT	American Aerolights Eagle	M. D. O'Brien	
G-MBNU	Hilander/Hiway Skytrike	D. Wilson & I. Williams	
G-MBNV	Sheffield Aircraft Skytrike	D. L. Buckley	
G-MBNW	Meagher Flexwing	P. R. Collier	
G-MBNX	Solar Storm	F. Kratky	
G-MBNY	Steer Terror Fledge II	M. J. Steer	
G-MBNZ	Hiway Skytrike Demon	E. Battersea & J. Paige	
G-MBOA	Flexiform Hilander	A. F. Stafford	
G-MBOB	American Aerolights Eagle	N. J. Oldacres	
G-MBOC	Ultrasports Tripacer 250	R. Lewis-Evans	
G-MBOD	American Aerolights Eagle	M. A. Ford & ptnrs	
G-MBOE	Solar Wing Typhoon Trike	W. Turner & C. Ferrie	
G-MBOF	Pakes Jackdaw	L. G. Pakes	
G-MBOG	Flexiform Sealander	M. J. B. Knapp	
G-MBOH	Microlight Engineering Mistral	N. A. Bell	
G-MBOI	Ultralight Flight Mirage II	H. I. Jones	
G-MBOJ	Pterodactyl Pfledgling	S. P. Dewhurst	
G-MBOK	Dunstable Microlight	W. E. Brooks	
G-MBOL	Pterodactyl Pfledgling 360	W. J. Neath	
G-MBOM	Hiway Hilander	P. H. Beaumont	
G-MBON	Eurowing Goldwing Canard	A. H. Dunlop	
G-MBOP	Hiway Demon Skytrike	R. E. Holden	
G-MBOR	Chotia 460B Weedhopper	D. J. Whysall	
G-MBOS	Hiway Super Scorpion	C. Montgomery	
G-MBOT	Hiway 250 Skytrike	M. I. Kaye	
G-MBOU	Wheeler Scout	T. Spiers	
G-MBOV	Southdown Lightning Trike	J. Messenger	
G-MBOW	Solar Wing Typhoon	R. Luke	
G-MBOX	American Aerolights Eagle	J. S. Paine	
G-MBPA	Weedhopper Srs 2	C. H. & P. B. Smith	
G-MBPB	Pterodactyl Ptraveller	P. E. Bailey	
G-MBPC	American Aerolights Eagle	Aerial Imaging Systems Ltd	
G-MBPD	American Aerolights Eagle	R. G. Harris & K. Hall	
G-MBPE	Ultrasports Trike	K. L. Turner	
G-MBPF	Southern Aerosports Scorpion	R. L. Wadley	
G-MBPG	Hunt Skytrike	A. F. Batcheler	
G-MBPI	MEA Mistral Trainer	M. J. Kenniston	
G-MBPJ	Moto-Delta	J. B. Jackson	
G-MBPL	Hiway Demon	B. J. Merrett	
G-MBPM	Eurowing Goldwing	A. B. Paton & ptnrs	
G-MBPN	American Aerolights Eagle	N. O. G. & P. C. Wooler	
G-MBPO	Volnik Arrow	N. A. Seymour	
G-MBPP	American Aerolights Eagle	R. C. Colbeck	
G-MBPR	American Aerolights Eagle	P. Kift	
G-MBPS	Gryphon Willpower	J. T. Meager	
G-MBPT	Hiway Demon	K. M. Simpson	
G-MBPU	Hiway Demon	L. F. Banham	
G-MBPW	Weedhopper	P. G. Walton	
G-MBPX	Eurowing Goldwing	W. R. Haworth & V. C. Cannon	
G-MBPY	Ultrasports Tripacer 330	P. A. Joyce	
G-MBPZ	Flexiform Striker	C. Harris	
G-MBRA	Eurowing Catto CP.16	J. Brown	
G-MBRB	Electraflyer Eagle 1	R. C. Bott	

Notes	Reg.	Type	Owner or Operator
	G-MBRC	Wheeler Scout Mk 3A	Skycraft (UK) Ltd
	G-MBRD	American Aerolights Eagle	D. G. Fisher
	G-MBRE	Wheeler Scout	R. G. Buck
	G-MBRF	Weedhopper 460C	L. R. Smith
	G-MBRH	Ultraflight Mirage Mk II	R. A. L. Hubbard
	G-MBRK	Huntair Pathfinder	F. M. Sharland
	G-MBRM	Hiway Demon	S. D. Hicks & ptnrs
	G-MBRN	Hiway Demon 175	G. J. Dunn
	G-MBRO	Hiway Skytrike 160	M. A. Saunders
	G-MBRP	American Aerolights Eagle	F. G. Rainbow
	G-MBRS	American Aerolights Eagle	R. W. Chatterton
	G-MBRU	Skyhook Cutlass	T. J. McLauchlan
	G-MBRV	Eurowing Goldwing	J. H. G. Lywood & A. A. Boyle
	G-MBRZ	Hiway Vulcan 250	D. J. Jackson
	G-MBSA	Ultraflight Mirage II	M. J. Laxton
	G-MBSB	Ultraflight Mirage II	Windsports Centre
	G-MBSC	Ultraflight Mirage II	M. E. Hollis
	G-MBSF	Ultraflight Mirage II	A. J. Horne
	G-MBSG	Ultraflight Mirage II	P. E. Owen
	G-MBSI	American Aerolights Eagle	M. Day
	G-MBSN	American Aerolights Eagle	D. Duckworth
	G-MBSS	Ultrasports Puma 2	Swancar
	G-MBSU	Ultraflight Mirage II	R. Lynn
	G-MBSW	Ultraflight Mirage II	G. Clare
	G-MBTA	UAS Storm Buggy 5 Mk 2	N. & D. McEwan
	G-MBTB	Davies Tri-Flyer S	P. J. Head
	G-MBTC	Weedhopper	P. C. Lovegrove
	G-MBTD	Solar Wings Cherokee 250 Trike	R. D. Yaxley
	G-MBTE	Hornet Dual Trainer Trike	A. B. Greenbank
	G-MBTF	Mainair Tri-Flyer Skytrike	Ministry of Defence (PE)
	G-MBTG	Mainair Tri-Flyer 2 Seat Skytrike	J. J. Thompson
	G-MBTH	Whittaker MW.4	MWA Flying Group
	G-MBTI	Hovey Whing Ding	A. Carr & R. Saddington
	G-MBTJ	Solar Wings Microlight	J. Swingler
	G-MBTK	Mainair Vortex 120P	R. Beer
	G-MBTL	Hiway Super Scorpion	C. S. Beer
	G-MBTN	Mitchell Wing B.10	N. F. James
	G-MBTO	Mainair Tri-Flyer 250	N. Huxtable
	G-MBTP	Hiway Demon	S. E. Huxtable
	G-MBTR	Skyhook Sailwings	R. Smith
	G-MBTS	Hovey WD-II Whing-Ding	T. G. Solomon
	G-MBTU	Cloudhopper Mk II	P. C. Lovegrove
	G-MBTV	Ultraflight Tomcat	M. C. Latham
	G-MBTW	Raven Vector 600	J. Spavins & A. L. Coleman
	G-MBTX	Hornet	A. F. Holdsworth
	G-MBTY	American Aerolights Eagle	Southall College of Technology
	G-MBTZ	Huntair Pathfinder	G. M. Hayden
	G-MBUA	Hiway Demon	R. J. Nicholson
	G-MBUB	Horne Sigma Skytrike	L. G. Horne
	G-MBUC	Huntair Pathfinder	Huntair Ltd
	G-MBUD	Wheeler Scout Mk III	R. J. Adams
	G-MBUE	MBA Tiger Cub 440	Herveport Ltd
	G-MBUG	Southern Aerosports Scorpion Twin	Flyflight Southeast
	G-MBUH	Hiway Skytrike	H. Glover
	G-MBUI	Wheeler Scout Mk I	G. C. Martin
	G-MBUJ	Rotec Rally 2B	L. T. Swallham
	G-MBUK	Mainair 330 Tri Pacer	J. D. Bridge
	G-MBUL	American Aerolights Eagle	Nottingham Offshore Marine
	G-MBUO	Southern Aerosports Scorpion	I. C. Vanner
	G-MBUP	Hiway Skytrike	P. Hamilton
	G-MBUS	MEA Mistral	F. G. Johnson Ltd
	G-MBUT	UAS Storm Buggy	J. N. Wrigley
	G-MBUU	Mainair Triflyer	G. E. Edwards
	G-MBUV	Huntair Pathfinder	G. H. Cork
	G-MBUW	Skyhook Sabre Trike	D. F. Soul
	G-MBUX	Pterodactyl Ptraveller	J. J. Harris
	G-MBUY	American Aerolights Eagle	Nottingham Offshore Marine
	G-MBUZ	Wheeler Scout Mk II	A. B. Cameron
	G-MBVA	Volmer Jensen VJ-23E	D. P. Eichorn
	G-MBVC	American Aerolights Eagle	E. M. Salt
	G-MBVE	Hiway 160 Valmet	T. J. Daly
	G-MBVF	Hornet	P. D. Hopkins
	G-MBVG	American Aerolights Eagle	Cipher Systems Ltd

Reg.	Type	Owner or Operator	Notes
G-MBVH	Mainair Triflyer Striker	M. A. Lomas	
G-MBVI	Hiway 250 Skytrike	D. L. B. Holliday	
G-MBVJ	Skyhook Trike	F. M. Ripley	
G-MBVK	Ultraflight Mirage II	R. Braxton	
G-MBVL	Southern Aerosports Scorpion	R. H. Wentham	
G-MBVM	Ultraflight Mirage II	Normalair Garrett Ltd	
G-MBVP	Mainair Triflyer 330 Striker	S. M. C. Kenton	
G-MBVR	Rotec Rally 2B	A. C. W. Day	
G-MBVS	Hiway Skytrike	M. A. Brown	
G-MBVT	American Aerolights Eagle	D. Cracknell	
G-MBVU	Flexiform Sealander Triflyer	B. Fallows	
G-MBVV	Hiway Skytrike	I. Shulver	
G-MBVW	Skyhook TR.2	Oban Divers Ltd	
G-MBVX	Tigair Power Fledge	D. G. Tigwell	
G-MBVY	Eipper Quicksilver MX	J. Moss	
G-MBVZ	Hornet Trike 250	R. F. Southcott	
G-MBWA	American Aerolights Eagle	S. Pizzey	
G-MBWB	Hiway Skytrike	C. K. Board	
G-MBWD	Rotec Rally 2B	A. Craw	
G-MBWE	American Aerolights Eagle	R. H. Tombs	
G-MBWF	Mainair Triflyer Striker	G. A. Archer	
G-MBWG	Huntair Pathfinder	R. V. Hogg	
G-MBWH	Designability Duet I	Designability Ltd	
G-MBWI	Microlight Lafayette Mk.1	F. W. Harrington	
G-MBWK	Mainair Triflyer	G. C. Weighwell	
G-MBWL	Huntair Pathfinder	D. A. Izod & R. C. Wright	
G-MBWM	American Aerolights Eagle	J. N. B. Mourant	
G-MBWN	American Aerolights Eagle	J. N. B. Mourant	
G-MBWO	Hiway Demon Skytrike	J. T. W. J. Edwards	
G-MBWP	Ultrasports Trike	M. R. Butterworth	
G-MBWR	Hornet	G. Edwards	
G-MBWT	Huntair Pathfinder	D. G. Gibson	
G-MBWU	Hiway Demon Skytrike	R. M. Lister	
G-MBWW	Southern Aerosports Scorpion	Twinflight Ltd	
G-MBWX	Southern Aerosports Scorpion	Twinflight Ltd	
G-MBWY	American Aerolights Eagle	C. Carber	
G-MBWZ	American Aerolights Eagle	B. Busby	
G-MBXA	Southern Aerosports Scorpion	Osprey Aviation Ltd	
G-MBXB	Southdown Sailwings Puma	C. Sharpe	
G-MBXC	Eurowing Goldwing	A. J. J. Bartak	
G-MBXD	Huntair Pathfinder	Border Aviation Ltd	
G-MBXE	Hiway Skytrike	T. A. Harlow	
G-MBXF	Hiway Skytrike	J. Robinson	
G-MBXG	Mainair Triflyer	R. Bailey	
G-MBXH	Southdown Sailwings Puma	Kingdom Prints (Cupar) Ltd	
G-MBXI	Hiway Skytrike	R. K. Parry	
G-MBXJ	Hiway Demon Skytrike	D. C. & J. M. Read	
G-MBXK	Ultrasports Puma	T. A. Brown	
G-MBXL	Eipper Quicksilver MX2	Flying Machines (Circa 1910) Ltd	
G-MBXM	American Aerolights Eagle	P. D. Schramm	
G-MBXN	Southdown Sailwings Lighting	T. W. Robinson	
G-MBXO	Sheffield Trident	M. I. Watson	
G-MBXP	Hornet Skytrike	G. Little	
G-MBXR	Hiway Skytrike 150	N. E. Smith	
G-MBXT	Eipper Quicksilver MX2	M. G. Edwards	
G-MBXW	Hiway Skytrike	R. M. Hydes	
G-MBXX	Ultraflight Mirage II	Newell Aircraft & Tool Co Ltd	
G-MBXY	Hornet	C. Leach	
G-MBXZ	Skyhook TR2	Dennar Engineering Ltd	
G-MBYA	Southern Aerosports Scorpion	Inkerman Microlight Sales Ltd	
G-MBYB	—	—	
G-MBYC	—	—	
G-MBYD	American Aerolights Eagle	J. M. Hutchinson	
G-MBYE	Eipper Quicksilver MX	M. J. Beeby	
G-MBYF	Skyhook TR2	M. L. J. Eichner	
G-MBYH	Maxair Hummer	A. Edwards	
G-MBYI	Ultraflight Lazair	A. M. Fleming	
G-MBYJ	Hiway Super Scorpion IIC	M. A. Gosden	
G-MBYK	Huntair Pathfinder	W. E. Lambert	
G-MBYL	Huntair Pathfinder 330	R. S. Peaks	
G-MBYM	Eipper Quicksilver MX	J. Wibberley	
G-MBYN	Livesey Super-Fly	D. M. Livesey	
G-MBYO	American Aerolights Eagle	B. J. & M. G. Ferguson	
G-MBYP	Hornet 440cc Flexwing Cutlass	T. J. B. Daly	

Notes	Reg.	Type	Owner or Operator
	G-MBYR	American Aerolights Eagle	F. Green & G. McCready
	G-MBYT	Ultraflight Mirage II	L. J. Perring
	G-MBYU	American Aerolights Eagle	F. L. Wiseman
	G-MBYV	Mainair Tri-Flyer 330	I. T. Ferguson
	G-MBYW	Levi Magpie	R. Levi
	G-MBYX	American Aerolights Eagle	N. P. Austen
	G-MBYY	Southern Aerosports Scorpion	D. J. Lovell
	G-MBZA	Ultrasports Tripacer 330	F. H. Cook
	G-MBZB	Hiway Skytrike	M. W. Hurst & B. Emery
	G-MBZD	Hiway Volmet 160cc	G. G. Williams
	G-MBZF	American Aerolights Eagle	G. Calder & A. C. Bernard
	G-MBZG	Twinflight Scorpion 2 seat	H. T. Edwards
	G-MBZH	Eurowing Goldwing	M. I. M. Smith
	G-MBZI	Eurowing Goldwing	R. C. Forsyth
	G-MBZJ	Ultrasports Puma	A. Barnish
	G-MBZK	Ultrasports Tripacer 250	R. Alistair
	G-MBZL	Weedhopper	A. R. Prior
	G-MBZM	UAS Storm Buggy	S. Comber & A. Crabtree
	G-MBZN	Ultrasports Puma	D. J. Cole
	G-MBZO	Mainair Triflyer 330	J. Baxendale
	G-MBZP	Skyhook TR2	Army Hang Gliding School
	G-MBZR	Eipper Quicksilver MX	R. Gill
	G-MBZS	Ultrasports Puma	T. Coughlan
	G-MBZT	Solarwings Skytrike	S. Hetherton
	G-MBZU	Skyhook Sabre C	G. N. Beyer-Kay
	G-MBZV	American Aerolights Eagle	M. D. Phillips
	G-MBZW	American Aerolights Eagle	M. J. Pugh
	G-MBZX	American Aerolights Eagle	M. J. Johnson
	G-MBZY	Waspair Tom Cat HM.81	A. C. Wendelken
	G-MBZZ	Southern Aerosports Scorpion	P. J. Harlow
	G-MJAA	Ultrasports Tripacer	M. J. Sinnett
	G-MJAB	Ultrasports Skytrike	I. W. Kemsley
	G-MJAC	American Aerolights Eagle 3	P. R. Fellden
	G-MJAD	Eipper Quicksilver MX	J. McCullough
	G-MJAE	American Aerolights Eagle	T. B. Wooley
	G-MJAF	Ultrasports Puma 440	A. B. Greenbank
	G-MJAG	Skyhook TR1	D. L. Aspinall
	G-MJAH	American Aerolights Eagle	R. L. Arscott
	G-MJAI	American Aerolights Eagle	Leisure Flight Ltd
	G-MJAJ	Eurowing Goldwing	J. S. R. Moodie
	G-MJAK	Hiway Demon	F. C. Potter
	G-MJAL	Wheeler Scout 3	D. Howe
	G-MJAM	Eipper Quicksilver MX	J. C. Larkin
	G-MJAN	Hiway Skytrike	G. M. Sutcliffe
	G-MJAO	Hiway Skytrike	J. Gallimore & ptnrs
	G-MJAP	Hiway 160	N. A. Bray
	G-MJAR	Chargus Titan	Quest Air Ltd
	G-MJAS	—	—
	G-MJAT	Hiway Demon Skytrike	P. H. Howell
	G-MJAU	Hiway Skytrike 244cc	A. P. Cross
	G-MJAV	Hiway Demon Skytrike 244cc	P. Shoemaker
	G-MJAW	Solar Wings Typhoon	M. R. Nicholls
	G-MJAX	American Aerolights Eagle	J. P. Simpson & C. W. Mellard
	G-MJAY	Eurowing Goldwing	J. F. White
	G-MJBA	Raven Vector 610	Raven Leisure Industries Ltd
	G-MJBB	Raven Vector 610	Raven Leisure Industries Ltd
	G-MJBC	Raven Vector 610	Raven Leisure Industries Ltd
	G-MJBD	Raven Vector 610	Raven Leisure Industries Ltd
	G-MJBE	Wheeler Scout X	Newell Aircraft & Tool Co Ltd
	G-MJBF	Southdown Puma 330	C. Jacobs
	G-MJBG	Mainair Solarwings Typhoon	A. J. M. Berry
	G-MJBH	American Aerolights Eagle	P. Smith
	G-MJBI	Eipper Quicksilver MX	J. I. Visser
	G-MJBJ	—	—
	G-MJBK	Swallow AeroPlane Swallow B	L. V. Strickland
	G-MJBL	American Aerolights Eagle	B. W. Olley
	G-MJBM	Eurowing CP.16	A. H. Milne
	G-MJBN	American Aerolights Eagle	D. Darke
	G-MJBO	Bell Microlight Type A	G. Bell
	G-MJBP	Eurowing Catto CP.16	I. Wilson
	G-MJBR	Eipper Quicksilver MX	J. Dilks
	G-MJBS	Ultralight Stormbuggy	G. I. Sargeant
	G-MJBT	Eipper Quicksilver MX	M. R. Grunwell

Reg.	Type	Owner or Operator	Notes
G-MJBV	American Aerolights Eagle	P. A. Ellis	
G-MJBW	American Aerolights Eagle	J. D. Penman	
G-MJBX	Pterodactyl Ptraveller	R. E. Hawkes	
G-MJBY	Rotec Rally 2B	B. Eastwood	
G-MJBZ	Huntair Pathfinder	J. C. Rose	
G-MJCA	Skyhook Sabre	B. G. Axworthy	
G-MJCB	Hornet 330	A. C. Aspden & ptnrs	
G-MJCC	Ultrasports Puma	D. J. Walter	
G-MJCD	Sigma Tetley Skytrike	N. L. Betts & B. Tetley	
G-MJCE	Ultrasports Tripacer	R. J. Osborne	
G-MJCF	Maxair Hummer	Southern Microlight Centre Ltd	
G-MJCG	S.M.C. Flyer Mk 1	E. N. Skinner	
G-MJCH	Ultraflight Mirage II	D. H. Brown	
G-MJCI	Kruchek Firefly 440	E. Kepka	
G-MJCJ	Hiway Spectrum	J. F. Mayes	
G-MJCK	Southern Aerosports Scorpion	S. L. Moss	
G-MJCL	Eipper Quicksilver MX	R. F. Witt	
G-MJCM	S.M.C. Flyer Mk 1	P. L. Gooch	
G-MJCN	S.M.C. Flyer Mk 1	C. W. Merriam	
G-MJCO	Striplin Lone Ranger	J. G. Wellans	
G-MJCP	—	—	
G-MJCR	American Aerolights Eagle	R. F. Hinton	
G-MJCS	EFS Pterodactyl	D. W. Evans	
G-MJCT	Hiway Skytrike	E. W. Barker	
G-MJCU	Tarjani	S. C. Goozes	
G-MJCV	Southern Flyer Mk 1	G. N. Harris	
G-MJCW	Hiway Super Scorpion	M. G. Sheppard	
G-MJCX	American Aerolights Eagle	S. C. Weston	
G-MJCY	Eurowing Goldwing	A. E. Dewdeswell	
G-MJCZ	Southern Aerosports Scorpion 2	C. Baldwin	
G-MJDA	Hornet Trike Executive	J. Hainsworth	
G-MJDB	Birdman Cherokee	J. K. Cook	
G-MJDC	Mainair Tri-Flyer Dual	A. C. Dommett	
G-MJDE	Huntair Pathfinder	E. H. Gould	
G-MJDF	Tripacer 250cc Striker	J. Hough	
G-MJDG	Hornet Supertrike	I. Roy	
G-MJDH	Huntair Pathfinder	Hewland Engineering Ltd	
G-MJDI	Southern Flyer Mk 1	N. P. Day	
G-MJDJ	Hiway Skytrike Demon	A. J. Cowan	
G-MJDK	American Aerolights Eagle	P. A. McPherson & ptnrs	
G-MJDL	American Aerolights Eagle	M. T. Edwards	
G-MJDM	Wheeler Scout Mk III	Skycraft (UK) Ltd	
G-MJDN	Skyhook Single Seat	G. Morgan	
G-MJDO	Southdown Puma 440	C. S. Beer	
G-MJDP	Eurowing Goldwing	P. B. Merritt	
G-MJDR	Hiway Demon Skytrike	P. J. Bullock	
G-MJDU	Eipper Quicksilver MX2	Microlight Airsport Services Ltd	
G-MJDV	Skyhook TR-1	R. Mason	
G-MJDW	Eipper Quicksilver MX	Remus International Ltd	
G-MJDX	Moyes Mega II	P. H. Davies	
G-MJDY	Ultrasports Solarwings	S. A. Barnes	
G-MJDZ	Chargus Cyclone	P. Barrow & B. C. Tolman	
G-MJEA	Flexiform Striker	S. J. O'Neill	
G-MJEC	Ultrasports Puma	F. W. Hartshorn	
G-MJED	Eipper Quicksilver MX	R. Haslam	
G-MJEE	Mainair Triflyer Trike	M. F. Eddington	
G-MJEF	Gryphon 180	F. C. Coulson	
G-MJEG	Eurowing Goldwing	G. J. Stamper	
G-MJEH	Rotec Rally 2B	J. G. Lindley	
G-MJEI	American Aerolights Eagle	A. Moss	
G-MJEJ	American Aerolights Eagle	J. Cole	
G-MJEK	Hiway Demon 330 Skytrike	J. M. Bain	
G-MJEL	GMD-01 Trike	G. M. Drinkell	
G-MJEM	Griffon 440 Trike	R. G. Griffin	
G-MJEN	Eurowing Catto CP.16	A. D. G. Wright	
G-MJEO	American Aerolights Eagle	A. M. Shaw	
G-MJEP	Pterodactyl Ptraveller	G. H. Liddle	
G-MJER	Flexiform Striker	D. S. Simpson	
G-MJES	Stratos Prototype 3 Axis 1	Stratos Aviation Ltd	
G-MJET	Stratos Prototype 3 Axis 1	Stratos Aviation Ltd	
G-MJEU	Hiway Skytrike	P. Best	
G-MJEV	Flexiform Striker	C. Scoble	
G-MJEW	Electraflyer Eagle	A. F. Keating	
G-MJEX	Eipper Quicksilver MX	M. J. Sundaram	

Notes	Reg.	Type	Owner or Operator
	G-MJEY	Southdown Lightning	P. M. Coppola
	G-MJEZ	Raven Vector 600	D. H. Handley
	G-MJFB	Flexiform Striker	J. Laleham
	G-MJFD	Ultrasports Tripacer	R. N. O. Kingsbury
	G-MJFE	Hiway Scorpion	N. A. Fisher
	G-MJFF	Huntair Pathfinder	S. R. L. Eversfield & ptnrs
	G-MJFG	Eurowing Goldwing	J. G. Aspinall & H. R. Marsden
	G-MJFH	Eipper Quicksilver MX	T. J. Drummond & ptnrs
	G-MJFI	Flexiform Striker	M. R. Parr
	G-MJFJ	Hiway Skytrike 250	S. Venus
	G-MJFK	Flexiform Skytrike Dual	J. Hollings
	G-MJFL	Mainair Tri-Flyer 440	J. Phillips
	G-MJFM	Huntair Pathfinder	C. Childs
	G-MJFN	Huntair Pathfinder	Times Newspapers Ltd
	G-MJFP	American Aerolights Eagle	D. A. Culpitt
	G-MJFR	American Aerolights Eagle	Southall College of Technology
	G-MJFS	American Aerolights Eagle	P. R. A. Elliston
	G-MJFT	American Aerolights Eagle	D. S. McMullen
	G-MJFV	Ultrasports Tripacer	Hatfield Polytechnic Students Union
	G-MJFW	Ultrasports Puma	J. McCarthy
	G-MJFX	Skyhook TR-1	Skyhook Sailwings Ltd
	G-MJFY	Hornet 250	H. Lang
	G-MJFZ	Hiway Demon Skytrike	J. A. Lowie
	G-MJGA	Hiway Skytrike 160	J. H. Wadsworth
	G-MJGB	American Aerolights Eagle	N. P. Day
	G-MJGC	Hornet	P. C. & S. J. Turnbull
	G-MJGD	Huntair Pathfinder	A. Carling
	G-MJGE	Eipper Quicksilver MX	D. Brown
	G-MJGF	Poisestar Aeolus Mk 1	Poisestar Ltd
	G-MJGG	Skyhook TR-1	R. Pritchard
	G-MJGH	Flexiform Skytrike	P. Newman
	G-MJGI	Eipper Quicksilver MX	J. M. Hayer & J. R. Wilman
	G-MJGJ	American Aerolights Eagle	B. J. Houlihan
	G-MJGK	Eurowing Goldwing	R. Haslam
	G-MJGM	Hiway Demon 195 Skytrike	J. M. Creasey
	G-MJGN	Greenslade Monotrike	P. G. Greenslade
	G-MJGO	Barnes Avon Skytrike	B. R. Barnes
	G-MJGP	Hiway Demon Skytrike	G. I. J. Thompson
	G-MJGR	Hiway Demon Skytrike	H. L. Clarke
	G-MJGS	American Aerolights Eagle	P. D. Griffiths
	G-MJGT	Skyhook Cutlass Trike	T. Silvester
	G-MJGU	Pterodactyl Mk 1	J. Pemberton
	G-MJGV	Eipper Quicksilver MX2	D. Beer
	G-MJGW	Solar Wings Trike	D. J. D. Beck
	G-MJGX	Ultrasports Puma 250	TDJ Flying Club
	G-MJGZ	Mainair Triflyer 330	A. Holt
	G-MJHA	Hiway Skytrike 250 Mk II	D. E. Oakley
	G-MJHB	AES Sky Ranger	J. H. L. B. Wijsmuller
	G-MJHC	Ultrasports Tripacer 330	G. van Der Gaag
	G-MJHD	Campbell-Jones Ladybird	M. A. Campbell-Jones
	G-MJHE	Hiway Demon Skytrike	G. Harrison
	G-MJHF	Skyhook Sailwing Trike	R. A. Watering
	G-MJHG	Huntair Pathfinder 330	A. Nice
	G-MJHH	Soleair Dactyl	C. N. Giddings
	G-MJHI	Soleair Dactyl	S. B. Giddings
	G-MJHJ	Redwing G.W.W.1	N. J. Mackay
	G-MJHK	Hiway Demon 195	B. Richardson
	G-MJHL	Mainair Triflyer Mk II	D. G. Jones
	G-MJHM	Ultrasports Trike	J. Richardson
	G-MJHN	American Aerolights Eagle	P. K. Ewens
	G-MJHO	Shilling Bumble Bee Srs 1	C. R. Shilling
	G-MJHP	American Aerolights Eagle	B. A. G. Scott & ptnrs
	G-MJHR	Southdown Lightning	G. N. Sugg
	G-MJHS	American Aerolights Eagle	R. M. Bacon
	G-MJHT	Eurowing Goldwing	J. D. Penman
	G-MJHU	Eipper Quicksilver MX	P. J. Hawcock & ptnrs
	G-MJHV	Hiway Demon 250	A. G. Griffiths
	G-MJHW	Ultrasports Puma 1	P. & C. Crayfourd
	G-MJHX	Eipper Quicksilver MX	G. J. Pill
	G-MJHY	American Aerolights Eagle	J. T. H. McAlpine
	G-MJHZ	Southdown Sailwings	D. Corke
	G-MJIA	Flexiform Striker	N. R. Beale
	G-MJIB	Hornet 250	S. H. Williams
	G-MJIC	Ultrasports Puma 330	A. E. Silvey

Reg.	Type	Owner or Operator	Notes
G-MJID	Southdown Sailwings Puma DS	J. W. Barber	
G-MJIE	Hornet 330	C. J. Dalby	
G-MJIF	Mainair Triflyer	D. P. Fiske	
G-MJIG	Hiway Demon Skytrike	W. A. Reynoldson	
G-MJIH	Ultrasports Tripacer	J. L. Bakewell	
G-MJII	American Aerolights Eagle	M. Flitman	
G-MJIJ	Ultrasports Tripacer 250	D. H. Targett	
G-MJIK	Southdown Sailwings Lightning	J. F. Chithalan	
G-MJIL	Bremner Mitchell B.10	D. S. & R. M. Bremner	
G-MJIM	Skyhook Cutlass	P. Rayner	
G-MJIN	Hiway Skytrike	P. W. Harding	
G-MJIO	American Aerolights Eagle	R. Apps & J. Marshall	
G-MJIO	Goldmark 250 Skytrike	W. E. Bray	
G-MJIP	Wheeler Scout Mk 33A	A. V. Wilson	
G-MJIR	Eipper Quicksilver MX	H. Feeney	
G-MJIS	American Aerolights Eagle	E. Gee	
G-MJIT	Hiway Skytrike	F. A. Mileham & D. W. B. Hatch	
G-MJIU	Eipper Quicksilver MX	O. W. A. Church	
G-MJIV	Pterodactyl Ptraveller	G. E. Fowles	
G-MJIX	Flexiform Hilander	D. Beer	
G-MJIY	Flexiform Voyager	C. & R. J. Sims	
G-MJIZ	Southdown Lightning	J. Stokes	
G-MJJA	Huntair Pathfinder	Quest Air Ltd	
G-MJJB	Eipper Quicksilver MX	J. W. V. Adkins	
G-MJJC	Eipper Quicksilver MX2	A. Brabiner	
G-MJJD	Birdman Cherokee	B. J. Sanderson	
G-MJJE	Douglas Type 1	R. A. Douglas	
G-MJJF	Sealey	L. G. Thomas & R. D. Thomasson	
G-MJJI	Mackinder Skyrider	R. H. Mackinder	
G-MJJJ	Moyes Knight	R. J. Broomfield	
G-MJJK	Eipper Quicksilver MX2	D. F. Gaughan	
G-MJJL	Solar Wings Storm	P. Wharton	
G-MJJM	Birdman Cherokee Mk 1	R. J. Wilson	
G-MJJN	Ultrasports Puma	J. Stokes	
G-MJJO	Flexiform Skytrike Dual	S. P. Slade & L. R. Mudge	
G-MJJP	American Aerolights Eagle	Flying Machines (Circa 1910) Ltd	
G-MJJR	Huntair Pathfinder 330	R. D. Beavan	
G-MJJS	Swallow AeroPlane Swallow B	D. Corrigan	
G-MJJT	Huntair Pathfinder	Macpara Ltd	
G-MJJU	Hiway Demon	M. R. Starling	
G-MJJV	Wheeler Scout	C. G. Johes	
G-MJJW	Chargus Kilmarnock	J. S. Potts	
G-MJJX	Hiway Skytrike	P. A. Mercer	
G-MJJY	Tirith Firefly	Tirith Microplane Ltd	
G-MJJZ	Hiway Demon 175 Skytrike	B. C. Williams	
G-MJKA	Skyhook Sabre Trike	E. James	
G-MJKB	Striplin Skyranger	A. P. Booth	
G-MJKC	Mainair Triflyer 330 Striker	W. H. Prince	
G-MJKD	—	—	
G-MJKE	Mainair Triflyer 330	R. E. D. Bailey	
G-MJKF	Hiway Demon	W. G. Reynolds	
G-MJKG	John Ivor Skytrike	R. C. Wright	
G-MJKH	Eipper Quicksilver MX II	Aerolite Aviation Co Ltd	
G-MJKI	Eipper Quicksilver MX	D. R. Gibbons	
G-MJKJ	Eipper Quicksilver MX	Aerolite Aviation Co Ltd	
G-MJKL	Ultrasports Puma	A. Tremer	
G-MJKM	Chargus Titan TS.440/38	Hiway Flight Services Ltd	
G-MJKN	Hiway Demon	Hiway Flight Services Ltd	
G-MJKO	Goldmark 250 Skytrike	M. J. Barry	
G-MJKP	Hiway Super Scorpion	M. Horsfall	
G-MJKR	Rotec Rally 2B	J. R. Darlow & J. D. Whitcock	
G-MJKS	Mainair Triflyer	P. Sutton	
G-MJKT	Hiway Super Scorpion	K. J. Morris	
G-MJKU	Hiway Demon 175	B. G. Staniscia & M. J. McCarthy	
G-MJKV	Hornet	C. Parkinson	
G-MJKW	Maxair Hummer TX	D. Roberts	
G-MJKX	Ultralight Skyrider Phantom	L. K. Fowler	
G-MJKY	Hiway Skytrike	N. R. Beale & W. H. Sherlock	
G-MJLA	Ultrasports Puma 2	J. F. Bickerstaffe	
G-MJLB	Ultrasports Puma 2	Breen Aviation Ltd	
G-MJLC	American Aerolights Double Eagle	Ardenco Ltd	
G-MJLD	Wheeler Scout Mk III	M. Buchanan-Jones	
G-MJLE	Lightwing Rooster 2 Type 5	J. Lee	

Notes	Reg.	Type	Owner or Operator
	G-MJLF	Southern Microlight Trike	P. A. Grimes
	G-MJLG	Hiway Skytrike Mk II	R. E. Neilson
	G-MJLH	American Aerolights Eagle 2	A. Cussins
	G-MJLI	Hiway Demon Skytrike	A. K. Coveney
	G-MJLJ	Flexiform Sealander	Questair Ltd
	G-MJLK	Dragonfly 250-II	G. Carter
	G-MJLL	Hiway Demon Skytrike	D. Hines
	G-MJLM	Mainair Triflyer 250	E. R. Fisher
	G-MJLN	Southern Microlight Gazelle	R. Rossiter
	G-MJLO	Goldmarque Skytrike	R. Knowles
	G-MJLR	Skyhook SK-1	T. Moore
	G-MJLS	Rotec Rally 2B	G. Messenger
	G-MJLT	American Aerolights Eagle	P. de Vere Hunt
	G-MJLU	Skyhook	C. A. Shayes
	G-MJLV	Eipper Quicksilver MX	W. Wade-Gery
	G-MJLW	Chargus Titan	C. Ellison
	G-MJLX	Rotec Rally 2B	J. Houldenshaw
	G-MJLY	American Aerolights Eagle	A. H. Read
	G-MJLZ	Hiway Demon Skytrike	K. James
	G-MJMA	Hiway Demon	T. & M. Bowyer
	G-MJMB	Weedhopper	J. E. Brown
	G-MJMC	Huntair Pathfinder	R. Griffiths
	G-MJMD	Hiway Demon Skytrike	D. Cussen
	G-MJME	Ultrasports Tripacer Mega II	J. Fleet
	G-MJMG	Weedhopper	S. Reynolds
	G-MJMH	American Aerolights Eagle	D. Crowson
	G-MJMI	Skyhook Sabre	W. P. Klotz
	G-MJMJ	Wheeler Scout III	R. Mitchell
	G-MJMK	Ultrasports Tripacer	M. F. J. Shipp
	G-MJML	Weedhopper D	V. Dixon
	G-MJMM	Chargus Vortex	D. Gwenin
	G-MJMN	Mainair Trike	D. Harrison
	G-MJMO	Lancashire Microlight Striker	N. Heap
	G-MJMP	Eipper Quicksilver MX	D. R. Peppercorn
	G-MJMR	Mainair Trike	T. Anderson
	G-MJMS	Hiway Skytrike	G. J. Foard
	G-MJMT	Hiway Demon Skytrike	R. Chiappa
	G-MJMU	Hiway Demon	J. Hall
	G-MJMV	Vulcan 2	R. Rawcliffe
	G-MJMW	Eipper Quicksilver MX2	S. E. Borrow
	G-MJMX	Ultrasports Tripacer	M. A. H. Milne
	G-MJMZ	Robertson Ultralight B1-RD	Southwest Aviation
	G-MJNA	Mainair Triflyer	M. T. Byrne
	G-MJNB	Hiway Skytrike	G. Hammond
	G-MJNC	Hiway Demon Skytrike	T. Gdaniec
	G-MJND	Mainair Triflyer	A. T. R. Nuttall
	G-MJNE	Hornet Supreme Dual Trike	Hornet Microlights
	G-MJNG	Eipper Quicksilver MX	R. Briggs-Price
	G-MJNH	Skyhook Cutlass Trike	B. M. Marsh
	G-MJNI	Hornet Sabre	T. M. Carter
	G-MJNJ	Gregory Typhoon	M. R. Gregory
	G-MJNK	Hiway Skytrike	S. J. Beecroft
	G-MJNL	American Aerolights Eagle	N. J. Williams
	G-MJNM	American Aerolights Double Eagle	E. G. Cullen
	G-MJNN	Ultraflight Mirage II	Breen Aviation Ltd
	G-MJNO	American Aerolights Double Eagle	R. S. Martin & J. L. May
	G-MJNP	American Aerolights Eagle	M. P. Harper & P. A. George
	G-MJNR	Ultralight Solar Buggy	D. J. Smith
	G-MJNS	Swallow AeroPlane Swallow B	—
	G-MJNT	Hiway Skytrike	P. R. Allery
	G-MJNU	Skyhook Cutlass	D. M. Camm
	G-MJNV	Eipper Quicksilver MX	W. Toulmin
	G-MJNW	Skyhook Silhouette	R. Hamilton
	G-MJNX	Eipper Quicksilver MX	R. Hurley
	G-MJNY	Skyhook Sabre Trike	P. Ratcliffe
	G-MJNZ	Skyhook Sabre Trike	R. Huthison
	G-MJOA	Chargus T.250 Vortex	R. J. Ridgway
	G-MJOB	Skyhook Cutlass CD Trike	J. M. Oliver
	G-MJOC	Huntair Pathfinder	R. Bowring
	G-MJOD	Rotec Rally 2B	A. J. Capel & K. D. Halsey
	G-MJOE	Eurowing Goldwing	B. C. Norris
	G-MJOF	Eipper Quicksilver MX	S. M. Wellband

Reg.	Type	Owner or Operator	Notes
G-MJOG	American Aerolights Eagle	J. B. Rush	
G-MJOH	Flexiform Striker	R. J. Butler	
G-MJOI	Hiway Demon	M. J. Coppel	
G-MJOJ	Flexiform Skytrike	D. Haynes	
G-MJOK	Mainair Triflyer 250	S. Pike & K. Fagan	
G-MJOL	Skyhook Cutlass	D. W. Nuttall	
G-MJOM	Southdown Puma 40F	J. G. Crawford	
G-MJON	Southdown Puma 40F	Peninsula Flight Ltd	
G-MJOO	Southdown Puma 40F	D. J. England	
G-MJOP	Southdown Puma 40F	Peninsula Flight Ltd	
G-MJOR	Solair Phoenix	Soleair Aviation	
G-MJOS	Southdown Lightning 170	R. C. Wright	
G-MJOT	Airwave Nimrod	W. G. Lamyman	
G-MJOU	Hiway Demon 175	H. Phipps	
G-MJOV	Solar Wings Typhoon	R. H. Lawson	
G-MJOW	Eipper Quicksilver MX	P. N. Haigh	
G-MJOX	Solar Wings Typhoon	L. Johnston	
G-MJOY	Eurowing CP.16	J. P. B. Chilton	
G-MJOZ	—	—	
G-MJPA	Rotec Rally 2B	A. Troughton	
G-MJPB	Manuel Ladybird	W. L. Manuel	
G-MJPC	American Aerolights Double Eagle	K. C. Wigley	
G-MJPD	Hiway Demon Skytrike	T. D. Adamson	
G-MJPE	Hiway Demon Skytrike	D. Hill	
G-MJPF	American Aerolights Eagle 430R	A. E. F. McClintock	
G-MJPG	American Aerolights Eagle 430R	C. J. W. Marriott	
G-MJPH	Huntair Pathfinder	A. J. Lambert	
G-MJPI	Flexiform Striker	M. Willan	
G-MJPJ	Flexiform Dual Trike 440	M. D. Phillips & ptnrs	
G-MJPK	Hiway Vulcan	R. G. Darcy	
G-MJPL	Birdman Cherokee	P. A. Leach	
G-MJPM	Huntair Pathfinder	Swift Systems Ltd	
G-MJPN	Mitchell B10	T. Willford	
G-MJPO	Eurowing Goldwing	M. Merryman	
G-MJPP	Hiway Super Scorpion	K. L. Mercer	
G-MJPR	Birdman Cherokee 250	R. J. Shaeffer	
G-MJPS	American Aerolights Eagle 430R	Peter Symonds & Co	
G-MJPT	Dragon	Fly-In Ltd	
G-MJPU	Solar Wings Typhoon	K. N. Dickinson	
G-MJPV	Eipper Quicksilver MX	J. B. Walker	
G-MJPW	Mainair Merlin	G. Deegan	
G-MJPX	Hiway Demon	R. Todd	
G-MJPY	American Aerolights Eagle	E. R. Brewster	
G-MJPZ	American Aerolights Eagle	A. T. Croy	
G-MJRA	Hiway Demon	P. Richardson & J. Martin	
G-MJRB	Eurowing Goldwing	J. S. Pyke	
G-MJRC	Eipper Quicksilver MX	R. W. Bunting	
G-MJRD	Gaze	M. Gaze	
G-MJRE	Hiway Demon	Elmstone Construction Ltd	
G-MJRG	Ultrasports Puma	J. Lakin	
G-MJRH	Hiway Skytrike	G. P. Foyle	
G-MJRI	American Aerolights Eagle	N. N. Brown	
G-MJRJ	Hiway Demon 175 Skytrike	M. Tomlinson	
G-MJRK	Flexiform Striker	P. J. & G. Long	
G-MJRL	Eurowing Goldwing	H. G. I. Goodheart	
G-MJRM	Dragon 150	Fly-In Ltd	
G-MJRN	Flexiform Striker	K. Handley	
G-MJRO	Eurowing Goldwing	I. D. Stokes	
G-MJRP	Mainair Triflyer 330	P. E. Blyth	
G-MJRR	Striplin Skyranger Srs 1	J. R. Reece	
G-MJRS	Eurowing Goldwing	J. G. Beesley	
G-MJRT	Southdown Lightning DS	R. Turner	
G-MJRU	MBA Tiger Cub 440	P. Johnson	
G-MJRV	Eurowing Goldwing	D. N. Williams	
G-MJRX	Ultrasports Puma II	E. M. Woods	
G-MJRY	MBA Super Tiger Cub 440	Vintage Displays & Training Services Ltd	
G-MJRZ	MBA Super Tiger Cub 440	Vintage Displays & Training Services Ltd	
G-MJSA	Mainair 2-Seat Trike	M. K. W. Hughes	

Notes	Reg.	Type	Owner or Operator
	G-MJSB	Eurowing Catto CP.16	Independent Business Forms (Scotland) Ltd
	G-MJSC	American Aerolights Eagle	E. McGuiness
	G-MJSD	Rotec Rally 2B Srs 1	K. J. Dickson
	G-MJSE	Skyrider Airsports Phantom	S. Montandon
	G-MJSF	Skyrider Airsports Phantom	Skyrider Airsports
	G-MJSH	American Aerolights Eagle	J. Walsom
	G-MJSI	Huntair Pathfinder	Huntair Ltd
	G-MJSK	Skyhook Sabre	G. E. Coole
	G-MJSL	Dragon 200	Dragon Light Aircraft Co Ltd
	G-MJSM	Weedhopper B	J. R. Bancroft
	G-MJSO	Hiway Skytrike	C. D. Hull
	G-MJSP	MBA Super Tiger Cub 440	J. W. E. Romain
	G-MJSR	Flexiform Micro-Trike II	R. J. S. Galley
	G-MJSS	American Aerolights Eagle	G. N. S. Farrant
	G-MJST	Pterodactyl Ptraveler	C. H. J. Goodwin
	G-MJSU	MBA Tiger Cub	R. J. Adams
	G-MJSV	MBA Tiger Cub	D. A. Izod
	G-MJSX	Simplicity Microlight	N. Smith
	G-MJSY	Eurowing Goldwing	A. J. Rex
	G-MJSZ	D.H. Wasp	D. Harker
	G-MJTA	Flexiform Striker	K. J. Regan
	G-MJTB	Eipper Quicksilver MX	R. F. G. King
	G-MJTC	Solar Wings Typhoon	J. Pemberton
	G-MJTD	Gardner T-M Scout	D. Gardner
	G-MJTE	Skyrider Airsports Phantom	R. C. Mark
	G-MJTF	Gryphon Wing	A. T. Armstrong
	G-MJTG	AES Sky Ranger	Aero & Engineering Services Ltd
	G-MJTH	S.M.D. Gazelle	R. B. Best
	G-MJTI	Huntair Pathfinder II	B. Gunn
	G-MJTJ	Weedhopper	M. J. Blanchard
	G-MJTK	American Aerolights Eagle	N. R. MacRae
	G-MJTL	Aerostructure Pipistrelle 2B	Southdown Aero Services Ltd
	G-MJTM	Aerostructure Pipistrelle 2B	Southdown Aero Services Ltd
	G-MJTN	Eipper Quicksilver MX	D. Stott
	G-MJTO	Jordan Duet Srs 1	J. R. Jordan
	G-MJTP	Flexiform Striker	A. J. Dawson & J. P. Hunt
	G-MJTR	Southdown Puma DS Mk 1	P. Soanes
	G-MJTU	Skyhook Cutlass 185	P. D. Wade
	G-MJTV	Chargus Titan 38	D. L. Harrison
	G-MJTW	Eurowing Trike	W. G. Lindsay
	G-MJTX	Skyrider Phantom	William Tomkins Ltd
	G-MJTY	Huntair Pathfinder	C. H. Smith
	G-MJTZ	Skyrider Airsports Phantom	J. H. Bakewell
	G-MJUA	MBA Super Tiger Cub	M. Ward
	G-MJUB	MBA Tiger Cub 440	C. C. Butt
	G-MJUC	MBA Tiger Cub 440	R. R. Hawkes
	G-MJUD	Southdown Puma 440	T. W. Jennings & R. A. H. Falla
	G-MJUE	Southdown Wild Cat II	W. M. A. Alladin
	G-MJUF	MBA Super Tiger Cub 440	M. P. Chetwyn-Talbot
	G-MJUH	MBA Tiger Cub 440	J. E. Johnes
	G-MJUI	Flexiform Striker	L. M. & R. E. Bailey
	G-MJUJ	Eipper Quicksilver MX II	M. Jones
	G-MJUK	Eipper Quicksilver MX II	P. Walker
	G-MJUL	Southdown Puma Sprint	S. M. Cook
	G-MJUM	Flexiform Striker	A. P. Smith
	G-MJUN	Hiway Skytrike	A. Donohue
	G-MJUO	Eipper Quicksilver MX II	Border Aviation Ltd
	G-MJUP	Weedhopper B	R. A. P. Cox
	G-MJUR	Skyrider Airsports Phantom	J. Hannibal
	G-MJUS	MBA Tiger Cub 440	B. Jenks
	G-MJUT	Eurowing Goldwing	D. L. Eite
	G-MJUU	Eurowing Goldwing	B. A. Akiens
	G-MJUV	Huntair Pathfinder	B. E. Francis
	G-MJUW	MBA Tiger Cub 440	G. R. Fountain
	G-MJUX	Skyrider Airsports Phantom	R. F. Fisher
	G-MJUY	Eurowing Goldwing	J. E. M. Barnatt-Millns
	G-MJUZ	Dragon Srs 150	J. R. Fairweather
	G-MJVA	Skyrider Airsports Phantom	Skyrider Airsports
	G-MJVB	Skyhook TR-2	Skyhook Sailwings Ltd
	G-MJVC	Hiway Skytrike	G. C. Martin
	G-MJVE	Hybred Skytrike	M. Dale
	G-MJVF	CFM Shadow	D. G. Cook
	G-MJVG	Hiway Skytrike	D. Bridges

Reg.	Type	Owner or Operator	Notes
G-MJVH	American Aerolights Eagle	R. G. Glenister	
G-MJVI	Lightwing Rooster 1 Srs 4	J. M. Lee	
G-MJVJ	Flexiform Striker	Hornet Microlights	
G-MJVL	Flexiform Striker	H. Phipps	
G-MJVM	Dragon 150	A. Fairweather	
G-MJVN	Ultrasports Puma 440	P. B. Robinson	
G-MJVP	Eipper Quicksilver MX II	D. F. Crowson	
G-MJVR	Flexiforn Striker	T. Dury	
G-MJVS	Hiway Super Scorpion	T. C. Harrold	
G-MJVT	Eipper Quicksilver MX	A. M. Reid	
G-MJVU	Eipper Quicksilver MX	B. J. Gordon	
G-MJVV	Hornet Supreme Dual	D. Atkinson & C. B. Mills	
G-MJVW	Airwave Nimrod	T. P. Mason	
G-MJVX	Skyrider Airsports Phantom	J. A. Grindley	
G-MJVY	Dragon Srs 150	M. J. Postlewaite	
G-MJVZ	Hiway Demon Tripacer	E. W. P. Van Zeller	
G-MJWA	Birdman Cherokee	R. Jakeway	
G-MJWB	Eurowing Goldwing	A. R. Slee	
G-MJWC	Paraglide Fabric Self Inflating Wing	O. W. Neumark	
G-MJWD	Solar Wings Typhoon XL	A. R. Hughes	
G-MJWE	Hiway Demon	M. D. Phillips & M. S. Henson	
G-MJWF	MBA Tiger Cub 440	B. R. Hunter	
G-MJWG	MBA Tiger Cub 440	D. H. Carter	
G-MJWH	—	—	
G-MJWI	Flexiform Striker	R. W. Twamley	
G-MJWJ	MBA Tiger Cub 440	H. A. Bromiley	
G-MJWK	Huntair Pathfinder	J. W. Keenan	
G-MJWL	Chargus Vortex T250	Pegasus Transport Systems	
G-MJWM	Chargus Vortex T250	Pegasus Transport Systems	
G-MJWN	Flexiform Striker	A. N. Baumber	
G-MJWO	Hiway Skytrike	W. Threlfall	
G-MJWR	MBA Tiger Cub 440	J. L. Burton	
G-MJWS	Eurowing Goldwing	J. W. Salter & R. J. Bell	
G-MJWT	American Aerolights Eagle	D. S. Baber	
G-MJWU	Maxair Hummer TX	D. Dugdale	
G-MJWV	Southdown Puma MS	P. A. C. Wheeler	
G-MJWW	MBA Super Tiger Cub 440	P. R. Colyer	
G-MJWX	Flexiform Striker	W. A. Bibby	
G-MJWY	Flexiform Striker	M. B. Horan	
G-MJWZ	Ultrasports Panther XL	T. V. Ward	
G-MJXA	Flexiform Striker	Chalteam Ltd	
G-MJXB	Eurowing Goldwing	A. W. Odell	
G-MJXD	MBA Tiger Cub 440	W. L. Rogers	
G-MJXE	Hiway Demon	H. Sykes	
G-MJXF	MBA Tiger Cub 440	E. J. Hadley	
G-MJXG	Flexiform Striker	D. W. Barnes	
G-MJXH	Mitchell Wing B10	M. M. Ruck	
G-MJXI	Flexiform Striker	A. P. Pearson	
G-MJXJ	MBA Tiger Cub 440	J. L. E. Griffiths	
G-MJXL	MBA Tiger Cub 440	M. J. Lister	
G-MJXM	Hiway Skytrike	G. S. & P. W. G. Carter	
G-MJXN	American Aerolights Eagle	C. H. Middleton	
G-MJXO	Middleton CM.5	C. H. Middleton	
G-MJXP	—	—	
G-MJXR	Huntair Pathfinder II	J. F. H. James	
G-MJXS	Huntair Pathfinder II	A. E. Sawyer	
G-MJXT	Phoenix Falcon 1	Phoenix Aircraft Co	
G-MJXU	MBA Tiger Cub 440	Radio West Ltd	
G-MJXV	Flexiform Striker	A. J. Doggett	
G-MJXW	Southdown Sigma	C. J. Tansley	
G-MJXX	Flexiform Striker	R. T. Lancaster	
G-MJXY	Hiway Demon Skytrike	N. Jackson & A. Dring	
G-MJXZ	Hiway Demon	R. P. Franks	
G-MJYA	Huntair Pathfinder	Ultrasports Ltd	
G-MJYB	Eurowing Goldwing	D. A. Farnworth	
G-MJYC	Ultrasports Panther XL Dual 440	J. Murphy	
G-MJYD	MBA Tiger Cub 440	M. L. Smith	
G-MJYE	Southdown Lightning Trike	G. Popplewell	
G-MJYF	—	—	
G-MJYG	Skyhook Orion Canard	Skyhook Sailwings Ltd	
G-MJYI	Mainair Triflyer	M. J. Johnson	
G-MJYJ	MBA Tiger Cub 440	M. F. Collett	
G-MJYL	Airwave Nimrod	R. Bull	

Notes	Reg.	Type	Owner or Operator
	G-MJYM	Southdown Puma Sprint	Breen Aviation Ltd
	G-MJYN	Mainair Triflyer 440	D. P. Fiske
	G-MJYO	Mainair Triflyer 330	P. Best
	G-MJYP	Mainair Triflyer 440	Mainair Sports Ltd
	G-MJYR	Catto CP.16	M. Hindley
	G-MJYS	Southdown Puma Sprint	G. Breen
	G-MJYT	Southdown Puma Sprint	G. Breen
	G-MJYV	Mainair Triflyer 2 Seat	D. A. McFadyean
	G-MJYW	Wasp Gryphon III	P. D. Lawrence
	G-MJYX	Mainair Triflyer	R. K. Birlison
	G-MJYY	Hiway Demon	N. H. Martin
	G-MJYZ	Flexiform Striker	R. Simpson & D. C. Davies
	G-MJZA	MBA Tiger Cub	C. R. Barsby
	G-MJZB	Flexiform Striker Dual	P. Howarth
	G-MJZC	MBA Tiger Cub 440	P. G. Walton
	G-MJZD	—	—
	G-MJZE	MBA Tiger Cub 440	D. Ridley & ptnrs
	G-MJZF	La Mouette Atlas 16	W. R. Crew
	G-MJZG	Mainair Triflyer 440	K. C. Bennett
	G-MJZH	Southdown Lightning 195	B. F. Crick
	G-MJZI	Eurowing Goldwing	A. J. Sharpe
	G-MJZJ	Hiway Cutlass Skytrike	G. D. H. Sandlin
	G-MJZK	Southdown Puma	Ministry of Defence (PE)
	G-MJZL	Eipper Quicksilver MX II	E. E. White
	G-MJZM	MBA Tiger Cub 440	F. M. Ward
	G-MJZN	Pterodactyl	C. J. Blundell
	G-MJZO	Flexiform Striker	P. Shelley
	G-MJZP	MBA Tiger Cub 440	Herts & Cambs Biplanes Ltd
	G-MJZR	Eurowing Zephyr 1	I. M. Vass
	G-MJZS	MMT Scorpion	C. Mowat
	G-MJZT	Flexiform Striker	J. Whitehouse
	G-MJZU	Flexiform Striker	G. J. Foard
	G-MJZV	Livesey Micro 5	D. M. Livesey
	G-MJZW	Eipper Quicksilver MX II	W. Smith & ptnrs
	G-MJZX	Maxair Hummer TX	K. T. G. Smith
	G-MJZZ	Skyhook Cutlass	J. Bradbury & D. F. Coles
	G-MMAE	Dragon Srs 150	I. Fleming
	G-MMAG	MBA Tiger Cub 440	W. R. Tull
	G-MMAH	Eipper Quicksilver MX II	T. E. McDonald
	G-MMAI	Dragon Srs 150	T. W. Dukes
	G-MMAK	MBA Tiger Cub 440	G. E. Heritage
	G-MMAL	Flexiform Striker Dual	D. J. Hand
	G-MMAM	MBA Tiger Cub 440	I. M. Bartlett
	G-MMAN	Flexiform Striker	E. Dean
	G-MMAO	Southdown Puma Sprint	R. A. Downham
	G-MMAP	Hummer TX	J. S. Millard
	G-MMAR	Southdown Puma Sprint MS	J. R. North
	G-MMAS	Southdown Sprint	Mainair Sports Ltd
	G-MMAT	Southdown Puma Sprint MS	Mainair Sports Ltd
	G-MMAU	Flexiform Rapier	T. P. Ord
	G-MMAV	American Aerolights Eagle	Aeri-Visual Ltd
	G-MMAW	Mainair Rapier	T. Green
	G-MMAX	Flexiform Striker	M. R. Thorne
	G-MMAY	Airwave Magic Nimrod	R. E. Patterson
	G-MMAZ	Southdown Puma Sprint	A. R. Smith
	G-MMBA	Hiway Super Scorpion	P. Dook
	G-MMBB	American Aerolights Eagle	Microlight Aviation (UK) Ltd
	G-MMBC	Hiway Super Scorpion	A. T. Grain
	G-MMBD	Spectrum 330	J. Hollings
	G-MMBE	MBA Tiger Cub 440	R. J. B. Jordan & R. W. Pearce
	G-MMBF	American Aerolights Eagle	N. V. Middleton
	G-MMBG	Chargus Cyclone	P. N. Long
	G-MMBH	MBA Super Tiger Cub 440	C. H. Jennings & J. F. Howesman
	G-MMBJ	Solar Wings Typhoon	R. F. Barber
	G-MMBK	American Aerolights Eagle	B. M. Quinn
	G-MMBL	Southdown Puma	A. J. M. Berry
	G-MMBM	La Mouette Azure	A. Christian
	G-MMBN	Eurowing Goldwing	J. E. Andrew
	G-MMBR	Hiway Demon 175	S. S. M. Turner
	G-MMBS	Flexiform Striker	A. A. Ridgway
	G-MMBT	MBA Tiger Cub 440	F. F. Chamberlain
	G-MMBU	Eipper Quicksilver MX II	C. Crawford
	G-MMBV	Huntair Pathfinder	M. P. Phillippe

Reg.	Type	Owner or Operator	Notes
G-MMBW	MBA Tiger Cub 440	J. C. Miles	
G-MMBX	MBA Tiger Cub 440	Fox Brothers Blackpool Ltd	
G-MMBY	Solar Wings Typhoon	D. Houghton	
G-MMCA	Solar Wings Storm	P. B. Currell	
G-MMCB	Huntair Pathfinder	S. Pizzey	
G-MMCC	American Aerolights Eagle	Microlight Aviation (UK) Ltd	
G-MMCD	Southdown Lightning DS	G. R. Bailie	
G-MMCE	MBA Tiger Cub 440	M. K. Dring	
G-MMCF	Solar Wings Panther 330	N. Birkin	
G-MMCG	Eipper Quicksilver MX I	R. W. Payne	
G-MMCH	Southdown Lightning Phase II	R. S. Andrew	
G-MMCI	Southdown Puma Sprint	D. M. Parsons	
G-MMCJ	Flexiform Striker	P. Hayes	
G-MMCK	Stewkie Aer-O-Ship LTA	K. Stewart	
G-MMCL	Stewkie Aer-O-Ship HAA	K. Stewart	
G-MMCM	Southdown Puma Sprint	C. Montgomery	
G-MMCN	Solar Wings Storm	A. P. S. Presland	
G-MMCO	Southdown Sprint	R. J. O. Walker	
G-MMCP	Southdown Lightning	J. McCallum	
G-MMCR	Eipper Quicksilver MX	T. L. & B. L. Holland	
G-MMCS	Southdown Puma Sprint	R. G. Calvert	
G-MMCT	Hiway Demon	R. G. Gray	
G-MMCV	Solar Wings Typhoon III	S. N. Pugh	
G-MMCW	Southdown Puma Sprint	S. Palmer	
G-MMCX	MBA Super Tiger Cub 440	D. Harkin	
G-MMCY	Flexiform Striker	A. P. White	
G-MMCZ	Flexiform Striker	A. A. Ridgeway	
G-MMDA	Mitchell Wing B-10	H. F. French	
G-MMDB	La Mouette Atlas	D. L. Bowtell	
G-MMDC	Eipper Quicksilver MXII	M. Risdale & C. Lamb	
G-MMDD	Huntair Pathfinder	Microlight Aviation (UK) Ltd	
G-MMDE	Solar Wings Typhoon	D. E. Smith	
G-MMDF	Southdown Lightning Phase II	P. Kelly	
G-MMDG	Eurowing Goldwing	Edgim Ltd	
G-MMDH	Manta Fledge 2B	R. G. Hooker	
G-MMDI	Hiway Super Scorpion	R. E. Hodge	
G-MMDJ	Solar Wings Typhoon	D. Johnson	
G-MMDK	Flexiform Striker	R. R. Wasson	
G-MMDL	Dragon Srs 150	Dragon Light Aircraft Co Ltd	
G-MMDM	MBA Tiger Cub 440	D. Marsh	
G-MMDN	Flexiform Striker	M. G. Griffiths	
G-MMDO	Southdown Sprint	E. Barfoot	
G-MMDP	Southdown Sprint	R. M. Strange	
G-MMDR	Huntair Pathfinder II	M. Shapland	
G-MMDS	Ultrasports Panther XLS	K. N. Dickinson	
G-MMDT	Flexiform Striker	P. G. Rawson	
G-MMDU	MBA Tiger Cub 440	P. Flynn	
G-MMDV	Ultrasports Panther	T. M. Evans	
G-MMDW	Pterodactyl Pfledgling	J. Fletcher	
G-MMDX	Solar Wings Typhoon	E. J. Lloyd	
G-MMDY	Southdown Puma Sprint	A. M. Brooks	
G-MMDZ	Flexiform Dual Strike	D. C. Seager-Thomas	
G-MMEA	MBA Tiger Cub 440	Border Aviation Ltd	
G-MMEB	Hiway Super Scorpion	A. A. Ridgway	
G-MMEC	Southdown Puma DS	A. E. Wilson	
G-MMED	Aeolus Mk 1	Aeolus Aviation	
G-MMEE	American Aerolights Eagle	G. R. Bell & J. D. Bailey	
G-MMEF	Hiway Super Scorpion	J. H. Cooling	
G-MMEG	Eipper Quicksilver MX	W. K. Harris	
G-MMEH	Ultrasports Panther	P. A. Harris	
G-MMEI	Hiway Demon	K. May	
G-MMEJ	Flexiform Striker	R. Calwood	
G-MMEK	Solar Wings Typhoon XL2	T. A. Baker	
G-MMEL	Solar Wings Typhoon XL2	D. Rigden	
G-MMEM	Solar Wings Typhoon XL2	Wyndham Wade Ltd	
G-MMEN	Solar Wings Typhoon XL2	I. M. Rapley	
G-MMEP	MBA Tiger Cub 440	P. M. Yeoman & D. Freestone-Barks	
G-MMER	—	—	
G-MMES	Southdown Puma Sprint	B. J. Sanderson	
G-MMET	Skyhook Sabre TR-1 Mk II	A. B. Greenbank	
G-MMEU	MBS Tiger Cub 440	R. Taylor	
G-MMEV	American Aerolights Eagle	J. G. Jennings	
G-MMEW	MBA Tiger Cub 440	V. N. Baker	
G-MMEX	Solar Wings Sprint	P. W. Robinson	

Notes	Reg.	Type	Owner or Operator
	G-MMEY	MBA Tiger Cub 440	M. G. Selley
	G-MMEZ	Southdown Puma Sprint	Southdown Sailwings
	G-MMFB	Flexiform Striker	Flexiform Sky Sails
	G-MMFC	Flexiform Striker	Flexiform Sky Sails
	G-MMFD	Flexiform Striker	B. J. Wood
	G-MMFE	Flexiform Striker	J. Ljustina
	G-MMFF	Flexiform Striker	Flexiform Sky Sails
	G-MMFG	Flexiform Striker	A. C. Donmett
	G-MMFH	Flexiform Striker	Flexiform Sky Sails
	G-MMFI	Flexiform Striker	C. H. P. Bell
	G-MMFJ	Flexiform Striker	G. Penson
	G-MMFK	Flexiform Striker	S. W. England
	G-MMFL	Flexiform Striker	J. G. McNally
	G-MMFM	Piranha Srs 200	G. A. Brown
	G-MMFN	MBA Tiger Cub 440	R. L. Barnett
	G-MMFP	MBA Tiger Cub 440	R. J. Adams
	G-MMFR	MBA Tiger Cub 440	R. J. Adams
	G-MMFS	MBA Tiger Cub 440	P. J. Hodgkinson
	G-MMFT	MBA Tiger Cub 440	E. Barfoot
	G-MMFV	Flexiform Dual Striker	R. A. Walton
	G-MMFW	Skyhook Cutlass	W. Chapel
	G-MMFX	MBA Tiger Cub 440	J. W. E. Romain
	G-MMFY	Flexiform Dual Striker	C. Sims
	G-MMFZ	AES Sky Ranger	H. A. Ward
	G-MMGA	Bass Gosling	G. J. Bass
	G-MMGB	Southdown Puma Sprint	G. Breen
	G-MMGC	Southdown Puma Sprint	Innovative Air Services Ltd
	G-MMGD	Southdown Puma Sprint	I. Hughes
	G-MMGE	Hiway Super Scorpion	N. R. D'Urso
	G-MMGF	MBA Tiger Cub 440	J. Ford-Dunn
	G-MMGG	Southdown Puma	J. D. Penman
	G-MMGH	Flexiform Dual Striker	J. Whitehouse
	G-MMGI	Flexiform Dual Striker	M. Hurtley
	G-MMGJ	MBA Tiger Cub 440	J. Laidler
	G-MMGK	Skyhook Silhouette	N. E. Smith
	G-MMGL	MBA Tiger Cub 440	A. R. Cornelius
	G-MMGN	Southdown Puma Sprint	H. Stieker
	G-MMGO	MBA Tiger Cub 440	T. J. Court
	G-MMGP	Southdown Puma Sprint	R. Coar
	G-MMGR	Flexiform Dual Striker	E. J. Richards
	G-MMGS	Solar Wings Panther Dual	D. J. Lewis
	G-MMGT	Solar Wings Typhoon	J. A. Hunt
	G-MMGU	Flexiform Sealander	C. J. Meadows
	G-MMGV	Sorcerer MW.5 Srs A	Microknight Aviation Ltd
	G-MMGW	Sorcerer MW.5 Srs B	Microknight Aviation Ltd
	G-MMGX	Southdown Puma	G. J. Sargent
	G-MMGY	Dean Piranha 1000	M. G. Dean
	G-MMGZ	Mitchell U2 Super Wing	R. A. Caudron
	G-MMHA	Skyhook TR-1 Pixie	L. Gabriels
	G-MMHB	Skyhook TR-1 Pixie	Skyhook Sailwings Ltd
	G-MMHC	American Aerolights Eagle	G. Davies
	G-MMHD	Hiway Demon 175	F. L. Allatt
	G-MMHE	Southdown Puma Sprint MS	R. Crosthwaite
	G-MMHF	Southdown Puma Sprint	D. M. Punnett
	G-MMHG	Solar Wings Storm	W. T. Price
	G-MMHH	Solar Wings Panther Dual	D. R. Beaumont
	G-MMHI	MBA Tiger Cub 440	R. W. Iddon
	G-MMHJ	Flexiform Hilander	A. N. Baggaley
	G-MMHK	Hiway Super Scorpion	R. Hemsworth
	G-MMHL	Hiway Super Scorpion	G. Ross
	G-MMHM	Goldmarque Gyr	G. J. Foard
	G-MMHN	MBA Tiger Cub 440	Gt Consall Copper Mines Co Ltd
	G-MMHO	MBA Tiger Cub 440	G. J. Foard
	G-MMHP	Hiway Demon	P. C. Collins
	G-MMHR	Southdown Puma Sprint DS	B. J. Bishop
	G-MMHS	SMD Viper	C. Scoble
	G-MMHT	Flexiform Striker	I. D. Swan
	G-MMHU	Flexiform Striker	P. R. Farnell
	G-MMHV	Chargus Vortex 120/T225	P. D. Larkin
	G-MMHW	Chargus Vortex 120/T225	P. D. Larkin
	G-MMHX	Hornet Invader 440	Hornet Microlights
	G-MMHY	Hornet Invader 440	W. Finlay
	G-MMHZ	Solar Wings Typhoon XL	A. T. Grain
	G-MMIA	Westwind Phoenix XP-3	Westwind Corporation Ltd

Reg.	Type	Owner or Operator	Notes
G-MMIB	MEA Mistral	D. Hines	
G-MMIC	Luscombe Vitality	Luscombe Aircraft Ltd	
G-MMID	Flexiform Dual Striker	D. C. North	
G-MMIE	MBA Tiger Cub 440	M. L. Philpott	
G-MMIF	Wasp Gryphon	F. Coulson	
G-MMIG	MBA Tiger Cub 440	R. F. Witt	
G-MMIH	MBA Tiger Cub 440	E. H. E. Nunn	
G-MMII	Southdown Puma Sprint 440	J. D. Hall	
G-MMIJ	Ultrasports Tripacer	R. W. Evans	
G-MMIK	Eipper Quicksilver MX II	Microlight Airsport Services Ltd	
G-MMIL	Eipper Quicksilver MX II	C. K. Brown	
G-MMIM	MBA Tiger Cub 440	D. A. Small	
G-MMIN	Luscombe Vitality	Luscombe Aircraft Ltd	
G-MMIO	Huntair Pathfinder II	C. Slater	
G-MMIP	Hiway Vulcan	C. J. Whittaker	
G-MMIR	Mainair Tri-Flyer 440	F. A. Prescott	
G-MMIS	Hiway Demon	M. P. Wing	
G-MMIT	Hiway Demon	D. M. Lyall	
G-MMIU	Southdown Puma Sprint	P. J. Bullock & P. Robinson	
G-MMIV	Southdown Puma Sprint	J. S. Walton	
G-MMIW	Southdown Puma Sprint	D. M. Hansell	
G-MMIX	MBA Tiger Cub 440	M. J. Butler & C. Bell	
G-MMIY	—	—	
G-MMIZ	Southdown Lightning II	G. A. Martin	
G-MMJA	Mitchell Wing B.10	J. Abbott	
G-MMJB	American Aerolights Eagle	M. R. Clarke	
G-MMJC	Southdown Sprint	P. G. Marshall	
G-MMJD	Southdown Puma Sprint	L. F. Kemmett	
G-MMJE	Southdown Puma Sprint	F. N. M. Sergeant	
G-MMJF	Ultrasports Panther Dual 440	D. H. Stokes	
G-MMJG	Mainair Tri-Flyer 440	J. G. Teague	
G-MMJH	Southdown Puma Sprint	A. R. Lawrence & T. J. Weston	
G-MMJI	Southdown Puma Sprint	W. G. Reynolds	
G-MMJJ	—	?	
G-MMJK	Hiway Demon	M. A. Baldwin	
G-MMJL	Flexiform 1+1 Sealander	J. S. Long	
G-MMJM	Southdown Puma Sprint	R. J. Sanger	
G-MMJN	Eipper Quicksilver MX II	R. M. Gunn	
G-MMJO	MBA Tiger Cub 440	R. J. Adams	
G-MMJP	—	—	
G-MMJR	MBA Tiger Cub 440	J. F. Ratcliffe	
G-MMJS	—	—	
G-MMJT	Southdown Puma Sprint MS	W. A. Reynoldson	
G-MMJU	Hiway Demon	D. Whiteside	
G-MMJV	MBA Tiger Cub 440	K. Bannister	
G-MMJW	Southdown Puma Sprint	D. J. Callender	
G-MMJX	Teman Mono-Fly	B. F. J. Hope	
G-MMJY	MBA Tiger Cub 440	Peterson Clarke Sports Ltd	
G-MMJZ	Skyhook Pixie	T. C. Harrold	
G-MMKA	Ultrasports Panther Dual	D. M. Lyall	
G-MMKB	Ultralight Flight Mirage II	B. K. Price	
G-MMKC	Southdown Puma Sprint MS	S. Farnsworth	
G-MMKD	Southdown Puma Sprint	L. W. Cload	
G-MMKE	Birdman Chinook WT-11	C. R. Gale & D. J. Royce	
G-MMKF	Ultrasports Panther Dual 440	G. H. Cork	
G-MMKG	Solar Wings Typhoon XL	D. Cassidy	
G-MMKH	Solar Wings Typhoon XL	D. E. Home & M. Baylis	
G-MMKI	Ultasports Panther 330	Lightflight Aviation	
G-MMKJ	Ultrasports Panther 330	A. Reynolds	
G-MMKK	Mainair Flash	P. D. Frain	
G-MMKL	Mainair Flash	Mainair Sports Ltd	
G-MMKM	Flexiform Dual Striker	M. R. Starling	
G-MMKN	Mitchell Wing B-10	R. A. Rumney	
G-MMKO	Southdown Puma Sprint	G. Breen	
G-MMKP	MBA Tiger Cub 440	C. O. Bibby	
G-MMKR	Southdown Lightning DS	C. Moore	
G-MMKS	Southdown Lightning 195	P. W. Fathers	
G-MMKT	MBA Tiger Cub 440	K. N. Townsend	
G-MMKU	Southdown Puma Sprint MS	R. S. T. Sears & R. A. Morris	
G-MMKV	Southdown Puma Sprint	J. Walsom	
G-MMKW	Solar Wings Storm	R. J. & J. T. Mowatt	
G-MMKY	Jordan Duet	C. H. Smith	
G-MMKZ	Ultrasports Puma 440	W. Anderson	
G-MMLA	American Aerolights Eagle	Ardenco Ltd	

Notes	Reg.	Type	Owner or Operator
	G-MMLB	MBA Tiger Cub 440	E. W. Scales
	G-MMLC	Scaled Composites 97M	Group Lotus Car Co Ltd
	G-MMLD	Solar Wings Typhoon S	N. P. Moran
	G-MMLE	Eurowing Goldwing SP	D. Lamberty
	G-MMLF	MBA Tiger Cub 440	J. R. Chichester-Constable
	G-MMLG	Solar Wings Typhoon S4 XL	B. Smith
	G-MMLH	Hiway Demon	P. M. Hendry & D. J. Lukery
	G-MMLI	Solar Wings Typhoon S	J. D. Grey
	G-MMLJ	—	—
	G-MMLK	MBA Tiger Cub 440	A. C. Barr
	G-MMLL	Midland Ultralights Sirocco	Midland Ultralights Ltd
	G-MMLM	MBA Tiger Cub 440	L. M. Campbell
	G-MMLN	Skyhook Pixie	J. F. Bishop
	G-MMLO	Skyhook Pixie	Skyhook Sailwings Ltd
	G-MMLP	Southdown Sprint	Aactron Equipment Co Ltd
	G-MMLR	Ultrasports Panther 330	Lightflight Aviation
	G-MMLU	—	—
	G-MMLV	—	—
	G-MMLW	—	—
	G-MMLX	Ultrasports Panther	R. Almond
	G-MMLY	—	—
	G-MMLZ	Mainair Tri-Flyer	A. Farnworth
	G-MMMA	Flexiform Dual Striker	N. P. Heap
	G-MMMB	Mainair Tri-Flyer	N. D. Turner & L. M. West
	G-MMMC	Southdown Puma SS	M. E. Hollis
	G-MMMD	Flexiform Dual Striker	K. P. Southwell & ptnr
	G-MMME	American Aerolights Eagle	C. Davies
	G-MMMF	American Aerolights Eagle	Aeronautical Logistics Ltd
	G-MMMG	Eipper Quicksilver MXL	P. Butler
	G-MMMH	Hadland Willow	M. J. Hadland
	G-MMMI	Southdown Lightning	A. Barnish
	G-MMMJ	Southdown Sprint	R. R. Wolfenden
	G-MMMK	Hornet Invader	R. R. Wolfenden
	G-MMML	Dragon 150	J. Doswell
	G-MMMN	Ultrasports Panther Dual 440	T. L. Travis
	G-MMMO	Solar Wings Typhoon	B. R. Underwood
	G-MMMP	Flexiform Dual Striker	K. P. Southwell
	G-MMMR	Flexiform Striker	M. A. Rigler
	G-MMMS	MBA Tiger Cub 440	M. H. D. Soltau
	G-MMMT	Hornet Sigma	R. Nay
	G-MMMU	Skyhook Cutlass CD	G. P. Wilkins
	G-MMMV	Skyhook Cutlass Dual	R. R. Wolfenden
	G-MMMW	Flexiform Striker	K. & M. Spedding
	G-MMMX	—	—
	G-MMMY	Hornet Nimrod	Bradford Motorcycle Ltd
	G-MMMZ	Southdown Puma Sprint MS	J. S. Potts
	G-MMNA	Eipper Quicksilver MX II	R. W. Payne
	G-MMNB	Eipper Quicksilver MX	N. J. Williams
	G-MMNC	Eipper Quicksilver MX	R. Gardner
	G-MMND	Eipper Quicksilver MX II-Q2	Southwest Airsports Ltd
	G-MMNE	Eipper Quicksilver MX II-Q2	Southwest Airsports Ltd
	G-MMNF	—	—
	G-MMNG	—	—
	G-MMNH	Dragon 150	H. & P. Neil Ltd
	G-MMNI	Solar Wings Typhoon S	S. Galley
	G-MMNJ	Hiway Skytrike	A. Helliwell
	G-MMNK	Solar Wings Typhoon S4	P. Jackson
	G-MMNL	Solar Wings Typhoon S4	P. Jackson
	G-MMNM	Hornet 330	D. C. Sanderson
	G-MMNN	Buzzard	E. W. Sherry
	G-MMNO	American Aerolights Eagle	P. J. Pentreath
	G-MMNP	Ultrasports Panther 250	R. Richardson
	G-MMNR	Dove	A. D. Wright
	G-MMNS	Mitchell U-2 Super Wing	D. J. Baldwin
	G-MMNT	Flexiform Striker	D. T. Kaberry
	G-MMNU	Ultrasports Panther	B. R. Underwood
	G-MMNV	Weedhopper	N. L. Rice
	G-MMNW	Mainair Tri-Flyer 330	T. Jackson
	G-MMNX	Solar Wings Panther XL	Ultrasports Ltd
	G-MMNY	Skyhook TR-1	N. H. Morley
	G-MMNZ	—	—
	G-MMOA	—	—
	G-MMOB	Southdown Sprint	E. Marsh
	G-MMOC	Huntair Pathfinder II	E. H. Gould

Reg.	Type	Owner or Operator	Notes
G-MMOD	MBA Tiger Cub 440	G. W. de Lancey Aitchison	
G-MMOE	Mitchell Wing B-10	T. Boyd	
G-MMOF	MBA Tiger Cub 440	Sunderland Microlights	
G-MMOG	Huntair Pathfinder	R. G. Maguire	
G-MMOH	Solar Wings Typhoon XL	T. H. Scott	
G-MMOI	MBA Tiger Cub 440	J. S. Smith & P. R. Talbot	
G-MMOJ	—	—	
G-MMOK	Solar Wings Panther XL	R. F. Foster	
G-MMOL	Skycraft Scout R3	P. D. G. Weller	
G-MMOM	Flexiform Striker	D. Haynes	
G-MMON	Microflight Monarch	Microflight	
G-MMOO	Southdown Storm	B. Berry	
G-MMOP	Solar Wings Panther Dual 440	J. Murphy	
G-MMOR	American Aerolights Eagle Cuyana	Southwest Air Sports Ltd	
G-MMOS	Eipper Quicksilver MX II	Southwest Air Sports Ltd	
G-MMOT	Solar Wings Typhoon XL	R. E. D. Bailey	
G-MMOU	American Aerolights Eagle	T. Crispin	
G-MMOV	Mainair Gemini Flash	R. C. Coles	
G-MMOW	Mainair Gemini Flash	G. R. Hillary	
G-MMOX	Mainair Gemini Flash	K. Handley	
G-MMOY	Mainair Gemini Sprint	Mainair Sports Ltd	
G-MMOZ	Mainair Tri-Flyer	Mainair Sports Ltd	
G-MMPB	Solar Wings Typhoon S	P. T. F. Bowden	
G-MMPC	Skyhook TR-1	J. S. Garvey	
G-MMPD	Mainair Tri-Flyer	A. R. J. Dorling	
G-MMPE	Eurowing Goldwing	J. Cuff	
G-MMPF	Eurowing Goldwing	J. Cuff	
G-MMPG	Southdown Puma	N. E. Asplin	
G-MMPH	Southdown Puma Sprint	L. H. Phillips & D. M. Mudie	
G-MMPI	Pterodactyl Ptraveller	Goodwins of Hanley Ltd	
G-MMPJ	Mainair Tri-Flyer 440	M. C. & S. J. Beaumont	
G-MMPK	Solar Wings Typhoon 1	P. J. D. Kerr	
G-MMPL	Flexiform Dual Striker	P. D. Lawrence	
G-MMPM	Ultrasports Puma 330	J. R. Moffatt	
G-MMPN	Chargus T250	S. M. Powrie	
G-MMPP	—	—	
G-MMPR	Dragon 150	P. N. B. Rosenfeld	
G-MMPS	American Aerolights Eagle	J. M. Tingle	
G-MMPT	SMD Gazelle	E. C. Poole	
G-MMPU	Ultrasports Tripacer 250	R. J. Heming	
G-MMPV	MBA Tiger Cub 440	R. Felton	
G-MMPW	Airwave Nimrod	P. W. Wisneiwski	
G-MMPX	Ultrasports Panther Dual 440	M. T. Jones	
G-MMPY	—	—	
G-MMPZ	Teman Mono-Fly	J. W. Highton	
G-MMRA	Mainair Tri-Flyer 250	S. R. Criddle	
G-MMRB	—	—	
G-MMRC	Southdown Lightning	R. T. Curant	
G-MMRD	Skyhook Cutlass CD	B. Barry	
G-MMRE	Maxair Hummer	R. D. Noble	
G-MMRF	MBA Tiger Cub 440	Fox Bros (Blackpool) Ltd	
G-MMRG	Eipper Quicksilver MX	L. P. Diede	
G-MMRH	Hiway Demon	J. S. McCaig	
G-MMRI	Skyhook Sabre	M. Salvini	
G-MMRJ	Solar Wings Panther XL	D. T. James	
G-MMRK	Ultrasports Panther XL	Enstone Microlight Centre	
G-MMRL	Solar Wings Panther XL	C. Smith	
G-MMRM	—	—	
G-MMRN	Southdown Puma Sprint	H. P. D. Tothill	
G-MMRO	—	—	
G-MMRP	Mainair Gemini	M. A. Pugh	
G-MMRR	Southdown Panther 250	D. D. & A. R. Young	
G-MMRS	Dragon 150	R. H. W. Strange	
G-MMRT	Southdown Puma Sprint	V. Brierley	
G-MMRU	Tirith Firebird FB-2	Tirith Microplane Ltd	
G-MMRV	MBA Tiger Cub 440	C. J. R. V. Baker	
G-MMRW	Flexiform Dual Striker	M. D. Hinge	
G-MMRX	Willmot J.W.1	N. J. Willmot	
G-MMRY	Chargus T.250	D. L. Edwards ⅛ ptnrs	
G-MMRZ	Ultrasports Panther Dual 440	Lancaster Partners (Holdings) Ltd	
G-MMSA	Ultrasports Panther XL	D. W. Taylor	
G-MMSB	Huntair Pathfinder II	S. R. Baugh	
G-MMSC	Mainair Gemini	A. B. Jones	

Notes	Reg.	Type	Owner or Operator
	G-MMSD	—	—
	G-MMSE	Eipper Quicksilver MX	S. Bateman
	G-MMSF	—	—
	G-MMSG	Solar Wings Typhoon XL	R. Simpson
	G-MMSH	Solar Wings Panther XL	C. Stallard
	G-MMSI	ParaPlane	International Fund for Animal Welfare
	G-MMSJ	ParaPlane	International Fund for Animal Welfare
	G-MMSK	—	—
	G-MMSL	Ultrasports Panther XLS	G. J. Slater
	G-MMSM	—	—
	G-MMSN	Mainair Gemini	P. G. Moore
	G-MMSO	—	—
	G-MMSP	Mainair Gemini Flash	B. D. Godden
	G-MMSR	MBA Tiger Cub 440	A. S. Reid
	G-MMSS	Solar Wings Panther 330	S. R. Stacey
	G-MMST	Southdown Puma Sprint	I. Davis
	G-MMSU	—	—
	G-MMSV	—	—
	G-MMSW	MBA Tiger Cub 440	D. R. Hemmings
	G-MMSX	—	—
	G-MMSY	Ultrasports Panther	R. W. Davies
	G-MMSZ	—	—
	G-MMTA	Ultrasports Panther XL	K. L. Turner
	G-MMTB	—	—
	G-MMTC	Ultrasports Panther Dual	A. D. Baker
	G-MMTD	Mainair Tri-Flyer 330	E. I. Armstrong
	G-MMTE	Mainair Gemini	R. Fyson
	G-MMTF	—	—
	G-MMTG	Mainair Gemini	R. P. W. Johnstone
	G-MMTH	Southdown Puma Sprint	J. I. Greenshields
	G-MMTI	Southdown Puma Sprint	K. B. Woods
	G-MMTJ	Southdown Puma Sprint	W. B. Cardew
	G-MMTK	Medway Hybred	Midway Microlights
	G-MMTL	Mainair Gemini	C. L. Ross
	G-MMTM	—	—
	G-MMTN	Hiway Skytrike	A. J. Blake
	G-MMTO	Mainair Tri-Flyer	R. G. Swales
	G-MMTP	Eurowing Goldwing	T. Crispin
	G-MMTR	Ultrasports Panther	K. N. Dickinson
	G-MMTS	—	—
	G-MMTT	Ultrasports Panther XL	W. Levinson
	G-MMTU	Flylite Super Scout	M. C. Drew
	G-MMTV	American Aerolights Eagle	P. J. Scott & ptnrs
	G-MMTW	American Aerolights Eagle	K. R. Gillett
	G-MMTX	—	—
	G-MMTY	Fisher FP.202U	B. E. Maggs
	G-MMTZ	Eurowing Goldwing	R. K. Young
	G-MMUA	Southdown Puma Sprint	J. T. Houghton & P. Scott
	G-MMUB	Ultrasports Tripacer 250	R. E. D. Bailey
	G-MMUC	Mainair Gemini 440	R. E. D. Bailey
	G-MMUD	Willmot Junior Cub	N. J. Willmot
	G-MMUE	Mainair Gemini Flash	Motor Services (Manchester) Ltd
	G-MMUF	Mainair Gemini	D. P. Fiske
	G-MMUG	Mainair Tri-Flyer	P. A. Leach
	G-MMUH	Mainair Tri-Flyer	J. P. Nicklin
	G-MMUI	—	—
	G-MMUJ	Southdown Puma Sprint 440	D. E. Richards
	G-MMUK	Mainair Tri-Flyer	B. R. Kirk
	G-MMUL	Ward Elf E.47	M. Ward
	G-MMUM	MBA Tiger Cub 440	N. C. Butcher
	G-MMUN	Ultrasports Panther Dual XL	K. S. Smith
	G-MMUO	Mainair Gemini Flash	Cloudbase
	G-MMUP	Airwave Nimrod 140	R. J. Bickham
	G-MMUR	Hiway Skytrike 250	W. J. Clayton
	G-MMUS	Mainair Gemini	P. Crossman
	G-MMUT	Mainair Tri-Flyer 440	A. Anderson
	G-MMUU	ParaPlane PM-1	Colt Balloons Ltd
	G-MMUV	Southdown Puma Sprint	J. T. W. J. Edwards
	G-MMUW	Mainair Gemini Flash	M. L. Smith
	G-MMUX	Mainair Gemini	S. E. Dollery
	G-MMUY	Mainair Gemini Flash	K. A. Fagan
	G-MMUZ	American Aerolights Eagle	A. C. Lowings
	G-MMVA	Southdown Puma Sprint	A. Ayre
	G-MMVB	Skyhook Pixie	C. L. S. Boswell

Reg.	Type	Owner or Operator	Notes
G-MMVC	Ultrasports Panther XL	E. R. Holton	
G-MMVE	—	—	
G-MMVF	Ultrasports Panther XL	Microlight Tuition & Sales Ltd	
G-MMVG	MBA Tiger Cub 440	C. W. Grant	
G-MMVH	Southdown Raven	Southdown International Ltd	
G-MMVI	Southdown Puma Sprint	Enstone Microlight Centre	
G-MMVJ	Southdown Puma Sprint	S. Baker	
G-MMVK	Sigh Wing ParaPlane	Sigh Wing Ltd	
G-MMVL	Ultrasports Panther XL-S	B. Milton	
G-MMVM	Whiteley Orion 1	P. N. Whiteley	
G-MMVN	Solar Wings Nomad 425F	P. George	
G-MMVO	Southdown Puma Sprint	G. Weighell	
G-MMVP	Mainair Gemini Flash	P. J. Head	
G-MMVR	Hiway Skytrike 1	F. Naylor	
G-MMVS	Skyhook Pixie	H. Smith	
G-MMVT	Mainair Gemini Flash	Aircraft Microlight Services	
G-MMVU	Mainair Gemini Flash	R. Wheeler	
G-MMVW	Skyhook Pixie	R. Keighley	
G-MMVX	Southdown Puma Sprint	T. W. Baxter	
G-MMVY	American Aerolights Eagle	R. Savva	
G-MMVZ	Southdown Puma Sprint	E. K. Battersea	
G-MMWA	Mainair Gemini Flash	W. L. Singleton	
G.MMWB	Huntair Pathfinder II	Brinhan Ltd	
G-MMWC	Eipper Quicksilver MXII	P. W. Cole & L. R. Mudge	
G-MMWD	Southdown Puma Sprint	B. J. Parker	
G-MMWE	Hiway Skytrike 250	R. Bailey	
G-MMWF	Hiway Skytrike 250	J. R. Du Plessis	
G-MMWG	Greenslade Mono-Trike	P. G. Greenslade	
G-MMWH	Southdown Puma Sprint 440	J. M. Thornton	
G-MMWI	Southdown Lightning	C. A. Crick	
G-MMWJ	Pterodactyl Ptraveler	P. Careless	
G-MMWK	Hiway Demon	J. A. Robinson	
G-MMWL	Eurowing Goldwing	D. J. White	
G-MMWM	—	—	
G-MMWN	Ultrasports Tripacer	N. Huxtable	
G-MMWO	Ultrasports Panther XL	P. M. Stoney	
G-MMWP	American Aerolights Eagle	R. A. V. Pendlebury	
G-MMWR	—	—	
G-MMWS	Mainair Tri-Flyer	S. E. Huxtable	
G-MMWT	CFM Shadow	K. H. Abel	
G-MMWU	Ultrasports Tripacer 250	J. R. Du Plessis	
G-MMWV	Flight Research Nomad 425F	R. J. B. S. Escott	
G-MMWX	Southdown Puma Sprint	D. O. Lewis & ptnrs	
G-MMWY	Skyhook Pixie	N. M. Cuthbertson	
G-MMWZ	Southdown Puma Sprint	C. I. D. H. Garrison	
G-MMXA	Mainair Gemini Flash	A. Wells	
G-MMXB	—	—	
G-MMXC	Mainair Gemini Flash	Aircraft Microlight Services	
G-MMXD	Mainair Gemini Flash	Aircraft Microlight Services	
G-MMXE	Mainair Gemini Flash	Aircraft Microlight Services	
G-MMXF	Mainair Gemini Flash	Aircraft Microlight Services	
G-MMXG	Mainair Gemini Flash	Allied Electrical	
G-MMXH	Mainair Gemini Flash	M. J. Starling	
G-MMXI	Horizon Prototype	Horizon Aerosails Ltd	
G-MMXJ	Mainair Gemini Flash	R. Meredith-Hardy	
G-MMXK	Mainair Gemini Flash	S. W. Harris	
G-MMXL	Mainair Gemini Flash	S. M. Cawthra	
G-MMXM	Mainair Gemini Flash	R. Perrett	
G-MMXN	Southdown Puma Sprint	B. Hill	
G-MMXO	Southdown Puma Sprint	C. Tomlin	
G-MMXP	Southdown Puma Sprint	D. M. Humphreys	
G-MMXR	Southdown Puma DS	R. P. Franks	
G-MMXS	Southdown Puma Sprint	G. Breen	
G-MMXT	Mainair Gemini Flash	S. J. Basey-Fisher	
G-MMXU	Mainair Gemini Flash	T. J. Franklin	
G-MMXV	—	—	
G-MMXW	Mainair Gemini	P. D. Judd	
G-MMXX	—	—	
G-MMXY	—	—	
G-MMXZ	Eipper Quicksilver MXII	Mangreen Holdings Ltd	
G-MMYA	Solar Wings Pegasus XL	Solar Wings Ltd	
G-MMYB	Solar Wings Pegasus XL	Solar Wings Ltd	
G-MMYC	Gryphon Cheetah	P. J. D. Kerr	
G-MMYD	CFM Shadow Srs B	CFM Metal-Fax Ltd	

Notes	Reg.	Type	Owner or Operator
	G-MMYE	—	—
	G-MMYF	—	—
	G-MMYG	—	—
	G-MMYH	—	—
	G-MMYI	Southdown Puma Sprint	A. Lawson
	G-MMYJ	Southdown Puma Sprint	D. Little
	G-MMYK	Southdown Puma Sprint	D. C. Fairbrass
	G-MMYL	Cyclone 70	Cyclone Hovercraft Ltd
	G-MMYM	—	—
	G-MMYN	Ultrasports Panther XL	M. J. Aubrey
	G-MMYO	Southdown Puma Sprint	K. J. Taylor
	G-MMYP	—	—
	G-MMYR	Eipper Quicksilver MXII	M. Reed
	G-MMYS	Southdown Puma Sprint	Enstone Microlight Centre
	G-MMYT	Southdown Puma Sprint	I. A. Stamp & T. P. Daniels
	G-MMYU	Southdown Puma Sprint	D. G. Hill
	G-MMYV	Webb Trike	J. D. Webb
	G-MMYW	Hiway Demon 115	T. J. Walsh
	G-MMYX	Mitchell U-2	R. A. Codling
	G-MMYY	Southdown Puma Sprint	J. C. & A. M. Rose
	G-MMYZ	Southdown Puma Sprint	L. H. Phillips & D. Mudie
	G-MMZA	Mainair Gemini Flash	R. E. D. Bailey
	G-MMZB	Mainair Gemini Flash	R. A. Guntrip
	G-MMZC	Mainair Gemini Flash	G. T. Johnston
	G-MMZD	Mainair Gemini Flash	R. Clegg
	G-MMZE	Mainair Gemini Flash	S. Saul
	G-MMZF	Mainair Gemini Flash	W. Myers
	G-MMZG	Ultrasports Panther XL-S	P. N. Long
	G-MMZH	—	—
	G-MMZI	Medway 130SX	P. M. Lang
	G-MMZJ	Mainair Gemini Flash	D. J. Griffiths
	G-MMZK	Mainair Gemini Flash	S. N. Inoff
	G-MMZL	Mainair Gemini Flash	P. R. Sexton
	G-MMZM	Mainair Gemini Flash	J. D. Hall
	G-MMZN	Mainair Gemini Flash	P. G. Kavanagh
	G-MMZO	Microflight Spectrum	Microflight Aircraft Ltd
	G-MMZP	Ultrasports Panther XL	H. Phipps
	G-MMZR	Southdown Puea Sprint	S. P. Milne
	G-MMZS	Eipper Quicksilver MX1	N. W. O'Brien
	G-MMZT	Ultrasports Tripacer	J. Bell
	G-MMZU	Southdown Puma DS	E. Clark
	G-MMZV	Mainair Gemini Flash	G. J. Pill
	G-MMZW	Southdown Puma Sprint	J. A. Carpenter
	G-MMZX	Southdown Puma Sprint	Midland Ultralights Ltd
	G-MMZY	Ultrasports Tripacer 330	K. M. Simpson
	G-MMZZ	Maxair Hummer	Microflight Ltd
	G-MNAA	Striplin Sky Ranger	Ingleby Microlight Flying Club
	G-MNAB	Ultrasports Panther XL	Scottish Microlights
	G-MNAC	Mainair Gemini Flash	D. Holliday
	G-MNAD	Mainair Gemini Flash	P. G. Moore
	G-MNAE	Mainair Gemini Flash	N. Jefferson
	G-MNAF	Solar Wings Panther XL	Pegasus Schools Ltd
	G-MNAG	Hiway Skytrike 1	R. J. Grogan
	G-MNAH	Solar Wings Panther XL	Pegasus Schools Ltd
	G-MNAI	Ultrasports Panther XL-S	J. C. Johnson
	G-MNAJ	Solar Wings Panther XL-S	D. M. Lyall
	G-MNAK	Solar Wings Panther XL-S	Windsports Centre Ltd
	G-MNAL	MBA Tiger Cub 440	Ace Aero Ltd
	G-MNAM	Solar Wings Panther XL-S	Solar Wings Ltd
	G-MNAN	Solar Wings Panther XL-S	Solar Wings Ltd
	G-MNAO	Solar Wings Panther XL-S	Solar Wings Ltd
	G-MNAP	—	—
	G-MNAR	Solar Wings Panther XL-S	Pegasus Schools Ltd
	G-MNAS	Solar Wings Pegasus XL-R	N. S. Payne
	G-MNAT	—	—
	G-MNAU	Solar Wings Pegasus XL-R	D. Corke
	G-MNAV	Southdown Puma Sprint	D. W. Barnes
	G-MNAW	Solar Wings Pegasus XL-R	A. W. Kowles
	G-MNAX	Solar Wings Pegasus XL-R	D. R. Gazey
	G-MNAY	Ultrasports Panther XL-S	Pegasus Schools Ltd
	G-MNAZ	Solar Wings Pegasus XL-R	D. Rogers
	G-MNBA	Solar Wings Pegasus XL-R	R. Moggridge
	G-MNBB	Solar Wings Pegasus XL-R	Microflight Ltd

Reg.	Type	Owner or Operator	Notes
G-MNBC	Solar Wings Pegasus XL-R	D. J. Lewis	
G-MNBD	—	—	
G-MNBE	Southdown Puma Sprint	P. J. Martin	
G-MNBF	Mainair Gemini Flash	J. B. Tate	
G-MNBG	—	—	
G-MNBH	Southdown Puma Sprint	J. A. Lancaster & ptnrs	
G-MNBI	Ultrasports Panther XL	Microflight	
G-MNBJ	Skyhook Pixie	G. M. Mansfield	
G-MNBK	Hiway Skytrike	R. Colquhoun	
G-MNBL	American Aerolights Z Eagle	J. H. Telford	
G-MNBM	Southdown Puma Sprint	M. W. Hurst	
G-MNBN	Mainair Gemini Flash	D. Adams	
G-MNBO	—	—	
G-MNBP	Mainair Gemini Flash	P. A. G. Harper	
G-MNBR	Mainair Gemini Flash	G. J. Stallard	
G-MNBS	Mainair Gemini Flash	R. Heathcote	
G-MNBT	Mainair Gemini Flash	W. A. Carlson	
G-MNBU	Mainair Gemini Flash	K. L. Turner	
G-MNBV	Mainair Gemini Flash	B. F. Hill	
G-MNBW	Mainair Gemini Flash	L. E. Comben	
G-MNBX	—	—	
G-MNBY	Mainair Gemini	A. C. Dommett	
G-MNBZ	Medway Half Pint	C. J. Draper	
G-MNCA	Adams Trike	D. Adams	
G-MNCB	Mainair Gemini Flash	K. B. O'Regan	
G-MNCC	Mainair Gemini	P. Browne	
G-MNCD	Harmsworth Trike	C. C. Harmsworth	
G-MNCE	Skyhook Pixie	C. F. Corke	
G-MNCF	Mainair Gemini Flash	A. Hetherington	
G-MNCG	Mainair Gemini Flash	J. K. Cross	
G-MNCH	Lancashire Micro Trike 330	C. F. Horsall	
G-MNCI	Southdown Puma Sprint	D. S. Anker	
G-MNCJ	Mainair Gemini Flash	J. R. North	
G-MNCK	Southdown Puma Sprint	D. J. Gibbs	
G-MNCL	Southdown Puma Sprint	Lasgo Exports Ltd	
G-MNCM	CFM Shadow Srs B	CFM Metal-Fax Ltd	
G-MNCN	Hiway Skytrike 250	N. J. Clemens	
G-MNCO	Eipper Quicksilver MXII	S. Lawton	
G-MNCP	Southdown Puma Sprint	J. G. Sealey	
G-MNCR	Flexiform Striker	I. M. Stamp	
G-MNCS	—	—	
G-MNCT	—	—	
G-MNCU	Medway Hybred	K. L. Fenn	
G-MNCV	Medway Typhoon XL	V. E. J. Smith	
G-MNCW	Hornet Dual Trainer	R. R. Wolfenden	
G-MNCX	Mainair Gemini Flash	Oban Divers Ltd	
G-MNCY	Skyhook Pixie	B. F. Johnson	
G-MNCZ	Solar Wings Pegasus XL-T	Solar Wings Ltd	
G-MNDA	Thruster TST	Thruster Aircraft (UK) Ltd	
G-MNDB	Southdown Puma Sprint	D. V. Brunt	
G-MNDC	Mainair Gemini Flash	S. E. H. Ellcombe	
G-MNDD	Mainair Scorcher Solo	Mainair Sports Ltd	
G-MNDE	Medway Half Pint	D. Thorpe	
G-MNDF	Mainair Gemini Flash	R. Meredith-Hardy	
G-MNDG	Southdown Puma Sprint	F. D. Bennett	
G-MNDH	Hiway Skytrike	N. R. Holloway	
G-MNDI	MBA Tiger Cub 440	F. Clarke	
G-MNDJ	—	—	
G-MNDK	Mainair Tri-Flyer 440	D. Kerr	
G-MNDL	Southdown Puma Sprint	L. F. Kemmett	
G-MNDM	Mainair Gemini Flash	J. P. McGuinness	
G-MNDN	Southdown Puma Sprint	C. H. Middleton	
G-MNDO	Mainair Flash	Solar Wings Ltd	
G-MNDP	Southdown Puma Sprint	R. Kimbell	
G-MNDR	—	—	
G-MNDS	—	—	
G-MNDT	—	—	
G-MNDU	Midland Sirocco 377GB	R. F. Bridgeland	
G-MNDW	Midland Sirocco 377GB	D. K. MacDonald	
G-MNDX	—	—	
G-MNDY	Southdown Puma Sprint	D. Young	
G-MNDZ	Southdown Puma Sprint	Aerotech International Ltd	
G-MNEA	Southern Airwolf	Southdown International Ltd	
G-MNEB	Southern Airwolf	Southdown International Ltd	

Notes	Reg.	Type	Owner or Operator
	G-MNEC	Southern Airwolf	Southdown International Ltd
	G-MNED	Skyhook Pixie	O. McCullogh
	G-MNEE	—	—
	G-MNEF	Mainair Gemini Flash	S. Meadowcroft
	G-MNEG	Mainair Gemini Flash	D. P. Moxon
	G-MNEH	Mainair Gemini Flash	I. Rawson
	G-MNEI	Medway Hybred 440	R. F. Miller
	G-MNEJ	Mainair Gemini Flash	I. A. Thomasson
	G-MNEK	Medway Half Pint	R. S. Peaks
	G-MNEL	Medway Half Pint	J. H. Brown
	G-MNEM	Solar Wings Pegasus Dual	C. Smith
	G-MNEN	Southdown Puma Sprint	A. J. Mann
	G-MNEO	Southdown Raven	Microflight
	G-MNEP	Aerostructure Pipstrelle P.2B	M. R. Guerard
	G-MNER	CFM Shadow Srs B	CFM Metal-Fax Ltd
	G-MNES	CFM Shadow Srs B	CFM Metal-Fax Ltd
	G-MNET	Mainair Gemini Flash	N. H. Martin
	G-MNEU	—	—
	G-MNEV	Mainair Gemini Flash	Alfasound Ltd
	G-MNEW	Mainair Tri-Flyer	M. A. Reeve
	G-MNEX	Mainair Gemini Flash	Airecraft Microlight Services Ltd
	G-MNEY	Mainair Gemini Flash	T. S. Elmhirst
	G-MNEZ	Skyhook TR1 Mk 2	Microflight
	G-MNFA	Solar Wings Typhoon	D. R. Joint
	G-MNFB	Southdown Puma Sprint	C. Lawrence
	G-MNFC	Midland Ultralights Sirocco 377GB	G. C. Sutton
	G-MNFD	—	—
	G-MNFE	Mainair Gemini Flash	G. M. Douglas
	G-MNFF	Mainair Gemini Flash	D. C. & J. A. Walker
	G-MNFG	Southdown Puma Sprint	A. Reynolds
	G-MNFH	Mainair Gemini Flash	E. J. Blyth
	G-MNFI	Medway Half Pint	C. Grainger
	G-MNFJ	Mainair Gemini Flash	L. A. Humphreys
	G-MNFK	Mainair Gemini Flash	Mainair Sports Ltd
	G-MNFL	AMF Chevron	AMF Microflight Ltd
	G-MNFM	Mainair Gemini Flash	R. Blenkey
	G-MNFN	Mainair Gemini Flash	J. D. Hall
	G-MNFO	—	—
	G-MNFP	Mainair Gemini Flash	A. W. Brown
	G-MNFR	Wright Tri-Flyer	R. L. Arscott
	G-MNFS	Ikarus Sherpa Dopplesitzer	U. Silvan
	G-MNFT	Mainair Gemini Flash	B. D. Golden
	G-MNFU	—	—
	G-MNFV	Ultrasports Trike	D. H. George
	G-MNFW	Medway Hybred 44XL	R. L. Wadley
	G-MNFX	Southdown Puma Sprint	A. M. Shaw
	G-MNFZ	Southdown Puma Sprint	M. D. Phillips
	G-MNGA	Aerial Arts Chasher 110SX	I. M. Grayland
	G-MNGB	—	—
	G-MNGB	Mainair Flash	Mainair Sports Ltd
	G-MNGC	CFM Shadow Srs B	C. R. Garner
	G-MNGD	Quest Air Services	P. R. Davey
	G-MNGE	Solar Wings Photon	Solar Wings Ltd
	G-MNGF	Solar Wings Pegasus	Solar Wings Ltd
	G-MNGG	Solar Wings Pegasus XL-R	G. H. Cork
	G-MNGH	Skyhook Pixie	N. A. Bray
	G-MNGI	—	—
	G-MNGJ	Skyhook Zipper	Skyhook Sailwings Ltd
	G-MNGK	Mainair Gemini Flash	P. C. Wilson
	G-MNGL	Mainair Gemini Flash	J. A. Hambleton
	G-MNGM	Mainair Gemini Flash	Airbourne Aviation Ltd
	G-MNGN	Mainair Gemini Flash	M. A. Dobson
	G-MNGO	Solar Wings Storm	S. Adams
	G-MNGP	—	—
	G-MNGR	Southdown Puma Sprint	A. K. Pomroy
	G-MNGS	Southdown Puma 330	G. J. Sargemt
	G-MNGT	Mainair Gemini Flash	G. C. Hobson
	G-MNGU	Mainair Gemini Flash	R. E. D. Bailey
	G-MNGV	—	—
	G-MNGW	Mainair Gemini Flash	M. A. Dobson
	G-MNGX	Southdown Puma Sprint	R. J. Morris
	G-MNGY	Hiway Skytrike 160	A. J. Wood
	G-MNGZ	Mainair Gemini Flash	N. Hatton

Reg.	Type	Owner or Operator	Notes
G-MNHA	Noble Hardman Snowbird	Noble Hardman Aviation Ltd	
G-MNHB	Solar Wings Pegasus XL-R	P. J. D. Kerr	
G-MNHC	Solar Wings Pegasus XL-R	Solar Wings Ltd	
G-MNHD	Solar Wings Pegasus XL-R	P. D. Stiles	
G-MNHE	Solar Wings Pegasus XL-R	A. Holley	
G-MNHF	Solar Wings Pegasus XL-R	Solar Wings Ltd	
G-MNHG	Solar Wings Pegasus XL-R	Solar Wings Ltd	
G-MNHH	Solar Wings Pegasus XL-S	D. J. Hampson	
G-MNHI	Solar Wings Pegasus XL-R	R. Thompson	
G-MNHJ	—	—	
G-MNHK	—	—	
G-MNHL	—	—	
G-MNHM	—	—	
G-MNHN	—	—	
G-MNHO	—	—	
G-MNHP	—	—	
G-MNHR	—	—	
G-MNHS	—	—	
G-MNHT	—	—	
G-MNHU	—	—	
G-MNHV	—	—	
G-MNHW	Medway Half Pint	R. Riley	
G-MNHX	Solar Wings Typhoon S4	R. Blackwell	
G-MNHY	Mainair Tri-Flyer 440	K. Wright	
G-MNHZ	Mainair Gemini Flash	R. J. Evans	
G-MNIA	Mainair Gemini Flash	R. J. Lamb	
G-MNIB	American Aerolights Eagle 215B	C. R. Cattell	
G-MNIC	MBA Tiger Cub 440	N. B. Kirby	
G-MNID	—	—	
G-MNIE	Mainair Gemini Flash	P. Eden & B. Weinrabe	
G-MNIF	Mainair Gemini Flash	D. Yarr	
G-MNIG	Mainair Gemini Flash	R. D. Noble	
G-MNIH	Mainair Gemini Flash	D. E. Richards	
G-MNII	Mainair Gemini Flash	R. F. Finnis	
G-MNIJ	—	—	
G-MNIK	Solar Wings Pegasus Photon	Solar Wings Ltd	
G-MNIL	Southdown Puma Sprint	G. C. Weighell	
G-MNIM	Maxair Hummer	P. J. Brookman	
G-MNIN	Designability Duet	S. Osmond	
G-MNIO	Mainair Gemini Flash	M. A. Hayward	
G-MNIP	Mainair Gemini Flash	G. S. Bulpitt	
G-MNIR	Skyhook Pixie 130	I. G. Cole	
G-MNIS	CFM Shadow Srs B	CFM Metal-Fax Ltd	
G-MNIT	—	—	
G-MNIU	Solar Wings Pegasus Photon	Solar Wings Ltd	
G-MNIV	—	—	
G-MNIW	Airwave Nimrod 165	G. E. Jewitt	
G-MNIX	Mainair Gemini Flash	B. J. Bishop	
G-MNIY	Skyhook Pixie Zipper	Skyhook Sailwings Ltd	
G-MNIZ	—	—	
G-MNJA	Southdown Lightning Skytrike	P. J. Hesketh	
G-MNJB	Southdown Raven	D. Millar	
G-MNJC	MBA Tiger Cub 440	J. G. Carpenter	
G-MNJD	Southdown Puma Sprint	J. B. Duffus	
G-MNJE	Southdown Puma Sprint	A. P. White	
G-MNJF	Dragon 150	L. R. Jillings	
G-MNJG	Mainair Tri-Flyer	J. V. George	
G-MNJH	Solar Wings Pegasus Flash	D. Gandle	
G-MNJI	Solar Wings Pegasus Flash	A. Barnish	
G-MNJJ	Solar Wings Pegasus Flash	R. Hill	
G-MNJK	Solar Wings Pegasus Flash	Pegasus Schools Ltd	
G-MNJL	Solar Wings Pegasus Flash	Pegasus Schools Ltd	
G-MNJM	Solar Wings Pegasus Flash	N. Jefferson	
G-MNJN	Solar Wings Pegasus Flash	D. J. Pay	
G-MNJO	Solar Wings Pegasus Flash	S. J. Farrant & Sons Ltd	
G-MNJP	—	—	
G-MNJR	—	—	
G-MNJS	Southdown Puma Sprint	C. E. Bates	
G-MNJT	Southdown Raven	P. J. D. Kerr	
G-MNJU	Mainair Gemini Flash	N. V. Wnekowski	
G-MNJV	Medway Half Pint	C. C. W. Mates	
G-MNJW	Mitchell Wing B10	J. D. Webb	
G-MNJX	Medway Hybred 44XL	H. A. Stewart	
G-MNJY	Medway Half Pint	P. M. Stoney	

Notes	Reg.	Type	Owner or Operator
	G-MNJZ	Aerial Arts Alpha 130SX	I. M. Grayland
	G-MNLR	Solar Wings Typhoon	B. J. Farrell
	G-MNMC	Southdown Puma MS	M. L. Coomber
	G-MNMF	Maxair Hummer TX	M. I. Smith
	G-MNMK	Solar Wings Pegasus XL-R	Knowles Transport Ltd
	G-MNML	Southdown Puma Sprint	Doncaster Aero Club Ltd
	G-MNMP	Pritchard Experimental Mk 1	M. H. Pritchard
	G-MNMS	Wheeler Scout	M. I. Smith
	G-MNPH	Flexiform Dual Striker	D. L. Aspinall
	G-MNPL	Ultrasports Panther 330	P. N. Long
	G-MNPR	Hiway Demon 175	P. Robinson
	G-MNPW	AMF Chevron	A.M.F. Microlight Ltd
	G-MNRD	Ultraflight Lazair	D. W. & M. F. Briggs
	G-MNSB	Southdown Puma Sprint	S. Baker
	G-MNUM	Southdown Puma Sprint MS	A. Sethi
	G-MWCR	Southdown Puma Sprint	C. R. Read
	G-MWFT	MBA Tiger Cub 440	W. F. Tremayne
	G-MWOW	CFM Shadow Srs B	Global Aviation Projects Ltd
	G-MWPL	MBA Tiger Cub 440	P. A. Lee
	G-MWRS	Ultravia Super Pelica	Embermere Ltd
	G-MWTF	Mainair Gemini	G. D. C. Buyers

G-ALOD Cessna 140.

G-ALIW DH.82A Tiger Moth.

G-AMRA Douglas Dakota 6 of Air Atlantique.

G-AXMU BAC One-Eleven 432 of Air UK. *Air UK*

G-BAKC Sikorsky S-61N. *A. S. Wright*

G-BEVV BN-2A-III-2 Trislander of Aviation West/Avon Aviation Services.

G-BEZB HPR-7 Herald of Channel Express.

G-BGTT Cessna 310R.

G-BIKN Boeing 757-236 of British Airways.

G-BKHG Piper J-3C-65 Cub. *A. S. Wright*

G-BKMF Short SC.7 Skyvan.

G-GAYL Learjet 35.

G-TYME Rockwell Commander 690B.

G-MJSU MBA Tiger Cub.

EC-DVN Boeing 736-2K2 of Hispania.

F-BPND Fokker F-27 Friendship Mk 500 of Air Inter.

F-BPVR Boeing 747-228F of Air France Cargo.

HB-ICO S.E.210 Caravelle 10R of C.T.A.

N187AT L.1011 TriStar 1 of American Trans Air.

OD-AFM Boeing 720-023B of Middle East Airlines.

PH-HAG GAF Nomad N24A of Holland Aero Lines.

PH-KJA BAe Jetstream 3102 of NetherLines.

TF-FLV Douglas DC-8-63 of Icelandair.

3B-NAG Boeing 747SP-44 of Air Mauritius.

AIRBAND MONITORS

FOR THE MOST COMPREHENSIVE RANGE IN THE NORTHWEST

BUY WITH CONFIDENCE FROM THE EXPERTS

Monitors by Signal, A.O.R., Regency, Revco.
Aerad Charts, Frequency Charts, Post Cards, Aerials, Accessories, Display Airliner Models.

Send ***40p*** *for Catalogue*

FAIRBOTHAM & CO LTD

58-62 LOWER HILLGATE, STOCKPORT SK1 3AN
Telephone: 061-480 4872

ACCESS AND VISA WELCOME

JAVIATION

CARLTON WORKS
CARLTON STREET
BRADFORD
WEST YORKSHIRE BD7 1DA
ENGLAND
Tel: (0274) 732146
Telex: 517382 JAVIA G

AIRBAND RADIOS AND ACCESSORIES

We keep a full range of receivers to suit all pockets from £12.95 to over £500, including the **AOR2002 VHF/UHF SYNTHESISED SCANNING RECEIVER** which gives you access to both **CIVIL AND MILITARY** airbands plus much more.

Please call, telephone, write (SAE please) or telex for further information.

SUPPLIERS OF AIRFRAME AND ENGINE COMPONENTS TO THE AIRLINES AND OPERATORS OF THE WORLD

Military to Civil Cross-Reference

Serial carried	*Civil identity*
04 (Luftwaffe)	G-WULF
14 (Luftwaffe)	G-BJZZ
26 (US)	G-BAVO
45 (Aeronavale)	G-BHFG
75	G-AFDX
92 (31-GW FrAF)	G-BJGW
103 (Aeronavale)	G-BHXJ
120 (Fr AF)	G-AZGC
152/17	G-ATJM
164 (USN)	G-BKGL
168	G-BFDE
385 (RCAF)	G-BGPB
422-15	G-AVJO
1049	G-BJCL
1076	G-AVEB
2345	G-ATVP
2807 (VE-111 USN)	G-BHTH
3066	G-AETA
4253/18	G-BFPL
5036	G-ATJX
5964	G-BFVH
7198/18	G-AANJ
7797 (USAAF)	G-BFAF
8449M	G-ASWJ
18393 (C.A.F.)	G-BCYK
2-7767	G-BIHW
115042 (TA-042 USAF)	G-BGHU
18-2001 (USAAF)	G-BIZV
315509 (USAAF)	G-BHUB
329417 (USAAF)	G-BDHK
329601 (D-44 USAAF)	G-AXHR
329934 (72-B USAAF)	G-BCPH
413048 (39-E USAAF)	G-BCXJ
454537 (04-J)	G-BFDL
461748	G-BHDK
472216	G-BIXL
479865 (A-44)	G-BHPK
480133 (44-B USAAF)	G-BDCD
485784 (YB-E)	G-BEDF
44-79609 (USAAF)	G-BHXY
44-80594 (USAAF)	G-BEDJ
53398 (2)	G-BFYO
56321 (RSwAF)	G-BKPY
542447	G-SCUB
542474 (R-184)	G-PCUB
51-15227 (USN)	G-BKRA
A16-199 (SF-R RAAF)	G-BEOX
A8226	G-BIDW
B1807	G-EAVX
B7270	G-BFCZ
D88	G-AWXZ
D5397/17	G-BFXL
C1701	G-AWYY
C1904	G-PFAP
D8096 (D)	G-AEPH
E-15 (RNethAF)	G-BIYU
E449	G-EBJE
F904	G-EBIA
F938	G-EBIC
F939 (6)	G-EBIB
F943	G-BIHF
F943	G-BKDT
F1425 (17)	G-BEFR
F5447	G-BKER
F5459	G-INNY
F8010	G-BDWJ
F8614	G-AWAU
G-29-1 (Class B)	G-APRJ
G-48-1 (Class B)	G-ALSX
H2311	G-ABAA
H5199	G-ADEV
J-108 (Swiss AF)	G-BJAX
J9941 (57)	G-ABMR
K-33 (USAAF)	G-BJLH
K123	G-EACN
K1786	G-AFTA
K2050	G-ASCM
K2059	G-PFAR
K2060	G-BKZM
K2567	G-MOTH
K2572	G-AOZH
K3215	G-AHSA
K4235	G-AHMJ
K5414	G-AENP
L2301	G-AIZG
L8032	G-AMRK
N1854	G-AIBE
N3788	G-AKPF
N4877 (VX-F)	G-AMDA
N5180	G-EBKY
N5182	G-APUP
N5195	G-ABOX
N5430	G-BHEW
N6452	G-BIAU
N6466	G-ANKZ
N6532	G-ANTS
N6847	G-APAL
N6848	G-BALX
N6985	G-AHMN
N9191	G-ALND
N9238	G-ANEL
N9389	G-ANJA
N9508	G-APCU
N9510	G-AOEL
P6382	G-AJRS
R1914	G-AHUJ
R4959	G-ARAZ
R5086	G-APIH
R7524	G-AIWA
S1287	G-BEYB
S3398 (2)	G-BFYO
T5424	G-AJOA
T5493	G-ANEF
T5854	G-ANKK
T6313	G-AHVU
T6645	G-AIIZ
T6818	G-ANKT
T7281	G-ARTL
T7404	G-ANMV
T7997	G-AOBH
T7909	G-ANON
T9707	G-AKKR
T9738	G-AKAT
U-0247	G-AGOY
U-142 (Swiss AF)	G-BONE
V3388	G-AHTW
V9281 (RU-M)	G-BCWL
V9441 (AR-A)	G-AZWT

Serial carried	*Civil identity*	*Serial carried*	*Civil identity*
Z2033	G-ASTL	TA719	G-ASKC
Z7015	G-BKTH	TJ569	G-AKOW
Z7197	G-AKZN	TW439	G-ANRP
Z7258	G-AHGD	TW641	G-ATDN
AP507 (KX-P)	G-ACWP	VI-3 (Finnish AF)	G-BAAY
AR213 (QG-A)	G-AIST	VL348	G-AVVO
AR501 (NN-D)	G-AWII	VL349	G-AWSA
BS676 (K-U)	G-KUKU	VM360	G-APHV
DE208	G-AGYU	VP955	G-DVON
DE363	G-ANFC	VP962	G-BLRB
DE623	G-ANFI	VR249	G-APIY
DE992	G-AXXV	VS356	G-AOLU
DF128 (RCO-U)	G-AOJJ	VS610	G-AOKL
DF130	G-BACK	VS623	G-AOKZ
DF155	G-ANFV	VZ728	G-AGOS
DF198	G-BBRB	WA576	G-ALSS
DG590	G-ADMW	WA577	G-ALST
DR613	G-AFJB	WB588	G-AOTD
EM720	G-AXAN	WB660	G-ARMB
EM903	G-APBI	WB763	G-BBMR
EX280	G-TEAC	WD321	G-BDCC
FE992	G-BDAM	WD363	G-BCIH
FS728	G-BAFM	WD379 (K)	G-APLO
FT239	G-BIWX	WD413	G-BFIR
FT323	G-AZSC	WE569	G-ASAJ
FT391	G-AZBN	WG307	G-BCYJ
FX301 (FD-NQ)	G-JUDI	WG348	G-BBMV
HB751	G-BCBL	WG350	G-BCYE
JV928 (Y)	G-BLSC	WG422 (16)	G-BFAX
KB976 (LQ-K)	G-BCOH	WG719	G-BRMA
KG874 (YS-L)	G-DAKS	WJ358	G-ARYD
LB312	G-AHXE	WJ945	G-BEDV
LB375	G-AHGW	WL626	G-BHDD
LS326	G-AJVH	WM167	G-LOSM
LZ766	G-ALCK	WP321 (750/CU)	G-BRFC
MD497	G-ANLW	WP788	G-BCHL
MH434 (ZD-B)	G-ASJV	WP790	G-BBNC
ML407 (OU-L)	G-LFIX	WP808	G-BDEU
ML417 (2I-T)	G-BJSG	WP857	G-BDRJ
MP425	G-AITB	WP977	G-BHRD
MT360	G-AKWT	WR149	G-BLKA
MT438	G-AREI	WT933	G-ALSW
MT818 (G-M)	G-AIDN	WV493	G-BDYG
MV293	G-SPIT	WV783	G-ALSP
MV370 (AV-L)	G-FXIV	WW397 (N-E)	G-BKHP
MW100	G-AGNV	WZ507	G-VTII
NF875 (603/CH)	G-AGTM	WZ711	G-AVHT
NH238 (D-A)	G-MKIX	XB733	G-ATBF
NJ695	G-AJXV	XF690	G-BGKA
NJ703	G-AKPI	XF785	G-ALBN
NM140	G-APGL	XF836 (J-G)	G-AWRY
NM181	G-AZGZ	XF877 (JX)	G-AWVF
NP181	G-AOAR	XG452	G-BRMB
NP184	G-ANYP	XG547	G-HAPR
NP303	G-ANZJ	XJ347	G-AMXT
NX611	G-ASXX	XJ348	G-NAVY
PG617	G-AYVY	XJ389	G-AJJP
PG651	G-AYUX	XJ407	G-BKHB
PL983	G-PRXI	XJ763	G-BKHA
RG333	G-AIEK	XK417	G-AVXY
RG333	G-AKEZ	XK655	G-AMXA
RH377	G-ALAH	XK896	G-RNAS
RH378	G-AJOE	XL717	G-AOXG
RL962	G-AHED	XM553	G-AWSV
RM221	G-ANXR	XM556	G-HELI
RM689 (AP-D)	G-ALGT	XM685	G-AYZJ
RR299 (HT-E)	G-ASKH	XN351	G-BKSC
SM832	G-WWII	XN437	G-AXWA
SM969	G-BRAF	XN441	G-BGKT
TA634	G-AWJV	XN637	G-BKOU

Serial carried	*Civil identity*	*Serial carried*	*Civil identity*
XP279	G-BWKK	F+IS (Luftwaffe)	G-BIRW
XP282	G-BGTC	BA+AY (Luftwaffe)	G-BAAY
XP328	G-BKHC	BU+CK (Luftwaffe)	G-BUCK
XP355	G-BEBC	RF+16 (Luftwaffe)	G-PTWO
XR240	G-BDFH	6J+PR (Luftwaffe)	G-AWHB
XR241	G-AXRR	7A+WN (Luftwaffe)	G-AZMH
XR267	G-BJXR	N8+AA (Luftwaffe)	G-BFHD
XR269	G-BDXY	N9+AA (Luftwaffe)	G-BECL
XR363	G-OHCA	ↀ (Russian AF)	G-KYAK
XS101	G-GNAT	CE (USAAF)	G-BICE

VP962 D.H.104 Devon C.2, civil identity G-BLRB.

XR269 Auster AOP.9, civil identity G-BDXY.

Overseas Airliner Registrations

(Aircraft included in this section are those most likely to be seen at UK and major European airports on scheduled or charter services.)

A2 (Botswana)

Reg.	*Type*	*Owner or Operator*	*Notes*
A2-ACA	L-100-30 Hercules	Air Botswana	

A40 (Oman)

A40-TP	L-1011-385 TriStar 100	Gulf Air	
A40-TR	L-1011-385 TriStar 100	Gulf Air	
A40-TS	L-1011-385 TriStar 100	Gulf Air	
A40-TT	L-1011-385 TriStar 200	Gulf Air	
A40-TV	L-1011-385 TriStar 200	Gulf Air	
A40-TW	L-1011-385 TriStar 200	Gulf Air	
A40-TX	L-1011-385 TriStar 200	Gulf Air	
A40-TY	L-1011-385 TriStar 200	Gulf Air	
A40-TZ	L-1011-385 TriStar 200	Gulf Air	

Note: Gulf Air also operates TriStar 200s N92TA and N92TB and Boeing 747 LN-AET.

AP (Pakistan)

AP-AWU	Boeing 707-373C	Pakistan International Airlines	
AP-AWY	Boeing 707-340C	Pakistan International Airlines	
AP-AXA	Boeing 707-340C	Pakistan International Airlines	
AP-AXG	Boeing 707-340C	Pakistan International Airlines	
AP-AYV	Boeing 747-282B	Pakistan International Airlines	
AP-AYW	Boeing 747-282B	Pakistan International Airlines	
AP-AZW	Boeing 707-351B	Pakistan International Airlines	
AP-BAA	Boeing 707-351B	Pakistan International Airlines	
AP-BAK	Boeing 747-240B	Pakistan International Airlines	
AP-BAT	Boeing 747-240B	Pakistan International Airlines	
AP-BBK	Boeing 707-323C	Pakistan International Airlines	
AP-	Boeing 747-217B	Pakistan International Airlines	
AP-	Boeing 747-217B	Pakistan International Airlines	
AP-	Boeing 747-217B	Pakistan International Airlines	
AP-	Boeing 747-217B	Pakistan International Airlines	

B (China/Taiwan)

B-198	Boeing 747-2R7F	China Airlines	
B-1862	Boeing 747SP-09	China Airlines	
B-1864	Boeing 747-209B	China Airlines	
B-1866	Boeing 747-209B	China Airlines	
B-1880	Boeing 747SP-09	China Airlines	
B-1886	Boeing 747-209B	China Airlines	
B-1888	Boeing 747-209B	China Airlines	
B-1894	Boeing 747-209F	China Airlines	
B-2402	Boeing 707-3J6B	CAAC	
B-2404	Boeing 707-3J6B	CAAC	

Notes	Reg.	Type	Owner or Operator
	B-2406	Boeing 707-3J6B	CAAC
	B-2408	Boeing 707-3J6B	CAAC
	B-2410	Boeing 707-3J6C	CAAC
	B-2412	Boeing 707-3J6C	CAAC
	B-2414	Boeing 707-3J6C	CAAC
	B-2416	Boeing 707-3J6C	CAAC
	B-2418	Boeing 707-3J6C	CAAC
	B-2420	Boeing 707-3J6C	CAAC
	B-2442	Boeing 747SP-J6	CAAC
	B-2444	Boeing 747SP-J6	CAAC
	B-2446	Boeing 747-2J6B	CAAC
	B-2448	Boeing 747-2J6B	CAAC

Note: CAAC also operates Boeing 747SPs N1301E and N1304E. China Airlines operates N4508H and N4522V, both Boeing 747SP-09s.

C9 (Mozambique)

Note: Linhas Aereas de Mocambique (LAM) operates DC-10-30 F-GDJK on lease from UTA.

Wardair Canada

AIR CANADA

C-F and C-G (Canada)

	C-FCPO	Douglas DC-8-63 (801)	Worldways Canada
	C-FCPP	Douglas DC-8-63 (802)	Worldways Canada
	C-FCPQ	Douglas DC-8-63 (803)	Worldways Canada
	C-FCPR	Douglas DC-8-63 (804)	Worldways Canada
	C-FCRA	Boeing 747-217B (741)	Canadian Pacific *Empress of Italy*
	C-FCRB	Boeing 747-217B (742)	Canadian Pacific *Empress of Canada*
	C-FCRD	Boeing 747-217B (743)	Canadian Pacific *Empress of Expo 86*
	C-FDJC	Boeing 747-1D1 (399)	Wardair Canada *Phil Garrett*
	C-FFUN	Boeing 747-1D1 (398)	Wardair Canada *Romeo Vachan*
	C-FTIK	Douglas DC-8-73AF (867)	Air Canada
	C-FTIO	Douglas DC-8-73AF (871)	Air Canada
	C-FTIP	Douglas DC-8-73AF (872)	Air Canada
	C-FTIQ	Douglas DC-8-73CF (873)	Air Canada
	C-FTIR	Douglas DC-8-73AF (874)	Air Canada
	C-FTIS	Douglas DC-8-73AF (875)	Air Canada
	C-FTIU	Douglas DC-8-73AF (877)	Air Canada
	C-FTNB	L.1011-385 TriStar 100 (502)	Air Canada
	C-FTND	L.1011-385 TriStar 100 (504)	Air Canada
	C-FTNH	L.1011-385 TriStar 100 (508)	Air Canada
	C-FTNI	L.1011-385 TriStar 100 (509)	Air Canada
	C-FTNJ	L.1011-385 TriStar 100 (510)	Air Canada
	C-FTNK	L.1011-385 TriStar 100 (511)	Air Canada
	C-FTNL	L.1011-385 TriStar 100 (512)	Air Canada
	C-FTOC	Boeing 747-133 (303)	Air Canada
	C-FTOD	Boeing 747-133 (304)	Air Canada
	C-FTOE	Boeing 747-133 (305)	Air Canada
	C-GAGA	Boeing 747-233B (306)	Air Canada
	C-GAGB	Boeing 747-233B (307)	Air Canada
	C-GAGF	L.1011-385 TriStar 500 (551)	Air Canada
	C-GAGG	L.1011-385 TriStar 500 (552)	Air Canada
	C-GAGH	L.1011-385 TriStar 500 (553)	Air Canada
	C-GAGI	L.1011-385 TriStar 500 (554)	Air Canada
	C-GAGJ	L.1011-385 TriStar 500 (555)	Air Canada
	C-GAGK	L.1011-385 TriStar 500 (556)	Air Canada
	C-GA	Boeing 767-233	Air Canada
	C-GA	Boeing 767-233	Air Canada
	C-GAVC	Boeing 767-233 (611)	Air Canada
	C-GAVF	Boeing 767-233 (612)	Air Canada
	C-GCPC	Douglas DC-10-30 (901)	CP *Empress of Amsterdam*
	C-GCPD	Douglas DC-10-30 (902)	CP *Empress of British Colombia*
	C-GCPE	Douglas DC-10-30 (903)	CP *Empress of Buenos Aires*
	C-GCPI	Douglas DC-10-30 (907)	CP *Empress of Auckland*

Reg.	Type	Owner or Operator	Notes
C-GCPJ	Douglas DC-10-30 (908)	Canadian Pacific *Empress of Rome*	
C-GCP-	Douglas DC-10-30	Canadian Pacific	
C-GCP-	Douglas DC-10-30	Canadian Pacific	
C-GCP-	Douglas DC-10-30	Canadian Pacific	
C-GCP-	Douglas DC-10-30	Canadian Pacific	
C-GFHX	Douglas DC-10-30 (103)	Wardair Canada *S. R. McMilland*	
C-GFLG	Boeing 707-365C (31)	Worldways Canada	
C-GHPW	L-100-30 Hercules (387)	North West Territorial Airways	
C-GIES	L.1011-385 TriStar 100	Worldways Canada	
C-GIFE	L.1011-385 TriStar 100	Worldways Canada	
C-GMXB	Douglas DC-8-61	Nationair	
C-GMXQ	Douglas DC-8-61	Nationair	
C-GMXR	Douglas DC-8-61	Nationair	
C-GMXY	Douglas DC-8-62	Nationair	
C-GQBA	Douglas DC-8-63	Quebecair	
C-GQBF	Douglas DC-8-63	Quebecair	
C-GXRA	Boeing 747-211B (397)	Wardair Canada *Herbert Hollick Kenyon*	
C-GXRB	Douglas DC-10-30 (101)	Wardair Canada *C. H. Punch Dickens*	
C-GXRC	Douglas DC-10-30 (102)	Wardair Canada *W. R. Wop May*	
C-GXRD	Boeing 747-211B (396)	Wardair Canada *H. A. Doc Oakes*	

Note: Airline fleet number carried on aircraft is shown in parenthesis. CP Air also operates DC-10-10s N1834U, N1836U and N1837U on lease from United Airlines.

CCCP (Russia)

All aircraft listed are operated by Aeroflot. The registrations are prefixed by CCCP in each case.

Reg.	Type	Notes
65020	Tu-134A	
65024	Tu-134A	
65027	Tu-134A	
65028	Tu-134A	
65035	Tu-134A	
65036	Tu-134A	
65038	Tu-134A	
65040	Tu-134A	
65042	Tu-134A	
65044	Tu-134A	
65048	Tu-134A	
65050	Tu-134A	
65051	Tu-134A	
65089	Tu-134A	
65107	Tu-134A	
65134	Tu-134A	
65135	Tu-134A	
65142	Tu-134A	
65601	Tu-134	
65602	Tu-134	
65603	Tu-134	
65604	Tu-134	
65605	Tu-134	
65606	Tu-134	
65607	Tu-134	
65608	Tu-134	
65609	Tu-134	
65610	Tu-134	
65611	Tu-134	
65612	Tu-134	
65613	Tu-134	
65614	Tu-134	
65615	Tu-134	
65616	Tu-134	
65617	Tu-134	
65618	Tu-134	
65619	Tu-134	
65620	Tu-134	
65621	Tu-134	
65625	Tu-134	
65627	Tu-134	
65628	Tu-134	
65629	Tu-134	
65630	Tu-134	
65631	Tu-134	
65632	Tu-134	
65633	Tu-134	
65634	Tu-134	
65635	Tu-134	
65636	Tu-134	
65637	Tu-134	
65639	Tu-134	
65642	Tu-134	
65643	Tu-134	
65644	Tu-134A	
65645	Tu-134A	
65646	Tu-134A	
65647	Tu-134A	
65648	Tu-134A	
65649	Tu-134A	
65650	Tu-134A	
65651	Tu-134A	
65652	Tu-134A	
65653	Tu-134A	
65654	Tu-134A	
65655	Tu-134A	
65656	Tu-134A	
65657	Tu-134A	
65658	Tu-134A	
65659	Tu-134A	
65660	Tu-134A	
65661	Tu-134A	
65662	Tu-134A	
65663	Tu-134A	
65664	Tu-134A	
65665	Tu-134A	
65666	Tu-134A	
65667	Tu-134A	
65669	Tu-134A	
65670	Tu-134A	

Notes	Reg.	Type
	65671	Tu-134A
	65672	Tu-134A
	65673	Tu-134A
	65674	Tu-134A
	65675	Tu-134A
	65676	Tu-134A
	65677	Tu-134A
	65678	Tu-134A
	65679	Tu-134A
	65680	Tu-134A
	65681	Tu-134A
	65683	Tu-134A
	65687	Tu-134A
	65689	Tu-134A
	65690	Tu-134A
	65691	Tu-134A
	65692	Tu-134A
	65694	Tu-134A
	65696	Tu-134A
	65697	Tu-134A
	65705	Tu-134A
	65706	Tu-134A
	65707	Tu-134A
	65711	Tu-134A
	65713	Tu-134A
	65714	Tu-134A
	65717	Tu-134A
	65718	Tu-134A
	65727	Tu-134A
	65728	Tu-134A
	65729	Tu-134A
	65730	Tu-134A
	65731	Tu-134A
	65732	Tu-134A
	65733	Tu-134A
	65734	Tu-134A
	65735	Tu-134A
	65739	Tu-134A
	65741	Tu-134A
	65742	Tu-134A
	65743	Tu-134A
	65744	Tu-134A
	65745	Tu-134A
	65746	Tu-134A
	65747	Tu-134A
	65748	Tu-134A
	65749	Tu-134A
	65753	Tu-134A
	65757	Tu-134A
	65758	Tu-134A
	65765	Tu-134A
	65769	Tu-134A
	65770	Tu-134A
	65777	Tu-134A
	65780	Tu-134A
	65781	Tu-134A
	65782	Tu-134A
	65783	Tu-134A
	65784	Tu-134A
	65785	Tu-134A
	65790	Tu-134A
	65791	Tu-134A
	65794	Tu-134A
	65795	Tu-134A
	65801	Tu-134A
	65802	Tu-134A
	65804	Tu-134A
	65806	Tu-134A
	65810	Tu-134A
	65811	Tu-134A
	65812	Tu-134A
	65815	Tu-134A
	65816	Tu-134A
	65817	Tu-134A

Notes	Reg.	Type
	65818	Tu-134A
	65820	Tu-134A
	65821	Tu-134A
	65822	Tu-134A
	65823	Tu-134A
	65825	Tu-134A
	65828	Tu-134A
	65829	Tu-134A
	65830	Tu-134A
	65831	Tu-134A
	65832	Tu-134A
	65833	Tu-134A
	65834	Tu-134A
	65836	Tu-134A
	65837	Tu-134A
	65839	Tu-134A
	65840	Tu-134A
	65841	Tu-134A
	65843	Tu-134A
	65844	Tu-134A
	65845	Tu-134A
	65848	Tu-134A
	65851	Tu-134A
	65852	Tu-134A
	65853	Tu-134A
	65854	Tu-134A
	65857	Tu-134A
	65861	Tu-134A
	65862	Tu-134A
	65863	Tu-134A
	65864	Tu-134A
	65865	Tu-134A
	65866	Tu-134A
	65867	Tu-134A
	65868	Tu-134A
	65869	Tu-134A
	65870	Tu-134A
	65871	Tu-134A
	65872	Tu-134A
	65873	Tu-134A
	65874	Tu-134A
	65877	Tu-134A
	65878	Tu-134A
	65879	Tu-134A
	65880	Tu-134A
	65881	Tu-134A
	65882	Tu-134A
	65883	Tu-134A
	65884	Tu-134A
	65886	Tu-134A
	65888	Tu-134A
	65890	Tu-134A
	65891	Tu-134A
	65892	Tu-134A
	65893	Tu-134A
	65894	Tu-134A
	65895	Tu-134A
	65898	Tu-134A
	65899	Tu-134A
	65903	Tu-134A
	65950	Tu-134A
	65951	Tu-134A
	65952	Tu-134A
	65953	Tu-134A
	65954	Tu-134A
	65955	Tu-134A
	65957	Tu-134A
	65960	Tu-134A
	65961	Tu-134A
	65962	Tu-134A
	65963	Tu-134A
	65964	Tu-134A
	65965	Tu-134A
	65967	Tu-134A

Reg.	*Type*	*Notes*
65969	Tu-134A	
65970	Tu-134A	
65971	Tu-134A	
65972	Tu-134A	
65973	Tu-134A	
65974	Tu-134A	
65975	Tu-134A	
65976	Tu-134A	
85001	Tu-154	
85002	Tu-154	
85003	Tu-154	
85004	Tu-154	
85005	Tu-154	
85006	Tu-154	
85007	Tu-154	
85008	Tu-154	
85009	Tu-154	
85010	Tu-154	
85011	Tu-154	
85012	Tu-154	
85013	Tu-154	
85014	Tu-154	
85016	Tu-154	
85017	Tu-154	
85018	Tu-154	
85019	Tu-154	
85020	Tu-154	
85021	Tu-154	
85022	Tu-154	
85024	Tu-154	
85025	Tu-154	
85028	Tu-154	
85029	Tu-154	
85030	Tu-154	
85031	Tu-154	
85032	Tu-154	
85033	Tu-154	
85034	Tu-154	
85035	Tu-154	
85037	Tu-154	
85038	Tu-154	
85039	Tu-154	
85040	Tu-154	
85041	Tu-154	
85042	Tu-154	
85043	Tu-154	
85044	Tu-154	
85049	Tu-154	
85050	Tu-154	
85051	Tu-154	
85052	Tu-154	
85053	Tu-154	
85054	Tu-154	
85055	Tu-154	
85057	Tu-154	
85059	Tu-154A	
85060	Tu-154A	
85061	Tu-154A	
85062	Tu-154A	
85063	Tu-154A	
85064	Tu-154A	
85065	Tu-154A	
85066	Tu-154A	
85067	Tu-154A	
85068	Tu-154A	
85069	Tu-154A	
85070	Tu-154A	
85071	Tu-154A	
85072	Tu-154A	
85074	Tu-154A	
85078	Tu-154A	
85079	Tu-154A	
85075	Tu-154A	
85076	Tu-154A	
85078	Tu-154A	
85079	Tu-154A	
65080	Tu-154A	
85081	Tu-154A	
85082	Tu-154A	
85083	Tu-154A	
85084	Tu-154A	
85085	Tu-154A	
85086	Tu-154A	
85087	Tu-154A	
85088	Tu-154A	
85090	Tu-154A	
85091	Tu-154A	
85092	Tu-154B-1	
85093	Tu-154A	
85094	Tu-154A	
85095	Tu-154A	
85096	Tu-154B-1	
85097	Tu-154B-1	
85098	Tu-154A	
85099	Tu-154A	
85100	Tu-154A	
85101	Tu-154A	
85102	Tu-154A	
85103	Tu-154A	
85104	Tu-154A	
85105	Tu-154A	
85106	Tu-154B	
85107	Tu-154A	
85108	Tu-154A	
85109	Tu-154B	
85110	Tu-154B	
85111	Tu-154A	
85112	Tu-154A	
85113	Tu-154A	
85114	Tu-154A	
85115	Tu-154A	
85116	Tu-154A	
85117	Tu-154A	
85118	Tu-154B	
85119	Tu-154A	
85120	Tu-154B	
85121	Tu-154B	
85122	Tu-154B	
85123	Tu-154B	
85124	Tu-154B	
85125	Tu-154B	
85126	Tu-154B	
85129	Tu-154B	
85130	Tu-154B	
85131	Tu-154B	
85132	Tu-154B	
85133	Tu-154B	
85134	Tu-154B	
85135	Tu-154B	
85136	Tu-154B	
85137	Tu-154B	
85138	Tu-154B	
85139	Tu-154B	
85140	Tu-154B	
85141	Tu-154B	
85142	Tu-154B	
85143	Tu-154B	
85145	Tu-154B	
85146	Tu-154B	
85147	Tu-154B	
85148	Tu-154B	
85149	Tu-154B	
85150	Tu-154B	
85151	Tu-154B	
85152	Tu-154B	
85153	Tu-154B	
85154	Tu-154B	
85155	Tu-154B	

Notes	Reg.	Type
	85156	Tu-154B
	85157	Tu-154B
	85158	Tu-154B
	85160	Tu-154B
	85162	Tu-154B
	85163	Tu-154B
	85164	Tu-154B
	85165	Tu-154B
	85166	Tu-154B
	85167	Tu-154B
	85168	Tu-154B
	85169	Tu-154B
	85170	Tu-154B
	85171	Tu-154B
	85172	Tu-154B
	85173	Tu-154B
	85174	Tu-154B
	85175	Tu-154B
	85176	Tu-154B
	85177	Tu-154B
	85178	Tu-154B
	85179	Tu-154B
	85180	Tu-154B
	85181	Tu-154B
	85182	Tu-154B
	85183	Tu-154B
	85184	Tu-154B
	85185	Tu-154B
	85186	Tu-154B
	85187	Tu-154B
	85188	Tu-154B
	85189	Tu-154B
	85190	Tu-154B
	85191	Tu-154B
	85192	Tu-154B
	85193	Tu-154B
	85194	Tu-154B
	85195	Tu-154B
	85196	Tu-154B
	85197	Tu-154B
	85198	Tu-154B
	85199	Tu-154B
	85200	Tu-154B
	85201	Tu-154B
	85202	Tu-154B
	85203	Tu-154B
	85204	Tu-154B
	85205	Tu-154B
	85206	Tu-154B
	85207	Tu-154B
	85210	Tu-154B
	85211	Tu-154B
	85212	Tu-154B
	85213	Tu-154B
	85214	Tu-154B
	85215	Tu-154B
	85216	Tu-154B
	85217	Tu-154B
	85218	Tu-154B
	85219	Tu-154B
	85220	Tu-154B
	85221	Tu-154B
	85222	Tu-154B
	85223	Tu-154B
	85226	Tu-154B
	85227	Tu-154B
	85228	Tu-154B
	85229	Tu-154B
	85230	Tu-154B
	85231	Tu-154B
	85232	Tu-154B
	85233	Tu-154B
	85234	Tu-154B
	85235	Tu-154B

Notes	Reg.	Type
	85236	Tu-154B
	85237	Tu-154B
	85238	Tu-154B
	85240	Tu-154B
	85241	Tu-154B
	85242	Tu-154B
	85243	Tu-154B
	85244	Tu-154B
	85245	Tu-154B
	85246	Tu-154B
	85247	Tu-154B
	85248	Tu-154B
	85249	Tu-154B
	85250	Tu-154B
	85251	Tu-154B
	85252	Tu-154B
	85253	Tu-154B
	85254	Tu-154B
	85255	Tu-154B
	85256	Tu-154B
	85257	Tu-154B
	85259	Tu-154B
	85260	Tu-154B
	85261	Tu-154B
	85263	Tu-154B
	85264	Tu-154B
	85265	Tu-154B
	85266	Tu-154B
	85267	Tu-154B
	85268	Tu-154B
	85269	Tu-154B
	85270	Tu-154B
	85271	Tu-154B
	85272	Tu-154B
	85273	Tu-154B
	85274	Tu-154B
	85275	Tu-154B
	85276	Tu-154B
	85277	Tu-154B
	85278	Tu-154B
	85279	Tu-154B
	85280	Tu-154B
	85281	Tu-154B
	85282	Tu-154B
	85283	Tu-154B
	85284	Tu-154B
	85285	Tu-154B
	85286	Tu-154B
	85287	Tu-154B
	85288	Tu-154B
	85289	Tu-154B
	85290	Tu-154B
	85291	Tu-154B
	85292	Tu-154B
	85293	Tu-154B
	85294	Tu-154B
	85295	Tu-154B
	85296	Tu-154B
	85297	Tu-154B
	85298	Tu-154B
	85299	Tu-154B
	85300	Tu-154B
	85301	Tu-154B
	85302	Tu-154B
	85303	Tu-154B
	85304	Tu-154B
	85305	Tu-154B
	85306	Tu-154B
	85307	Tu-154B
	85308	Tu-154B
	85309	Tu-154B
	85310	Tu-154B
	85311	Tu-154B
	85312	Tu-154B

Reg.	Type	Notes
85313	Tu-154B	
85314	Tu-154B	
85315	Tu-154B	
85316	Tu-154B	
85317	Tu-154B	
85318	Tu-154B	
85319	Tu-154B	
85321	Tu-154B	
85322	Tu-154B	
85323	Tu-154B	
85324	Tu-154B	
85325	Tu-154B	
85328	Tu-154B	
85329	Tu-154B	
85330	Tu-154B	
85331	Tu-154B	
85332	Tu-154B	
85333	Tu-154B-2	
85334	Tu-154B	
85335	Tu-154B	
85336	Tu-154B	
85337	Tu-154B	
85338	Tu-154B	
85339	Tu-154B	
85340	Tu-154B	
85344	Tu-154B-2	
85346	Tu-154B	
85347	Tu-154B	
85348	Tu-154B-2	
85349	Tu-154B	
85350	Tu-154B	
85351	Tu-154B-2	
85352	Tu-154B-2	
85353	Tu-154B	
85354	Tu-154B	
85355	Tu-154B	
85356	Tu-154B	
85357	Tu-154B	
85358	Tu-154B	
85359	Tu-154B	
85360	Tu-154B-2	
85361	Tu-154B-2	
85362	Tu-154B	
85363	Tu-154B	
85364	Tu-154B	
85365	Tu-154B	
85366	Tu-154B	
85367	Tu-154B	
85368	Tu-154B	
85369	Tu-154B-2	
85370	Tu-154B-2	
85371	Tu-154B-2	
85372	Tu-154B	
85374	Tu-154B	
85375	Tu-154B	
85376	Tu-154B	
85377	Tu-154B	
85378	Tu-154B	
85379	Tu-154B	
85380	Tu-154B-2	
85381	Tu-154B	
85382	Tu-154B	
85383	Tu-154B-2	
85384	Tu-154B-2	
85385	Tu-154B	
85386	Tu-154B-2	
85387	Tu-154B-2	
85388	Tu-154B-2	
85389	Tu-154B-2	
85390	Tu-154B	
85392	Tu-154B-2	
85395	Tu-154B	
85396	Tu-154B	
85397	Tu-154B	

Reg.	Type	Notes
85398	Tu-154B	
85399	Tu-154B	
85400	Tu-154B	
85402	Tu-154B	
85403	Tu-154B-2	
85404	Tu-154B-2	
85405	Tu-154B-2	
85406	Tu-154B-2	
85407	Tu-154B	
85409	Tu-154B	
85410	Tu-154B	
85411	Tu-154B	
85412	Tu-154B	
85413	Tu-154B	
85414	Tu-154B	
85415	Tu-154B-2	
85416	Tu-154B-2	
85418	Tu-154B	
85423	Tu-154B	
85424	Tu-154B	
85431	Tu-154B	
85432	Tu-154B	
85433	Tu-154B-2	
85434	Tu-154B-2	
85435	Tu-154B-2	
85436	Tu-154B-2	
85437	Tu-154B	
85438	Tu-154B	
85441	Tu-154B	
85442	Tu-154B-2	
85443	Tu-154B-2	
85444	Tu-154B-2	
85445	Tu-154B-2	
85446	Tu-154B-2	
85447	Tu-154B-2	
85448	Tu-154B-2	
85449	Tu-154B-2	
85450	Tu-154B-2	
85451	Tu-154B-2	
85452	Tu-154B-2	
85453	Tu-154B-2	
85454	Tu-154B-2	
85455	Tu-154B	
85459	Tu-154B	
85460	Tu-154B	
85462	Tu-154B	
85472	Tu-154B	
85476	Tu-154B	
85478	Tu-154B	
85479	Tu-154B	
85486	Tu-154B	
85490	Tu-154B	
85491	Tu-154B	
85494	Tu-154B	
85495	Tu-154B	
85496	Tu-154B	
85497	Tu-154B	
85498	Tu-154B	
85499	Tu-154B	
85503	Tu-154B	
85504	Tu-154B	
85510	Tu-154B-2	
85513	Tu-154B	
85514	Tu-154B-2	
85515	Tu-154B-2	
85516	Tu-154B-2	
85518	Tu-154B	
85519	Tu-154B	
85525	Tu-154B	
85526	Tu-154B	
85530	Tu-154B-2	
85531	Tu-154B-2	
85532	Tu-154B-2	
85533	Tu-154B-2	

Notes	Reg.	Type
	85534	Tu-154B-2
	85535	Tu-154B
	85542	Tu-154B-2
	85544	Tu-154B-2
	85545	Tu-154B
	85546	Tu-154B-2
	85547	Tu-154B-2
	85548	Tu-154B-2
	85549	Tu-154B-2
	85550	Tu-154B-2
	85551	Tu-154B-2
	85552	Tu-154B-2
	85553	Tu-154B
	85554	Tu-154B
	85555	Tu-154B
	85556	Tu-154B
	85557	Tu-154B
	85558	Tu-154B
	85559	Tu-154B
	85560	Tu-154B
	85561	Tu-154B
	85562	Tu-154B
	85563	Tu-154B
	85564	Tu-154B
	85565	Tu-154B
	85566	Tu-154B
	85567	Tu-154B
	85568	Tu-154B
	85569	Tu-154B
	85570	Tu-154B
	85571	Tu-154B
	85572	Tu-154B
	85573	Tu-154B
	85574	Tu-154B
	85575	Tu-154B
	85576	Tu-154B
	85577	Tu-154B
	85578	Tu-154B
	85579	Tu-154B
	85580	Tu-154B
	85584	Tu-154B-2
	85585	Tu-154B-2
	85588	Tu-154B-2
	85590	Tu-154B-2
	85595	Tu-154B-2
	85596	Tu-154B-2
	85600	Tu-154B-2
	85601	Tu-154B-2
	85602	Tu-154M-2
	85606	Tu-154M-2
	85609	Tu-154B-2
	85610	Tu-154B-2
	86000	IL-86
	86002	IL-86
	86003	IL-86
	86004	IL-86
	86005	IL-86
	86006	IL-86
	86007	IL-86
	86008	IL-86
	86009	IL-86
	86010	IL-86
	86011	IL-86
	86012	IL-86
	86015	IL-86
	86016	IL-86
	86022	IL-86
	86025	IL-86
	86050	IL-86
	86054	IL-86
	86058	IL-86
	86059	IL-86
	86065	IL-86
	86066	IL-86

Notes	Reg.	Type
	86067	IL-86
	86068	IL-86
	86069	IL-86
	86070	IL-86
	86072	IL-86
	86074	IL-86
	86075	IL-86
	86450	IL-62
	86451	IL-62
	86452	IL-62M
	86453	IL-62M
	86454	IL-62M
	86455	IL-62M
	86456	IL-62M
	86457	IL-62M
	86458	IL-62M
	86459	IL-62M
	86460	IL-62
	86461	IL-62
	86462	IL-62M
	86463	IL-62M
	86464	IL-62M
	86465	IL-62M
	86469	IL-62M
	86471	IL-62M
	86472	IL-62M
	86473	IL-62M
	86474	IL-62M
	86475	IL-62M
	86476	IL-62M
	86477	IL-62M
	86478	IL-62M
	86479	IL-62M
	86480	IL-62M
	86481	IL-62M
	86482	IL-62M
	86483	IL-62M
	86484	IL-62M
	86485	IL-62M
	86486	IL-62M
	86487	IL-62M
	86488	IL-62M
	86489	IL-62M
	86490	IL-62M
	86491	IL-62M
	86492	IL-62M
	86493	IL-62M
	86494	IL-62M
	86497	IL-62M
	86498	IL-62M
	86499	IL-62M
	86500	IL-62M
	86501	IL-62M
	86502	IL-62M
	86503	IL-62M
	86504	IL-62M
	86506	IL-62M
	86507	IL-62M
	86508	IL-62M
	86509	IL-62M
	86510	IL-62M
	86511	IL-62M
	86512	IL-62M
	86513	IL-62M
	86514	IL-62M
	86516	IL-62M
	86517	IL-62M
	86518	IL-62M
	86519	IL-62M
	86520	IL-62M
	86521	IL-62M
	86522	IL-62M

Reg.	Type	Notes
86523	IL-62M	
86524	IL-62M	
86530	IL-62M	
86531	IL-62M	
86532	IL-62M	
86533	IL-62M	
86534	IL-62M	
86535	IL-62M	
86536	IL-62M	
86605	IL-62	
86606	IL-62	
86607	IL-62M	
86608	IL-62	
86609	IL-62	
86610	IL-62	
86611	IL-62	
86612	IL-62	
86613	IL-62	
86615	IL-62	
86616	IL-62	
86617	IL-62	
86618	IL-62M	
86619	IL-62	
86620	IL-62M	
86621	IL-62M	
86622	IL-62M	
86623	IL-62M	
86624	IL-62	
86648	IL-62	
86649	IL-62	
86650	IL-62	
86652	IL-62	
86653	IL-62	
86654	IL-62	
86655	IL-62	
86656	IL-62M	
86657	IL-62	
86658	IL-62M	
86661	IL-62	
86662	IL-62	
86663	IL-62	

Reg.	Type	Notes
86664	IL-62	
86665	IL-62	
86666	IL-62	
86667	IL-62	
86668	IL-62	
86669	IL-62	
86670	IL-62	
86672	IL-62	
86673	IL-62M	
86674	IL-62	
86675	IL-62	
86676	IL-62	
86677	IL-62	
86678	IL-62	
86679	IL-62	
86680	IL-62	
86681	IL-62	
86682	IL-62	
86683	IL-62	
86684	IL-62	
86685	IL-62	
86686	IL-62	
86687	IL-62	
86688	IL-62	
86689	IL-62	
86690	IL-62	
86691	IL-62	
86692	IL-62M	
86693	IL-62M	
86694	IL-62	
86695	IL-62	
86696	IL-62	
86697	IL-62	
86698	IL-62	
86699	IL-62	
86700	IL-62M	
86701	IL-62M	
86702	IL-62M	
86703	IL-62	
86704	IL-62	
86705	IL-62M	

CN (Morocco)

royal air maroc

Notes	Reg.	Type	Owner or Operator
	CN-CCF	Boeing 727-2B6	Royal Air Maroc *Fez*
	CN-CCG	Boeing 727-2B6	Royal Air Maroc *l'Oiseau de la Providence*
	CN-CCH	Boeing 727-2B6	Royal Air Maroc *Marrakesh*
	CN-CCW	Boeing 727-2B6	Royal Air Maroc *Agadir*
	CN-RMB	Boeing 707-351C	Royal Air Maroc *Tangier*
	CN-RMC	Boeing 707-351C	Royal Air Maroc *Casablanca*
	CN-RME	Boeing 747-2B6B	Royal Air Maroc
	CN-RMI	Boeing 737-2B6	Royal Air Maroc *El Ayoun*
	CN-RMJ	Boeing 737-2B6	Royal Air Maroc *Oujda*
	CN-RMK	Boeing 737-2B6	Royal Air Maroc *Smara*
	CN-RML	Boeing 737-2B6	Royal Air Maroc
	CN-RMM	Boeing 737-2B6C	Royal Air Maroc
	CN-RMN	Boeing 737-2B6C	Royal Air Maroc
	CN-RMO	Boeing 727-2B6	Royal Air Maroc
	CN-RMP	Boeing 727-2B6	Royal Air Maroc
	CN-RMQ	Boeing 727-2B6	Royal Air Maroc
	CN-RMR	Boeing 727-2B6	Royal Air Maroc
	CN-RMS	Boeing 747SP-44	Royal Air Maroc

CS (Portugal)

	CS-TBA	Boeing 707-382B	Air Atlantis
	CS-TBC	Boeing 707-382B	TAP — Air Portugal *Cidade de Luanda*
	CS-TBD	Boeing 707-382B	TAP — Air Portugal *Mocambique*
	CS-TBG	Boeing 707-382B	TAP — Air Portugal *Fernao de Magalhaes*
	CS-TBJ	Boeing 707-373C	TAP — Air Portugal *Lisboa*
	CS-TBK	Boeing 727-82	Air Atlantis
	CS-TBL	Boeing 727-82	Air Atlantis
	CS-TBM	Boeing 727-82	TAP — Air Portugal *Algarve*
	CS-TBO	Boeing 727-82QC	TAP — Air Portugal *Costa do Estoril*
	CS-TBS	Boeing 727-282	TAP — Air Portugal *Lisboa*
	CS-TBT	Boeing 707-3F5C	TAP — Air Portugal *Humberto Delgado*
	CS-TBU	Boeing 707-3F5C	TAP — Air Portugal *Jaime Cortesao*
	CS-TBW	Boeing 727-282	TAP — Air Portugal *Coimbra*
	CS-TBX	Boeing 727-282	TAP — Air Portugal *Faro*
	CS-TBY	Boeing 727-282	TAP — Air Portugal *Amadora*
	CS-TEA	L.1011-385 TriStar 500	TAP — Air Portugal *Luis de Camoes*
	CS-TEB	L.1011-385 TriStar 500	TAP — Air Portugal *Infante D. Henrique*
	CS-TEC	L.1011-385 TriStar 500	TAP — Air Portugal *Gago Coutinho*
	CS-TED	L.1011-385 TriStar 500	TAP — Air Portugal *Bartolomeu de Gusmao*
	CS-TEE	L.1011-385 TriStar 500	TAP — Air Portugal *St Antonio Lisboa*
	CS-TEK	Boeing 737-282	TAP — Air Portugal *Ponta Delgada*
	CS-TEL	Boeing 737-282	TAP — Air Portugal *Funchal*
	CS-TEM	Boeing 737-282	TAP — Air Portugal *Setubal*
	CS-TEN	Boeing 737-282	TAP — Air Portugal *Braga*
	CS-TEO	Boeing 737-282	TAP — Air Portugal *Evora*
	CS-TEP	Boeing 737-282	TAP — Air Portugal *Oporto*
	CS-TEQ	Boeing 737-282C	TAP — Air Portugal *Vila Real*

CU (Cuba)

	CU-T1208	Ilyushin IL-62M	Cubana *Capt Wifredo Perez*
	CU-T1209	Ilyushin IL-62M	Cubana
	CU-T1215	Ilyushin IL-62M	Cubana
	CU-T1216	Ilyushin IL-62M	Cubana
	CU-T1217	Ilyushin IL-62M	Cubana
	CU-T1218	Ilyushin IL-62M	Cubana
	CU-T1225	Ilyushin IL-62M	Cubana
	CU-T1226	Ilyushin IL-62M	Cubana
	CU-T1252	Ilyushin IL-62M	Cubana

D (German Federal Republic)

LTU   Hapag-Lloyd   Lufthansa   Condor Flugdienst

Reg.	Type	Owner or Operator	Notes
D-AAST	S.E.210 Caravelle 10R	Aero Lloyd	
D-ABAK	S.E.210 Caravelle 10R	Aero Lloyd	
D-ABAP	S.E.210 Caravelle 10-R	SAT Flug	
D-ABAV	S.E.210 Caravelle 10-R	SAT Flug	
D-ABAW	S.E.210 Caravelle 10-R	SAT Flug	
D-ABCI	Boeing 727-230	Lufthansa *Karlsruhe*	
D-ABDI	Boeing 727-230	Lufthansa *Lübeck*	
D-ABFA	Boeing 737-230	Lufthansa *Regensburg*	
D-ABFB	Boeing 737-230	Lufthansa *Flensburg*	
D-ABFC	Boeing 737-230	Lufthansa *Würzburg*	
D-ABFD	Boeing 737-230	Lufthansa *Bamberg*	
D-ABFF	Boeing 737-230	Lufthansa *Gelsenkirchen*	
D-ABFH	Boeing 737-230	Lufthansa *Pforzheim*	
D-ABFI	Boeing 727-230	Lufthansa *Münster*	
D-ABFK	Boeing 737-230	Lufthansa *Wuppertal*	
D-ABFL	Boeing 737-230	Lufthansa *Coburg*	
D-ABFM	Boeing 737-230	Lufthansa *Osnabrück*	
D-ABFN	Boeing 737-230	Lufthansa *Kempton*	
D-ABFP	Boeing 737-230	Lufthansa *Offenbach*	
D-ABFR	Boeing 737-230	Lufthansa *Solingen*	
D-ABFS	Boeing 737-230	Lufthansa *Oldenburg*	
D-ABFT	Boeing 737-230	Condor Flugdienst	
D-ABFU	Boeing 737-230	Lufthansa *Mülheim a.d.R*	
D-ABFW	Boeing 737-230	Lufthansa *Wolfsberg*	
D-ABFX	Boeing 737-230	Lufthansa *Tübingen*	
D-ABFY	Boeing 737-230	Lufthansa *Göttingen*	
D-ABFZ	Boeing 737-230	Lufthansa *Wilhelmshaven*	
D-ABGI	Boeing 727-230	Lufthansa *Leverkusen*	
D-ABHA	Boeing 737-230	Lufthansa *Koblenz*	
D-ABHB	Boeing 737-230	Lufthansa *Goslar*	
D-ABHC	Boeing 737-230	Lufthansa *Friedrichshafen*	
D-ABHD	Boeing 737-230	Condor Flugdienst	
D-ABHE	Boeing 737-230C	Lufthansa	
D-ABHF	Boeing 737-230	Lufthansa *Heilbronn*	
D-ABHH	Boeing 737-230	Lufthansa *Marburg*	
D-ABHI	Boeing 727-230	Lufthansa *Mönchengladbach*	
D-ABHK	Boeing 737-230	Lufthansa *Bayreuth*	
D-ABHL	Boeing 737-230	Lufthansa *Worms*	
D-ABHM	Boeing 737-230	Lufthansa *Landshut*	
D-ABHN	Boeing 737-230	Lufthansa *Trier*	
D-ABHP	Boeing 737-230	Lufthansa *Erlangen*	
D-ABHR	Boeing 737-230	Lufthansa *Darmstadt*	
D-ABHS	Boeing 737-230	Lufthansa *Remscheid*	
D-ABHT	Boeing 737-230	Condor Flugdienst	
D-ABHU	Boeing 737-230	Lufthansa *Konstanz*	
D-ABHW	Boeing 737-230	Lufthansa *Baden Baden*	
D-ABHX	Boeing 737-230	Condor Flugdienst	
D-ABKA	Boeing 727-230	Lufthansa *Heidelberg*	
D-ABKB	Boeing 727-230	Lufthansa *Augsburg*	
D-ABKC	Boeing 727-230	Lufthansa *Braunschweig*	
D-ABKD	Boeing 727-230	Lufthansa *Freiburg*	
D-ABKE	Boeing 727-230	Lufthansa *Mannheim*	
D-ABKF	Boeing 727-230	Lufthansa *Saarbrücken*	
D-ABKG	Boeing 727-230	Lufthansa *Kassel*	
D-ABKH	Boeing 727-230	Lufthansa *Kiel*	
D-ABKI	Boeing 727-230	Lufthansa *Bremerhaven*	
D-ABKJ	Boeing 727-230	Lufthansa *Wiesbaden*	
D-ABKK	Boeing 727-230	Condor Flugdienst	
D-ABKL	Boeing 727-230	Condor Flugdienst	
D-ABKM	Boeing 727-230	Lufthansa *Hagen*	
D-ABKN	Boeing 727-230	Lufthansa *Ulm*	
D-ABKP	Boeing 727-230	Lufthansa *Krefeld*	
D-ABKQ	Boeing 727-230	Lufthansa *Mainz*	
D-ABKR	Boeing 727-230	Lufthansa *Bielefeld*	
D-ABKS	Boeing 727-230	Lufthansa *Oberhausen*	

Notes	Reg.	Type	Owner or Operator
	D-ABKT	Boeing 727-230	Lufthansa *Aachen*
	D-ABLI	Boeing 727-230	Lufthansa *Ludwigshafen a.Rh.*
	D-ABMA	Boeing 737-230	Lufthansa *Idar-doerst*
	D-ABMB	Boeing 737-230	Lufthansa *Ingolstadt*
	D-ABMC	Boeing 737-230	Lufthansa *Norderstedt*
	D-ABMD	Boeing 737-230	Lufthansa *Paderborn*
	D-ABME	Boeing 737-230	Lufthansa *Schweinfurth*
	D-ABMF	Boeing 737-230	Lufthansa *Verden*
	D-ABMI	Boeing 727-230	Condor Flugdienst
	D-ABNI	Boeing 727-230	Condor Flugdienst
	D-ABPI	Boeing 727-230	Condor Flugdienst
	D-ABQI	Boeing 727-230	Lufthansa *Hildesheim*
	D-ABRI	Boeing 727-230	Lufthansa *Esslingen*
	D-ABSI	Boeing 727-230	Lufthansa *Hof*
	D-ABTI	Boeing 727-230	Condor Flugdienst
	D-ABVI	Boeing 727-230	Condor Flugdienst
	D-ABWI	Boeing 727-230	Condor Flugdienst
	D-ABXA	Boeing 737-330	Lufthansa
	D-ABXB	Boeing 737-330	Lufthansa
	D-ABXC	Boeing 737-330	Lufthansa
	D-ABXD	Boeing 737-330	Lufthansa
	D-ABXE	Boeing 737-330	Lufthansa
	D-ABXF	Boeing 737-330	Lufthansa
	D-ABXH	Boeing 737-330	Lufthansa
	D-ABXI	Boeing 737-330	Lufthansa
	D-ABXK	Boeing 737-330	Lufthansa
	D-ABXL	Boeing 737-330	Lufthansa
	D-ABYJ	Boeing 747-230B	Lufthansa *Hessen*
	D-ABYK	Boeing 747-230B	Lufthansa *Rheinland-Pfalz*
	D-ABYL	Boeing 747-230B	Lufthansa *Saarland*
	D-ABYM	Boeing 747-230B	Lufthansa *Schleswig-Holstein*
	D-ABYN	Boeing 747-230B	Lufthansa *Baden-Wurttemberg*
	D-ABYO	Boeing 747-230F	Lufthansa *America*
	D-ABYP	Boeing 747-230B	Lufthansa *Niedersachen*
	D-ABYQ	Boeing 747-230B	Lufthansa *Bremen*
	D-ABYR	Boeing 747-230B	Lufthansa *Nordrhein-Westfalen*
	D-ABYS	Boeing 747-230B	Lufthansa *Bayern*
	D-ABYT	Boeing 747-230B	Lufthansa *Hamburg*
	D-ABYU	Boeing 747-230F	Lufthansa *Asia*
	D-ABYW	Boeing 747-230B	Lufthansa *Berlin*
	D-ABYX	Boeing 747-230B	Lufthansa *Köln*
	D-ABYY	Boeing 747-230B	Lufthansa *München*
	D-ABYZ	Boeing 747-230B	Lufthansa *Frankfurt*
	D-ABZA	Boeing 747-230B	Lufthansa *Dusseldorf*
	D-ABZB	Boeing 747-230F	Lufthansa
	D-ABZC	Boeing 747-230B	Lufthansa
	D-ABZD	Boeing 747-230B	Lufthansa
	D-ABZE	Boeing 747-330	Lufthansa
	D-ABZF	Boeing 747-330	Lufthansa
	D-ACVK	S.E.210 Caravelle 10R	Aero Lloyd
	D-ADAO	Douglas DC-10-30	Lufthansa *Düsseldorf*
	D-ADBO	Douglas DC-10-30	Lufthansa *Bochum*
	D-ADCO	Douglas DC-10-30	Lufthansa *Frankfurt*
	D-ADDO	Douglas DC-10-30	Lufthansa *Duisburg*
	D-ADFO	Douglas DC-10-30	Lufthansa *Fürth*
	D-ADGO	Douglas DC-10-30	Lufthansa *Bonn*
	D-ADHO	Douglas DC-10-30	Lufthansa *Hannover*
	D-ADJO	Douglas DC-10-30	Lufthansa *Essen*
	D-ADKO	Douglas DC-10-30	Lufthansa *Stuttgart*
	D-ADLO	Douglas DC-10-30	Lufthansa *Nurnberg*
	D-ADMO	Douglas DC-10-30	Lufthansa *Dortmund*
	D-ADPO	Douglas DC-10-30	Condor Flugdienst
	D-ADQO	Douglas DC-10-30	Condor Flugdienst
	D-ADSO	Douglas DC-10-30	Condor Flugdienst
	D-ADUA	Douglas DC-8-73CF	German Cargo
	D-ADUC	Douglas DC-8-73CF	Condor Flugdienst/German Cargo
	D-ADUE	Douglas DC-8-73CF	German Cargo
	D-ADUI	Douglas DC-8-73CF	German Cargo
	D-ADUO	Douglas DC-8-73CF	German Cargo
	D-AERE	L.1011-385 TriStar 1	LTU
	D-AERI	L.1011-385 TriStar 1	LTU
	D-AERL	L.1011-385 TriStar 500	LTU
	D-AERM	L.1011-385 TriStar 1	LTU

Reg.	Type	Owner or Operator	Notes
D-AERN	L.1011-385 TriStar 1	LTU	
D-AERP	L.1011-385 TriStar 1	LTU	
D-AERT	L.1011-385 TriStar 500	LTU	
D-AERU	L.1011-385 TriStar 100	LTU	
D-AHLB	A.300C4 Airbus	Hapag-Lloyd	
D-AHLC	A.300B4 Airbus	Hapag-Lloyd	
D-AHLG	Boeing 737-2K5	Hapag-Lloyd	
D-AHLH	Boeing 737-2K5	Hapag-Lloyd	
D-AHLI	Boeing 737-2K5	Hapag-Lloyd	
D-AHLJ	A.300B4 Airbus	Hapag-Lloyd	
D-AHLK	A.300B4 Airbus	Hapag-Lloyd	
D-AHLM	Boeing 727-81	Hapag-Lloyd	
D-AHLN	Boeing 727-81	Hapag-Lloyd/SAT Flug	
D-AHLS	Boeing 727-89	Hapag-Lloyd/SAT Flug	
D-AHLT	Boeing 727-2K5	Hapag-Lloyd	
D-AHLU	Boeing 727-2K5	Hapag-Lloyd	
D-AHLZ	A.300B4 Airbus	Hapag-Lloyd	
D-AHSA	H.S.748 Srs 2B	D.L.T.	
D-AHSB	H.S.748 Srs 2B	D.L.T.	
D-AHSC	H.S.748 Srs 2B	D.L.T.	
D-AHSD	H.S.748 Srs 2B	D.L.T.	
D-AHSE	H.S.748 Srs 2B	D.L.T.	
D-AHSF	H.S.748 Srs 2B	D.L.T.	
D-AIBA	A.300B4 Airbus	Lufthansa	
D-AIBB	A.300B4 Airbus	Lufthansa	
D-AIBC	A.300B4 Airbus	Lufthansa *Lindau/Bodensee*	
D-AIBD	A.300B4 Airbus	Lufthansa *Erbach/Odenwald*	
D-AIBF	A.300B4 Airbus	Lufthansa *Kronberg/Taunus*	
D-AICA	A.310-203 Airbus	Lufthansa *Neustadt an der Weinstrausse*	
D-AICB	A.310-203 Airbus	Lufthansa *Garmisch-Partenkirchen*	
D-AICC	A.310-203 Airbus	Lufthansa *Kaiserslauten*	
D-AICD	A.310-203 Airbus	Lufthansa *Detmold*	
D-AICF	A.310-203 Airbus	Lufthansa *Rüdesheim am Rhein*	
D-AICH	A.310-203 Airbus	Lufthansa *Lüneburg*	
D-AICK	A.310-203 Airbus	Lufthansa *Westerland-Sylt*	
D-AICL	A.310-203 Airbus	Lufthansa *Rothenburg ob der Tauber*	
D-AICM	A.310-203 Airbus	Condor Flugdienst	
D-AICN	A.310-203 Airbus	Condor Flugdienst	
D-AICP	A.310-203 Airbus	Condor Flugdienst	
D-AICR	A.310-203 Airbus	Lufthansa	
D-AICS	A.310-203 Airbus	Lufthansa	
D-	A.310-203 Airbus	Lufthansa	
D-AJAA	Boeing 727-81	Jetair *Heinrich der Löive*	
D-AJAB	Boeing 727-81	Jetair *Kurfurst Max Emanuel*	
D-ALLA	Douglas DC-9-32	Aero Lloyd	
D-ALLB	Douglas DC-9-32	Aero Lloyd	
D-ALLC	Douglas DC-9-32	Aero Lloyd	
D-AL	Douglas DC-9-83	Aero Lloyd	
D-AMAP	A.300B4 Airbus	Hapag-Lloyd	
D-AMAX	A.300B4 Airbus	Hapag-Lloyd	
D-AMAY	A.300B4 Airbus	Hapag-Lloyd	
D-AMUR	Boeing 757-2G5	Luftransport Sud (LTS)	
D-AMUS	Boeing 757-2G5	Luftransport Sud (LTS)	
D-BAKA	F.27 Friendship Mk 100	WDL	
D-BAKI	F.27 Friendship Mk 100	WDL	
D-BAKU	F.27 Friendship Mk 200	WDL	
D-B	Fokker 50	D.L.T.	
D-B	Fokker 50	D.L.T.	
D-CABA	Swearingen SA226AC Metro III	Nürnberger Flugdienst (NFD)	
D-CABB	Swearingen SA226AC Metro III	Nürnberger Flugdienst (NFD)	
D-CABD	Swearingen SA226AC Metro III	Nürnberger Flugdienst (NFD)	
D-CABE	Swearingen SA226AC Metro III	Nürnberger Flugdienst (NFD)	
D-CABF	Swearingen SA226AC Metro III	Nürnberger Flugdienst (NFD)	
D-CABH	Swearingen SA226AC Metro III	Nürnberger Flugdienst (NFD)	
D-C	EMB-120 Brasilia	D.L.T.	
D-C	EMB-120 Brasilia	D.L.T.	
D-C	EMB-120 Brasilia	D.L.T.	
D-C	EMB-120 Brasilia	D.L.T.	
D-C	EMB-120 Brasilia	D.L.T.	
D-C	EMB-120 Brasilia	D.L.T.	
D-CONE	BAe Jetstream 3102	Contactair	
D-CONI	BAe Jetstream 3102	Contactair	

Notes	Reg.	Type	Owner or Operator
	D-CONU	BAe Jetstream 3102	Contactair
	D-IASN	Swearingen SA226TC Metro II	Delta-Air
	D-IASX	Dornier Do.228-200	Delta Air
	D-IBCF	Swearingen SA226TC Metro II	D.L.T.
	D-ICRJ	Swearingen SA.226TC Metro II	R.F.G.
	D-IDWT	DHC-6 Twin Otter 310	D.L.T.
	D-IDOM	Dornier Do.228-100	Holiday Express
	D-IHCW	Swearingen SA.226TC Metro II	R.F.G.
	D-IHUC	Swearingen SA226TC Metro II	Delta Air

DDR (German Democratic Republic)

	DDR-SCB	Tupolev Tu-134	Interflug
	DDR-SCE	Tupolev Tu-134	Interflug
	DDR-SCF	Tupolev Tu-134	Interflug
	DDR-SCG	Tupolev Tu-134	Interflug
	DDR-SCH	Tupolev Tu-134	Interflug
	DDR-SCI	Tupolev Tu-134A	Interflug
	DDR-SCK	Tupolev Tu-134A	Interflug
	DDR-SCL	Tupolev Tu-134A	Interflug
	DDR-SCN	Tupolev Tu-134A	Interflug
	DDR-SCO	Tupolev Tu-134A	Interflug
	DDR-SCP	Tupolev Tu-134A	Interflug
	DDR-SCR	Tupolev Tu-134A	Interflug
	DDR-SCS	Tupolev Tu-134A	Interflug
	DDR-SCT	Tupolev Tu-134A	Interflug
	DDR-SCU	Tupolev Tu-134A	Interflug
	DDR-SCV	Tupolev Tu-134A	Interflug
	DDR-SCW	Tupolev Tu-134A	Interflug
	DDR-SCX	Tupolev Tu-134A	Interflug
	DDR-SCY	Tupolev Tu-134A	Interflug
	DDR-SCZ	Tupolev Tu-134	Interflug
	DDR-SDC	Tupolev Tu-134A	Interflug
	DDR-SDE	Tupolev Tu-134A	Interflug
	DDR-SDF	Tupolev Tu-134A	Interflug
	DDR-SDG	Tupolev Tu-134A	Interflug
	DDR-SDU	Tupolev Tu-134A	Interflug
	DDR-SEB	Ilyushin IL-62	Interflug
	DDR-SEC	Ilyushin IL-62	Interflug
	DDR-SEF	Ilyushin IL-62	Interflug
	DDR-SEG	Ilyushin IL-62	Interflug
	DDR-SEH	Ilyushin IL-62	Interflug
	DDR-SEI	Ilyushin IL-62M	Interflug
	DDR-SEK	Ilyushin IL-62M	Interflug
	DDR-SEL	Ilyushin IL-62M	Interflug
	DDR-SEM	Ilyushin IL-62M	Interflug
	DDR-SEO	Ilyushin IL-62M	Interflug
	DDR-SEP	Ilyushin IL-62M	Interflug
	DDR-STA	Ilyushin IL-18D	Interflug
	DDR-STB	Ilyushin IL-18D	Interflug
	DDR-STC	Ilyushin IL-18D	Interflug
	DDR-STD	Ilyushin IL-18D	Interflug
	DDR-STE	Ilyushin IL-18D	Interflug
	DDR-STF	Ilyushin IL-18D	Interflug
	DDR-STG	Ilyushin IL-18D	Interflug
	DDR-STH	Ilyushin IL-18D	Interflug
	DDR-STI	Ilyushin IL-18D	Interflug
	DDR-STK	Ilyushin IL-18D	Interflug
	DDR-STM	Ilyushin IL-18D	Interflug
	DDR-STN	Ilyushin IL-18D	Interflug
	DDR-STO	Ilyushin IL-18D	Interflug
	DDR-STP	Ilyushin IL-18D	Interflug

EC (Spain)

Reg.	Type	Owner or Operator	Notes
EC-BIG	Douglas DC-9-32	Iberia *Villa de Madrid*	
EC-BIH	Douglas DC-9-32	Aviaco *Ciudad de Barcelona*	
EC-BIJ	Douglas DC-9-32	Iberia *Santa Cruz de Tenerife*	
EC-BIK	Douglas DC-9-32	Aviaco *Castillo de Guanapa*	
EC-BIL	Douglas DC-9-32	Iberia *Ciudad de Zaragoza*	
EC-BIM	Douglas DC-9-32	Iberia *Ciudad de Santander*	
EC-BIN	Douglas DC-9-32	Iberia *Palma de Mallorca*	
EC-BIO	Douglas DC-9-32	Iberia *Villa de Bilbao*	
EC-BIP	Douglas DC-9-32	Aviaco *Castillo de Monteagudo*	
EC-BIQ	Douglas DC-9-32	Aviaco *Castillo de Argueso*	
EC-BIR	Douglas DC-9-32	Iberia *Ciudad de Valencia*	
EC-BIS	Douglas DC-9-32	Iberia *Ciudad de Alicante*	
EC-BIT	Douglas DC-9-32	Iberia *Ciudad de San Sebastian*	
EC-BIU	Douglas DC-9-32	Iberia *Ciudad de Oviedo*	
EC-BJD	CV-990A Coronado	Spantax	
EC-BPF	Douglas DC-9-32	Iberia *Ciudad de Almeria*	
EC-BPG	Douglas DC-9-32	Iberia *Ciudad de Vigo*	
EC-BPH	Douglas DC-9-32	Iberia *Ciudad de Gerona*	
EC-BQA	CV-990A Coronado	Spantax	
EC-BQQ	CV-990A Coronado	Spantax	
EC-BQT	Douglas DC-9-32	Iberia *Ciudad de Murcia*	
EC-BQU	Douglas DC-9-32	Iberia *Ciudad de la Coruna*	
EC-BQV	Douglas DC-9-32	Iberia *Ciudad de Ibiza*	
EC-BQX	Douglas DC-9-32	Iberia *Ciudad de Valladolid*	
EC-BQY	Douglas DC-9-32	Aviaco *Ciudad de Cordoba*	
EC-BQZ	Douglas DC-9-32	Iberia *Ciudad de Santa Cruz de la Palma*	
EC-BRQ	Boeing 747-256B	Iberia *Calderon de la Barca*	
EC-BTE	CV-990A Coronado	Spantax	
EC-BXI	CV-990A Coronado	Spantax	
EC-BYD	Douglas DC-9-32	Iberia *Ciudad de Arrecife de Lanzarote*	
EC-BYE	Douglas DC-9-32	Iberia *Ciudad de Mahon*	
EC-BYF	Douglas DC-9-32	Iberia *Ciudad de Granada*	
EC-BYG	Douglas DC-9-32	Iberia *Ciudad de Pamplona*	
EC-BYH	Douglas DC-9-32	Aviaco *Castillo de Butron*	
EC-BYI	Douglas DC-9-32	Iberia *Ciudad de Vitoria*	
EC-BYJ	Douglas DC-9-32	Iberia *Ciudad de Salamanca*	
EC-BYK	Douglas DC-9-33RC	Iberia *Ciudad de Badajoz*	
EC-BYL	Douglas DC-9-33RC	Iberia *Cuidad de Albacete*	
EC-BYM	Douglas DC-9-33RC	Iberia *Ciudad de Cangas de Onis*	
EC-BYN	Douglas DC-9-33RC	Iberia *Ciudad de Caceres*	
EC-BZO	CV-990A Coronado	Spantax	
EC-BZP	CV-990A Coronado	Spantax	
EC-CAI	Boeing 727-256	Iberia *Castilla la Neuva*	
EC-CAJ	Boeing 727-256	Iberia *Cataluna*	
EC-CAK	Boeing 727-256	Iberia *Aragon*	
EC-CBA	Boeing 727-256	Iberia *Vascongadas*	
EC-CBB	Boeing 727-256	Iberia *Valencia*	
EC-CBC	Boeing 727-256	Iberia *Navarra*	
EC-CBD	Boeing 727-256	Iberia *Murcia*	
EC-CBE	Boeing 727-256	Iberia *Leon*	
EC-CBF	Boeing 727-256	Iberia *Gran Canaria*	
EC-CBG	Boeing 727-256	Iberia *Extremadura*	
EC-CBH	Boeing 727-256	Iberia *Galicia*	
EC-CBI	Boeing 727-256	Iberia *Asturias*	
EC-CBJ	Boeing 727-256	Iberia *Andalucia*	
EC-CBK	Boeing 727-256	Iberia *Baleares*	
EC-CBL	Boeing 727-256	Iberia *Tenerife*	
EC-CBM	Boeing 727-256	Iberia *Castilla la Vieja*	
EC-CBO	Douglas DC-10-30	Iberia *Costa del Sol*	
EC-CBP	Douglas DC-10-30	Iberia *Costa Dorada*	
EC-CEZ	Douglas DC-10-30	Iberia *Costa del Azahar*	
EC-CFA	Boeing 727-256	Iberia *Jerez Xeres Sherry*	
EC-CFB	Boeing 727-256	Iberia *Rioja*	
EC-CFC	Boeing 727-256	Iberia *Tarragona*	
EC-CFD	Boeing 727-256	Iberia *Montilla Moriles*	
EC-CFE	Boeing 727-256	Iberia *Penedes*	
EC-CFF	Boeing 727-256	Iberia *Valdepenas*	

Notes	Reg.	Type	Owner or Operator
	EC-CFG	Boeing 727-256	Iberia *La Mancha*
	EC-CFH	Boeing 727-256	Iberia *Priorato*
	EC-CFI	Boeing 727-256	Iberia *Carinena*
	EC-CFK	Boeing 727-256	Iberia *Riberio*
	EC-CGN	Douglas DC-9-32	Aviaco *Martin Alonso Pinzon*
	EC-CGO	Douglas DC-9-32	Aviaco *Pedro Alonso Nino*
	EC-CGP	Douglas DC-9-32	Aviaco *Juan Sebastian Elcano*
	EC-CGQ	Douglas DC-9-32	Aviaco *Alonso de Ojeda*
	EC-CGR	Douglas DC-9-32	Aviaco *Francisco de Orellana*
	EC-CID	Boeing 727-256	Iberia *Malaga*
	EC-CIE	Boeing 727-256	Iberia *Esparragosa*
	EC-CIZ	S.E.210 Caravelle 10R	Hispania
	EC-CLB	Douglas DC-10-30	Iberia *Costa Blanca*
	EC-CLD	Douglas DC-9-32	Aviaco *Hernando de Soto*
	EC-CLE	Douglas DC-9-32	Aviaco *Jaun Ponce de Leon*
	EC-CNF	CV-990A Coronado	Spantax
	EC-CNH	CV-990A Coronado	Spantax
	EC-CPI	S.E.210 Caravelle 10R	Hispania
	EC-CSJ	Douglas DC-10-30	Iberia *Costa de la Luz*
	EC-CSK	Douglas DC-10-30	Iberia *Cornisa Cantabrica*
	EC-CTR	Douglas DC-9-34CF	Aviaco *Hernan Cortes*
	EC-CTS	Douglas DC-9-34CF	Aviaco *Francisco Pizarro*
	EC-CTT	Douglas DC-9-34CF	Aviaco *Pedro de Valdivia*
	EC-CTU	Douglas DC-9-34CF	Aviaco *Pedro de Alvarado*
	EC-CYI	S.E.210 Caravelle 10R	Hispania
	EC-CZE	Douglas DC-8-61	Spantax
	EC-DCC	Boeing 727-256	Iberia *Albarino*
	EC-DCD	Boeing 727-256	Iberia *Chacoli*
	EC-DCE	Boeing 727-256	Iberia *Mentrida*
	EC-DCN	S.E.210 Caravelle 10R	Hispania
	EC-DDV	Boeing 727-256	Iberia *Acueducto de Segovia*
	EC-DDX	Boeing 727-256	Iberia *Monasterio de Poblet*
	EC-DDY	Boeing 727-256	Iberia *Cuevas de Altamira*
	EC-DDZ	Boeing 727-256	Iberia *Murallas de Avila*
	EC-DEA	Douglas DC-10-30	Iberia *Rias Gallegas*
	EC-DGB	Douglas DC-9-34	Aviaco *Castillo de Javier*
	EC-DGC	Douglas DC-9-34	Aviaco *Castillo de Sotomayor*
	EC-DGD	Douglas DC-9-34	Aviaco *Castillo de Arcos*
	EC-DGE	Douglas DC-9-34	Aviaco *Castillo de Bellver*
	EC-DHZ	Douglas DC-10-30	Iberia *Costas Canarias*
	EC-DIA	Boeing 747-256B	Iberia *Tirso de Molina*
	EC-DIB	Boeing 747-256B	Iberia *Cervantes*
	EC-DLC	Boeing 747-256B SCD	Iberia *Francisco de Quevedo*
	EC-DLD	Boeing 747-256B SCD	Iberia *Lupe de Vega*
	EC-DLE	A.300B4 Airbus	Iberia *Doana*
	EC-DLF	A.300B4 Airbus	Iberia *Canadas del Teide*
	EC-DLG	A.300B4 Airbus	Iberia *Tablas de Daimiel*
	EC-DLH	A.300B4 Airbus	Iberia *Aigues Tortes*
	EC-DNP	Boeing 747-256B	Iberia *Juan Ramon Jimenez*
	EC-DNQ	A.300B4 Airbus	Iberia *Islas Cies*
	EC-DNR	A.300B4 Airbus	Iberia *Ordesa*
	EC-DTR	Boeing 737-2K5	Spantax
	EC-DUB	Boeing 737-2K5	Spantax
	EC-DUG	Douglas DC-10-30	Spantax
	EC-DVB	Douglas DC-8-61	Spantax
	EC-DVC	Douglas DC-8-61	Spantax
	EC-DVE	Boeing 737-204	Spantax (*leased from Britannia*)
	EC-DVN	Boeing 737-2K2	Hispania (*leased from Transavia*)
	EC-DXK	Boeing 737-204	Spantax
	EC-DXV	Boeing 737-2L9	Hispania

EI (Republic of Ireland)

Including complete current Irish Civil Register

Reg.	*Type*	*Owner or Operator*	*Notes*
EI-ADV	PA-12 Super Cruiser	R. E. Levis	
EI-AFF	B.A. Swallow 2 ★	J. McCarthy	
EI-AFN	B.A. Swallow 2 ★	J. McCarthy	
EI-AGB	Miles M.38 Messenger 4 ★	J. McLoughlin	
EI-AGD	Taylorcraft Plus D ★	H. Wolf	
EI-AGJ	J/I Autocrat	W. G. Rafter	
EI-AHA	D.H.82A Tiger Moth ★	J. H. Maher	
EI-AHR	D.H.C.1 Chipmunk 22 ★	C. Lane	
EI-AKM	Piper J-3C-65 Cub	Setanta Flying Group	
EI-ALH	Taylorcraft Plus D	N. Reilly	
EI-ALP	Avro 643 Cadet	J. C. O'Loughlin	
EI-ALU	Avro 631 Cadet	M. P. Cahill	
EI-AMK	J/I Autocrat	Irish Aero Club	
EI-AMO	J/IB Aiglet	R. Hassett	
EI-AND	Cessna 175A	Jack Braithwaite (Ireland) Ltd	
EI-ANE	BAC One-Eleven 208AL	Aer Lingus *St Mel*	
EI-ANF	BAC One-Eleven 208AL	Aer Lingus *St Malachy*	
EI-ANG	BAC One-Eleven 208AL	Aer Lingus *St Declan*	
EI-ANH	BAC One-Eleven 208AL	Aer Lingus *St Ronan*	
EI-ANT	Champion 7ECA Citabria	W. Kennedy	
EI-AOB	PA-28 Cherokee 140	Oscar Bravo Flying Group Ltd	
EI-AOD	Cessna 182J Skylane	Oscar Delta Flying Training Co Ltd	
EI-AOK	Cessna F.172G	R. J. Cloughley & N. J. Simpson	
EI-AOO	Cessna 150E	R. Hassett	
EI-AOP	D.H.82A Tiger Moth	Dublin Tiger Group	
EI-AOS	Cessna 310B	Joyce Aviation Ltd	
EI-APF	Cessna F.150F	L. O. Kennedy	
EI-APS	Schleicher ASK 14	G. W. Connolly & M. Slazenger	
EI-ARH	Currie Wot/S.E.5 Replica	L. Garrison	
EI-ARM	Currie Wot/S.E.5 Replica	L. Garrison	
EI-ARW	Jodel D.R.1050	M. Mannion	
EI-ASA	Boeing 737-248	Aer Lingus *St Jarlath*	
EI-ASB	Boeing 737-248	Aer Lingus *St Albert*	
EI-ASC	Boeing 737-248C	Aer Lingus *St Macartan*	
EI-ASD	Boeing 737-248C	Aer Lingus *St Ide*	
EI-ASE	Boeing 737-248C	Aer Lingus *St Fachtna*	
EI-ASF	Boeing 737-248	Aer Lingus *St Nathy*	
EI-ASG	Boeing 737-248	Aer Lingus *St Cormack*	
EI-ASH	Boeing 737-248	Aer Lingus *St Eugene* (leased Air Tara)	
EI-ASI	Boeing 747-148	Aer Lingus *St Colmcille*	
EI-ASJ	Boeing 747-148	Aer Lingus *St Patrick*	
EI-ASL	Boeing 737-248C	Aer Lingus *St Killian*	
EI-ASO	Boeing 707-349C	Aer Lingus *St Canice*	
EI-AST	Cessna F.150H	Liberty Flying Group	
EI-ATH	Cessna F.150J	Hibernian Flying Club	
EI-ATJ	B.121 Pup I	Wexford Aero Club	
EI-ATK	PA-28 Cherokee 140	Mayo Flying Club Ltd	
EI-ATS	M.S.880B Rallye Club	O. Bruton & G. Farrar	
EI-AUC	Cessna FA.150K Aerobat	Flying Fifteen Aero Club Ltd	
EI-AUD	M.S.880B Rallye Club	Kilkenny Flying Club Ltd	
EI-AUE	M.S.880B Rallye Club	Kilkenny Flying Club Ltd	
EI-AUG	M.S.894 Rallye Minerva 220	Weston Ltd	
EI-AUJ	M.S.880B Rallye Club	P. Mulhall	
EI-AUM	J/I Autocrat	J. G. Rafter	
EI-AUO	Cessna FA.150K Aerobat	Kerry Aero Club	
EI-AUP	M.S.880B Rallye Club	Limerick Flying Club	
EI-AUS	J/5F Aiglet Trainer	T. Stephens & T. Lennon	
EI-AUT	Forney F-1A Aircoupe	Joyce Aviation Ltd	
EI-AUV	PA-23 Aztec 250	Shannon Executive Aviation	
EI-AUY	Morane-Saulnier M.S.502	Historical Aircraft Preservation Group	
EI-AVB	Aeronca 7AC Champion	G. G. Bracken	
EI-AVC	Cessna F.337F	337 Flying Group	
EI-AVM	Cessna F.150L	G. Farrar	
EI-AVN	Hughes 369HM	Helicopter Maintenance Ltd	
EI-AWE	Cessna F.150M	Third Flight Group	

Notes	Reg.	Type	Owner or Operator
	EI-AWH	Cessna 210J	Southern Air Ltd
	EI-AWP	D.H.82A Tiger Moth	A. Lyons
	EI-AWR	Malmo MFI-9 Junior	G. Fawcett
	EI-AWU	M.S.880B Rallye Club	Longford Aviation Ltd
	EI-AWW	Cessna 414	Ace Coin (Ireland) Ltd
	EI-AYA	M.S.880B Rallye Club	Dundalk Aero Club Ltd
	EI-AYB	GY-80 Horizon 180	Westwing Flying Group
	EI-AYD	AA-5 Traveler	P. Howick & ptnrs
	EI-AYF	Cessna FRA.150L	Garda Flying Club
	EI-AYI	M.S.880B Rallye Club	Irish Air Training Group
	EI-AYK	Cessna F.172M	J. Martyn & J. Hession
	EI-AYL	A.109 Airedale	J. Ronan
	EI-AYN	BN-2A Islander	Aer Arran
	EI-AYO	Douglas DC-3A ★	Science Museum, Wroughton
	EI-AYR	Schleicher ASK-16	Kilkenny Airport Ltd
	EI-AYS	PA-22 Colt 108	Messrs Skelly & Hall
	EI-AYT	M.S.894A Rallye Minerva	R. C. Cunningham
	EI-AYV	M.S.892A Rallye Commodore 150	P. Murtagh
	EI-AYW	PA-23 Aztec 250	Chutewell International Ltd
	EI-AYY	Evans VP-1	M. Donoghue
	EI-BAB	M.S.894E Rallye Minerva	J. Phelan
	EI-BAF	Thunder Ax6-56 balloon	W. G. Woollett
	EI-BAJ	Stampe SV-4C	Dublin Tiger Group
	EI-BAO	Cessna F.172G	Kingdom Air Ltd
	EI-BAR	Thunder Ax8-105 balloon	J. Burke & V. Hourihane
	EI-BAS	Cessna F.172M	Falcon Aviation Ltd
	EI-BAT	Cessna F.150M	20th Air Training Co Ltd
	EI-BAU	Stampe SV.4C	S. P. O'Carroll
	EI-BAV	PA-22 Colt 108	J. P. Montcalm
	EI-BAY	Cameron V-77 balloon	F. N. Lewis
	EI-BBC	PA-28 Cherokee 180C	Rathcoole Flying Group
	EI-BBD	Evans VP-1	Volksplane Group
	EI-BBE	7FC Tri-Traveler (tailwheel)	Aeronca Flying Group
	EI-BBF	EAA P-2 Biplane	B. Feeley
	EI-BBG	M.S.880B Rallye Club	Weston Ltd
	EI-BBI	M.S.892 Rallye Commodore	Kilkenny Airport Ltd
	EI-BBJ	M.S.880B Rallye Club	Weston Ltd
	EI-BBK	A.109 Airedale	H. S. Igoe
	EI-BBL	R. Turbo 690A Commander	The Earl of Granard
	EI-BBM	Cameron O-65 balloon	Dublin Ballooning Club
	EI-BBN	Cessna F.150M	Sligo N.W. Aero Club
	EI-BBO	M.S.893E Rallye 180GT	J. G. Lacey & ptnrs
	EI-BBV	Piper J-3C-65 Cub	F. Cronin
	EI-BBW	M.S.894A Rallye Minerva	J. J. Ladbrook
	EI-BCE	BN-2A-26 Islander	Aer Arann
	EI-BCF	Bensen B.8M	T. A. Brennan
	EI-BCH	M.S.892A Rallye Commodore 150	The Condor Group
	EI-BCJ	F.8L Falco 1 Srs 3	D. Kelly
	EI-BCK	Cessna F.172K	Iona National Airways
	EI-BCL	Cessna 182P	Iona National Airways
	EI-BCM	Piper J-3C-65 Cub	Kilmoon Flying Group
	EI-BCN	Piper J-3C-65 Cub	Snowflake Flying Group
	EI-BCO	Piper J-3C-65 Cub	J. Molloy
	EI-BCR	Boeing 737-281	Aer Lingus *St Oliver Plunkett*
	EI-BCS	M.S.880B Rallye Club	J. Murphy
	EI-BCT	Cessna 411A	Air Surveys International
	EI-BCU	M.S.880B Rallye Club	Weston Ltd
	EI-BCV	Cessna F.150M	Hibernian Flying Club Ltd
	EI-BCW	M.S.880B Rallye Club	H. Clarke
	EI-BCY	Beech 200 Super King Air (232)	Minister of Defence
	EI-BDH	M.S.880B Rallye Club	Munster Wings Ltd
	EI-BDK	M.S.880B Rallye Club	Limerick Flying Club Ltd
	EI-BDL	Evans VP-2	J. Duggan
	EI-BDM	PA-23 Aztec 250D	Executive Air Services
	EI-BDO	Cessna F.152	Iona National Airways
	EI-BDP	Cessna 182P	182 Flying Group
	EI-BDR	PA-28 Cherokee 180	Cork Flying Club

Reg.	Type	Owner or Operator	Notes
EI-BDY	Boeing 737-2E1	Aer Lingus Teo *St Brigid*	
EI-BEA	M.S.880B Rallye 100ST	Weston Ltd	
EI-BEB	Boeing 737-248	Aer Lingus Teo *St Eunan*	
EI-BEC	Boeing 737-248	Aer Lingus Teo *St Fiacre*	
EI-BED	Boeing 747-130	Aer Linte Eireann Teo *St Kieran*	
EI-BEE	Boeing 737-281	Aer Tara Ltd	
EI-BEI	—	Aer Lingus Teo	
EI-BEJ	—	Aer Lingus Teo	
EI-BEK	Short SD3-60	Aer Lingus Teo *St Eithne*	
EI-BEL	Short SD3-60	Aer Lingus Teo	
EI-BEM	Short SD3-60	Aer Lingus Teo (crashed 31 Jan 86)	
EI-BEN	Piper J-3C-65 Cub	Capt J. J. Sullivan	
EI-BEO	Cessna 310Q	Iona National Airways	
EI-BEP	M.S.892A Rallye Commodore	H. Lynch & J. O'Leary	
EI-BEY	Naval N3N-3	Huntley & Huntley Ltd	
EI-BFB	M.S.880B Rallye 100ST	Weston Ltd	
EI-BFC	Boeing 737-2H4	Air Tara Ltd (leased to Nigeria Airways)	
EI-BFE	Cessna F.150G	Joyce Aviation Ltd	
EI-BFF	Beech A.23 Musketeer	A. Cody	
EI-BFH	Bell 212	Irish Helicopters Ltd	
EI-BFI	M.S.880B Rallye 100ST	J. O'Neill	
EI-BFJ	Beech A.200 Super King Air (234)	Minister of Defence	
EI-BFM	M.S.893E Rallye 235GT	NDM Aviation	
EI-BFO	Piper J-3C-90 Cub	M. Slattery	
EI-BFP	M.S.800B Rallye 100ST	Weston Ltd	
EI-BFR	M.S.880B Rallye 100ST	Galway Flying Club	
EI-BFS	FRED Srs 2	G. J. McGlennon	
EI-BFV	M.S.880B Rallye 100ST	Ormond Flying Club	
EI-BGA	SOCATA Rallye 100T	T. Daly	
EI-BGB	M.S.880B Rallye Club	G. N. Atkinson	
EI-BGC	M.S.880B Rallye Club	P. Moran	
EI-BGD	M.S.880B Rallye Club	N. Kavanagh	
EI-BGG	M.S.893 Rallye 180GT	C. Weldon	
EI-BGH	Cessna F.172N	Iona National Airways	
EI-BGJ	Cessna F.152	Kerry Aero Club	
EI-BGK	Cessna P206D	B. A. Carpenter	
EI-BGO	Canadair CL-44D-4J	Aer Turas Teo *City of Dublin*	
EI-BGP	Cessna 414A	Iona National Airways	
EI-BGS	M.S.893B Rallye 180GT	M. Farrelly	
EI-BGT	Colt 77A balloon	K. Haugh	
EI-BGU	M.S.880B Rallye Club	M. F. Neary	
EI-BGV	AA-5 Traveler	J. Crowe	
EI-BHA	Beech A200 Super King Air	*Stored*	
EI-BHB	M.S.887 Rallye 125	Hotel Bravo Flying Club	
EI-BHC	Cessna F.177RG	P. J. McGuire & B. Palfrey	
EI-BHD	M.S.893E Rallye 180GT	Epic Flying Group	
EI-BHF	M.S.892A Rallye Commodore 150	B. Mullen	
EI-BHI	Bell 206B JetRanger 2	J. Mansfield	
EI-BHK	M.S.880B Rallye Club	J. Lawlor & B. Lyons	
EI-BHL	Beech E90 King Air	Stewart Singlam Fabrics Ltd	
EI-BHM	Cessna 337E	The Ross Flying Group	
EI-BHN	M.S.893A Rallye Commodore	K. O'Driscoll & ptnrs	
EI-BHO	Sikorsky S-61N	Irish Helicopters Ltd	
EI-BHP	M.S.893A Rallye Commodore	G. Atkinson	
EI-BHT	Beech 77 Skipper	M. Cullinane	
EI-BHV	Champion 7EC Traveler	Condor Group	
EI-BHW	Cessna F.150F	Shannon Executive Aviation	
EI-BHY	M.S.892E Rallye Commodore	D. Killian	
EI-BIB	Cessna F.152	Galway Flying Club	
EI-BIC	Cessna F.172N	Oriel Flying Group Ltd	
EI-BID	PA-18 Super Cub 95	D. MacCarthy	
EI-BIE	Cessna FA.152	D. F. McEllin	
EI-BIF	M.S.894 Rallye Minerva 235	Empire Enterprises Ltd	
EI-BIG	Zlin 526	P. von Lonkhuyzen	
EI-BIJ	AB-206B JetRanger 2	Irish Helicopters Ltd	
EI-BIK	PA-18-180 Super Cub	Dublin Gliding Club	
EI-BIL	Beech A35 Bonanza	W. J. Phelan	
EI-BIM	M.S.880B Rallye Club	D. Millar	
EI-BIO	Piper J-3C-65 Cub	Monasterevin Flying Club	

Notes	Reg.	Type	Owner or Operator
	EI-BIR	Cessna F.172M	P. O'Reilly
	EI-BIS	Robin R.1180TD	Robin Aiglon Group
	EI-BIT	M.S.887 Rallye 125	Tango Flying Group
	EI-BIU	Robin R.2112A	Bruton Aircraft Engineering Ltd
	EI-BIV	Bellanca 8KCAB Citabria	Aerocrats Flying Group
	EI-BIW	M.S.880B Rallye Club	E. J. Barr
	EI-BJA	Cessna FRA.150L	Blackwater Flying Group
	EI-BJC	Aeronca 7AC Champion	R. J. Bentley
	EI-BJE	Boeing 737-275	Air Tara Ltd
	EI-BJF	AA-5 Traveler	P. Mercer & ptnrs
	EI-BJG	Robin R.1180	N. Hanley
	EI-BJJ	Aeronca 15AC Sedan	A. A. Alderdice & S. H. Boyd
	EI-BJK	M.S.880B Rallye 110ST	JK Group
	EI-BJL	Cessna 550 Citation II	Helicopter Maintenance Ltd
	EI-BJM	Cessna A.152	Leinster Aero Club
	EI-BJN	Cessna 500 Citation	Tool & Mould Steel (Ireland) Ltd
	EI-BJO	Cessna R.172K	P. Hogan & G. Ryder
	EI-BJP	Boeing 737-275C	Air Tara Ltd (leased to Nigeria Airways)
	EI-BJS	AA-5B Tiger	A. Killian & C. Pearce
	EI-BJT	PA-38-112 Tomahawk	Shannon Executive Aviation
	EI-BJV	AB-206B JetRanger 3	J. Kelly
	EI-BJW	D.H.104 Dove 6	S. J. Filhol Ltd
	EI-BKC	Aeronca 115AC Sedan	G. Treacy
	EI-BKD	Mooney M.20J	Limerick Warehousing Ltd
	EI-BKE	M.S.885 Super Rallye	C. Brady & G. Groom
	EI-BKF	Cessna F.172H	M. & M. C. Veale
	EI-BKL	Cessna FR.172F	J. Sullivan
	EI-BKM	Zenith CH.200	B. McGann
	EI-BKN	M.S.880B Rallye 100ST	Weston Ltd
	EI-BKP	Zenith CH.200	L. McEnteggart
	EI-BKS	Eipper Quicksilver	Irish Microlight Ltd
	EI-BKT	AB-206B JetRanger 3	Irish Helicopters Ltd
	EI-BKU	M.S.892A Rallye Commodore	T. Maguire
	EI-BLA	PA-23 Aztec 250	National Aluminium Ltd
	EI-BLB	Stampe SV-4C	J. E. Hutchinson & R. A. Stafford
	EI-BLD	Bolkow Bo 105C	Irish Helicopters Ltd
	EI-BLE	Eipper Microlight	R. P. St George-Smith
	EI-BLF	Hiway Demon	F. Warren
	EI-BLG	AB.206B JetRanger 3	Anglo Irish Meat Co Ltd
	EI-BLJ	Cessna T.210H	R. Neeson & ptnrs
	EI-BLK	PA-28-181 Archer II	J. J. Sundival
	EI-BLL	Cessna F.172P	J. J. Spollen
	EI-BLM	Hiway Skytrike II/MSD	R. Hudson
	EI-BLN	Eipper Quicksilver MX	O. J. Conway & B. Daffy
	EI-BLO	Catto CP.16	R. W. Hall
	EI-BLR	PA-34-200T Seneca II	R. Paris
	EI-BLU	Evans VP-1	A. Bailey
	EI-BLW	PA-23 Aztec 250	Shannon Executive Aviation
	EI-BLY	Sikorsky S-61N	Irish Helicopters Ltd
	EI-BMA	M.S.880B Rallye Club	Trinity Aviation Ltd
	EI-BMB	M.S.880B Rallye 100T	Clyde Court Development Ltd
	EI-BMC	Hiway Demon Skytrike	S. Pallister
	EI-BMD	Eagle Microlight	E. Fitzgerald
	EI-BMF	F.8L Falco	M. Slazenger
	EI-BMH	M.S.880B Rallye Club	N. J. Bracken
	EI-BMI	SOCATA TB.9 Tampico	Weston Ltd
	EI-BMJ	M.S.893A Rallye Club	Galway Flying Club
	EI-BMK	Cessna 310Q	Iona National Airways Ltd
	EI-BML	PA-23 Aztec 250	Bruton Aircraft Engineering Ltd
	EI-BMM	Cessna F.152 II	Iona National Airways Ltd
	EI-BMN	Cessna F.152 II	Iona National Airways Ltd
	EI-BMO	Robin R.2160	The Robin Group
	EI-BMR	Southdown Puma	R. Hudson
	EI-BMS	Cessna F.177RG	A. M. Smyth
	EI-BMU	Monnet Sonerai II	P. Forde & D. Connaire
	EI-BMV	AA-5 Traveler	D. M. Lummes
	EI-BMW	Vulcan Air Trike	L. Maddock
	EI-BMY	Boeing 737-2L9	Air Tara Ltd
	EI-BNA	Douglas DC-8-63CF	Aer Turas
	EI-BNB	Lake LA-4-200 Buccaneer	L. McNamara & M. Ledwith
	EI-BNC	Cessna F.152	Iona National Airlines

Reg.	Type	Owner or Operator	Notes
EI-BND	Conroy CL-44-0	HeavyLift Cargo Airlines Ltd/Stansted	
EI-BNF	Goldwing Canard	T. Morelli	
EI-BNG	M.S.892A Rallye Commodore	Shannon Executive Aviation	
EI-BNH	Hiway Skytrike	M. Martin	
EI-BNI	Bell 412	Vellare Ltd	
EI-BNJ	Evans VP-2	G. A. Cashman	
EI-BNK	Cessna U.206F	Irish Parachute Club Ltd	
EI-BNL	Rand KR-2	K. Hayes	
EI-BNN	SC.7 Skyvan	Shannon Executive Aviation	
EI-BNP	Rotorway 133	R. L. Renfroe	
EI-BNR	AA-5 Traveler	Victor Mike Flying Group	
EI-BNT	Cvjetkovic CA-65	B. Tobin & P. G. Ryan	
EI-BNU	M.S.880B Rallye Club	P. A. Doyle	
EI-BNV	PA-23 Aztec 250	Epic Flying Group Ltd	
EI-BNY	SNIAS SN.601 Corvette	Flightline Ltd	
EI-BOA	Pterodactyl Ptraveller	A. Murphy	
EI-BOE	SOCATA TB.10 Tobago	E. L. Symmons	
EI-BOF	SOCATA TB.10 Tobago	J. Condron	
EI-BOH	Eipper Quicksilver	L. Leech	
EI-BOK	PA-23 Aztec 250	K. A. O'Connor	
EI-BOM	Boeing 737-2T4	Air Tara Ltd (*leased out*)	
EI-BON	Boeing 737-2T4	Air Tara Ltd	
EI-BOO	PA-23 Aztec 250	McAuliffe's Photographic Labs	
EI-BOP	M.S.892A Rallye 150	Limerick Flying Club	
EI-BOR	Bell 222	V. O'Brien	
EI-BOT	AS.350B Ecureuil	J. Kelly	
EI-BOV	Rand KR-2	G. O'Hara & G. Callan	
EI-BOW	Cessna 182M	Safety Plane Ltd	
EI-BOX	Duet	K. Riccius	
EI-BOY	Murphy Sprite	A. Murphy	
EI-BPA	PA-23 Aztec 250E	D. M. Hillary	
EI-BPB	PA-28R Cherokee Arrow 200	G. P. O'Gorman	
EI-BPC	Rockwell 690B	Flightline Ltd	
EI-BPD	Short SD3-60	Aer Lingus Teo *St Gall*	
EI-BPE	Viking Dragonfly	G. Bracken	
EI-BPH	Boeing 747-133	Air Tara Ltd (*leased in US*)	
EI-BPI	EMB-110P1 Bandeirante	Ryan Air	
EI-BPJ	Cessna 182A	J. Matthews & V. McCarthy	
EI-BPL	Cessna F.172K	Phoenix Flying	
EI-BPM	AS.350B Ecureuil	Helicopter Maintenance Ltd	
EI-BPO	Puma Skytrike	A. Morelli	
EI-BPP	Quicksilver MX	J. A. Smith	
EI-BPR	Boeing 737-2S3	Air Tara Ltd (*leased to Air Cal*)	
EI-BPS	PA-30 Twin Comanche 160	Group Air	
EI-BPT	Skyhook Sabre	T. McGrath	
EI-BPU	Hiway Demon	J. T. Haycock	
EI-BPV	Boeing 737-2T5	Air Tara Ltd (*leased to Haiti Air*)	
EI-BPW	Boeing 737-2S3	Air Tara Ltd (*leased to Midway*)	
EI-BPY	Boeing 737-2S3	Air Tara Ltd	
EI-BRB	Boeing 737-2S3	Air Tara Ltd	
EI-BRC	PA-31-350 Navajo Chieftain	Comfhorbairt	
EI-BRD	Boeing 727-247	Air Tara Ltd (*leased in US*)	
EI-BRF	Boeing 727-264	Air Tara Ltd (*leased in US*)	
EI-BRG	H.S.125 Srs 600	Anglo-Irish Meats	
EI-BRH	Mainair Gemini Flash	F. Warren & T. McGrath	
EI-BRI	Swearingen SA.226TC Metro II	Shannon Executive Aviation	
EI-BRK	Flexiform Trike	L. Maddock	
EI-BRM	Cessna 172Q	Iona National Airways	
EI-BRN	Boeing 737-2T4	Air Tara Ltd	
EI-BRO	Cessna F.152	Iona National Airways	
EI-BRP	Canadair CL-44J	Aer Turas	
EI-BRR	Boeing 747-133	Air Tara Ltd	
EI-BRS	Cessna P.172D	D. & M. Hilary	
EI-BRT	Flexwing M17727	S. Pallister	
EI-BRU	Evans VP-1	C. Quinn	
EI-BRV	Hiway Demon	M. Garvey & C. Tully	
EI-BRW	Ultralight Deltabird	A. & E. Aerosport	
EI-BRX	—	—	
EI-BRY	—	—	
EI-BSA	Learjet 55	S. M. McDaniel	
EI-BSB	Jodel D.112	S. Bruton	

EL (Liberia)

Notes	Reg.	Type	Owner or Operator
	EL-AJA	Boeing 707-321C	Liberia World Airlines
	EL-AJK	Douglas DC-8-54F	Trans Sahel Airlines

EP (Iran)

Notes	Reg.	Type	Owner or Operator
	EP-IAA	Boeing 747SP-86	Iran Air *Fars*
	EP-IAB	Boeing 747SP-86	Iran Air *Kurdistan*
	EP-IAC	Boeing 747SP-86	Iran Air *Khuzestan*
	EP-IAD	Boeing 747SP-86	Iran Air
	EP-IAG	Boeing 747-286B	Iran Air *Azarabadegan*
	EP-IAH	Boeing 747-286B	Iran Air *Khorasan*
	EP-IAM	Boeing 747-186B	Iran Air
	EP-ICB	Boeing 747-2J9F	Iran Air
	EP-ICC	Boeing 747-2J9F	Iran Air
	EP-IRK	Boeing 707-321C	Iran Air
	EP-IRL	Boeing 707-386C	Iran Air *Apadana*
	EP-IRM	Boeing 707-386C	Iran Air *Ekbatana*
	EP-IRN	Boeing 707-386C	Iran Air *Pasargad*
	EP-NHD	Boeing 747-131F	Iran Air
	EP-NHJ	Boeing 747-131	Iran Air
	EP-NHK	Boeing 747-131F	Iran Air
	EP-NHN	Boeing 747-2J9F	Iran Air
	EP-NHS	Boeing 747-131	Iran Air

ET (Ethiopia)

Notes	Reg.	Type	Owner or Operator
	ET-AAH	Boeing 720-060B	Ethiopian Airlines *White Nile*
	ET-ABP	Boeing 720-060B	Ethiopian Airlines
	ET-ACQ	Boeing 707-379C	Ethiopian Airlines
	ET-AFK	Boeing 720-024B	Ethiopian Airlines
	ET-AIE	Boeing 767-260ER	Ethiopian Airlines
	ET-AIF	Boeing 767-260ER	Ethiopian Airlines
	ET-AIV	Boeing 707-327C	Ethiopian Airlines

F (France)

Notes	Reg.	Type	Owner or Operator
	F-BEIG	Douglas DC-3	Stellair
	F-BIUK	F.27 Friendship Mk 100	Uni-Air
	F-BJEN	S.E.210 Caravelle 10B	Air Charter/E.A.S.
	F-BJTU	S.E.210 Caravelle 10B	Air Charter/E.A.S.
	F-BLHX	Nord 262A	Compagnie Aérienne du Languedoc
	F-BMKS	S.E.210 Caravelle 10B	Air Charter/E.A.S.
	F-BNKG	S.E.210 Caravelle III	Altair
	F-BNOG	S.E.210 Caravelle 12	Air Inter
	F-BNOH	S.E.210 Caravelle 12	Air Inter
	F-BOJA	Boeing 727-228	Air France
	F-BOJB	Boeing 727-228	Air France
	F-BOJC	Boeing 727-228	Air France
	F-BOJD	Boeing 727-228	Air France
	F-BOJE	Boeing 727-228	Air France

Reg.	Type	Owner or Operator	Notes
F-BOJF	Boeing 727-228	Air France	
F-BOLF	Douglas DC-8-62	Union de Transports Aériens (UTA)	
F-BPJG	Boeing 727-228	Air France	
F-BPJH	Boeing 727-228	Air France	
F-BPJI	Boeing 727-228	Air France	
F-BPJJ	Boeing 727-228	Air France	
F-BPJK	Boeing 727-228	Air France	
F-BPJL	Boeing 727-228	Air France	
F-BPJM	Boeing 727-228	Air France	
F-BPJN	Boeing 727-228	Air France	
F-BPJO	Boeing 727-228	Air France	
F-BPJP	Boeing 727-228	Air France	
F-BPJQ	Boeing 727-228	Air France	
F-BPJR	Boeing 727-228	Air France	
F-BPJS	Boeing 727-228	Air France	
F-BPJT	Boeing 727-228	Air France	
F-BPJU	Boeing 727-214	Air Charter	
F-BPJV	Boeing 727-214	Air Charter	
F-BPNA	F.27 Friendship Mk 500	Air Inter	
F-BPNB	F.27 Friendship Mk 500	Air Inter	
F-BPNC	F.27 Friendship Mk 500	Air Inter	
F-BPND	F.27 Friendship Mk 500	Air Inter	
F-BPNE	F.27 Friendship Mk 500	Air Inter	
F-BPNG	F.27 Friendship Mk 500	Brit Air	
F-BPNH	F.27 Friendship Mk 500	Air Inter	
F-BPNI	F.27 Friendship Mk 500	Brit Air	
F-BPNJ	F.27 Friendship Mk 500	Air Inter	
F-BPPA	Aero Spacelines Guppy-201	Airbus Industrie *Airbus Skylink 2*	
F-BPUA	F.27 Friendship Mk 500	Air France	
F-BPUB	F.27 Friendship Mk 500	Air France	
F-BPUC	F.27 Friendship Mk 500	Air France	
F-BPUD	F.27 Friendship Mk 500	Air France	
F-BPUE	F.27 Friendship Mk 500	Air France	
F-BPUF	F.27 Friendship Mk 500	Air France	
F-BPUG	F.27 Friendship Mk 500	Air France	
F-BPUH	F.27 Friendship Mk 500	Air France	
F-BPUI	F.27 Friendship Mk 500	Air France	
F-BPUJ	F.27 Friendship Mk 500	Air France	
F-BPUK	F.27 Friendship Mk 500	Air France	
F-BPUL	F.27 Friendship Mk 500	Air France	
F-BPVA	Boeing 747-128	Air France	
F-BPVB	Boeing 747-128	Air France	
F-BPVC	Boeing 747-128	Air France	
F-BPVD	Boeing 747-128	Air France	
F-BPVE	Boeing 747-128	Air France	
F-BPVF	Boeing 747-128	Air France	
F-BPVG	Boeing 747-128	Air France	
F-BPVH	Boeing 747-128	Air France	
F-BPVJ	Boeing 747-128	Air France	
F-BPVK	Boeing 747-128	Air France	
F-BPVL	Boeing 747-128	Air France	
F-BPVM	Boeing 747-128	Air France	
F-BPVN	Boeing 747-128	Air France	
F-BPVO	Boeing 747-228F	Air France	
F-BPVP	Boeing 747-128	Air France	
F-BPVQ	Boeing 747-128	Air France	
F-BPVR	Boeing 747-228F	Air France	
F-BPVS	Boeing 747-228B	Air France	
F-BPVT	Boeing 747-228B	Air France	
F-BPVU	Boeing 747-228B	Air France	
F-BPVV	Boeing 747-228F	Air France	
F-BPVX	Boeing 747-228B	Air France	
F-BPVY	Boeing 747-228B	Air France	
F-BPVZ	Boeing 747-228F	Air France	
F-BRGU	S.E.210 Caravelle VI-N	Minerve	
F-BRNI	Beech 70 Queen Air	Lucas Air Transport	
F-BRUN	Beech 99	T.A.T.	
F-BSGT	Boeing 707-321B	Pointair *Bernard Audebourg*	
F-BSUM	F.27 Friendship Mk 500	Air France	
F-BSUN	F.27 Friendship Mk 500	Air France	
F-BSUO	F.27 Friendship Mk 500	Air France	
F-BTDB	Douglas DC-10-30	Union de Transports Aériens (UTA)	
F-BTDC	Douglas DC-10-30	Union de Transports Aériens (UTA)	

Notes	Reg.	Type	Owner or Operator
	F-BTDD	Douglas DC-10-30	Union de Transports Aériens (UTA)
	F-BTDE	Douglas DC-10-30	Union de Transports Aériens (UTA)
	F-BTDG	Boeing 747-2B3B	Union de Transports Aériens (UTA)
	F-BTDH	Boeing 747-2B3B	Union de Transports Aériens (UTA)
	F-BTGV	Aero Spacelines Guppy-201	Airbus Industrie *Airbus Skylink 1*
	F-BTMA	Beech 99	T.A.T.
	F-BTMD	Mercure 100	Air-Inter
	F-BTMJ	Beech 99	T.A.T.
	F-BTMK	Beech 99	T.A.T.
	F-BTOA	S.E.210 Caravelle 12	Air Inter
	F-BTOB	S.E.210 Caravelle 12	Air Inter
	F-BTOC	S.E.210 Caravelle 12	Air Inter
	F-BTOD	S.E.210 Caravelle 12	Air Inter
	F-BTOE	S.E.210 Caravelle 12	Air Inter
	F-BTSD	Concorde 101	Air France
	F-BTTA	Mercure 100	Air Inter
	F-BTTB	Mercure 100	Air Inter
	F-BTTC	Mercure 100	Air Inter
	F-BTTD	Mercure 100	Air Inter
	F-BTTE	Mercure 100	Air Inter
	F-BTTF	Mercure 100	Air Inter
	F-BTTG	Mercure 100	Air Inter
	F-BTTH	Mercure 100	Air Inter
	F-BTTI	Mercure 100	Air Inter
	F-BTTJ	Mercure 100	Air Inter
	F-BUAE	A.300B2 Airbus	Air Inter
	F-BUAF	A.300B2 Airbus	Air Inter
	F-BUAG	A.300B2 Airbus	Air Inter
	F-BUAH	A.300B2 Airbus	Air Inter
	F-BUAI	A.300B2 Airbus	Air Inter
	F-BUAJ	A.300B2 Airbus	Air Inter
	F-BUAK	A.300B2 Airbus	Air Inter
	F-BUAL	A.300B4 Airbus	Air Inter
	F-BUAM	A.300B2 Airbus	Air Inter
	F-BUAN	A.300B2 Airbus	Air Inter
	F-BUFP	Transall C-160P	Air France
	F-BUFQ	Transall C-160P	Air France
	F-BUFR	Transall C-160P	Air France
	F-BUFS	Transall C-160P	Air France
	F-BUOR	Douglas DC-8-55F	SFAIR
	F-BUTI	F-28 Fellowship 1000	T.A.T./Air France
	F-BUZC	S.E.210 Caravelle VI-R	Minerve
	F-BVFA	Concorde 101	Air France
	F-BVFB	Concorde 101	Air France
	F-BVFC	Concorde 101	Air France
	F-BVFF	Concorde 101	Air France
	F-BVFG	Nord 262A	Air Limousin *Lac de Vassivière*
	F-BVFH	Nord 262A	Air Limousin
	F-BVFI	Nord 262A	Air Limousin
	F-BVFJ	Nord 262A	Air Limousin *Lac de Briance*
	F-BVFP	HPR-7 Herald 214	Trans Azur Aviation
	F-BVGA	A.300B2 Airbus	Air France
	F-BVGB	A.300B2 Airbus	Air France
	F-BVGC	A.300B2 Airbus	Air France
	F-BVGD	A.300B2 Airbus	Air Inter
	F-BVGE	A.300B2 Airbus	Air Inter
	F-BVGF	A.300B2 Airbus	Air Inter
	F-BVGG	A.300B4 Airbus	Air France
	F-BVGH	A.300B4 Airbus	Air France
	F-BVGI	A.300B4 Airbus	Air France
	F-BVGJ	A.300B4 Airbus	Air France
	F-BVGL	A.300B4 Airbus	Air France
	F-BVGM	A.300B4 Airbus	Air France
	F-BVGN	A.300B4 Airbus	Air France
	F-BVGO	A.300B4 Airbus	Air France
	F-BVGP	A.300B4 Airbus	Air France
	F-BVGQ	A.300B4 Airbus	Air France
	F-BVGR	A.300B4 Airbus	Air France
	F-BVGS	A.300B4 Airbus	Air France
	F-BVGT	A.300B4 Airbus	Air France
	F-BVJL	Beech 99	T.A.T.
	F-BVPZ	S.E.210 Caravelle VI-N	Corse Air *Golfe du Valinco*
	F-BVSF	S.E.210 Caravelle VI-N	Corse Air *Golfe de Porto Vecchio*

Reg.	Type	Owner or Operator	Notes
F-BYAB	F.27 Friendship Mk 400	Air Jet	
F-BYAG	D.H.C.-6 Twin Otter 310	Air Littoral	
F-BYAP	F.27 Friendship Mk 100	Uni-Air	
F-BYCD	S.E.210 Caravelle VI-N	Corse Air *Ajaccio*	
F-BYCJ	BN-2A-IIIA Trislander	Europe Aero Fret	
F-BYCU	Douglas DC-3	Stellair	
F-BYFM	Douglas DC-8-53	Minerve	
F-GAOT	F.27 Friendship Mk 100	Uni-Air	
F-GATS	EMB-110P2 Bandeirante	Air Littoral	
F-GATZ	S.E.210 Caravelle VI-N	Minerve	
F-GBBR	F.28 Fellowship 1000	T.A.T./Air France	
F-GBBS	F.28 Fellowship 1000	T.A.T./Air France	
F-GBBT	F.28 Fellowship 1000	T.A.T./Air France	
F-GBBX	F.28 Fellowship 1000	T.A.T./Air France	
F-GBEA	A.300B2 Airbus	Air Inter	
F-GBEB	A.300B2 Airbus	Air France	
F-GBEC	A.300B2 Airbus	Air France	
F-GBEI	Nord 262A	Compagnie Aérienne du Languedoc	
F-GBEJ	Nord 262A	Compagnie Aérienne du Langeudoc	
F-GBEK	Nord 262A	Air Littoral	
F-GBGA	EMB-110P2 Bandeirante	Brit Air	
F-GBLE	EMB-110P2 Bandeirante	Brit Air	
F-GBME	EMB-110P2 Bandeirante	Air Littoral	
F-GBMF	EMB-110P2 Bandeirante	Cie Aérienne du Languedoc/Air Littoral	
F-GBMG	EMB-110P2 Bandeirante	Brit Air	
F-GBMQ	EMB-110P2 Bandeirante	Air Littoral	
F-GBOX	Boeing 747-2B3F	Union de Transports Aériens (UTA)	
F-GBRM	EMB-110P2 Bandeirante	Brit Air	
F-GBRQ	FH.227B Friendship	T.A.T.	
F-GBTO	SA.226TC Metro II	Cie Aèrienne du Languedoc	
F-GBYA	Boeing 737-228	Air France	
F-GBYB	Boeing 737-228	Air France	
F-GBYC	Boeing 737-228	Air France	
F-GBYD	Boeing 737-228	Air France	
F-GBYE	Boeing 737-228	Air France	
F-GBYF	Boeing 737-228	Air France	
F-GBYG	Boeing 737-228	Air France	
F-GBYH	Boeing 737-228	Air France	
F-GBYI	Boeing 737-228	Air France	
F-GBYJ	Boeing 737-228	Air France	
F-GBYK	Boeing 737-228	Air France	
F-GBYL	Boeing 737-228	Air France	
F-GBYM	Boeing 737-228	Air France	
F-GCBA	Boeing 747-228B	Air France	
F-GCDA	Boeing 727-228	Air France	
F-GCDB	Boeing 727-228	Air France	
F-GCDC	Boeing 727-228	Air France	
F-GCDD	Boeing 727-228	Air France	
F-GCDE	Boeing 727-228	Air France	
F-GCDF	Boeing 727-228	Air France	
F-GCDG	Boeing 727-228	Air France	
F-GCDH	Boeing 727-228	Air France	
F-GCDI	Boeing 727-228	Air France	
F-GCFC	FH.227B Friendship	T.A.T.	
F-GCFE	SA.226TC Metro II	Cie Aérienne du Langudoc	
F-GCGH	FH.227B Friendship	T.A.T.	
F-GCGQ	Boeing 727-227	Europe Aero Service *Normandie*	
F-GCJL	Boeing 737-222	Euralair	
F-GCJO	FH.227B Friendship	T.A.T.	
F-GCJT	S.E.210 Caravelle 10B	Europe Aero Service *Bretagne*	
F-GCJV	F.27 Friendship Mk 400	Air Jet	
F-GCLA	EMB-110P2 Bandeirante	Lucas Air Transport	
F-GCLL	Boeing 737-222	Euralair	
F-GCLM	FH.227B Friendship	T.A.T.	
F-GCLN	FH.227B Friendship	T.A.T.	
F-GCLO	FH.227B Friendship	T.A.T.	
F-GCLQ	FH.227B Friendship	T.A.T.	
F-GCMQ	EMB-110P2 Bandeirante	Air Littoral	
F-GCMV	Boeing 727-2X3	Air Charter	
F-GCMX	Boeing 727-2X3	Air Charter	
F-GCPG	SA.226TC Metro II	Cie Aérienne du Languedoc	
F-GCPS	FH.227B Friendship	T.A.T.	
F-GCPT	FH.227B Friendship	T.A.T.	

Notes	Reg.	Type	Owner or Operator
	F-GCPU	FH.227B Friendship	T.A.T.
	F-GCPV	FH.227B Friendship	T.A.T.
	F-GCPX	FH.227B Friendship	T.A.T.
	F-GCPY	FH.227B Friendship	T.A.T.
	F-GCPZ	FH.227B Friendship	T.A.T.
	F-GCSL	Boeing 737-222	Euralair
	F-GCTE	SA.226TC Metro II	Cie Aérienne du Languedoc
	F-GCVI	S.E.210 Caravelle 12	Air Inter
	F-GCVJ	S.E.210 Caravelle 12	Air Inter
	F-GCVK	S.E.210 Caravelle 12	Air Inter
	F-GCVL	S.E.210 Caravelle 12	Air Inter
	F-GCVM	S.E.210 Caravelle 12	Air Inter
	F-GDAQ	L.100 Hercules	SFAIR
	F-GDAR	L.100 Hercules	SFAIR
	F-GDCI	EMB-110P2 Bandeirante	Air Littoral
	F-GDFC	F.28 Fellowship 4000	T.A.T./Air France
	F-GDFD	F.28 Fellowship 4000	T.A.T./Air France
	F-GDFY	S.E.210 Caravelle 10B	Air Charter
	F-GDFZ	S.E.210 Caravelle 10B	Air Charter
	F-GDJK	Douglas DC-10-30	Lineas Aereas de Mocambique (LAM)
	F-GDJM	Douglas DC-8-62CF	Minerve
	F-GDJU	S.E.210 Caravelle 10B	Europe Aero Service *Lorraine*
	F-GDMR	SA.226TC Metro II	Cie Aérienne du Languedoc
	F-GDPM	Douglas DC-8-53	Minerve
	F-GDPP	Douglas DC-3	Transvalair
	F-GDPS	Douglas DC-8-61	Pointair *Les Trois Voltas*
	F-GDSG	UTA Super Guppy	Airbus Industrie *Airbus Skylink 3*
	F-GDSK	F.28 Fellowship 4000	T.A.T.
	F-GDUY	F.28 Fellowship 4000	T.A.T./Air France
	F-GDUZ	F.28 Fellowship 4000	T.A.T./Air France
	F-GEAI	UTA Super Guppy	Airbus Industrie *Airbus Skylink 4*
	F-GECK	F.28 Fellowship 1000	T.A.T./Air France
	F-GEGC	Aeritalia ATR.42	Air Littoral
	F-GEGD	Aeritalia ATR.42	Air Littoral
	F-GEMA	A.310-203 Airbus	Air France
	F-GEMB	A.310-203 Airbus	Air France
	F-GEMC	A.310-203 Airbus	Air France
	F-GEMD	A.310-203 Airbus	Air France
	F-GEME	A.310-203 Airbus	Air France
	F-GEOM	Douglas DC-3	Stellair
	F-GEPC	S.E.210 Caravelle 10B	Corse Air *Golfe de Porto Marina*
	F-GETA	Boeing 747-3B3	Union de Transports Aériens (UTA)
	F-GFAE	DHC-6 Twin Otter 310	T.A.T.
	F-GFAF	DHC-6 Twin Otter 310	T.A.T.
	F-GFAG	DHC-6 Twin Otter 310	T.A.T.
	F-GFAR	L-100-30 Hercules	SFAIR
	F-G	Douglas DC-8-73	Minerve
	F-GMOL	F.28 Fellowship 1000	T.A.T.
	F-GPAN	Boeing 747-2B3F	UTA/Saudia

Note: Air France also operates five more Boeing 747s which retain the US registrations N1289E, N1305E, N4506H, N4508E and N4544F. UTA also operates two DC-10-30s registered N54629 and N54649, while Minerva uses a DC-8-73 which retains N4805J.

HA (Hungary)

MALĒV

	HA-LBE	Tupolev Tu-134	Malev
	HA-LBF	Tupolev Tu-134	Malev
	HA-LBG	Tupolev Tu-134	Malev
	HA-LBH	Tupolev Tu-134	Malev
	HA-LBI	Tupolev Tu-134A	Malev
	HA-LBK	Tupolev Tu-134A	Malev
	HA-LBN	Tupolev Tu-134A	Malev
	HA-LBO	Tupolev Tu-134A	Malev
	HA-LBP	Tupolev Tu-134A	Malev
	HA-LBR	Tupolev Tu-134A	Malev

Reg.	Type	Owner or Operator	Notes
HA-LCA	Tupolev Tu-154B	Malev	
HA-LCB	Tupolev Tu-154B	Malev	
HA-LCE	Tupolev Tu-154B	Malev	
HA-LCG	Tupolev Tu-154B	Malev	
HA-LCH	Tupolev Tu-154B	Malev	
HA-LCM	Tupolev Tu-154B	Malev	
HA-LCN	Tupolev Tu-154B	Malev	
HA-LCO	Tupolev Tu-154B	Malev	
HA-LCP	Tupolev Tu-154B	Malev	
HA-LCR	Tupolev Tu-154B	Malev	

HB (Switzerland)

swissair

BALAIR

HB-AHA	Saab-Fairchild SF.340A	Crossair	
HB-AHB	Saab-Fairchild SF.340A	Crossair	
HB-AHC	Saab-Fairchild SF.340A	Crossair	
HB-AHD	Saab-Fairchild SF.340A	Crossair	
HB-AHE	Saab-Fairchild SF.340A	Crossair	
HB-AHF	Saab-Fairchild SF.340A	Crossair	
HB-AHG	Saab-Fairchild SF.340A	Crossair	
HB-AHH	Saab-Fairchild SF.340A	Crossair	
HB-AHI	Saab-Fairchild SF.340A	Crossair	
HB-AHK	Saab-Fairchild SF.340A	Crossair	
HB-ICI	S.E.210 Caravelle 10-R	CTA	
HB-ICN	S.E.210 Caravelle 10-R	CTA *Ville de Genève*	
HB-ICO	S.E.210 Caravelle 10-R	CTA *Romandie*	
HB-ICQ	S.E.210 Caravelle 10-R	CTA	
HB-IDO	Douglas DC-9-32	Swissair *Genève-Cointrin*	
HB-IDP	Douglas DC-9-32	Swissair *Basel-Land*	
HB-IFH	Douglas DC-9-32	Swissair *Baden*	
HB-IFU	Douglas DC-9-32	Swissair *Chur*	
HB-IFV	Douglas DC-9-32	Swissair *Bülach*	
HB-IFZ	Douglas DC-9-33F	Balair	
HB-IGC	Boeing 747-357	Swissair *Bern*	
HB-IGD	Boeing 747-357	Swissair *Basel*	
HB-IHC	Douglas DC-10-30	Swissair *Luzern*	
HB-IHD	Douglas DC-10-30	Swissair *Thurgau*	
HB-IHE	Douglas DC-10-30	Swissair *Vaud*	
HB-IHF	Douglas DC-10-30	Swissair *Nidwalden*	
HB-IHG	Douglas DC-10-30	Swissair *Graubünden*	
HB-IHH	Douglas DC-10-30	Swissair *Schaffhausen*	
HB-IHI	Douglas DC-10-30	Swissair *Fribourg*	
HB-IHK	Douglas DC-10-30	Balair	
HB-IHL	Douglas DC-10-30ER	Swissair *Ticino*	
HB-IHM	Douglas DC-10-30ER	Swissair *Valais-Wallis*	
HB-IHN	Douglas DC-10-30ER	Swissair *St Gallen*	
HB-IHO	Douglas DC-10-30ER	Swissair *Uri*	
HB-IKK	Douglas DC-9-82	Alisarda (Italy)	
HB-IKL	Douglas DC-9-82	Alisarda (Italy)	
HB-INA	Douglas DC-9-81	Swissair *Obwalden*	
HB-INB	Douglas DC-9-82	Balair	
HB-INC	Douglas DC-9-81	Swissair *Lugano*	
HB-IND	Douglas DC-9-81	Swissair *Zug*	
HB-INE	Douglas DC-9-81	Swissair *Rümlang*	
HB-INF	Douglas DC-9-81	Swissair *Appenzell a.Rh.*	
HB-ING	Douglas DC-9-81	Swissair *Glarus*	
HB-INH	Douglas DC-9-81	Swissair *Winterthur*	
HB-INI	Douglas DC-9-81	Swissair *Kloten*	
HB-INK	Douglas DC-9-81	Swissair *Opfikon*	
HB-INL	Douglas DC-9-81	Swissair *Jura*	
HB-INM	Douglas DC-9-81	Swissair *Lausanne*	
HB-INN	Douglas DC-9-81	Swissair *Appenzell i.Rh.*	
HB-INO	Douglas DC-9-81	Swissair *Bellinzona*	
HB-INP	Douglas DC-9-81	Swissair *Oberglatt*	
HB-INR	Douglas DC-9-82	Balair	
HB-INS	Douglas DC-9-81	Swissair	

Notes	Reg.	Type	Owner or Operator
	HB-INT	Douglas DC-9-81	Swissair
	HB-INU	Douglas DC-9-81	Swissair
	HB-INV	Douglas DC-9-81	Swissair
	HB-IPA	A.310-221 Airbus	Swissair *Aargau*
	HB-IPB	A.310-221 Airbus	Swissair *Neuchatel*
	HB-IPC	A.310-221 Airbus	Swissair *Schwyz*
	HB-IPD	A.310-221 Airbus	Swissair *Solothurn*
	HB-IPE	A.310-221 Airbus	Swissair *Basel-Land*
	HB-IPF	A.310-322 Airbus	Swissair *Glarus*
	HB-IPG	A.310-322 Airbus	Swissair *Zug*
	HB-IPH	A.310-322 Airbus	Swissair *Appenzell*
	HB-IPI	A.310-322 Airbus	Swissair *Luzern*
	HB-IPK	A.310-322 Airbus	Balair
	HB-ISK	Douglas DC-9-51	Swissair *Höri*
	HB-ISL	Douglas DC-9-51	Swissair *Köniz*
	HB-ISM	Douglas DC-9-51	Swissair *Wettingen*
	HB-ISN	Douglas DC-9-51	Swissair *Sion*
	HB-ISO	Douglas DC-9-51	Swissair *Bienne*
	HB-ISV	Douglas DC-9-51	Swissair *Winkel*
	HB-ISW	Douglas DC-9-51	Swissair *Dubendorf*
	HB-IVA	Fokker 100	Swissair
	HB-IVB	Fokker 100	Swissair
	HB-IVC	Fokker 100	Swissair
	HB-IVD	Fokker 100	Swissair
	HB-IVE	Fokker 100	Swissair
	HB-IVF	Fokker 100	Swissair
	HB-IVG	Fokker 100	Swissair
	HB-IVH	Fokker 100	Swissair
	HB-LLF	SA.227AC Metro III	Crossair
	HB-LNA	SA.227AC Metro III	Crossair
	HB-LNB	SA.227AC Metro III	Crossair
	HB-LNC	SA.227AC Metro III	Crossair
	HB-LNE	SA.227AC Metro III	Crossair
	HB-LNO	SA.227AC Metro III	Crossair

Note: Swissair also operates three Boeing 747-357s which retain their US registrations N221GE, N221GF and N221GG.

HK (Colombia)

	HK-2900X	Boeing 747-124F	Avianca
	HK-2980X	Boeing 747-259B	Avianca *Cartagena de Indias*

Note: Avianca also operates Boeing 747-123 registered N9664.

HL (Korea)

	HL7406	Boeing 707-3B5C	Korean Air
	HL7427	Boeing 707-321C	Korean Air
	HL7431	Boeing 707-321C	Korean Air
	HL7432	Boeing 707-338C	Korean Air
	HL7433	Boeing 707-338C	Korean Air
	HL7435	Boeing 707-321B	Korean Air
	HL7440	Boeing 747-230B	Korean Air
	HL7441	Boeing 747-230F	Korean Air
	HL7443	Boeing 747-2B5B	Korean Air
	HL7447	Boeing 747-230B	Korean Air
	HL7451	Boeing 747-2B5F	Korean Air
	HL7452	Boeing 747-2B5F	Korean Air
	HL7454	Boeing 747-2B5B	Korean Air
	HL7458	Boeing 747-2B5B	Korean Air
	HL7459	Boeing 747-2B5F	Korean Air
	HL7463	Boeing 747-2B5B	Korean Air
	HL7464	Boeing 747-2B5B	Korean Air
	HL7468	Boeing 747-3B5	Korean Air
	HL7469	Boeing 747-3B5	Korean Air
	HL7471	Boeing 747-273C	Korean Air/Saudia

HS (Thailand)

Reg.	Type	Owner or Operator	Notes
HS-TGA	Boeing 747-2D7B	Thai Airways International *Visuthakasatriya*	
HS-TGB	Boeing 747-2D7B	Thai Airways International *Sirisobhakya*	
HS-TGC	Boeing 747-2D7B	Thai Airways International *Dararasmi*	
HS-TGF	Boeing 747-2D7B	Thai Airways International *Phimara*	
HS-TGG	Boeing 747-2D7B	Thai Airways International *Sriwanna*	
HS-TGS	Boeing 747-2D7B	Thai Airways International *Chainarai*	

HZ (Saudi Arabia)

HZ-AHA	L.1011-385 TriStar 200	Saudia — Saudi Arabian Airlines	
HZ-AHB	L.1011-385 TriStar 200	Saudia — Saudi Arabian Airlines	
HZ-AHC	L.1011-385 TriStar 200	Saudia — Saudi Arabian Airlines	
HZ-AHD	L.1011-385 TriStar 200	Saudia — Saudi Arabian Airlines	
HZ-AHE	L.1011-385 TriStar 200	Saudia — Saudi Arabian Airlines	
HZ-AHF	L.1011-385 TriStar 200	Saudia — Saudi Arabian Airlines	
HZ-AHG	L.1011-385 TriStar 200	Saudia — Saudi Arabian Airlines	
HZ-AHH	L.1011-385 TriStar 200	Saudia — Saudi Arabian Airlines	
HZ-AHI	L.1011-385 TriStar 200	Saudia — Saudi Arabian Airlines	
HZ-AHJ	L.1011-385 TriStar 200	Saudia — Saudi Arabian Airlines	
HZ-AHL	L.1011-385 TriStar 200	Saudia — Saudi Arabian Airlines	
HZ-AHM	L.1011-385 TriStar 200	Saudia — Saudi Arabian Airlines	
HZ-AHN	L.1011-385 TriStar 200	Saudia — Saudi Arabian Airlines	
HZ-AHO	L.1011-385 TriStar 200	Saudia — Saudi Arabian Airlines	
HZ-AHP	L.1011-385 TriStar 200	Saudia — Saudi Arabian Airlines	
HZ-AHQ	L.1011-385 TriStar 200	Saudia — Saudi Arabian Airlines	
HZ-AHR	L.1011-385 TriStar 200	Saudia — Saudi Arabian Airlines	
HZ-AIA	Boeing 747-168B	Saudia — Saudi Arabian Airlines	
HZ-AIB	Boeing 747-168B	Saudia — Saudi Arabian Airlines	
HZ-AIC	Boeing 747-168B	Saudia — Saudi Arabian Airlines	
HZ-AID	Boeing 747-168B	Saudia — Saudi Arabian Airlines	
HZ-AIE	Boeing 747-168B	Saudia — Saudi Arabian Airlines	
HZ-AIF	Boeing 747SP-68	Saudia — Saudi Arabian Airlines	
HZ-AIG	Boeing 747-168B	Saudia — Saudi Arabian Airlines	
HZ-AIH	Boeing 747-168B	Saudia — Saudi Arabian Airlines	
HZ-AII	Boeing 747-168B	Saudia — Saudi Arabian Airlines	
HZ-AIJ	Boeing 747SP-68	Saudia — Saudi Arabian Airlines	
HZ-AIK	Boeing 747-368	Saudia — Saudi Arabian Airlines	
HZ-AIL	Boeing 747-368	Saudia — Saudi Arabian Airlines	
HZ-AIM	Boeing 747-368	Saudia — Saudi Arabian Airlines	
HZ-AIN	Boeing 747-368	Saudia — Saudi Arabian Airlines	
HZ-AIO	Boeing 747-368	Saudia — Saudi Arabian Airlines	
HZ-AIP	Boeing 747-368	Saudia — Saudi Arabian Airlines	
HZ-AIQ	Boeing 747-368	Saudia — Saudi Arabian Airlines	
HZ-AIR	Boeing 747-368	Saudia — Saudi Arabian Airlines	
HZ-AIS	Boeing 747-368	Saudia — Saudi Arabian Airlines	
HZ-AIT	Boeing 747-368	Saudia — Saudi Arabian Airlines	
HZ-AJA	A.300-620 Airbus	Saudia — Saudi Arabian Airlines	
HZ-AJB	A.300-620 Airbus	Saudia — Saudi Arabian Airlines	
HZ-AJC	A.300-620 Airbus	Saudia — Saudi Arabian Airlines	
HZ-AJD	A.300-620 Airbus	Saudia — Saudi Arabian Airlines	
HZ-AJE	A.300-620 Airbus	Saudia — Saudi Arabian Airlines	
HZ-AJF	A.300-620 Airbus	Saudia — Saudi Arabian Airlines	
HZ-AJG	A.300-620 Airbus	Saudia — Saudi Arabian Airlines	
HZ-AJH	A.300-620 Airbus	Saudia — Saudi Arabian Airlines	
HZ-AJI	A.300-620 Airbus	Saudia — Saudi Arabian Airlines	
HZ-AJJ	A.300-620 Airbus	Saudia — Saudi Arabian Airlines	
HZ-AJK	A.300-620 Airbus	Saudia — Saudi Arabian Airlines	

Note: Saudia also operates DC-8s leased from National Airlines and Boeing 747s HL7471 and N743TV leased from Korean Air and Transamerica respectively.

Notes	Reg.	Type	Owner or Operator
	I-ATIE	Douglas DC-9-32	Aero Trasporti Italiani (ATI) *Toscana*
	I-ATIH	Douglas DC-9-32	Aero Trasporti Italiani (ATI)
			Lido degli Estensi
	I-ATIJ	Douglas DC-9-32	Aero Trasporti Italiani (ATI)
	I-ATIK	Douglas DC-9-32	Aero Trasporti Italiani (ATI) *Sardegna*
	I-ATIO	Douglas DC-9-32	Aero Trasporti Italiani (ATI)
			Emilia Romagna
	I-ATIQ	Douglas DC-9-32	Aero Trasporti Italiani (ATI) *Sila*
	I-ATIU	Douglas DC-9-32	Aero Trasporti Italiani (ATI)
	I-ATIW	Douglas DC-9-32	Aero Trasporti Italiani (ATI) *Lazio*
	I-ATIY	Douglas DC-9-32	Aero Trasporti Italiani (ATI) *Lombardia*
	I-ATJA	Douglas DC-9-32	Aero Trasporti Italiani (ATI) *Sicilia*
	I-ATJB	Douglas DC-9-32	Aero Trasporti Italiani (ATI)
			Riviera de Conero
	I-BUSB	A.300B4 Airbus	Alitalia *Tiziano*
	I-BUSC	A.300B4 Airbus	Alitalia *Botticelli*
	I-BUSD	A.300B4 Airbus	Alitalia *Caravaggio*
	I-BUSF	A.300B4 Airbus	Alitalia *Tintoretto*
	I-BUSG	A.300B4 Airbus	Alitalia *Canaletto*
	I-BUSH	A.300B4 Airbus	Alitalia *Mantegua*
	I-BUSJ	A.300B4 Airbus	Alitalia *Tiepolo*
	I-BUSL	A.300B4 Airbus	Alitalia *Pinturicchia*
	I-DAVA	Douglas DC-9-82	Aero Trasporti Italiani (ATI) *Cuneo*
	I-DAVB	Douglas DC-9-82	Aero Trasporti Italiani (ATI) *Ferrara*
	I-DAVC	Douglas DC-9-82	Aero Trasporti Italiani (ATI) *Lucca*
	I-DAVD	Douglas DC-9-82	Aero Trasporti Italiani (ATI) *Mantova*
	I-DAVF	Douglas DC-9-82	Aero Trasporti Italiani (ATI) *Oristano*
	I-DAVG	Douglas DC-9-82	Aero Trasporti Italiani (ATI) *Pesaro*
	I-DAVH	Douglas DC-9-82	Aero Trasporti Italiani (ATI) *Salerno*
	I-DAWA	Douglas DC-9-82	Alitalia *Roma*
	I-DAWB	Douglas DC-9-82	Alitalia *Cagliari*
	I-DAWC	Douglas DC-9-82	Alitalia *Campobasso*
	I-DAWD	Douglas DC-9-82	Alitalia *Catanzaro*
	I-DAWE	Douglas DC-9-82	Alitalia *Milano*
	I-DAWF	Douglas DC-9-82	Alitalia *Firenze*
	I-DAWG	Douglas DC-9-82	Alitalia *L'Aquila*
	I-DAWH	Douglas DC-9-82	Alitalia
	I-DAWI	Douglas DC-9-82	Alitalia *Ancona*
	I-DAWJ	Douglas DC-9-82	Alitalia *Napoli*
	I-DAWL	Douglas DC-9-82	Alitalia *Perugia*
	I-DAWM	Douglas DC-9-82	Alitalia *Potenza*
	I-DAWO	Douglas DC-9-82	Alitalia *Bari*
	I-DAWP	Douglas DC-9-82	Alitalia *Torino*
	I-DAWQ	Douglas DC-9-82	Alitalia *Trieste*
	I-DAWR	Douglas DC-9-82	Alitalia *Venezia*
	I-DAWS	Douglas DC-9-82	Alitalia *Aosta*
	I-DAWT	Douglas DC-9-82	Aero Trasporti Italiani (ATI) *Napoli*
	I-DAWU	Douglas DC-9-82	Alitalia *Bologna*
	I-DAWV	Douglas DC-9-82	Aero Trasporti Italiani (ATI) *Trento*
	I-DAWW	Douglas DC-9-82	Aero Trasporti Italiani (ATI) *Riace*
	I-DAWY	Douglas DC-9-82	Aero Trasporti Italiani (ATI) *Agrigento*
	I-DAWZ	Douglas DC-9-82	Aero Trasporti Italiani (ATI) *Avellino*
	I-DEMC	Boeing 747-243B	Alitalia *Taormina*
	I-DEMD	Boeing 747-243B	Alitalia *Cortina d'Ampezzo*
	I-DEMF	Boeing 747-243B	Alitalia *Portofino*
	I-DEMG	Boeing 747-243B	Alitalia *Cervinia*
	I-DEML	Boeing 747-243B	Alitalia *Sorrento*
	I-DEMN	Boeing 747-243B	Alitalia *Portocervo*
	I-DEMP	Boeing 747-243B	Alitalia *Capri*
	I-DEMR	Boeing 747-243B	Alitalia *Stresa*
	I-DEMS	Boeing 747-243B	Alitalia *Monte Argentario*
	I-DEMT	Boeing 747-243B	Alitalia *Monte Catini*
	I-DEMV	Boeing 747-243B	Alitalia *Sestriere*

Reg.	Type	Owner or Operator	Notes
I-DIBC	Douglas DC-9-32	Alitalia *Isola di Lampedusa*	
I-DIBD	Douglas DC-9-32	Alitalia *Isola di Montecristo*	
I-DIBJ	Douglas DC-9-32	Alitalia *Isola della Capraia*	
I-DIBN	Douglas DC-9-32	Alitalia *Isola della Palmaria*	
I-DIBO	Douglas DC-9-32	Aero Trasporti Italiani (ATI) *Conca d'Ora*	
I-DIBQ	Douglas DC-9-32	Alitalia *Isola di Pianosa*	
I-DIKM	Douglas DC-9-32	Aero Trasporti Italiani (ATI) *Positano*	
I-DIKP	Douglas DC-9-32	Alitalia	
I-DIKR	Douglas DC-9-32	Aero Trasporti Italiani (ATI) *Piemonte*	
I-DIKS	Douglas DC-9-32	Aero Trasporti Italiani (ATI) *Isola di Filicudi*	
I-DIKT	Douglas DC-9-32	Aero Trasporti Italiani (ATI) *Isola d'Ustica*	
I-DIKV	Douglas DC-9-32	Alitalia *Isola di Vulcano*	
I-DIKY	Douglas DC-9-32	Aero Trasporti Italiani (ATI) *Puglia*	
I-DIKZ	Douglas DC-9-32	Alitalia *Isola di Linosa*	
I-DIZA	Douglas DC-9-32	Alitalia *Isola di Palmarola*	
I-DIZB	Douglas DC-9-32	Aero Trasporti Italiani (ATI) *Umbria*	
I-DIZC	Douglas DC-9-32	Aero Trasporti Italiani (ATI)	
I-DIZE	Douglas DC-9-32	Aero Trasporti Italiani (ATI)	
I-DIZF	Douglas DC-9-32	Aero Trasporti Italiani (ATI)	
I-DIZI	Douglas DC-9-32	Aero Trasporti Italiani (ATI)	
I-DIZO	Douglas DC-9-32	Aero Trasporti Italiani (ATI) *Liguria*	
I-DIZU	Douglas DC-9-32	Aero Trasporti Italiani (ATI)	
I-GISE	S.E.210 Caravelle III	Altair *Citta di Civago*	
I-GISO	S.E.210 Caravelle 10B	Altair *Carol*	
I-GISU	S.E.210 Caravelle 10B	Altair	
I-SMEA	Douglas DC-9-51	Alisarda	
I-SMEI	Douglas DC-9-51	Alisarda	
I-SMEU	Douglas DC-9-51	Alisarda	
I-TIDB	F.28 Fellowship 1000	Unifly	
I-TIDI	F-28 Fellowship 1000	Unifly	

Note: Alisarda also uses DC-9s which retain their Swiss registrations HB-IKK and HB-IKL. Alitalia is in the process of changing its DC-9 fleet. Ten of its DC-9-32s have been reregistered N516MD, N901DC, N902DC, N903DC, N904DC, N905DC, N906DC, N2786S, N2786T and N43265. Similarly ATI uses DC-9-32 N515MD. All the US registered aircraft will gradually be replaced by Srs 82.

J2 (Djibouti)

Note: Air Dijbouti leases a Boeing 737 from Sobelair or Sabena.

JA (Japan)

JA8036	Douglas DC-8-62AF	Japan Air Lines	
JA8044	Douglas DC-8-62AF	Japan Air Lines	
JA8055	Douglas DC-8-62AF	Japan Air Lines	
JA8101	Boeing 747-146	Japan Air Lines	
JA8104	Boeing 747-246B	Japan Air Lines	
JA8105	Boeing 747-246B	Japan Air Lines	
JA8106	Boeing 747-246B	Japan Air Lines	
JA8107	Boeing 747-146A	Japan Air Lines	
JA8108	Boeing 747-246B	Japan Air Lines	
JA8110	Boeing 747-246B	Japan Air Lines	
JA8111	Boeing 747-246B	Japan Air Lines	
JA8112	Boeing 747-146A	Japan Air Lines	
JA8113	Boeing 747-246B	Japan Air Lines	
JA8114	Boeing 747-246B	Japan Air Lines	
JA8115	Boeing 747-146A	Japan Air Lines	
JA8116	Boeing 747-146A	Japan Air Lines	
JA8122	Boeing 747-246B	Japan Air Lines	
JA8123	Boeing 747-246F	Japan Air Lines	
JA8125	Boeing 747-246B	Japan Air Lines	

Notes	Reg.	Type	Owner or Operator
	JA8127	Boeing 747-246B	Japan Air Lines
	JA8128	Boeing 747-146A	Japan Air Lines
	JA8129	Boeing 747-246B	Japan Air Lines
	JA8130	Boeing 747-246B	Japan Air Lines
	JA8131	Boeing 747-246B	Japan Air Lines
	JA8132	Boeing 747-246F	Japan Air Lines
	JA8140	Boeing 747-246B	Japan Air Lines
	JA8141	Boeing 747-246B	Japan Air Lines
	JA8142	Boeing 747-146B	Japan Air Lines
	JA8143	Boeing 747-146B	Japan Air Lines
	JA8144	Boeing 747-246F	Japan Air Lines
	JA8149	Boeing 747-246B	Japan Air Lines
	JA8150	Boeing 747-246B	Japan Air Lines
	JA8151	Boeing 747-246F	Japan Air Lines
	JA8154	Boeing 747-246B	Japan Air Lines
	JA8155	Boeing 747-246B	Japan Air Lines
	JA8160	Boeing 747-221F	Japan Air Lines
	JA8161	Boeing 747-246B	Japan Air Lines
	JA8162	Boeing 747-246B	Japan Air Lines
	JA8163	Boeing 747-346	Japan Air Lines
	JA8164	Boeing 747-146B	Japan Air Lines
	JA8165	Boeing 747-221F	Japan Air Lines
	JA8166	Boeing 747-346	Japan Air Lines
	JA8169	Boeing 747-246B	Japan Air Lines
	JA8170	Boeing 747-146B	Japan Air Lines
	JA8171	Boeing 747-246F	Japan Air Lines
	JA8234	Boeing 747-346	Japan Air Lines
	JA8235	Boeing 747-346	Japan Air Lines
	JA8236	Boeing 747-346	Japan Air Lines
	JA8534	Douglas DC-10-40	Japan Air Lines
	JA8535	Douglas DC-10-40	Japan Air Lines
	JA8538	Douglas DC-10-40	Japan Air Lines
	JA8539	Douglas DC-10-40	Japan Air Lines
	JA8541	Douglas DC-10-40	Japan Air Lines
	JA8542	Douglas DC-10-40	Japan Air Lines
	JA8543	Douglas DC-10-40	Japan Air Lines
	JA8544	Douglas DC-10-40	Japan Air Lines
	JA8545	Douglas DC-10-40	Japan Air Lines
	JA8547	Douglas DC-10-40	Japan Air Lines

Note: Japan Air Lines also operates a Boeing 747-221F which retains its US registration N211JL and two 747-346s N212JL and N213JL.

JY (Jordan)

	JY-ADP	Boeing 707-3D3C	Alia — The Royal Jordanian Airline *City of Amman*
	JY-AEB	Boeing 707-384C	Alia/Sierra Leone Airlines
	JY-AEC	Boeing 707-384C	Alia — The Royal Jordanian Airline *Um Qais*
	JY-AES	Boeing 707-321C	Alia — The Royal Jordanian Airline
	JY-AFA	Boeing 747-2D3B	Alia — The Royal Jordanian Airline *Prince Ali*
	JY-AFS	Boeing 747-2D3B	Alia — The Royal Jordanian Airline *Prince Hamzah*
	JY-AGA	L.1011-385 TriStar 500	Alia — The Royal Jordanian Airline *Abas Bin Firnas*
	JY-AGB	L.1011-385 TriStar 500	Alia — The Royal Jordanian Airline *Ibn Batouta*
	JY-AGC	L.1011-385 TriStar 500	Alia — The Royal Jordanian Airline *Al Jawaheri*
	JY-AGD	L.1011-385 TriStar 500	Alia — The Royal Jordanian Airline *Ibn Sina*
	JY-AGE	L.1011-385 TriStar 500	Alia — The Royal Jordanian Airline *Al Biruni*
	JY-AGH	L.1011-385 TriStar 500	Alia — The Royal Jordanian Airline
	JY-AGI	L.1011-385 TriStar 500	Alia — The Royal Jordanian Airline
	JY-AGJ	L.1011-385 TriStar 500	Alia — The Royal Jordanian Airline

Note: Arab Air Cargo now use the prefix 4YB- instead of JY.

LN (Norway)

FRED OLSEN AIRTRANSPORT

Reg.	*Type*	*Owner or Operator*	*Notes*
LN-AEO	Boeing 747-283B	S.A.S. *Ivar Viking*	
LN-AET	Boeing 747-283B	Gulf Air	
LN-AKA	F.27 Friendship Mk 200	Busy Bee	
LN-AKB	F.27 Friendship Mk 200	Busy Bee	
LN-AKC	F.27 Friendship Mk 200	Busy Bee	
LN-AKD	F.27 Friendship Mk 200	Busy Bee	
LN-BSC	F.27J Friendship	Swedair	
LN-BSD	F.27F Friendship	Swedair	
LN-BWG	Convair 580	Partnair	
LN-BWN	Convair 580	Partnair	
LN-FOG	L-188AF Electra	Fred Olsen Airtransport	
LN-FOH	L-188AF Electra	Fred Olsen Airtransport	
LN-FOI	L-188CF Electra	Fred Olsen Airtransport	
LN-KLK	Convair 440	Norsk Metropolitan Fly Klubb	
LN-MOF	Douglas DC-8-63	S.A.S. *Bue Viking*	
LN-MOW	Douglas DC-8-62	S.A.S. *Roald Viking*	
LN-NPB	Boeing 737-2R4C	Busy Bee	
LN-NPC	F-27 Friendship Mk 100	Busy Bee (*for sale*)	
LN-NPD	F-27 Friendship Mk.100	Busy Bee (*for sale*)	
LN-NPH	F.27 Friendship Mk 300	Busy Bee	
LN-NPI	F.27 Friendship Mk 100	Busy Bee (*for sale*)	
LN-NPM	F.27 Friendship Mk 100	Busy Bee (*for sale*)	
LN-RCA	A.300B4 Airbus	Scanair *Snorre Viking*	
LN-RKA	Douglas DC-10-30	S.A.S. *Olav Viking*	
LN-RKC	Douglas DC-10-30	S.A.S. *Leif Viking*	
LN-RLA	Douglas DC-9-41	S.A.S. *Are Viking*	
LN-RLB	Douglas DC-9-41	S.A.S. *Arne Viking*	
LN-RLC	Douglas DC-9-41	S.A.S. *Gunnar Viking*	
LN-RLD	Douglas DC-9-41	S.A.S. *Torleif Viking*	
LN-RLE	Douglas DC-9-81	S.A.S. *Trygve Viking*	
LN-RLF	Douglas DC-9-82	S.A.S. *Finn Viking*	
LN-RLH	Douglas DC-9-41	S.A.S. *Einar Viking*	
LN-RLJ	Douglas DC-9-41	S.A.S. *Stein Viking*	
LN-RLK	Douglas DC-9-41	S.A.S. *Erling Viking*	
LN-RLL	Douglas DC-9-21	S.A.S. *Guttorm Viking*	
LN-RLN	Douglas DC-9-41	S.A.S. *Halldor Viking*	
LN-RLO	Douglas DC-9-21	S.A.S. *Gunder Viking*	
LN-RLP	Douglas DC-9-41	S.A.S. *Froste Viking*	
LN-RLS	Douglas DC-9-41	S.A.S. *Asmund Viking*	
LN-RLT	Douglas DC-9-41	S.A.S. *Audun Viking*	
LN-RLU	Douglas DC-9-41	S.A.S. *Eivind Viking*	
LN-RLW	Douglas DC-9-33AF	S.A.S. *Rand Viking*	
LN-RLX	Douglas DC-9-41	S.A.S. *Sote Viking*	
LN-RLZ	Douglas DC-9-41	S.A.S. *Bodvar Viking*	
LN-RNY	F.27 Friendship Mk 600	S.A.S.	
LN-RNZ	F.27 Friendship Mk 600	S.A.S. *Vemund Viking*	
LN-SUA	Boeing 737-205	Braathens SAFE	
LN-SUB	Boeing 737-205	Braathens SAFE *Magnus Den Gode*	
LN-SUC	F.28 Fellowship 1000	Braathens SAFE *Olav Kyrre*	
LN-SUD	Boeing 737-205	Braathens SAFE *Olav Tryggvason*	
LN-SUE	F.27 Friendship Mk 100	Busy Bee	
LN-SUF	F.27 Friendship Mk 100	Busy Bee	
LN-SUG	Boeing 737-205	Braathens SAFE *Harald Gille*	
LN-SUH	Boeing 737-205	Braathens SAFE *Sigurd Jorsalfar*	
LN-SUI	Boeing 737-205	Braathens SAFE *Haakon den Gode*	
LN-SUJ	Boeing 737-205	Braathens SAFE	
LN-SUK	Boeing 737-205	Braathens SAFE *Magnus Erlingsson*	
LN-SUL	F.27 Friendship Mk 100	Busy Bee	

Note: S.A.S. also operates two Boeing 747-283Bs which retain their US registrations N4501Q and N4502R.

Notes	Reg.	Type	Owner or Operator
	LN-SUM	Boeing 737-205	Braathens SAFE *Magnus Lagaboter*
	LN-SUN	F.28 Fellowship 1000	Braathens SAFE *Haakon Sverresson*
	LN-SUO	F.28 Fellowship 1000	Braathens SAFE *Magnus Barfot*
	LN-SUP	Boeing 737-205	Braathens SAFE *Haakon IV Hakonsson*
	LN-SUQ	Boeing 737-205	Braathens SAFE
	LN-SUS	Boeing 737-205	Braathens SAFE *Haakon V Magnusson*
	LN-SUT	Boeing 737-205	Braathens SAFE *Oystein Magnusson*
	LN-SUU	Boeing 737-205	Braathens SAFE
	LN-SUV	Boeing 737-205	Braathens SAFE
	LN-SUW	Boeing 767-205	Braathens *Olav den Hellige*
	LN-SUX	F.28 Fellowship 1000	Braathens SAFE *Harald Hardrade*
	LN-SUZ	Boeing 737-205	Braathens SAFE

LV (Argentina)

	LV-MLO	Boeing 747-287B	Flying Tiger Line
	LV-MLP	Boeing 747-287B	Aerolineas Argentinas
	LV-MLR	Boeing 747-287B	Aerolineas Argentinas
	LV-OEP	Boeing 747-287B	Aerolineas Argentinas
	LV-OHV	Boeing 747SP-27	Aerolineas Argentinas
	LV-OOZ	Boeing 747-287B	Aerolineas Argentinas
	LV-OPA	Boeing 747-287B	Aerolineas Argentinas

Note: Services to the UK are suspended.

LX (Luxembourg)

	LX-DCV	Boeing 747-2R7F	Cargolux *City of Luxembourg*
	LX-GCV	Douglas DC-8-54C	Cargolux
	LX-LGA	F.27 Friendship Mk 100	Luxair *Prince Henri*
	LX-LGB	F.27 Friendship Mk 100	Luxair *Prince Jean*
	LX-LGD	F.27 Friendship Mk 600	Luxair *Princess Margaretha*
	LX-LGH	Boeing 737-2C9	Luxair *Prince Guillaume*
	LX-LGI	Boeing 737-2C9	Luxair *Princess Marie-Astrid*
	LX-LGJ	F.27 Friendship Mk 200	Luxair
	LX-LGK	F.27 Friendship Mk 200	Luxair
	LX-LGP	A.300B4 Airbus	Luxair
	LX-TAP	Boeing 747-254F	Cargolux

LZ (Bulgaria)

	LZ-BEA	Ilyushin IL-18D	Balkan Bulgarian Airlines
	LZ-BEK	Ilyushin IL-18V	Balkan Bulgarian Airlines
	LZ-BEL	Ilyushin IL-18V	Balkan Bulgarian Airlines
	LZ-BEO	Ilyushin IL-18D	Balkan Bulgarian Airlines
	LZ-BEP	Ilyushin IL-18V	Balkan Bulgarian Airlines
	LZ-BET	Ilyushin IL-18D	Balkan Bulgarian Airlines
	LZ-BEU	Ilyushin IL-18V	Balkan Bulgarian Airlines
	LZ-BEV	Ilyushin IL-18V	Balkan Bulgarian Airlines
	LZ-BTA	Tupolev Tu-154B	Balkan Bulgarian Airlines
	LZ-BTC	Tupolev Tu-154B	Balkan Bulgarian Airlines
	LZ-BTD	Tupolev Tu-154B	Balkan Bulgarian Airlines
	LZ-BTE	Tupolev Tu-154B	Balkan Bulgarian Airlines
	LZ-BTF	Tupolev Tu-154B	Balkan Bulgarian Airlines
	LZ-BTG	Tupolev Tu-154B	Balkan Bulgarian Airlines
	LZ-BTI	Tupolev Tu-154M	Balkan Bulgarian Airlines
	LZ-BTJ	Tupolev Tu-154B	Balkan Bulgarian Airlines
	LZ-BTK	Tupolev Tu-154B	Balkan Bulgarian Airlines
	LZ-BTL	Tupolev Tu-154B	Balkan Bulgarian Airlines
	LZ-BTM	Tupolev Tu-154B	Balkan Bulgarian Airlines
	LZ-BTO	Tupolev Tu-154B	Balkan Bulgarian Airlines

Reg.	Type	Owner or Operator	Notes
LZ-BTP	Tupolev Tu-154B	Balkan Bulgarian Airlines	
LZ-BTR	Tupolev Tu-154B	Balkan Bulgarian Airlines	
LZ-BTS	Tupolev Tu-154B	Balkan Bulgarian Airlines	
LZ-BTT	Tupolev Tu-154B	Balkan Bulgarian Airlines	
LZ-BTU	Tupolev Tu-154B	Balkan Bulgarian Airlines	
LZ-BTV	Tupolev Tu-154B	Balkan Bulgarian Airlines	
LZ-BTW	Tupolev Tu-154B	Balkan Bulgarian Airlines	
LZ-TUA	Tupolev Tu-134	Balkan Bulgarian Airlines	
LZ-TUC	Tupolev Tu-134	Balkan Bulgarian Airlines	
LZ-TUD	Tupolev Tu-134	Balkan Bulgarian Airlines	
LZ-TUE	Tupolev Tu-134	Balkan Bulgarian Airlines	
LZ-TUF	Tupolev Tu-134	Balkan Bulgarian Airlines	
LZ-TUG	Tupolev Tu-134A	Balkan Bulgarian Airlines	
LZ-TUK	Tupolev Tu-134A	Balkan Bulgarian Airlines	
LZ-TUL	Tupolev Tu-134A	Balkan Bulgarian Airlines	
LZ-TUM	Tupolev Tu-134A	Balkan Bulgarian Airlines	
LZ-TUN	Tupolev Tu-134A	Balkan Bulgarian Airlines	
LZ-TUO	Tupolev Tu-134A	Balkan Bulgarian Airlines	
LZ-TUP	Tupolev Tu-134A	Balkan Bulgarian Airlines	
LZ-TUS	Tupolev Tu-134A	Balkan Bulgarian Airlines	
LZ-TUT	Tupulev Tu-134A	Balkan Bulgarian Airlines	
LZ-TUU	Tupolev Tu-134A	Balkan Bulgarian Airlines	
LZ-TUV	Tupolev Tu-134A	Balkan Bulgarian Airlines	
LZ-TUZ	Tupolev Tu-134A	Balkan Bulgarian Airlines	

N (USA)

N10ST	L-100-30 Hercules	Transamerica Airlines	
N11ST	L-100-30 Hercules	Transamerica Airlines	
N12ST	L-100-30 Hercules	Transamerica Airlines	
N15ST	L-100-30 Hercules	Transamerica Airlines	
N16ST	L-100-30 Hercules	Transamerica Airlines	
N18ST	L-100-30 Hercules	Transamerica Airlines	
N19ST	L-100-30 Hercules	Transamerica Airlines	
N20ST	L-100-30 Hercules	Transamerica Airlines	
N21ST	L-100-30 Hercules	Transamerica Airlines	
N23ST	L-100-30 Hercules	Transamerica Airlines	
N37ST	L-100-30 Hercules	Transamerica Airlines	
N38ST	L-100-30 Hercules	Transamerica Airlines	
N39ST	L-100-30 Hercules	Transamerica Airlines	
N63AF	Boeing 737-222	Pan Am *Clipper Schoneberg*	
N64AF	Boeing 737-222	Pan Am *Clipper Spandau*	
N64RD	Douglas DC-8-61	Airlift International	
N67AF	Boeing 737-222	Pan Am *Clipper Templehof*	
N68AF	Boeing 737-222	Pan Am *Clipper Zehlendorf*	
N69AF	Boeing 737-222	Pan Am *Clipper Charlottenburg*	
N92TA	L-1011-385 TriStar 100	Gulf Air	
N92TB	L-1011-385 TriStar 100	Gulf Air	
N103WA	Douglas DC-10-30CF	World Airways	
N104WA	Douglas DC-10-30CF	World Airways	
N105WA	Douglas DC-10-30CF	World Airways	
N106WA	Douglas DC-10-30CF	World Airways	
N107WA	Douglas DC-10-30CF	World Airways	
N108AK	L-100-30 Hercules	Markair	
N108WA	Douglas DC-10-30CF	World Airways	
N109WA	Douglas DC-10-30CF	World Airways	
N112WA	Douglas DC-10-30CF	World Airways	

Notes	Reg.	Type	Owner or Operator
	N116KB	Boeing 747-312B	Singapore Airlines
	N117KC	Boeing 747-312B	Singapore Airlines
	N118KD	Boeing 747-312B	Singapore Airlines
	N119KE	Boeing 747-312B	Singapore Airlines
	N120KF	Boeing 747-312B	Singapore Airlines
	N121AE	Canadair CL-44D-4	Aeron International *City of Stamford*
	N121KG	Boeing 747-312B	Singapore Airlines
	N122AE	Canadair CL-44D-4	Aeron International *Dixie*
	N122KH	Boeing 747-312B	Singapore Airlines
	N123KJ	Boeing 747-312B	Singapore Airlines
	N124KK	Boeing 747-312B	Singapore Airlines
	N125KL	Boeing 747-312B	Singapore Airlines
	N133TW	Boeing 747-156	Trans World Airlines
	N134TW	Boeing 747-156	Trans World Airlines
	N136AA	Douglas DC-10-30	American Airlines
	N137AA	Douglas DC-10-30	American Airlines
	N138AA	Douglas DC-10-30	American Airlines
	N139AA	Douglas DC-10-30	American Airlines
	N140AA	Douglas DC-10-30	American Airlines
	N141AA	Douglas DC-10-30	American Airlines
	N142AA	Douglas DC-10-30	American Airlines
	N143AA	Douglas DC-10-30	American Airlines
	N144AA	Douglas DC-10-30	American Airlines
	N163AA	Douglas DC-10-30	American Airlines
	N184AT	Douglas DC-10-40	American Trans Air
	N185AT	L-1011 TriStar 1	American Trans Air
	N186AT	L.1011 TriStar 1	American Trans Air
	N187AT	L.1011 TriStar 1	American Trans Air
	N188AT	L.1011 TriStar 1	American Trans Air
	N189AT	L.1011 TriStar 1	American Trans Air
	N190AT	L.1011 TriStar 1	American Trans Air
	N191AT	L.1011 TriStar 1	American Trans Air
	N192AT	L.1011 TriStar 1	American Trans Air
	N193AT	L.1011 TriStar 1	American Trans Air
	N202PA	A.300B4 Airbus	Pan Am *Clipper America*
	N203PA	A.300B4 Airbus	Pan Am *Clipper New York*
	N204PA	A.300B4 Airbus	Pan Am *Clipper Washington*
	N205PA	A.300B4 Airbus	Pan Am *Clipper Miami*
	N206PA	A.300B4 Airbus	Pan Am *Clipper Tampa*
	N207PA	A.300B4 Airbus	Pan Am *Clipper Los Angeles*
	N208PA	A.300B4 Airbus	Pan Am *Clipper San Francisco*
	N209PA	A.300B4 Airbus	Pan Am *Clipper Boston*
	N210PA	A.300B4 Airbus	Pan Am *Clipper Dallas*
	N211JL	Boeing 747-221F	Japan Air Lines
	N211PA	A.300B4 Airbus	Pan Am *Clipper Houston*
	N212JL	Boeing 747-346	Japan Air Lines
	N212PA	A.300B4 Airbus	Pan Am *Clipper Detroit*
	N213JL	Boeing 747-346	Japan Air Lines
	N213PA	A.300B4 Airbus	Pan Am *Clipper Chicago*
	N221GE	Boeing 747-357	Swissair *Geneve*
	N221GF	Boeing 747-357	Swissair *Zurich*
	N250SF	L-100-30 Hercules	Southern Air Transport
	N251SF	L-100-30 Hercules	Southern Air Transport
	N301FE	Douglas DC-10-30	Federal Express
	N301TW	Boeing 747-282B	Trans World Airlines
	N302FE	Douglas DC-10-30	Federal Express
	N302TW	Boeing 747-282B	Trans World Airlines
	N303FE	Douglas DC-10-30	Federal Express
	N303TW	Boeing 747-257B	Trans World Airlines
	N304FE	Douglas DC-10-30	Federal Express
	N304TW	Boeing 747-257B	Trans World Airlines
	N305FE	Douglas DC-10-30	Federal Express
	N305TW	Boeing 747-284B	Trans World Airlines
	N306FE	Douglas DC-10-30	Federal Express
	N307FE	Douglas DC-10-30	Federal Express
	N308FE	Douglas DC-10-30	Federal Express
	N309FE	Douglas DC-10-30	Federal Express
	N310FE	Douglas DC-10-30	Federal Express
	N311FE	Douglas DC-10-30	Federal Express
	N319AA	Boeing 767-223ER	American Airlines
	N320AA	Boeing 767-223ER	American Airlines
	N321AA	Boeing 767-223ER	American Airlines
	N322AA	Boeing 767-223ER	American Airlines

Reg.	Type	Owner or Operator	Notes
N323AA	Boeing 767-223ER	American Airlines	
N324AA	Boeing 767-223ER	American Airlines	
N325AA	Boeing 767-223ER	American Airlines	
N345HC	Douglas DC-10-30ER	Finnair	
N380PA	Boeing 737-275	Pan Am *Clipper Neukölln*	
N381PA	Boeing 737-275	Pan Am *Clipper Wedding*	
N382PA	Boeing 737-214	Pan Am *Clipper Kreuzberg*	
N383PA	Boeing 737-2A9C	Pan Am *Clipper Steglitz*	
N385PA	Boeing 737-2Q9	Pan Am *Clipper Berlin*	
N387PA	Boeing 737-296	Pan Am *Clipper Tiergarten*	
N388PA	Boeing 737-296	Pan Am *Clipper Reinickendorf*	
N389PA	Boeing 737-296	Pan Am *Clipper Frankfurt*	
N391EA	Douglas DC-10-30	Eastern Air Lines	
N392EA	Douglas DC-10-30	Eastern Air Lines	
N393EA	Douglas DC-10-30	Eastern Air Lines	
N498GA	Boeing 707-321B	Global International Airways	
N515MD	Douglas DC-9-32	Alitalia	
N516MD	Douglas DC-9-32	Alitalia	
N529PA	Boeing 747SP-27	Pan Am *Clipper America*	
N530PA	Boeing 747SP-21	Pan Am *Clipper Mayflower*	
N531PA	Boeing 747SP-21	Pan Am *Clipper Freedom*	
N532PA	Boeing 747SP-21	Pan Am *Clipper Constitution*	
N533PA	Boeing 747SP-21	Pan Am *Clipper New Horizons*	
N534PA	Boeing 747SP-21	Pan Am *Clipper Great Republic*	
N536PA	Boeing 747SP-21	Pan Am *Clipper Lindbergh*	
N537PA	Boeing 747SP-21	Pan Am *Clipper Washington*	
N538PA	Boeing 747SP-21	Pan Am *Clipper Plymouth Rock*	
N539PA	Boeing 747SP-21	Pan Am *Clipper Liberty Bell*	
N540PA	Boeing 747SP-21	Pan Am *China Clipper*	
N601BN	Boeing 747-127	Tower Air	
N601TW	Boeing 767-231	Trans World Airlines	
N601US	Boeing 747-151	Northwest Orient	
N602FF	Boeing 747-	Tower Air	
N602TW	Boeing 767-231	Trans World Airlines	
N602US	Boeing 747-151	Northwest Orient	
N603PE	Boeing 747-143	People Express	
N603TW	Boeing 767-231	Trans World Airlines	
N603US	Boeing 747-151	Northwest Orient	
N604PE	Boeing 747-243B	People Express	
N604TW	Boeing 767-231	Trans World Airlines	
N604US	Boeing 747-151	Northwest Orient	
N605PE	Boeing 747-243B	People Express	
N605TW	Boeing 767-231	Trans World Airlines	
N605US	Boeing 747-151	Northwest Orient	
N606PE	Boeing 747-143	People Express	
N606TW	Boeing 767-231	Trans World Airlines	
N606US	Boeing 747-151	Northwest Orient	
N607PE	Boeing 747-238B	People Express	
N607TW	Boeing 767-231	Trans World Airlines	
N607US	Boeing 747-151	Northwest Orient	
N608PE	Boeing 747-238B	People Express	
N608TW	Boeing 767-231	Trans World Airlines	
N608US	Boeing 747-151	Northwest Orient	
N609PE	Boeing 747-	People Express	
N609TW	Boeing 767-231	Trans World Airlines	
N609US	Boeing 747-151	Northwest Orient	
N610TW	Boeing 767-231	Trans World Airlines	
N610US	Boeing 747-151	Northwest Orient	
N611US	Boeing 747-251B	Northwest Orient	
N612US	Boeing 747-251B	Northwest Orient	
N613US	Boeing 747-251B	Northwest Orient	
N614US	Boeing 747-251B	Northwest Orient	
N615US	Boeing 747-251B	Northwest Orient	
N616US	Boeing 747-251F	Northwest Orient	
N617US	Boeing 747-251F	Northwest Orient	
N618US	Boeing 747-251F	Northwest Orient	
N619US	Boeing 747-251F	Northwest Orient	
N620US	Boeing 747-135	Northwest Orient	
N621US	Boeing 747-135	Northwest Orient	
N622US	Boeing 747-251B	Northwest Orient	
N623US	Boeing 747-251B	Northwest Orient	
N624US	Boeing 747-251B	Northwest Orient	
N625US	Boeing 747-251B	Northwest Orient	

Notes	Reg.	Type	Owner or Operator
	N626US	Boeing 747-251B	Northwest Orient
	N627US	Boeing 747-251B	Northwest Orient
	N628US	Boeing 747-251B	Northwest Orient
	N629US	Boeing 747-251F	Northwest Orient
	N630US	Boeing 747-2J9F	Northwest Orient
	N631US	Boeing 747-251B	Northwest Orient
	N632US	Boeing 747-251B	Northwest Orient
	N633US	Boeing 747-227B	Northwest Orient
	N634US	Boeing 747-227B	Northwest Orient
	N635US	Boeing 747-227B	Northwest Orient
	N636US	Boeing 747-251B	Northwest Orient
	N637US	Boeing 747-251B	Northwest Orient
	N638US	Boeing 747-251B	Northwest Orient
	N652PA	Boeing 747-121	Pan Am *Clipper Mermaid*
	N653PA	Boeing 747-121	Pan Am *Clipper Pride of the Ocean*
	N655PA	Boeing 747-121	Pan Am *Clipper Sea Serpent*
	N656PA	Boeing 747-121	Pan Am *Clipper Empress of the Seas*
	N657PA	Boeing 747-121	Pan Am *Clipper Seven Seas*
	N659PA	Boeing 747-121	Pan Am *Clipper Romance of the Seas*
	N670PA	Boeing 747-123	Pan Am *Clipper Empress of the Skies*
	N674PA	Boeing 747-123	Pan Am *Clipper Beacon Light*
	N707ZS	Boeing 707-309C	Jet Cargo *Miritza*
	N724DA	L.1011 TriStar 200	Delta Air Lines
	N724PA	Boeing 747-212B	Pan Am
	N725DA	L.1011 TriStar 100	Delta Air Lines
	N725PA	Boeing 747-132	Pan Am *Clipper Mandarin*
	N726PA	Boeing 747-212B	Pan Am *Clipper Cathay*
	N727PA	Boeing 747-212B	Pan Am *Clipper Belle of the Skies*
	N728PA	Boeing 747-212B	Pan Am *Clipper Water Witch*
	N729PA	Boeing 747-212B	Pan Am *Clipper Wild Wave*
	N730PA	Boeing 747-212B	Pan Am *Clipper Gem of the Ocean*
	N731PA	Boeing 747-121	Pan Am *Clipper Ocean Express*
	N732PA	Boeing 747-121	Pan Am *Clipper Ocean Telegraph*
	N733PA	Boeing 747-121	Pan Am *Clipper Pride of the Seas*
	N734PA	Boeing 747-121	Pan Am *Clipper Champion of the Seas*
	N735PA	Boeing 747-121	Pan Am *Clipper Spark of the Ocean*
	N737PA	Boeing 747-121	Pan Am *Clipper Ocean Herald*
	N739PA	Boeing 747-121	Pan Am *Clipper Maid of the Seas*
	N740PA	Boeing 747-121	Pan Am *Clipper Ocean Pearl*
	N741PA	Boeing 747-121	Pan Am *Clipper Sparkling Wave*
	N741PR	Boeing 747-2F6B	Philippine Airlines
	N741TV	Boeing 747-271C	Transamerica Airlines
	N742PA	Boeing 747-121	Pan Am *Clipper Neptune's Car*
	N742PR	Boeing 747-2F6B	Philippine Airlines
	N742TV	Boeing 747-271C	Transamerica Airlines
	N743PA	Boeing 747-121	Pan Am *Clipper Black Sea*
	N743PR	Boeing 747-2F6B	Philippine Airlines
	N743TV	Boeing 747-271C	Transamerica Airlines/Saudia
	N744PA	Boeing 747-121	Pan Am *Clipper Ocean Spray*
	N744PR	Boeing 747-2F6B	Philippine Airlines
	N747PA	Boeing 747-121	Pan Am *Clipper Juan J. Trippe*
	N747WA	Boeing 747-273C	World Airways
	N747WR	Boeing 747-273C	Flying Tiger Line
	N748PA	Boeing 747-121	Pan Am *Clipper Crest of the Wave*
	N748WA	Boeing 747-273C	National Airlines (*leased Korean Air*)
	N749PA	Boeing 747-121	Pan Am *Clipper Dashing Wave*
	N749WA	Boeing 747-273C	World Airways/Flying Tiger
	N750PA	Boeing 747-121	Pan Am *Clipper Neptune's Favorite*
	N751DA	L-1011-385 TriStar 500	Delta Air Lines
	N751PA	Boeing 747-121	Pan Am *Clipper Gem of the Seas*
	N752DA	L-1011-385 TriStar 500	Delta Air Lines
	N753DA	L-1011-385 TriStar 500	Delta Air Lines
	N753PA	Boeing 747-121	Pan Am *Clipper Queen of the Seas*
	N754DL	L.1011 TriStar 500	Delta Air Lines
	N754PA	Boeing 747-121	Pan Am *Clipper Ocean Rover*
	N755DL	L.1011 TriStar 500	Delta Air Lines
	N755PA	Boeing 747-121	Pan Am *Clipper Sovereign of the Seas*
	N756DR	L.1011 TriStar 500	Delta Air Lines
	N762BE	L.1011 TriStar 1	Hawaiian Air
	N763BE	L.1011 TriStar 1	Hawaiian Air
	N764BE	L.1011 TriStar 1	Hawaiian Air
	N765BE	L.1011 TriStar 1	Hawaiian Air
	N770PA	Boeing 747-121	Pan Am *Clipper Queen of the Pacific*

Reg.	Type	Owner or Operator	Notes
N791FT	Douglas DC-8-73CF	Emery Worldwide	
N792FT	Douglas DC-8-73CF	Emery Worldwide	
N795FT	Douglas DC-8-73CF	Emery Worldwide	
N796FT	Douglas DC-8-73CF	Emery Worldwide	
N801PA	A.310-221 Airbus	Pan Am *Clipper Berlin*	
N801WA	Douglas DC-8-73CF	Emery Worldwide	
N802PA	A.310-221 Airbus	Pan Am *Clipper Frankfurt*	
N803FT	Boeing 747-132F	Flying Tiger Line	
N803PA	A.310-221 Airbus	Pan Am *Clipper Munich*	
N804FT	Boeing 747-132F	Flying Tiger Line	
N804PA	A.310-221 Airbus	Pan Am *Clipper Hamburg*	
N805FT	Boeing 747-132F	Flying Tiger Line	
N805PA	A.310-222 Airbus	Pan Am	
N806FT	Boeing 747-249F	Flying Tiger Line *Robert W. Prescott*	
N806PA	A.310-222 Airbus	Pan Am	
N807FT	Boeing 747-249F	Flying Tiger Line *Thomas Haywood*	
N807PA	A.310-222 Airbus	Pan Am	
N808FT	Boeing 747-249F	Flying Tiger Line *William E. Bartlett*	
N810FT	Boeing 747-249F	Flying Tiger Line *Clifford G. Groh*	
N811FT	Boeing 747-245F	Flying Tiger Line	
N812FT	Boeing 747-245F	Flying Tiger Line	
N813FT	Boeing 747-245F	Flying Tiger Line	
N814FT	Boeing 747-245F	Flying Tiger Dine	
N815EV	Douglas DC-8-73CF	Evergreen International Airlines/ Air India	
N815FT	Boeing 747-245F	Flying Tiger Line *W. Henry Renniger*	
N816EV	Douglas DC-8-73CF	Evergreen International Airlines	
N816FT	Boeing 747-245F	Flying Tiger Line *Henry L. Heguy*	
N817FT	Boeing 747-121F	Flying Tiger Line	
N818FT	Boeing 747-121F	Flying Tiger Line	
N819FT	Boeing 747-121F	Flying Tiger Line	
N820FT	Boeing 747-121F	Flying Tiger Line	
N865F	Douglas DC-8-63	Emery Worldwide	
N870TV	Douglas DC-8-73	Emery Worldwide	
N872TV	Douglas DC-8-73CF	Transamerica Airlines	
N897U	Douglas DC-8-62	Arrow Air	
N901PA	Boeing 747-123F	Pan Am *Clipper Telegraph*	
N902DC	Douglas DC-9-32	Alitalia *Isola d'Elba*	
N902JW	Douglas DC-10-10	World Airways	
N902PA	Boeing 747-132	Pan Am *Clipper Seaman's Bridge*	
N902R	Douglas DC-8-55	National Airlines	
N903DC	Douglas DC-9-32	Alitalia *Isola di Murano*	
N904DC	Douglas DC-9-32	Alitalia *Isola di Pantellaria*	
N904WA	Douglas DC-10-10	Arrow Air	
N905DC	Douglas DC-9-32	Alitalia *Isola d'Ischia*	
N906DC	Douglas DC-9-32	Alitalia *Isola del Giglio*	
N906R	Douglas DC-8-63CF	National Airlines	
N910R	Douglas DC-8-55	Saudia — Saudi Arabian Airlines	
N915R	Douglas DC-8-61	National Airlines	
N916JW	Douglas DC-10-10	World Airways	
N917JW	Douglas DC-10-10	World Airways	
N917R	Douglas DC-8-71	National Airlines/Icelandair	
N921R	Douglas DC-8-63CF	Emery Worldwide	
N941JW	Douglas DC-8-63	Arrow Air	
N950R	Douglas DC-8-63	Emery Worldwide	
N951R	Douglas DC-8-63F	Emery Worldwide	
N954R	Douglas DC-8-61	National Airlines	
N957R	Douglas DC-8-63CF	Emery Worldwide	
N959R	Douglas DC-8-63CF	Emery Worldwide	
N961R	Douglas DC-8-73	Emery Worldwide	
N964R	Douglas DC-8-63CF	Emery Worldwide	
N1289E	Boeing 747-228B	Air France	
N1295E	Boeing 747-306	K.L.M. *The Ganges*	
N1298E	Boeing 747-306	K.L.M. *The Indus*	
N1301E	Boeing 747SP-27	CAAC	
N1304E	Boeing 747SP-J6	CAAC	
N1305E	Boeing 747-228B	Air France	
N1309E	Boeing 747-206B	K.L.M. *Admiral Richard E. Byrd*	
N1803	Douglas DC-8-62	Arrow Air	
N1804	Douglas DC-8-62	Rich International Airways	
N1805	Douglas DC-8-62	Rich International Airways	
N1808E	Douglas DC-8-62	Rich International Airways	
N1809E	Douglas DC-8-62	Arrow Air	

Notes	Reg.	Type	Owner or Operator
	N1834U	Douglas DC-10-10	CP Air
	N1836U	Douglas DC-10-10	CP Air
	N1837U	Douglas DC-10-10	CP Air
	N2674U	Douglas DC-8-73CF	Emery Worldwide
	N2786S	Douglas DC-9-32	Alitalia *Isola di Giannutri*
	N2786T	Douglas DC-9-32	Alitalia *Isola di Panarea*
	N2941W	Boeing 737-2K5	Air Berlin
	N3016Z	Douglas DC-10-30	Zambia Airways
	N3140D	L.1011 TriStar 500	B.W.I.A.
	N4501Q	Boeing 747-283B	S.A.S. *Dan Viking*
	N4502R	Boeing 747-283B	S.A.S. *Huge Viking*
	N4506H	Boeing 747-228B	Air France
	N4508E	Boeing 747-228F	Air France
	N4508H	Boeing 747SP-09	China Airways
	N4522V	Boeing 747SP-09	China Airways
	N4544F	Boeing 747-228B	Air France
	N4548M	Boeing 747-306	K.L.M. *Sir Frank Whittle*
	N4551N	Boeing 747-306	K.L.M. *Sir Geoffrey de Havilland*
	N4574P	Douglas DC-8-63	Arrow Air
	N4703U	Boeing 747-122	United Airlines *William M. Allen*
	N4717U	Boeing 747-122	United Airlines *Edward E. Carlson*
	N4720U	Boeing 747-122	United Airlines
	N4726U	Boeing 747-122	United Airlines
	N4727U	Boeing 747-122	United Airlines
	N4805J	Douglas DC-8-73	Minerve
	N4864T	Douglas DC-8-73CF	Transamerica Airlines
	N4865T	Douglas DC-8-73CF	Transamerica Airlines
	N4866T	Douglas DC-8-73CF	Transamerica Airlines
	N4867T	Douglas DC-8-73CF	Transamerica Airlines
	N4868T	Douglas DC-8-73CF	Transamerica Airlines
	N4869T	Douglas DC-8-73CF	Transamerica Airlines
	N4902W	Boeing 737-210C	Pan Am *Clipper Wilmersdorf*
	N6161A	Douglas DC-8-63CF	Arrow Air
	N7035T	L.1011 TriStar 100	Trans World Airlines
	N7036T	L.1011 TriStar 100	Trans World Airlines
	N7777G	L.749A Constellataon ★	Science Museum/Wroughton
	N7984S	L-100-20 Hercules	Southern Air Transport
	N8034T	L.1011 TriStar 100	Trans World Airlines
	N8764	Douglas DC-8-61	Airlift International
	N8766	Douglas DC-8-61	Airlift International
	N8968U	Douglas DC-8-62	Arrow Air
	N8969U	Douglas DC-8-62	Arrow Air
	N8970U	Douglas DC-8-62	Arrow Air
	N8973U	Douglas DC-8-62	Arrow Air
	N8974U	Douglas DC-8-62	Arrow Air
	N9232R	L-100-30 Hercules	Southern Air Transport
	N9266R	L-100-20 Hercules	Southern Air Transport
	N9664	Boeing 747-123	Avianca
	N12061	Douglas DC-10-30	Continental Airlines
	N14062	Douglas DC-10-30	Continental Airlines
	N14063	Douglas DC-10-30	Continental Airlines
	N17125	Boeing 747-136	Trans World Airlines
	N17126	Boeing 747-136	Trans World Airlines
	N31018	L.1011-385 TriStar 50	Trans World Airlines
	N31019	L.1011-385 TriStar 50	Trans World Airlines
	N31021	L.1011-385 TriStar 50	Trans World Airlines
	N31022	L.1011-385 TriStar 50	Trans World Airlines
	N31023	L.1011-385 TriStar 50	Trans World Airlines
	N31024	L.1011-385 TriStar 50	Trans World Airlines
	N31029	L.1011-385 TriStar 100	Trans World Airlines
	N31030	L.1011-385 TriStar 100	Trans World Airlines
	N31031	L.1011-385 TriStar 100	Trans World Airlines
	N31032	L.1011-385 TriStar 100	Trans World Airlines
	N31033	L.1011-385 TriStar 100	Trans World Airlines
	N39307	Douglas DC-8-62	Sea & Sun Aviation
	N41020	L.1011-385 TriStar 50	Trans World Airlines
	N43265	Douglas DC-9-32	Alitalia *Isola di Lipari*
	N46965	L-100-30 Hercules	Southern Air Transport
	N53110	Boeing 747-131	Trans World Airlines
	N53116	Boeing 747-131	Trans World Airlines
	N54629	Douglas DC-10-30	U.T.A.
	N54649	Douglas DC-10-30	U.T.A.
	N68060	Douglas DC-10-30	Continental Airlines *Robert F. Six*

Reg.	Type	Owner or Operator	Notes
N70723	Boeing 737-297	Pan Am *Clipper Luftikus*	
N70724	Boeing 737-297	Pan Am *Clipper Spreeathen*	
N81025	L.1011-385 TriStar 100	Trans World Airlines	
N81026	L.1011-385 TriStar 100	Trans World Airlines	
N81027	L.1011-385 TriStar 50	Trans World Airlines	
N81028	L.1011-385 TriStar 100	Trans World Airlines	
N93104	Boeing 747-131	Trans World Airlines	
N93105	Boeing 747-131	Trans World Airlines	
N93106	Boeing 747-131	Trans World Airlines	
N93107	Boeing 747-131	Trans World Airlines	
N93108	Boeing 747-131	Trans World Airlines	
N93109	Boeing 747-131	Trans World Airlines	
N93115	Boeing 747-131	Trans World Airlines	
N93117	Boeing 747-131	Trans World Airlines	
N93119	Boeing 747-131	Trans World Airlines	

OD (Lebanon)

OD-AFD	Boeing 707-3B4C	Middle East Airlines	
OD-AFE	Boeing 707-3B4C	Middle East Airlines	
OD-AFL	Boeing 720-023B	Middle East Airlines	
OD-AFM	Boeing 720-023B	Middle East Airlines	
OD-AFN	Boeing 720-023B	Middle East Airlines	
OD-AFY	Boeing 707-327C	Trans Mediterranean Airways	
OD-AFZ	Boeing 720-023B	Middle East Airlines	
OD-AGB	Boeing 720-023B	Middle East Airlines	
OD-AGD	Boeing 707-323C	Trans Mediterranean Airways	
OD-AGF	Boeing 720-047B	Middle East Airlines	
OD-AGJ	Boeing 747-2B4B	Middle East Airlines	
OD-AGO	Boeing 707-321C	Trans Mediterranean Airways	
OD-AGP	Boeing 707-321C	Trans Mediterranean Airways	
OD-AGQ	Boeing 720-047B	Middle East Airlines	
OD-AGS	Boeing 707-331C	Trans Mediterranean Airways	
OD-AGU	Boeing 707-347C	Middle East Airlines	
OD-AGV	Boeing 707-347C	Middle East Airlines	
OD-AGX	Boeing 707-327C	Trans Mediterranean Airways	
OD-AGY	Boeing 707-327C	Trans Mediterranean Airways	
OD-AGZ	Boeing 707-327C	Trans Mediterranean Airways	
OD-AHB	Boeing 707-323C	Middle East Airlines	
OD-AHC	Boeing 707-323C	Middle East Airlines	
OD-AHD	Boeing 707-323C	Middle East Airlines	
OD-AHE	Boeing 707-323C	Middle East Airlines	

OE (Austria)

OE-FGW	Boeing 737-3Z9	Lauda-Air	
OE-HLS	D.H.C. 7-102 Dash Seven	Tyrolean Airways *Stadt Innsbruck*	
OE-HLT	D.H.C. 7-102 Dash Seven	Tyrolean Airways *Stadt Wien*	
OE-HLR	D.H.C.8 Dash Eight	Tyrolean Airways	
OE-LAA	A.310-322 Airbus	Austrian Airlines	
OE-LAI	A.310-322 Airbus	Austrian Airlines	
OE-LDF	Douglas DC-9-32	Austrian Airlines	
OE-LDG	Douglas DC-9-32	Austrian Airlines	
OE-LDH	Douglas DC-9-32	Austrian Airlines	
OE-LDI	Douglas DC-9-32	Austrian Airlines	
OE-LDJ	Douglas DC-9-32	Austrian Airlines *Bregenz*	
OE-LDP	Douglas DC-9-81	Austrian Airlines	
OE-LDR	Douglas DC-9-81	Austrian Airlines *Wien*	
OE-LDS	Douglas DC-9-81	Austrian Airlines *Niederösterreich*	
OE-LDT	Douglas DC-9-81	Austrian Airlines *Burgenland*	
OE-LDU	Douglas DC-9-81	Austrian Airlines *Kärnten*	
OE-LDV	Douglas DC-9-81	Austrian Airlines *Steiermark*	
		Austrian Airlines *Oberösterreich*	

Notes	Reg.	Type	Owner or Operator
	OE-LDW	Douglas DC-9-81	Austrian Airlines *Salzburg*
	OE-LDX	Douglas DC-9-81	Austrian Airlines *Tirol*
	OE-LDY	Douglas DC-9-81	Austrian Airlines *Vorarlberg*
	OE-LDZ	Douglas DC-9-81	Austrian Airlines *Bregenz*
	OE-LMA	Douglas DC-9-81	Austrian Airlines
	OE-LMB	Douglas DC-9-81	Austrian Airlines
	OE-LMC	Douglas DC-9-81	Austrian Airlines
	OE-	Douglas DC-9-87	Austrian Airlines
	OE-	Douglas DC-9-87	Austrian Airlines
	OE-	Douglas DC-9-87	Austrian Airlines
	OE-	Douglas DC-9-87	Austrian Airlines

Note: Austrian Airlines is in the process of replacing its DC-9-32/51s with Series 81s.

OH (Finland)

KAR-AIR FINNAIR

	OH-LFZ	Douglas DC-8-62	Kar Air
	OH-LHA	Douglas DC-10-30ER	Finnair *Iso Antti*
	OH-LHB	Douglas DC-10-30ER	Finnair
	OH-LHD	Douglas DC-10-30ER	Finnair
	OH-LMN	Douglas DC-9-82	Finnair
	OH-LMO	Douglas DC-9-82	Finnair
	OH-LMP	Douglas DC-9-82	Finnair
	OH-LMR	Douglas DC-9-83	Finnair
	OH-LMS	Douglas DC-9-83	Finnair
	OH-LNB	Douglas DC-9-41	Finnair
	OH-LNC	Douglas DC-9-41	Finnair
	OH-LND	Douglas DC-9-41	Finnair
	OH-LNE	Douglas DC-9-41	Finnair
	OH-LNF	Douglas DC-9-41	Finnair
	OH-LYH	Douglas DC-9-15MC	Finnair
	OH-LYN	Douglas DC-9-51	Finnair
	OH-LYO	Douglas DC-9-51	Finnair
	OH-LYP	Douglas DC-9-51	Finnair
	OH-LYR	Douglas DC-9-51	Finnair
	OH-LYS	Douglas DC-9-51	Finnair
	OH-LYT	Douglas DC-9-51	Finnair
	OH-LYU	Douglas DC-9-51	Finnair
	OH-LYV	Douglas DC-9-51	Finnair
	OH-LYW	Douglas DC-9-51	Finnair
	OH-LYX	Douglas DC-9-51	Finnair
	OH-LYY	Douglas DC-9-51	Finnair
	OH-LYZ	Douglas DC-9-51	Finnair
	OH-	Douglas DC-9-87	Finnair
	OH-	Douglas DC-9-87	Finnair
	OH-	Douglas DC-9-87	Finnair
	OH-	Douglas DC-9-87	Finnair
	OH-	Douglas DC-9-87	Finnair
	OH-	Douglas DC-9-87	Finnair
	OH-	Douglas DC-9-87	Finnair
	OH-	Douglas DC-9-87	Finnair

Note: Finnair also operates a DC-10-30 which retains its US registration N345HC.

OK (Czechoslovakia)

CSA

	OK-ABD	Ilyushin IL-62	Ceskoslovenske Aerolinie *Kosice*
	OK-AFA	Tupolev Tu-134A	Ceskoslovenske Aerolinie
	OK-AFB	Tupolev Tu-134A	Ceskoslovenske Aerolinie
	OK-CFC	Tupolev Tu-134A	Ceskoslovenske Aerolinie
	OK-CFE	Tupolev Tu-134A	Ceskoslovenske Aerolinie
	OK-CFF	Tupolev Tu-134A	Ceskoslovenske Aerolinie
	OK-CFG	Tupolev Tu-134A	Ceskoslovenske Aerolinie
	OK-CFH	Tupolev Tu-134A	Ceskoslovenske Aerolinie
	OK-DBE	Ilyushin IL-62	Ceskoslovenske Aerolinie *Brno*

Reg.	*Type*	*Owner or Operator*	*Notes*
OK-DFI	Tupolev Tu-134A	Ceskoslovenske Aerolinie	
OK-EBG	Ilyushin IL-62	Ceskoslovenske Aerolinie *Banska Bystrica*	
OK-EFJ	Tupolev Tu-134A	Ceskoslovenske Aerolinie	
OK-EFK	Tupolev Tu-134A	Ceskoslovenske Aerolinie	
OK-FBF	Ilyushin IL-62	Ceskoslovenske Aerolinie	
OK-GBH	Ilyushin IL-62	Ceskoslovenske Aerolinie *Usti Nad Labem*	
OK-HFL	Tupolev Tu-134A	Ceskoslovenske Aerolinie	
OK-HFM	Tupolev Tu-134A	Ceskoslovenske Aerolinie	
OK-IFN	Tupolev Tu-134A	Ceskoslovenske Aerolinie	
OK-JBI	Ilyushin IL-62M	Ceskoslovenske Aerolinie *Plzen*	
OK-JBJ	Ilyushin IL-62M	Ceskoslovenske Aerolinie *Hradec Kralové*	
OK-KBK	Ilyushin IL-62M	Ceskoslovenske Aerolinie *Ceske Budejovice*	
OK-OBL	Ilyushin IL-62M	Ceskoslovenske Aerolinie	
OK-PBM	Ilyushin IL-62M	Ceskoslovenske Aerolinie	
OK-YBA	Ilyushin IL-62	Ceskoslovenske Aerolinie *Praha*	
OK-YBB	Ilyushin IL-62	Ceskoslovenske Aerolinie *Bratislava*	
OK-ZBC	Ilyushin IL-62	Ceskoslovenske Aerolinie *Ostrava*	

OO (Belgium)

OO-DTA	FH-227B Friendship	Delta Air Transport	
OO-DTB	FH-227B Friendship	Delta Air Transport	
OO-DTC	FH-227B Friendship	Delta Air Transport	
OO-DTD	FH-227B Friendship	Delta Air Transport	
OO-DTE	FH-227B Friendship	Delta Air Transport	
OO-JPA	Swearingen SA226AT Metro II	European Air Transport	
OO-JPI	Swearingen SA226TC Metro II	European Air Transport	
OO-JPK	Swearingen SA226TC Metro II	European Air Transport	
OO-JPN	Swearingen SA226AT Metro II	European Air Transport	
OO-JPY	Swearingen SA226TC Metro II	European Air Transport	
OO-PLH	Boeing 737-247	Air Belgium	
OO-	Boeing 737-347	Air Belgium	
OO-SBQ	Boeing 737-229	Sobelair	
OO-SBS	Boeing 737-229	Sobelair	
OO-SBT	Boeing 737-229	Sobelair	
OO-SBU	Boeing 707-373C	Sobelair	
OO-SCA	A.310-221 Airbus	Sabena	
OO-SCB	A.310-221 Airbus	Sabena	
OO-SCC	A.310-221 Airbus	Sabena	
OO-SDA	Boeing 737-229	Sabena	
OO-SDB	Boeing 737-229	Sabena	
OO-SDC	Boeing 737-229	Sabena	
OO-SDD	Boeing 737-229	Sabena	
OO-SDE	Boeing 737-229	Sabena	
OO-SDF	Boeing 737-229	Sabena	
OO-SDG	Boeing 737-229	Sabena	
OO-SDJ	Boeing 737-229C	Sabena	
OO-SDK	Boeing 737-229C	Sabena	
OO-SDL	Boeing 737-229	Sabena	
OO-SDM	Boeing 737-229	Sabena	
OO-SDN	Boeing 737-229	Sabena	
OO-SDO	Boeing 737-229	Sabena	
OO-SDP	Boeing 737-229C	Sabena	
OO-SDR	Boeing 737-229C	Sabena	
OO-SGA	Boeing 747-129A	Sabena	
OO-SGB	Boeing 747-129A	Sabena	
OO-SGC	Boeing 747-329	Sabena	
OO-SJM	Boeing 707-329C	Sobelair	
OO-SLA	Douglas DC-10-30CF	Sabena	
OO-SLB	Douglas DC-10-30CF	Sabena	
OO-SLC	Douglas DC-10-30CF	Sabena	
OO-SLD	Douglas DC-10-30CF	Sabena	
OO-SLE	Douglas DC-10-30CF	Sabena	

Notes	Reg.	Type	Owner or Operator
	OO-TED	Boeing 707-131	Trans European Airways *Rena*
	OO-TEF	A.300B1 Airbus	Trans European Airways
	OO-TEH	Boeing 737-2M8	Trans European Airways *Marcus Johannes*
	OO-TEK	Boeing 737-2Q9	Trans European Airways
	OO-TEL	Boeing 737-2M8	Trans European Airways *Antwerpen*
	OO-TEO	Boeing 737-2M8	Trans European Airways *Jonathan*
	OO-TYC	Boeing 707-328B	Trans European Airways
	OO-WAY	Beech 99	Sabena/Publi-Air

OY (Denmark)

	OY-APE	F.27 Friendship Mk 600	Alkair
	OY-APP	Boeing 737-2L9	Maersk Air
	OY-APU	Boeing 720-051B	Conair
	OY-APV	Boeing 720-051B	Conair
	OY-APW	Boeing 720-051B	Conair
	OY-APY	Boeing 720-051B	Conair
	OY-APZ	Boeing 720-051B	Conair
	OY-CCK	F.27 Friendship Mk 600	Alkair
	OY-DSP	Boeing 720-025	Conair
	OY-GAW	Swearingen SA227AC Metro III	Metro Airways
	OY-KAC	F-27 Friendship Mk 600	S.A.S.
	OY-KAD	F.27 Friendship Mk 600	S.A.S.
	OY-KDA	Douglas DC-10-30	S.A.S. *Gorm Viking*
	OY-KDB	Douglas DC-10-30	S.A.S. *Frode Viking*
	OY-KGA	Douglas DC-9-41	S.A.S. *Heming Viking*
	OY-KGB	Douglas DC-9-41	S.A.S. *Toste Viking*
	OY-KGC	Douglas DC-9-41	S.A.S. *Helge Viking*
	OY-KGD	Douglas DC-9-21	S.A.S. *Ubbe Viking*
	OY-KGE	Douglas DC-9-21	S.A.S. *Orvar Viking*
	OY-KGF	Douglas DC-9-21	S.A.S. *Rolf Viking*
	OY-KGG	Douglas DC-9-41	S.A.S. *Sune Viking*
	OY-KGH	Douglas DC-9-41	S.A.S. *Eiliv Viking*
	OY-KGI	Douglas DC-9-41	S.A.S. *Bent Viking*
	OY-KGK	Douglas DC-9-41	S.A.S. *Ebbe Viking*
	OY-KGL	Douglas DC-9-41	S.A.S. *Angantyr Viking*
	OY-KGM	Douglas DC-9-41	S.A.S. *Arnfinn Viking*
	OY-KGN	Douglas DC-9-41	S.A.S. *Gram Viking*
	OY-KGO	Douglas DC-9-41	S.A.S. *Holte Viking*
	OY-KGP	Douglas DC-9-41	S.A.S. *Torbern Viking*
	OY-KGR	Douglas DC-9-41	S.A.S. *Holger Viking*
	OY-KGS	Douglas DC-9-41	S.A.S. *Hall Viking*
	OY-KTF	Douglas DC-8-63	Scanair *Mette Viking*
	OY-KTG	Douglas DC-8-63	S.A.S. *Torodd Viking*
	OY-MBC	D.H.C.-7 Dash Seven	Maersk Air
	OY-MBD	D.H.C.-7 Dash Seven	Maersk Air
	OY-MBE	D.H.C.-7 Dash Seven	Maersk Air
	OY-MBV	Boeing 737-2L9	Maersk Air
	OY-MBZ	Boeing 737-2L9	Maersk Air
	OY-MMK	Boeing 737-3L9	Maersk Air
	OY-MML	Boeing 737-3L9	Maersk Air
	OY-SBE	Boeing 727-2J4	Sterling Airways
	OY-SBF	Boeing 727-2J4	Sterling Airways
	OY-SBG	Boeing 727-2J4	Sterling Airways
	OY-SBK	Douglas DC-8-63	Sterling Airways
	OY-SBL	Douglas DC-8-63	Sterling Airways
	OY-SBM	Douglas DC-8-63	Sterling Airways
	OY-STC	S.E.210 Caravelle 10B	Sterling Airways
	OY-STD	S.E.210 Caravelle 10B	Sterling Airways
	OY-STF	S.E.210 Caravelle 10B	Sterling Airways
	OY-STH	S.E.210 Caravelle 10B	Sterling Airways
	OY-STI	S.E.210 Caravelle 10B	Sterling Airways
	OY-STM	S.E.210 Caravelle 10B	Sterling Airways

Note: S.A.S. also operates two Boeing 747-283Bs which retain their U.S. registrations N4501Q and N4502R.

PH (Netherlands)

Reg.	Type	Owner or Operator	Notes
PH-AGA	A.310-203 Airbus	K.L.M. *Rembrandt*	
PH-AGB	A.310-203 Airbus	K.L.M. *Jeroen Bosch*	
PH-AGC	A.310-203 Airbus	K.L.M. *Albert Cuyp*	
PH-AGD	A.310-203 Airbus	K.L.M. *Marinus Ruppert*	
PH-AGE	A.310-203 Airbus	K.L.M. *Nicolaas Maes*	
PH-AGF	A.310-203 Airbus	K.L.M. *Jan Steen*	
PH-AGG	A.310-203 Airbus	K.L.M. *Vincent van Gogh*	
PH-AGH	A.310-203 Airbus	K.L.M. *Peiter de Hoogh*	
PH-AGI	A.310-203 Airbus	K.L.M. *Jan Toorop*	
PH-AGK	A.310-203 Airbus	K.L.M. *Johannes Vermeer*	
PH-AHB	Boeing 727-2H3	Air Holland	
PH-AHD	Boeing 727-2H3	Air Holland	
PH-ATR	Aeritalia ATR-42	Holland Aero Lines	
PH-BDA	Boeing 737-306	K.L.M. *Willem Barentz*	
PH-BDB	Boeing 737-306	K.L.M. *Olivier van Noort*	
PH-BDC	Boeing 737-306	K.L.M. *Cornelis De Houteman*	
PH-BDD	Boeing 737-306	K.L.M. *Anthony van Diemen*	
PH-BDE	Boeing 737-306	K.L.M. *Abel J. Tasman*	
PH-BDG	Boeing 737-306	K.L.M. *Michiel A. D. Ruyter*	
PH-BDH	Boeing 737-306	K.L.M. *Petrus Plancius*	
PH-BDI	Boeing 737-306	K.L.M. *Maarten H. Tromp*	
PH-BDK	Boeing 737-306	K.L.M. *Jan H. van Linschoten*	
PH-BDL	Boeing 737-306	K.L.M. *Piet Heyn*	
PH-BUA	Boeing 747-206B	K.L.M. *The Mississippi*	
PH-BUB	Boeing 747-206B	K.L.M. *The Danube*	
PH-BUC	Boeing 747-206B	K.L.M. *The Amazon*	
PH-BUD	Boeing 747-206B	K.L.M. *The Nile*	
PH-BUE	Boeing 747-206B	K.L.M. *Rio de la Plata*	
PH-BUG	Boeing 747-206B	K.L.M. *The Orinoco*	
PH-BUH	Boeing 747-306	K.L.M. *Dr Albert Plesman*	
PH-BUI	Boeing 747-306	K.L.M. *Wilbur Wright*	
PH-BUK	Boeing 747-306	K.L.M. *Louis Blèriot*	
PH-BUL	Boeing 747-306	K.L.M. *Charles A. Lindbergh*	
PH-BUM	Boeing 747-206B	K.L.M. *Sir Charles E. Kingsford-Smith*	
PH-BUN	Boeing 747-306	K.L.M. *Anthony H. G. Fokker*	
PH-BUO	Boeing 747-306	K.L.M. *Missouri*	
PH-BUW	Boeing 747-306	K.L.M. *Leonardo da Vinci*	
PH-CHB	F.28 Fellowship 4000	N.L.M. *City of Birmingham*	
PH-CHD	F.28 Fellowship 4000	N.L.M. *City of Maastricht*	
PH-CHF	F.28 Fellowship 4000	N.L.M. *Island of Guernsey*	
PH-CHN	F.28 Fellowship 4000	N.L.M.	
PH-DDA	Douglas DC-3	Dutch Dakota Association	
PH-DNC	Douglas DC-9-15	K.L.M. *City of Luxembourg*	
PH-DNG	Douglas DC-9-32	K.L.M. *City of Rotterdam*	
PH-DNH	Douglas DC-9-32	K.L.M. *City of Zurich*	
PH-DNI	Douglas DC-9-32	K.L.M. *City of Istanbul*	
PH-DNK	Douglas DC-9-32	K.L.M. *City of Copenhagen*	
PH-DNL	Douglas DC-9-32	K.L.M. *City of London*	
PH-DNM	Douglas DC-9-33RC	K.L.M. *City of Madrid*	
PH-DNN	Douglas DC-9-33RC	K.L.M. *City of Vienna*	
PH-DNO	Douglas DC-9-33RC	K.L.M. *City of Oslo*	
PH-DNP	Douglas DC-9-33RC	K.L.M. *City of Athens*	
PH-DNR	Douglas DC-9-33RC	K.L.M. *City of Stockholm*	
PH-DNS	Douglas DC-9-32	K.L.M. *City of Arnhem*	
PH-DNT	Douglas DC-9-32	K.L.M. *City of Lisbon*	
PH-DNV	Douglas DC-9-32	K.L.M. *City of Warsaw*	
PH-DNW	Douglas DC-9-32	K.L.M. *City of Moscow*	
PH-DNY	Douglas DC-9-33RC	K.L.M. *City of Paris*	
PH-DOA	Douglas DC-9-32	K.L.M. *City of Utrecht*	
PH-DOB	Douglas DC-9-32	K.L.M. *City of Santa Monica*	
PH-DTA	Douglas DC-10-30	K.L.M. *Johann Sebastian Bach*	
PH-DTB	Douglas DC-10-30	K.L.M. *Ludwig van Beethoven*	
PH-DTC	Douglas DC-10-30	K.L.M. *Frédéric Francçois Chopin*	
PH-DTD	Douglas DC-10-30	K.L.M. *Maurice Ravel*	
PH-DTL	Douglas DC-10-30	K.L.M. *Edvard Hagerup Grieg*	
PH-FKT	F-27 Friendship Mk 600	XP Parcel Service	
PH-HAF	GAF Nomad N24A	Holland Aero Lines	
PH-HAG	GAF Nomad N24A	Holland Aero Lines	

Notes	Reg.	Type	Owner or Operator
	PH-HAL	GAF Nomad N24A	Holland Aero Lines
	PH-HVF	Boeing 737-3K2	Transavia
	PH-HVG	Boeing 737-3K2	Transavia
	PH-IFH	Aeritalia ATR-42	Holland Aero Lines
	PH-KFD	F.27 Friendship Mk 200	N.L.M. *Jan Moll*
	PH-KFE	F.27 Friendship Mk 600	N.L.M. *Jan Dellaert*
	PH-KFG	F.27 Friendship Mk 200	N.L.M. *Koos Abspoel*
	PH-KFI	F.27 Friendship Mk 500	N.L.M. *Bremen*
	PH-KFK	F.27 Friendship Mk 500	N.L.M. *Zestienhoven*
	PH-KFL	F.27 Friendship Mk 500	N.L.M.
	PH-KJA	BAe Jetstream 3102	Nether Lines
	PH-KJB	BAe Jetstream 3102	Nether Lines
	PH-KJC	BAe Jetstream 3102	Nether Lines
	PH-KJD	BAe Jetstream 3102	Nether Lines
	PH-KJE	BAe Jetstream 3102	Nether Lines
	PH-KJF	BAe Jetstream 3102	Nether Lines
	PH-KLC	Fokker 100	K.L.M. *Gerard Mercator*
	PH-KLD	Fokker 100	K.L.M. *Jan A. Leeghwater*
	PH-KLE	Fokker 100	K.L.M. *Gerard J. Leeuwenhoek*
	PH-KLG	Fokker 100	K.L.M. *Johannes Blaeu*
	PH-KLH	Fokker 100	K.L.M. *Christiaan Huygens*
	PH-KLI	Fokker 100	K.L.M. *Antonie van Leeuwenhoek*
	PH-KLK	Fokker 100	K.L.M. *Chris H. D. Buys Ballot*
	PH-KLL	Fokker 100	K.L.M. *Hendrick A. Lorentz*
	PH-KLN	Fokker 100	K.L.M. *Pieter Zeeman*
	PH-KLO	Fokker 100	K.L.M. *Jan H. Oort*
	PH-MAX	Douglas DC-9-32	K.L.M. *City of Rome*
	PH-MBG	Douglas DC-10-30CF	Martinair *Kohoutek*
	PH-MBN	Douglas DC-10-30CF	Martinair *Anthony Ruys*
	PH-MBP	Douglas DC-10-30CF	Martinair *Hong Kong*
	PH-MBT	Douglas DC-10-30CF	Martinair
	PH-MBZ	Douglas DC-9-82	Martinair *Prinses Juiliana*
	PH-MCA	A.310-202 Airbus	Martinair
	PH-MCB	A.310-202CF Airbus	Martinair
	PH-MCD	Douglas DC-9-82	Martinair *Lucien Ruys*
	PH-SAD	F.27 Friendship Mk 200	N.L.M. *Evert van Dijk*
	PH-TVC	Boeing 737-2K2C	Transavia *Richard Gordon*
	PH-TVD	Boeing 737-2K2C	Transavia *Charles Conrad*
	PH-TVE	Boeing 737-2K2C	Transavia *Alan Bean*
	PH-TVH	Boeing 737-222	Transavia *Neil Armstrong*
	PH-TVR	Boeing 737-2K2	Transavia
	PH-TVU	Boeing 737-2K2	Transavia

Note: K.L.M. also operates Boeing 747-206B, N1309E and Boeing 747-306s N1295E, N1298E, N4548M and N4551N. Ten series 206 are to be converted to series 306. Sixteen DC-9s have been sold in America for delivery spread over two years starting in August 1986.

PK (Indonesia)

	PK-GSA	Boeing 747-2U3B	Garuda Indonesian Airways *City of Jakarta*
	PK-GSB	Boeing 747-2U3B	Garuda Indonesian Airways *City of Bandung*
	PK-GSC	Boeing 747-2U3B	Garuda Indonesian Airways *City of Medan*
	PK-GSD	Boeing 747-2U3B	Garuda Indonesian Airways *City of Surabaya*
	PK-GSE	Boeing 747-2U3B	Garuda Indonesian Airways *City of Yogyakarte*
	PK-GSF	Boeing 747-2U3B	Garuda Indonesian Airways *City of Denpasar*

PP (Brazil)

Reg.	Type	Owner or Operator	Notes
PP-VJH	Boeing 707-320C	VARIG	
PP-VJK	Boeing 707-379C	VARIG	
PP-VJS	Boeing 707-341C	VARIG	
PP-VJX	Boeing 707-345C	VARIG	
PP-VJY	Boeing 707-345C	VARIG	
PP-VLI	Boeing 707-385C	VARIG	
PP-VLK	Boeing 707-324C	VARIG	
PP-VLL	Boeing 707-324C	VARIG	
PP-VLM	Boeing 707-324C	VARIG	
PP-VLN	Boeing 707-324C	VARIG	
PP-VLO	Boeing 707-324C	VARIG	
PP-VLP	Boeing 707-323C	VARIG	
PP-VMA	Douglas DC-10-30	VARIG	
PP-VMB	Douglas DC-10-30	VARIG	
PP-VMD	Douglas DC-10-30	VARIG	
PP-VMQ	Douglas DC-10-30	VARIG	
PP-VMS	Douglas DC-10-30	VARIG	
PP-VMT	Douglas DC-10-30	VARIG	
PP-VMU	Douglas DC-10-30	VARIG	
PP-VMV	Douglas DC-10-30	VARIG	
PP-VMW	Douglas DC-10-30	VARIG	
PP-VMX	Douglas DC-10-30	VARIG	
PP-VMY	Douglas DC-10-30	VARIG	
PP-VMZ	Douglas DC-10-30	VARIG	
PP-VNA	Boeing 747-2L5B	VARIG	
PP-VNB	Boeing 747-2L5B	VARIG	
PP-VNC	Boeing 747-2L5B	VARIG	
PP-VNH	Boeing 747-341	VARIG	
PP-VNI	Boeing 747-341	VARIG	

RP (Philippines)

Note: Philippine Airlines operates four Boeing 747s which retain their U.S. registrations N741PR, N742PR, N743PR and N744PR.

S2 (Bangladesh)

S2-ABN	Boeing 707-351C	Bangladesh Biman *City of Shah Jalal*	
S2-ACA	Boeing 707-351C	Bangladesh Biman *Khan Jahan Ali*	
S2-ACE	Boeing 707-351C	Bangladesh Biman *City of Tokyo*	
S2-ACF	Boeing 707-351C	Bangladesh Biman *City of Hazrat Shah Balkhi*	
S2-ACK	Boeing 707-321B	Bangladesh Biman *City of Kuwait*	
S2-ACO	Douglas DC-10-30	Bangladesh Biman *City of Hazrat-Shar Makhdoom (R.A.)*	
S2-ACP	Douglas DC-10-30	Bangladesh Biman *City of Dhaka*	
S2-ACQ	Douglas DC-10-30	Bangladesh Biman *City of Hz Shah Jalal (R.A.)*	

S7 (Seychelles)

S7-SIS	Douglas DC-8-63	Seychelles International	
S7-	A.310-300 Airbus	Air Seychelles	

SE (Sweden)

Notes	Reg.	Type	Owner or Operator
	SE-DAK	Douglas DC-9-41	S.A.S. *Ragnvald Viking*
	SE-DAL	Douglas DC-9-41	S.A.S. *Algot Viking*
	SE-DAM	Douglas DC-9-41	S.A.S. *Starkad Viking*
	SE-DAN	Douglas DC-9-41	S.A.S. *Alf Viking*
	SE-DAO	Douglas DC-9-41	S.A.S. *Asgaut Viking*
	SE-DAP	Douglas DC-9-41	S.A.S. *Torgils Viking*
	SE-DAR	Douglas DC-9-41	S.A.S. *Agnar Viking*
	SE-DAS	Douglas DC-9-41	S.A.S. *Garder Viking*
	SE-DAT	Douglas DC-9-41	S.A.S. *Gissur Viking*
	SE-DAU	Douglas DC-9-41	S.A.S. *Hadding Viking*
	SE-DAW	Douglas DC-9-41	S.A.S. *Gotrik Viking*
	SE-DAX	Douglas DC-9-41	S.A.S. *Helsing Viking*
	SE-DBG	Douglas DC-8-62	S.A.S. *Jorund Viking*
	SE-DBI	Douglas DC-8-62CF	Thai International
	SE-DBK	Douglas DC-8-63	Scanair *Sigyn Viking*
	SE-DBL	Douglas DC-8-63	Scanair *Bodil Viking*
	SE-DBM	Douglas DC-9-41	S.A.S. *Ossur Viking*
	SE-DBN	Douglas DC-9-33AF	S.A.S. *Sigtrygg Viking*
	SE-DBO	Douglas DC-9-21	S.A.S. *Siger Viking*
	SE-DBP	Douglas DC-9-21	S.A.S. *Rane Viking*
	SE-DBR	Douglas DC-9-21	S.A.S. *Skate Viking*
	SE-DBS	Douglas DC-9-21	S.A.S. *Svipdag Viking*
	SE-DBT	Douglas DC-9-41	S.A.S. *Agne Viking*
	SE-DBU	Douglas DC-9-41	S.A.S. *Hjalmar Viking*
	SE-DBW	Douglas DC-9-41	S.A.S. *Adils Viking*
	SE-DBX	Douglas DC-9-41	S.A.S. *Arnljot Viking*
	SE-DDP	Douglas DC-9-41	S.A.S. *Brun Viking*
	SE-DDR	Douglas DC-9-41	S.A.S. *Atle Viking*
	SE-DDS	Douglas DC-9-41	S.A.S. *Alrik Viking*
	SE-DDT	Douglas DC-9-41	S.A.S. *Amund Viking*
	SE-DDU	Douglas DC-8-62	S.A.S. *Knud Viking*
	SE-DEH	S.E.210 Caravelle 10B	Transwede
	SE-DFD	Douglas DC-10-30	S.A.S. *Dag Viking*
	SE-DFE	Douglas DC-10-30	S.A.S. *Sverker Viking*
	SE-DFF	Douglas DC-10-30	S.A.S. *Solve Viking*
	SE-DFG	Douglas DC-10-30	S.A.S.
	SE-DFK	A.300B4 Airbus	Scanair *Sven Viking*
	SE-DFL	A.300B4 Airbus	Scanair *Ingemar Viking*
	SE-DFS	Douglas DC-9-82	S.A.S. *Gaut Viking*
	SE-DFT	Douglas DC-9-82	S.A.S. *Assur Viking*
	SE-	Douglas DC-9-81	S.A.S.
	SE-	Douglas DC-9-81	S.A.S.
	SE-DFZ	Boeing 747-283B	S.A.S. *Knut Viking*
	SE-DGA	F.28 Fellowship 1000	Linjeflyg
	SE-DGB	F.28 Fellowship 1000	Linjeflyg
	SE-DGC	F.28 Fellowship 1000	Linjeflyg
	SE-DGD	F.28 Fellowship 4000	Linjeflyg
	SE-DGE	F.28 Fellowship 4000	Linjeflyg
	SE-DGF	F.28 Fellowship 4000	Linjeflyg
	SE-DGG	F.28 Fellowship 4000	Linjeflyg
	SE-DGH	F.28 Fellowship 4000	Linjeflyg
	SE-DGI	F.28 Fellowship 4000	Linjeflyg
	SE-DGK	F.28 Fellowship 4000	Linjeflyg
	SE-DGL	F.28 Fellowship 4000	Linjeflyg
	SE-DGM	F.28 Fellowship 4000	Linjeflyg
	SE-DGN	F.28 Fellowship 4000	Linjeflyg
	SE-DGO	F.28 Fellowship 4000	Linjeflyg
	SE-DGP	F.28 Fellowship 4000	Linjeflyg
	SE-DGR	F.28 Fellowship 4000	Linjeflyg
	SE-DGS	F.28 Fellowship 4000	Linjeflyg
	SE-DGT	F.28 Fellowship 4000	Linjeflyg
	SE-IEG	F.27 Friendship	Swedair
	SE-IEY	Convair 580	ScanBee
	SE-INA	F.27 Friendship	Swedair
	SE-IPE	F.27J Friendship	Swedair

Reg.	Type	Owner or Operator	Notes
SE-IRF	F.27 Friendship Mk 600	S.A.S. *Vinge Viking*	
SE-IRG	F.27 Friendship Mk 600	S.A.S. *Vigge Viking*	
SE-ITH	F.27 Friendship Mk 600	S.A.S. *Visbur Viking*	
SE-ITI	F.27 Friendship Mk 600	S.A.S. *Vidar Viking*	

Note: S.A.S. also operates two Boeing 747-283Bs, which retain their U.S. registrations N4501Q and N4502R. Swedair operates two Fairchild F.27s registered LN-BSC and LN-BSD.

SP (Poland)

SP-LBA	Ilyushin IL-62M	Polskie Linie Lotnicze (LOT) *Juliusz Sowacki*	
SP-LBB	Ilyushin IL-62M	Polskie Linie Lotnicze (LOT) *Jgnacy Paderewski*	
SP-LBC	Ilyushin IL-62M	Polskie Linie Lotnicze (LOT) *Joseph Conrad-Korzeniowski*	
SP-LBD	Ilyushin IL-62M	Polskie Linie Lotnicze (LOT)	
SP-LBE	Ilyushin IL-62M	Polskie Linie Lotnicze (LOT)	
SP-LBF	Ilyushin IL-62M	Polskie Linie Lotnicze (LOT)	
SP-LBG	Ilyushin IL-62M	Polskie Linie Lotnicze (LOT)	
SP-LHA	Tupolev Tu-134A	Polskie Linie Lotnicze (LOT)	
SP-LHB	Tupolev Tu-134A	Polskie Linie Lotnicze (LOT)	
SP-LHC	Tupolev Tu-134A	Polskie Linie Lotnicze (LOT)	
SP-LHD	Tupolev Tu-134A	Polskie Linie Lotnicze (LOT)	
SP-LHE	Tupolev Tu-134A	Polskie Linie Lotnicze (LOT)	
SP-LHF	Tupolev Tu-134A	Polskie Linie Lotnicze (LOT)	
SP-LHG	Tupolev Tu-134A	Polskie Linie Lotnicze (LOT)	
SP-LSA	Ilyushin IL-18V (Cargo)	Polskie Linie Lotnicze (LOT)	
SP-LSB	Ilyushin IL-18V	Polskie Linie Lotnicze (LOT)	
SP-LSC	Ilyushin IL-18V (Cargo)	Polskie Linie Lotnicze (LOT)	
SP-LSD	Ilyushin IL-18V	Polskie Linie Lotnicze (LOT)	
SP-LSE	Ilyushin IL-18V	Polskie Linie Lotnicze (LOT)	
SP-LSF	Ilyushin IL-18E	Polskie Linie Lotnicze (LOT)	
SP-LSG	Ilyushin IL-18E	Polskie Linie Lotnicze (LOT)	
SP-LSH	Ilyushin IL-18V	Polskie Linie Lotnicze (LOT)	
SP-LSI	Ilyushin IL-18D	Polskie Linie Lotnicze (LOT)	
SP-	Tupolev Tu-154B	Polskie Linie Lotnicze (LOT)	
SP-	Tupolev Tu-154B	Polskie Linie Lotnicze (LOT)	

Note: LOT also operates two IL-62s registered CCCP-86611 and YR-IRE.

ST (Sudan)

ST-AFA	Boeing 707-3J8C	Sudan Airways	
ST-AFB	Boeing 707-3J8C	Sudan Airways	
ST-AIX	Boeing 707-369C	Sudan Airways	
ST-AJD	Douglas DC-8-55F	Trans Arabian Air Transport	
ST-AJR	Douglas DC-8-55F	Trans Arabian Air Transport	

SU (Egypt)

SU-AOU	Boeing 707-366C	EgyptAir *Khopho*	
SU-APD	Boeing 707-366C	EgyptAir *Khafrah*	
SU-AVX	Boeing 707-366C	EgyptAir *Tutankhamun*	
SU-AVY	Boeing 707-366C	EgyptAir *Akhenaton*	
SU-AVZ	Boeing 707-366C	EgyptAir *Mena*	
SU-AXK	Boeing 707-366C	EgyptAir *Seti I*	
SU-BCA	A.300B4 Airbus	EgyptAir *Horus*	

Notes	Reg.	Type	Owner or Operator
	SU-BCB	A.300B4 Airbus	EgyptAir *Osiris*
	SU-BCC	A.300B4 Airbus	EgyptAir *Nout*
	SU-BDF	A.300B4 Airbus	EgyptAir *Hathor*
	SU-BDG	A.300B4 Airbus	EgyptAir *Aton*
	SU-DAA	Boeing 707-351C	Zakani Aviation Services
	SU-DAB	Boeing 707-328C	Zakani Aviation Services
	SU-DAC	Boeing 707-336C	Zakani Aviation Services
	SU-DAD	Boeing 707-336C	Zakani Aviation Services
	SU-EAA	Boeing 707-351C	Misr Overseas Airways
	SU-FAA	Boeing 707-138B	Misr Overseas Airways
	SU-FAB	Boeing 707-138B	Misr Overseas Airways
	SU-FAC	Boeing 707-323C	Misr Overseas Airways
	SU-GAA	A.300B4 Airbus	EgyptAir *Isis*
	SU-GAB	A.300B4 Airbus	EgyptAir *Amun*
	SU-GAC	A.300B4 Airbus	EgyptAir *Bennou*
	SU-GAH	Boeing 767-266ER	Egyptair *Nefertiti*
	SU-GAI	Boeing 767-266ER	Egyptair *Nefertari*
	SU-GAJ	Boeing 767-266ER	Egyptair *Tiye*

SX (Greece)

	SX-BCA	Boeing 737-284	Olympic Airlines *Apollo*
	SX-BCB	Boeing 737-284	Olympic Airlines *Hermes*
	SX-BCC	Boeing 737-284	Olympic Airlines *Hercules*
	SX-BCD	Boeing 737-284	Olympic Airlines *Hephaestus*
	SX-BCE	Boeing 737-284	Olympic Airlines *Dionysus*
	SX-BCF	Boeing 737-284	Olympic Airlines *Poseidon*
	SX-BCG	Boeing 737-284	Olympic Airlines *Phoebus*
	SX-BCH	Boeing 737-284	Olympic Airlines *Triton*
	SX-BCI	Boeing 737-284	Olympic Airlines *Proteus*
	SX-BCK	Boeing 737-284	Olympic Airlines *Nereus*
	SX-BCL	Boeing 737-284	Olympic Airlines *Isle of Thassos*
	SX-BEB	A.300B4 Airbus	Olympic Airways *Odysseus*
	SX-BEC	A.300B4 Airbus	Olympic Airways *Achilleus*
	SX-BED	A.300B4 Airbus	Olympic Airways *Telemachos*
	SX-BEE	A.300B4 Airbus	Olympic Airways *Nestor*
	SX-BEF	A.300B4 Airbus	Olympic Airways *Ajax*
	SX-BEG	A.300B4 Airbus	Olympic Airways *Diamedes*
	SX-BEH	A.300B4 Airbus	Olympic Airways *Peleus*
	SX-BEI	A.300B4 Airbus	Olympic Airways *Neoptolemos*
	SX-CBA	Boeing 727-284	Olympic Airways *Mount Olympus*
	SX-CBB	Boeing 727-284	Olympic Airways *Mount Pindos*
	SX-CBC	Boeing 727-284	Olympic Airways *Mount Parnassus*
	SX-CBD	Boeing 727-284	Olympic Airways *Mount Helicon*
	SX-CBE	Boeing 727-284	Olympic Airways *Mount Athos*
	SX-CBF	Boeing 727-284	Olympic Airways *Mount Taygetus*
	SX-DBC	Boeing 707-384C	Olympic Airways *City of Knossos*
	SX-DBD	Boeing 707-384C	Olympic Airways *City of Sparta*
	SX-DBE	Boeing 707-384B	Olympic Airways *City of Pella*
	SX-DBF	Boeing 707-384B	Olympic Airways *City of Mycenae*
	SX-DBO	Boeing 707-351C	Olympic Airways *City of Lindos*
	SX-DBP	Boeing 707-351C	Olympic Airways *City of Thebes*
	SX-OAB	Boeing 747-284B	Olympic Airways *Olympic Eagle*
	SX-OAC	Boeing 747-212B	Olympic Airways *Olympic Spirit*
	SX-OAD	Boeing 747-212B	Olympic Airways *Olympic Flame*
	SX-OAE	Boeing 747-212B	Olympic Airways

TC (Turkey)

	TC-JAB	Douglas DC-9-32	Turk Hava Yollari (THY) *Bagazici*
	TC-JAD	Douglas DC-9-32	Turk Hava Yollari (THY) *Andadolu*
	TC-JAE	Douglas DC-9-32	Turk Hava Yollari (THY) *Trakya*
	TC-JAF	Douglas DC-9-32	Turk Hava Yollari (THY) *Ege*
	TC-JAG	Douglas DC-9-32	Turk Hava Yollari (THY) *Akdeniz*
	TC-JAK	Douglas DC-9-32	Turk Hava Yollari (THY) *Karadeniz*

Reg.	Type	Owner or Operator	Notes
TC-JAL	Douglas DC-9-32	Turk Hava Yollari (THY) *Halic*	
TC-JAU	Douglas DC-10-10	Türk Hava Yollari (THY) *Istanbul*	
TC-JAY	Douglas DC-10-10	Türk Hava Yollari (THY) *Izmir*	
TC-JBF	Boeing 727-2F2	Türk Hava Yollari (THY) *Adana*	
TC-JBG	Boeing 727-2F2	Türk Hava Yollari (THY) *Ankara*	
TC-JBJ	Boeing 727-2F2	Türk Hava Yollari (THY) *Diyarbakir*	
TC-JBK	Douglas DC-9-32	Turk Hava Yollari (THY) *Aydin*	
TC-JBL	Douglas DC-9-32	Turk Hava Yollari (THY) *Gediz*	
TC-JBM	Boeing 727-2F2	Türk Hava Yollari (THY) *Menderes*	
TC-JCA	Boeing 727-2F2	Türk Hava Yollari (THY) *Edirne*	
TC-JCB	Boeing 727-2F2	Türk Hava Yollari (THY) *Kars*	
TC-JCC	Boeing 707-321C	Türk Hava Yollari (THY) *Kervan I*	
TC-JCD	Boeing 727-2F2	Türk Hava Yollari (THY) *Sinop*	
TC-JCE	Boeing 727-2F2	Türk Hava Yollari (THY) *Hatay*	
TC-JCF	Boeing 707-321C	Türk Hava Yollari (THY) *Kervan II*	
TC-JCK	Boeing 727-243	Türk Hava Yollari (THY)	
TC-JCL	A310-203 Airbus	Türk Hava Yollari (THY) *Seyhan*	
TC-JCM	A310-203 Airbus	Türk Hava Yollari (THY) *Ceyhan*	
TC-JCN	A310-203 Airbus	Türk Hava Yollari (THY) *Dicle*	
TC-JCO	A310-203 Airbus	Türk Hava Yollari (THY) *Firat*	
TC-JCP	A310-203 Airbus	Türk Hava Yollari (THY)	
TC-JCR	A310-203 Airbus	Türk Hava Yollari (THY)	
TC-JCS	A310-203 Airbus	Türk Hava Yollari (THY)	

TF (Iceland)

TF-AEA	Boeing 707-	Air Arctic Icelandic	
TF-AEB	Boeing 707-338C	Air Arctic Icelandic	
TF-AEC	Boeing 707-338C	Air Arctic Icelandic	
TF-FLC	Douglas DC-8-63CF	Saudia — Saudi Arabian Airlines	
TF-FLG	Boeing 727-185C	Icelandair *Heim Fari*	
TF-FLI	Boeing 727-208	Icelandair *Fronfari*	
TF-FLT	Douglas DC-8-63	Icelandair	
TF-FLU	Douglas DC-8-63	Icelandair *Austurfari*	
TF-FLV	Douglas DC-8-63	Icelandair *Vesturfari*	
TF-GCV	Douglas DC-8-54F	Cargolux	
TF-ISA	Douglas DC-8-61	Eagle Air	
TF-VLJ	Boeing 707-324C	Eagle Air	
TF-VLT	Boeing 737-205C	Eagle Air	
TF-VLV	Boeing 707-382B	Eagle Air	
TF-VLW	Douglas DC-8-61	Eagle Air	

Note: Icelandair also operates DC-8s on lease.

TJ (Cameroon)

TJ-CAA	Boeing 707-3H7C	Cameroon Airlines *La Sanaga*	
TJ-CAB	Boeing 747-2H7B	Cameroon Airlines *Mont Cameroun*	

TR (Gabon)

TR-LVK	Douglas DC-8-55F	Air Gabon Cargo/Affretair	

Note: Air Gabon operates Boeing 747-2Q2B F-ODJG on lease.

TS (Tunisia)

TS-IEB	Boeing 737-2L9	Tunis Air	
TS-IMA	A.300B4 Airbus	Tunis-Air *Amilcar*	
TS-	A.300B4 Airbus	Tunis-Air	

Notes	Reg.	Type	Owner or Operator
	TS-	A.300B4 Airbus	Tunis-Air
	TS-IOC	Boeing 737-2H3	Tunis-Air *Salammbo*
	TS-IOD	Boeing 737-2H3C	Tunis-Air *Bulla Regia*
	TS-IOE	Boeing 737-2H3	Tunis-Air *Zarzis*
	TS-IOF	Boeing 737-2H3	Tunis-Air *Sousse*
	TS-JHN	Boeing 727-2H3	Tunis-Air *Carthago*
	TS-JHQ	Boeing 727-2H3	Tunis-Air *Tozeur-Nefta*
	TS-JHR	Boeing 727-2H3	Tunis-Air *Bizerte*
	TS-JHS	Boeing 727-2H3	Tunis-Air *Kairouan*
	TS-JHT	Boeing 727-2H3	Tunis-Air *Sidi Bousaid*
	TS-JHU	Boeing 727-2H3	Tunis-Air *Hannibal*
	TS-JHV	Boeing 727-2H3	Tunis-Air *Jugurtha*
	TS-JHW	Boeing 727-2H3	Tunis-Air *Ibn Khaldoun*

TU (Ivory Coast)

	TU-TAL	Douglas DC-10-30	Air Afrique *Libreville*
	TU-TAM	Douglas DC-10-30	Air Afrique
	TU-TAN	Douglas DC-10-30	Air Afrique *Niamey*
	TU-TAO	A.300B4-203 Airbus	Air Afrique *Nouackchott*
	TU-TAS	A.300B4-203 Airbus	Air Afrique *Bangui*
	TU-TAT	A.300B4-203 Airbus	Air Afrique

TZ (Mali)

	TZ-ADL	Boeing 737-2D6	Air Mali/Britannia Airways
	TZ-ADR	Boeing 727-173C	Air Mali

VH (Australia)

	VH-EBG	Boeing 747-238B	Qantas Airways *City of Hobart*
	VH-EBH	Boeing 747-238B	Qantas Airways *City of Newcastle*
	VH-EBI	Boeing 747-238B	Qantas Airways *City of Darwin*
	VH-EBJ	Boeing 747-238B	Qantas Airways *City of Geelong*
	VH-EBL	Boeing 747-238B	Qantas Airways *City of Townsville*
	VH-EBM	Boeing 747-238B	Qantas Airways *City of Parramatta*
	VH-EBN	Boeing 747-238B	Qantas Airways *City of Albury*
	VH-EBO	Boeing 747-238B	Qantas Airways *City of Elizabeth*
	VH-EBP	Boeing 747-238B	Qantas Airways *City of Freemantle*
	VH-EBQ	Boeing 747-238B	Qantas Airways *City of Bunbury*
	VH-EBR	Boeing 747-238B	Qantas Airways *City of Dubbo*
	VH-EBS	Boeing 747-238B	Qantas Airways *City of Longreach*
	VH-EBT	Boeing 747-338	Qantas Airways *City of Canberra*
	VH-EBU	Boeing 747-338	Qantas Airways *City of Sydney*
	VH-EBV	Boeing 747-338	Qantas Airways *City of Melbourne*

VR-H (Hong Kong)

	VR-HIA	Boeing 747-267B	Cathay Pacific Airways
	VR-HIB	Boeing 747-267B	Cathay Pacific Airways
	VR-HIC	Boeing 747-267B	Cathay Pacific Airways
	VR-HID	Boeing 747-267B	Cathay Pacific Airways
	VR-HIE	Boeing 747-267B	Cathay Pacific Airways
	VR-HIF	Boeing 747-267B	Cathay Pacific Airways
	VR-HIH	Boeing 747-367	Cathay Pacific Airways
	VR-HII	Boeing 747-367	Cathay Pacific Airways
	VR-HIK	Boeing 747-367	Cathay Pacific Airways

Reg.	Type	Owner or Operator	Notes
VR-H	Boeing 747-367	Cathay Pacific Airways	
VR-HKG	Boeing 747-267B	Cathay Pacific Airways	
VR-HVY	Boeing 747-236F	Cathay Pacific Airways	

VT (India)

Reg.	Type	Owner or Operator	Notes
VT-DPM	Boeing 707-337B	Air-India *Dhaulagiri*	
VT-DSI	Boeing 707-337B	Air-India *Lhotse*	
VT-DVA	Boeing 707-337B	Air-India *Annapoorna*	
VT-DVB	Boeing 707-337C	Air-India *Kamet*	
VT-DXT	Boeing 707-337C	Air-India *Trishul*	
VT-EBE	Boeing 747-237B	Air-India *Emperor Shahjehan*	
VT-EBN	Boeing 747-237B	Air-India *Emperor Rajendra Chola*	
VT-EBO	Boeing 747-237B	Air-India *Emperor Nikramaditya*	
VT-EDU	Boeing 747-237B	Air-India *Emperor Akbar*	
VT-EFJ	Boeing 747-237B	Air-India *Emperor Chandragupta*	
VT-EFO	Boeing 747-237B	Air-India *Emperor Kanishka*	
VT-EFU	Boeing 747-237B	Air-India *Emperor Krishna Deva*	
VT-EGA	Boeing 747-237B	Air-India *Emperor Samudra Gupto*	
VT-EGB	Boeing 747-237B	Air-India *Emperor Mahendra Varman*	
VT-EGC	Boeing 747-237B	Air India *Emperor Harsha Vardhuma*	
VT-EJG	A.310-304 Airbus	Air India *Vamuna*	
VT-EJH	A.310-304 Airbus	Air India *Tista*	
VT-EJI	A.310-304 Airbus	Air India *Saraswati*	
VT-EJJ	A.310-304 Airbus	Air India	
VT-EJK	A.310-304 Airbus	Air India	
VT-EJL	A.310-304 Airbus	Air India	

Note: Air-India Cargo operates Douglas DC-8s on lease from various airlines.

YI (Iraq)

Reg.	Type	Owner or Operator	Notes
YI-AGE	Boeing 707-370C	Iraqi Airways	
YI-AGG	Boeing 707-370C	Iraqi Airways	
YI-AGN	Boeing 747-270C	Iraqi Airways	
YI-AGO	Boeing 747-270C	Iraqi Airways	
YI-AGP	Boeing 747-270C	Iraqi Airways	
YI-AKO	Ilyushin IL-76M	Iraqi Airways	
YI-AKP	Ilyushin IL-76M	Iraqi Airways	
YI-AKQ	Ilyushin IL-76M	Iraqi Airways	
YI-AKS	Ilyushin IL-76M	Iraqi Airways	
YI-AKT	Ilyushin IL-76M	Iraqi Airways	
YI-AKU	Ilyushin IL-76M	Iraqi Airways	
YI-AKV	Ilyushin IL-76M	Iraqi Airways	
YI-AKW	Ilyushin IL-76M	Iraqi Airways	
YI-AKX	Ilyushin IL-76M	Iraqi Airways	
YI-ALL	Ilyushin IL-76M	Iraqi Airways	
YI-ALM	Boeing 747SP-70	Iraqi Airways *Al Qadissiya*	
YI-ALO	Ilyushin IL-76M	Iraqi Airways	
YI-ALP	Ilyushin IL-76M	Iraqi Airways	
YI-ALQ	Ilyushin IL-76MD	Iraqi Airways	
YI-ALR	Ilyushin IL-76MD	Iraqi Airways	
YI-ALS	Ilyushin IL-76MD	Iraqi Airways	
YI-ALT	Ilyushin IL-76MD	Iraqi Airways	

YK (Syria)

SYRIAN ARAB AIRLINES

Reg.	Type	Owner or Operator	Notes
YK-AGA	Boeing 727-29A	Syrian Arab Airlines *October 6*	
YK-AGB	Boeing 727-294	Syrian Arab Airlines *Damascus*	

Notes	Reg.	Type	Owner or Operator
	YK-AGC	Boeing 727-294	Syrian Arab Airlines *Palmyra*
	YK-AHA	Boeing 747SP-94	Syrian Arab Airlines *16 Novembre*
	YK-AHB	Boeing 747SP-94	Syrian Arab Airlines *Arab Solidarity*
	YK-AIA	Tupolev Tu-154M	Syrian Arab Airlines
	YK-AIB	Tupolev Tu-154M	Syrian Arab Airlines
	YK-AIC	Tupolev Tu-154M	Syrian Arab Airlines
	YK-ATA	Ilyushin IL-76M	Syrian Arab Airlines
	YK-ATB	Ilyushin IL-76M	Syrian Arab Airlines
	YK-ATC	Ilyushin IL-76T	Syrian Arab Airlines
	YK-ATD	Ilyushin IL-76T	Syrian Arab Airlines

YR (Romania)

	YR-ABA	Boeing 707-3K1C	Tarom
	YR-ABC	Boeing 707-3K1C	Tarom
	YR-ABM	Boeing 707-321C	Tarom
	YR-ABN	Boeing 707-321C	Tarom
	YR-BCB	BAC One-Eleven 424EU	Tarom
	YR-BCC	BAC One-Eleven 424EU	Liniile Aeriene Romane (LAR)
	YR-BCD	BAC One-Eleven 424EU	Liniile Aeriene Romane (LAR)
	YR-BCE	BAC One-Eleven 424EU	Tarom
	YR-BCF	BAC One-Eleven 424EU	Liniile Aeriene Romane (LAR)
	YR-BCG	BAC One-Eleven 401AK	Tarom
	YR-BCH	BAC One-Eleven 402AP	Tarom
	YR-BCI	BAC One-Eleven 525FT	Tarom
	YR-BCJ	BAC One-Eleven 525FT	Tarom
	YR-BCK	BAC One-Eleven 525FT	Tarom
	YR-BCL	BAC One-Eleven 525FT	Tarom
	YR-BCM	BAC One-Eleven 525FT	Tarom
	YR-BCN	BAC One-Eleven 525FT	Tarom
	YR-BCO	BAC One-Eleven 525FT	Tarom
	YR-BCR	BAC One-Eleven 487GK	Tarom
	YR-BRA	Rombac One-Eleven 560	Tarom
	YR-BRB	Rombac One-Eleven 560	Tarom
	YR-BRC	Rombac One-Eleven 560	Tarom
	YR-BRD	Rombac One-Eleven 560	Tarom
	YR-BRE	Rombac One-Eleven 560	Tarom
	YR-BRF	Rombac One-Eleven 560	Tarom
	YR-IMC	Ilyushin IL-18V	Tarom
	YR-IMD	Ilyushin IL-18V	Tarom
	YR-IMG	Ilyushin IL-18V	Tarom
	YR-IMH	Ilyushin IL-18V	Tarom
	YR-IMJ	Ilyushin IL-18D	Tarom
	YR-IML	Ilyushin IL-18D	Tarom
	YR-IMM	Ilyushin IL-18D	Tarom
	YR-IMZ	Ilyushin IL-18V	Tarom
	YR-IRA	Ilyushin IL-62	Tarom
	YR-IRB	Ilyushin IL-62	Tarom
	YR-IRC	Ilyushin IL-62	Tarom
	YR-IRD	Ilyushin IL-62M	Tarom
	YR-IRE	Ilyushin IL-62M	Tarom
	YR-TPA	Tupolev Tu-154B	Tarom
	YR-TPB	Tupolev Tu-154B	Tarom
	YR-TPC	Tupolev Tu-154B	Tarom
	YR-TPD	Tupolev Tu-154B	Tarom
	YR-TPE	Tupolev Tu-154B	Tarom
	YR-TPF	Tupolev Tu-154B	Tarom
	YR-TPG	Tupolev Tu-154B	Tarom
	YR-TPI	Tupolev Tu-154B	Tarom
	YR-TPJ	Tupolev Tu-154B	Tarom
	YR-TPK	Tupolev Tu-154B	Tarom
	YR-TPL	Tupolev Tu-154B	Tarom

YU (Yugoslavia)

INEX ADRIA

AVIOGENEX

Reg.	*Type*	*Owner or Operator*	*Notes*
YU-AGE	Boeing 707-340C	Jugoslovenski Aerotransport	
YU-AGG	Boeing 707-340C	Jugoslovenski Aerotransport	
YU-AGI	Boeing 707-351C	Jugoslovenski Aerotransport	
YU-AGJ	Boeing 707-351C	Jugoslovenski Aerotransport	
YU-AHJ	Douglas DC-9-32	Inex Adria Ariways *Ljubljana*	
YU-AHL	Douglas DC-9-32	Jugoslovenski Aerotransport	
YU-AHM	Douglas DC-9-32	Jugoslovenski Aerotransport *Tivat*	
YU-AHN	Douglas DC-9-32	Jugoslovenski Aerotransport	
YU-AHO	Douglas DC-9-32	Jugoslovenski Aerotransport	
YU-AHP	Douglas DC-9-32	Jugoslovenski Aerotransport	
YU-AHU	Douglas DC-9-32	Jugoslovenski Aerotransport	
YU-AHV	Douglas DC-9-32	Jugoslovenski Aerotransport	
YU-AHW	Douglas DC-9-33CF	Inex Adria Airways *Sarajevo*	
YU-AHX	Tupolev Tu-134A	Aviogenex *Beograd*	
YU-AHY	Tupolev Tu-134A	Aviogenex *Zagreb*	
YU-AJA	Tupolev Tu-134A	Aviogenex *Titograd*	
YU-AJD	Tupolev Tu-134A	Aviogenex *Skopje*	
YU-AJF	Douglas DC-9-32	Inex Adria Airways	
YU-AJH	Douglas DC-9-32	Jugoslovenski Aerotransport	
YU-AJI	Douglas DC-9-32	Jugoslovenski Aerotransport	
YU-AJJ	Douglas DC-9-32	Jugoslovenski Aerotransport	
YU-AJK	Douglas DC-9-32	Jugoslovenski Aerotransport	
YU-AJL	Douglas DC-9-32	Jugoslovenski Aerotransport	
YU-AJM	Douglas DC-9-32	Jugoslovenski Aerotransport	
YU-AJT	Douglas DC-9-51	Inex Adria Airways	
YU-AJU	Douglas DC-9-51	Inex Adria Airways *Maribor*	
YU-AJV	Tupolev Tu-134A	Aviogenex *Mostar*	
YU-AJW	Tupolev Tu-134A	Aviogenex *Pristina*	
YU-AJZ	Douglas DC-9-81	Inex Adria Airways	
YU-AKA	Boeing 727-2H9	Jugoslovenski Aerotransport	
YU-AKB	Boeing 727-2H9	Jugoslovenski Aerotransport	
YU-AKD	Boeing 727-2L8	Aviogenex *Zagreb*	
YU-AKE	Boeing 727-2H9	Jugoslovenski Aerotransport	
YU-AKF	Boeing 727-2H9	Jugoslovenski Aerotransport	
YU-AKG	Boeing 727-2H9	Jugoslovenski Aerotransport	
YU-AKH	Boeing 727-2L8	Aviogenex *Beograd*	
YU-AKI	Boeing 727-2H9	Jugoslovenski Aerotransport	
YU-AKJ	Boeing 727-2H9	Jugoslovenski Aerotransport	
YU-AKK	Boeing 727-2H9	Jugoslovenski Aerotransport	
YU-AKL	Boeing 727-2H9	Jugoslovenski Aerotransport	
YU-AKM	Boeing 727-243	Aviogenex *Pula*	
YU-AMA	Douglas DC-10-30	Jugoslovenski Aerotransport *Nikola Tesla*	
YU-AMB	Douglas DC-10-30	Jugoslovenski Aerotransport *Edvard Rusijan*	
YU-ANB	Douglas DC-9-82	Inex Adria Airways	
YU-ANC	Douglas DC-9-82	Inex Adria Airways	
YU-AND	Boeing 737-3H9	Jugoslovenski Aerotransport	
YU-ANE	Tupolev Tu-134A	Aviogenex *Novi Sad*	
YU-ANF	Boeing 737-3H9	Jugoslovenski Aerotransport	
YU-ANG	Douglas DC-9-82	Inex Adria Airways	
YU-ANH	Boeing 737-3H9	Jugoslovenski Aerotransport	
YU-ANI	Boeing 737-3H9	Jugoslovenski Aerotransport	
YU-ANJ	Tupolev Tu-134A	Aviogenex	

YV (Venezuela)

YV-134C	Douglas DC-10-30	Viasa	
YV-135C	Douglas DC-10-30	Viasa	
YV-136C	Douglas DC-10-30	Viasa	
YV-137C	Douglas DC-10-30	Viasa	
YV-138C	Douglas DC-10-30	Viasa	
YV-	Douglas DC-10-30	Viasa	

Z (Zimbabwe)

Notes	Reg.	Type	Owner or Operator
	Z-WKR	Boeing 707-330B	Air Zimbabwe
	Z-WKS	Boeing 707-330B	Air Zimbabwe
	Z-WKT	Boeing 707-330B	Air Zimbabwe
	Z-WKU	Boeing 707-330B	Air Zimbabwe
	Z-WKV	Boeing 707-330B	Air Zimbabwe
	Z-WMJ	Douglas DC-8-54F	Affretair *Captain Jack Malloch*

Note: Affretair also operates DC-8-55 TR-LVK.

ZK (New Zealand)

air new zealand
The Ritz of the Skies

	ZK-NZU	Boeing 747-219B	Air New Zealand
	ZK-NZV	Boeing 747-219B	Air New Zealand *Aotea*
	ZK-NZW	Boeing 747-219B	Air New Zealand *Tainui*
	ZK-NZX	Boeing 747-219B	Air New Zealand *Takitimu*
	ZK-NZY	Boeing 747-219B	Air New Zealand *Te Arawa*
	ZK-NZZ	Boeing 747-219B	Air New Zealand *Tokomaru*

ZS (South Africa)

	ZS-SAL	Boeing 747-244B	South African Airways *Tafelberg*
	ZS-SAM	Boeing 747-244B	South African Airways *Drakensberg*
	ZS-SAN	Boeing 747-244B	South African Airways *Lebombo*
	ZS-SAO	Boeing 747-244B	South African Airways *Magaliesberg*
	ZS-SAP	Boeing 747-244B	South African Airways *Swartberg*
	ZS-SAR	Boeing 747-244B	South African Airways *Waterberg*
	ZS-SAS	Boeing 747-244B	South African Airways *Helderberg*
	ZS-SAT	Boeing 747-344	South African Airways *Johannesburg*
	ZS-SAU	Boeing 747-344	South African Airways *Cape Town*
	ZS-SPA	Boeing 747SP-44	South African Airways *Matroosberg*
	ZS-SPB	Boeing 747SP-44	South African Airways *Outeniqua*
	ZS-SPE	Boeing 747SP-44	South African Airways *Hantam*
	ZS-SPF	Boeing 747SP-44	South African Airways *Soutpansberg*

3B (Mauritius)

	3B-NAE	Boeing 707-344B	Air Mauritius *City of Port Louis*
	3B-NAF	Boeing 707-344B	Air Mauritius *Ville de Curepipe*
	3B-NAG	Boeing 747SP-44	Air Mauritius *Chateau du Reduit*

3X (Guinea)

	3X-GAZ	Boeing 707-351C	Air Guinee

4R (Sri Lanka)

	4R-ULC	L.1011-385 TriStar 100	Air Lanka *City of Anuradhapu*
	4R-ULD	L.1011-385 TriStar 100	Air Lanka *City of Galle*
	4R-ULE	L.1011-385 TriStar 1	Air Lanka *City of Ratnapura*
	4R-ULF	Boeing 747-238B	Air Lanka
	4R-ULG	Boeing 747-238B	Air Lanka *King Tissa*

4W (Yemen)

Reg.	Type	Owner or Operator	Notes
4W-ACF	Boeing 727-2N8	Yemen Airways	
4W-ACG	Boeing 727-2N8	Yemen Airways	
4W-ACH	Boeing 727-2N8	Yemen Airways	
4W-ACI	Boeing 727-2N8	Yemen Airways	
4W-ACJ	Boeing 727-2N8	Yemen Airways	

4X (Israel)

4X-ABN	Boeing 737-258	El Al	
4X-ABO	Boeing 737-258	El Al	
4X-ATB	Boeing 707-458	El Al	
4X-ATD	Boeing 707-331B	El Al/Arkia	
4X-ATR	Boeing 707-358B	El Al	
4X-ATS	Boeing 707-358B	El Al	
4X-ATT	Boeing 707-358B	El Al	
4X-ATX	Boeing 707-358C	El Al	
4X-ATY	Boeing 707-358C	Sun d'Or International Airlines	
4X-AXA	Boeing 747-258B	El Al	
4X-AXB	Boeing 747-258B	El Al	
4X-AXC	Boeing 747-258B	El Al	
4X-AXD	Boeing 747-258C	El Al	
4X-AXF	Boeing 747-258C	El Al	
4X-AXG	Boeing 747-258F	El Al	
4X-AXH	Boeing 747-258B	El Al	
4X-AXZ	Boeing 747-124F	El Al	
4X-EAA	Boeing 767-258	El Al	
4X-EAB	Boeing 767-258	El Al	
4X-EAC	Boeing 767-258ER	El Al	
4X-EAD	Boeing 767-258ER	El Al	

4YB (Jordan)

4YB-CAB	Boeing 707-321C	Arab Air Cargo	
4YB-CAC	Boeing 707-370C	Arab Air Cargo	

5A (Libya)

5A-DAI	Boeing 727-224	Libyan Arab Airlines	
5A-DAK	Boeing 707-3L5C	Libyan Arab Airlines	
5A-DIA	Boeing 727-2L5	Libyan Arab Airlines	
5A-DIB	Boeing 727-2L5	Libyan Arab Airlines	
5A-DIC	Boeing 727-2L5	Libyan Arab Airlines	
5A-DID	Boeing 727-2L5	Libyan Arab Airlines	
5A-DIE	Boeing 727-2L5	Libyan Arab Airlines	
5A-DIF	Boeing 727-2L5	Libyan Arab Airlines	
5A-DIG	Boeing 727-2L5	Libyan Arab Airlines	
5A-DIH	Boeing 727-2L5	Libyan Arab Airlines	
5A-DII	Boeing 727-2L5	Libyan Arab Airlines	
5A-DIK	Boeing 707-328C	Libyan Arab Airlines	
5A-DJM	Boeing 707-321B	Libyan Arab Airlines	
5A-DLT	Boeing 707-328B	Libyan Arab Airlines	

5B (Cyprus)

Notes	Reg.	Type	Owner or Operator
	5B-DAG	BAC One Eleven 537GF	Cyprus Airways
	5B-DAH	BAC One Eleven 537GF	Cyprus Airways
	5B-DAJ	BAC One Eleven 537GF	Cyprus Airways
	5B-DAL	Boeing 707-123B	Cyprus Airways
	5B-DAO	Boeing 707-123B	Cyprus Airways
	5B-DAP	Boeing 707-123B	Cyprus Airways
	5B-DAQ	A.310-203 Airbus	Cyprus Airways
	5B-DAR	A.310-203 Airbus	Cyprus Airways
	5B-DAS	A.310-203 Airbus	Cyprus Airways

5N (Nigeria)

	5N-ABJ	Boeing 707-3F9C	Nigeria Airways
	5N-ABK	Boeing 707-3F9C	Nigeria Airways
	5N-ANN	Douglas DC-10-30	Nigeria Airways
	5N-ANO	Boeing 707-3F9C	Nigeria Airways
	5N-ANR	Douglas DC-10-30	Nigeria Airways
	5N-AON	Douglas DC-8-62	Okada Air
	5N-AOQ	Boeing 707-355C	Okada Air
	5N-ARQ	Boeing 707-338C	G.A.S. Air Cargo
	5N-ASY	Boeing 707-351C	United Air Services
	5N-AUE	A.310-221 Airbus	Nigeria Airways
	5N-AUF	A.310-221 Airbus	Nigeria Airways
	5N-AUG	A.310-221 Airbus	Nigeria Airways
	5N-AUH	A.310-221 Airbus	Nigeria Airways
	5N-AVR	Douglas DC-8-52	Intercontinental Airlines
	5N-AVS	Douglas DC-8-52	Intercontinental Airlines (*stored*)

5R (Madagascar)

	5R-MFT	Boeing 747-2B2B	Air Madagascar *Tolom Piavotana*

5X (Uganda)

	5X-UAC	Boeing 707-351C	Uganda Airlines
	5X-UBC	Boeing 707-338C	Uganda Airlines *Pearl of Africa*
	5X-UCF	Lockheed L382G Hercules	Uganda Airlines

5Y (Kenya)

Kenya Airways

	5Y-BBI	Boeing 707-351B	Kenya Airlines
	5Y-BBJ	Boeing 707-351B	Kenya Airlines
	5Y-BBK	Boeing 707-351B	Kenya Airlines
	5Y-BBX	Boeing 720-047B	Kenya Airlines
	5Y-	A.310-322 Airbus	Kenya Airlines
	5Y-	A.310-322 Airbus	Kenya Airlines

6O (Somalia)

Reg.	*Type*	*Owner or Operator*	*Notes*
6O-SBS	Boeing 707-330B	Somali Airlines	
6O-SBT	Boeing 707-330B	Somali Airlines	

6Y (Jamaica)

Note: Air Jamaica operates its UK services jointly with British Airways.

7T (Algeria)

7T-VEA	Boeing 727-2D6	Air Algerie *Tassili*	
7T-VEB	Boeing 727-2D6	Air Algerie *Hoggar*	
7T-VED	Boeing 737-2D6C	Air Algerie *Atlas Saharien*	
7T-VEE	Boeing 737-2D6C	Air Algerie *Oasis*	
7T-VEF	Boeing 737-2D6	Air Algerie *Saoura*	
7T-VEG	Boeing 737-2D6	Air Algerie *Monts des Ouleds Neils*	
7T-VEH	Boeing 727-2D6	Air Algerie *Lalla Khadidja*	
7T-VEI	Boeing 727-2D6	Air Algerie *Djebel Amour*	
7T-VEJ	Boeing 737-2D6	Air Algerie *Chrea*	
7T-VEK	Boeing 737-2D6	Air Algerie *Edough*	
7T-VEL	Boeing 737-2D6	Air Algerie *Akfadou*	
7T-VEM	Boeing 727-2D6	Air Algerie *Mont du Ksall*	
7T-VEN	Boeing 737-2D6	Air Algerie *La Soummam*	
7T-VEO	Boeing 737-2D6	Air Algerie *La Titteri*	
7T-VEP	Boeing 727-2D6	Air Algerie *Mont du Tessala*	
7T-VEQ	Boeing 737-2D6	Air Algerie *Le Zaccar*	
7T-VER	Boeing 737-2D6	Air Algerie *Le Souf*	
7T-VES	Boeing 737-2D6C	Air Algerie *Le Tadmaït*	
7T-VET	Boeing 727-2D6	Air Algerie *Georges du Rhumel*	
7T-VEU	Boeing 727-2D6	Air Algerie	
7T-VEV	Boeing 727-2D6	Air Algerie	
7T-VEW	Boeing 727-2D6	Air Algerie	
7T-VEX	Boeing 727-2D6	Air Algerie	
7T-VEY	Boeing 737-2D6	Air Algerie *Rhoufi*	
7T-VEZ	Boeing 737-2T4	Air Algerie *Monts du Daia*	
7T-VJA	Boeing 737-2T4	Air Algerie *Monts des Babors*	
7T-VJB	Boeing 737-2T4	Air Algerie *Monts des Bibons*	
7T-VJC	A.310-203 Airbus	Air Algerie	
7T-	A.310-203 Airbus	Air Algerie	

9G (Ghana)

9G-ADB	Boeing 707-331C	West Coast Airlines	
9G-ANA	Douglas DC-10-30	Ghana Airways	

9H (Malta)

9H-AAK	Boeing 720-047B	Air Malta	
9H-AAL	Boeing 720-047B	Air Malta	
9H-AAO	Boeing 720-047B	Air Malta	
9H-ABA	Boeing 737-2Y5	Air Malta	
9H-ABB	Boeing 737-2Y5	Air Malta	
9H-ABC	Boeing 737-2Y5	Air Malta	

9J (Zambia)

Zambia Airways

Notes	Reg.	Type	Owner or Operator
	9J-ADY	Boeing 707-349C (Cargo)	Zambia Airways
	9J-AEB	Boeing 707-351C	Zambia Airways
	9J-AEL	Boeing 707-338C	Zambia Airways
	9J-AEQ	Boeing 707-321C (Cargo)	Zambia Airways

Note: Zambia Airways also operates DC-10-30 N3016Z which is expected to be allocated a 9J registration.

9K (Kuwait)

	9K-ADA	Boeing 747-269B	Kuwait Airways *Al Sabahiya*
	9K-ADB	Boeing 747-269B	Kuwait Airways *Al Jaberiya*
	9K-ADC	Boeing 747-269B	Kuwait Airways *Al Murbarakiya*
	9K-ADD	Boeing 747-269B	Kuwait Airways *Al Salmiya*
	9K-AHA	A.310-222 Airbus	Kuwait Airways
	9K-AHB	A.310-222 Airbus	Kuwait Airways
	9K-AHD	A.310-222 Airbus	Kuwait Airways *Failaka*
	9K-AHE	A.310-222 Airbus	Kuwait Airways
	9K-AHF	A.300-620C Airbus	Kuwait Airways
	9K-AHG	A.300-620C Airbus	Kuwait Airways
	9K-AHI	A.300-620C Airbus	Kuwait Airways
	9K-AIA	Boeing 767-269	Kuwait Airways
	9K-AIB	Boeing 767-269	Kuwait Airways
	9K-AIC	Boeing 767-269	Kuwait Airways

9L (Sierra Leone)

Sierra Leone Airways' services between Freetown and London are operated by using Boeing 707 JY-AEB leased from Alia.

9M (Malaysia)

mas
malaysian airline system

	9M-MHI	Boeing 747-236B	Malaysian Airline System
	9M-MHJ	Boeing 747-236B	Malaysian Airline System

9Q (Zaïre)

	9Q-CKQ	Canadair CL-44-6	Virunga Air Cargo
	9Q-CLI	Douglas DC-10-30	Air Zaïre *Mont Ngaliema*
	9Q-CQU	Canadair CL-44D4	Virunga Air Cargo
	9Q-CVG	Boeing 707-329C	Katale Aero Transport
	9Q-CZF	Boeing 707-344	Inter-Frêt Transport Aérien
	9Q-CZK	Boeing 707-321F	Inter-Frêt Transport Aérien

9V (Singapore)

Reg.	Type	Owner or Operator	Notes
9V-SKA	Boeing 747-312	Singapore Airlines	
9V-SKI	Boeing 747-312	Singapore Airlines	
9V-SKM	Boeing 747-312	Singapore Airlines	
9V-SKN	Boeing 747-312	Singapore Airlines	
9V-SQK	Boeing 747-212B	Singapore Airlines	
9V-SQL	Boeing 747-212B	Singapore Airlines	
9V-SQM	Boeing 747-212B	Singapore Airlines	
9V-SQN	Boeing 747-212B	Singapore Airlines	
9V-SQO	Boeing 747-212B	Singapore Airlines	
9V-SQP	Boeing 747-212B	Singapore Airlines	
9V-SQQ	Boeing 747-212B	Singapore Airlines	
9V-SQR	Boeing 747-212B	Singapore Airlines	
9V-SQS	Boeing 747-212B	Singapore Airlines	

Note: Singapore Airlines also operates Boeing 747-312Bs N116KB, N117KC, N118KD, N119KE, N120KF, N121KG, N122KH, N123KJ, N124KK and N125KL.

9XR (Rwanda)

9XR-JA	Boeing 707-328C	Air Rwanda	

9Y (Trinidad and Tobago)

9Y-TGJ	L.1011 TriStar 500	B.W.I.A. *Flamingo*	
9Y-TGN	L.1011 TriStar 500	B.W.I.A.	
9Y-THA	L.1011 TriStar 500	B.W.I.A.	

Note: B.W.I.A. also operates a TriStar 500 which retains its US registration N3140D.

Radio Frequencies

The frequencies used by the larger airfields/airports are listed below. Abbreviations used: TWR — Tower, APP — Approach, A/G — Air-ground advisory. It is possible for changes to be made from time to time with the frequencies allocated which are all quoted in Megahertz (MHz).

Airfield	*TWR*	*APP*	*A/G*
Aberdeen	118.1	120.4	
Aldergrove	118.3	120.0	
Alderney	123.6		
Andrewsfield			130.55
Barton			122.7
Barrow			123.2
Bembridge			123.25
Biggin Hill	129.4	118.42	
Birmingham	118.3	120.5	
Blackbushe			122.3
Blackpool	118.4	118.4	
Bodmin			122.7
Booker			121.15
Bourn			129.8
Bournemouth	125.6	118.65	
Bristol	120.55	127.75	
Cambridge	122.2	123.6	
Cardiff	121.2	125.85	
Carlisle			123.6
Compton Abbas			122.7
Coventry	119.25	119.25	
Cranfield	123.2	122.85	
Crowland			122.6
Denham			130.72
Doncaster			122.9
Dundee	122.9	122.9	
Dunkeswell			123.5
Dunsfold	130.0	122.55	
Duxford			123.5
East Midlands	124.0	119.65	
Edinburgh	118.7	121.2	
Elstree			122.4
Exeter	119.8	128.15	
Fairoaks			123.42
Felthorpe			123.5
Fenland			123.05
Filton	124.95	130.85	
Gamston			130.47
Gatwick	124.22	119.6	
Glasgow	118.8	119.1	
Goodwood	119.7	122.45	
Guernsey	119.95	128.65	
Halfpenny Green			123.0
Hatfield	130.8	123.35	
Haverfordwest			122.2
Hawarden	124.95	123.35	
Hayes Heliport			123.65
Headcorn			122.0
Heathrow	118.7	119.2	
	121.0	119.5	
Hethal			122.35
Hucknall			130.8
Humberside	118.55	123.15	
Ingoldmells			130.45
Inverness	122.6	122.6	
Ipswich	123.25		
Jersey	119.45	120.3	
Kidlington	119.8	130.3	
Land's End			122.3
Leavesden	122.15		
Leeds	120.3	123.75	
Leicester			122.25
Little Snoring			122.4
Liverpool	118.1	119.85	
Long Marston			130.1
Luton	120.2	129.55	
Lydd	131.0	120.7	
Manchester	118.7	119.4	
Manston	124.9	126.35	
Netherthorpe			123.5
Newcastle	119.7	126.35	
North Denes			120.45
Norwich	118.9	119.35	
Panshanger			120.25
Perth	119.8	122.3	
Plymouth	122.6	123.2	
Popham			129.8
Prestwick	118.15	120.55	
Redhill			123.22
Rochester			122.25
Ronaldsway	118.9	120.85	
Sandown			123.5
Seething			122.6
Sherburn			122.6
Shobdon			123.5
Shoreham	125.4	123.15	
Sibson			122.3
Sleap			122.45
Southampton	118.2	128.85	
Southend	119.7	128.95	
Stansted	118.15	126.95	
Stapleford			122.8
Staverton	125.65		
Sumburgh	118.25	123.15	
Swansea	119.7		
Swanton Morley	123.5		
Sywell			122.7
Tees-side	119.8	118.85	
Thruxton			130.45
Tollerton			122.8
Wellesbourne			130.45
Weston	122.5		
White Waltham	122.6		
Wick	119.7		
Wickenby			122.45
Woodford	122.5	130.05	
Yeovil	125.4	130.8	

Airline Codes

Two character codes are used by airlines to prefix flight numbers in timetables, airport movement boards, etc. Those listed below identify both U.K. and overseas airlines appearing in the book.

AA	American A/L	N
AC	Air Canada	C
AE	Air Europe	G
AF	Air France	F
AH	Air Algerie	7T
AI	Air India	VT
AK	Air Bridge	G
AO	Aviaco	EC
AR	Aerolineas Argentinas	LV
AT	Royal Air Maroc	CN
AX	Connectair	G
AY	Finnair	OH
AZ	Alitalia	I
BA	British Airways	G
BB	Balair	HB
BD	British Midland	G
BG	Bangladesh Biman	S2
BM	ATI	I
BR	British Caledonian	G
BS	Busy Bee	LN
BU	Braathens	LN
BW	B.W.I.A.	9Y
BX	Spantax	EC
BY	Britannia	G
BZ	Brown Air	G
CA	CAAC	B
CC	Air Freight Egypt	SU
CO	Continental A/L	N
CP	CP Air	C
CS	Corse Air	F
CU	Cubana	CU
CV	Cargolux	LX
CX	Cathay Pacific	VR-H
CY	Cyprus A/W	5B
DA	Dan-Air	G
DB	Brit Air	F
DE	Delta Air Transport	OO
DF	Condor	D
DG	Air Atlantique	G
DK	Scanair	SE
DL	Delta A/L	N
DM	Maersk	OY
DQ	Air Limousin	F
DW	DLT	D
EA	Eastern A/L	N
EC	Air Ecosse	G
EI	Aer Lingus	EI
EL	Euralair	F
EO	Euroflite	G
ER	Sun d'Or Intl A/L	4X
ET	Ethiopian A/L	ET
EY	Europe Aero Service	F
EZ	Euroair	G
EZ	Evergreen Intl	N
FC	Fairflight	G
FD	Ford	G
FG	Ariana	YA
FI	Icelandair	TF
FO	Fred Olsen	LN
FQ	Minerve	F
FT	Flying Tiger	N
FZ	Altair	I
GA	Garuda	PK
GE	Guernsey A/L	G
GE	German Cargo	D
GF	Gulf Air	A40
GG	Air London	G
GH	Ghana A/W	9G
GI	Air Guinee	3X
GM	Swedair	SE
GR	Aurigny A/S	G
GT	GB Airways	G
GX	Global Intl	N
HA	Hawaiian Air	N
HE	Trans European A/W	OO
HF	Hapag-Lloyd	D
HI	Hispania	EC
HN	N.L.M.	PH
HV	Transavia	PH
HZ	Thurston Aviation	G
IA	Iraq A/W	YI
IB	Iberia	EC
IF	Interflug	DDR
IG	Alisarda	I
IK	Tradewinds	G
IO	TAT	F
IR	Iran Air	EP
IT	Air Inter	F
IY	Yemen A/W	4W
JA	Jetair	D
JE	Manx Airlines	G
JJ	Aviogenex	YU
JL	Japan A/L	JA
JP	Inex Adria	YU
JU	JAT	YU
JW	Arrow Air	N
JY	Jersey European	G
KB	Burnthills Aviation	G
KD	British Island A/W	G
KG	Orion A/W	G
KL	K.L.M.	PH
KM	Air Malta	9H
KQ	Kenya A/W	5Y
KR	Kar-Air	OH
KT	British Airtours	G
KU	Kuwait A/W	9K
KY	W. Africa Aircargo	9G
LC	Loganair	G
LF	Linjeflyg	SE
LG	Luxair	LX
LH	Lufthansa	D
LJ	Sierra Leone A/W	9L
LK	Lucas A/T	F
LL	Aero Lloyd	D
LN	Libyan Arab A/L	5A
LO	Polish A/L (LOT)	SP
LP	Air Alpes	F
LS	Channel Express	G
LT	LTU	D
LX	Crossair	HB
LY	El Al	4X
LZ	Bulgarian A/L	LZ
MA	Malev	HA
MB	Federal Express	N
ME	Middle East A/L	OD
MH	Malaysian A/L	9M
MK	Air Mauritius	3B
MO	Misr Overseas	SU
MP	Martinair	PH
MS	Egyptair	SU
NB	Sterling A/W	OY
NP	Heavy Lift	G
NQ	NW Territorial A/W	C
NV	Northern Executive	G
NW	Northwest Orient	N
OA	Olympic A/W	SX
OK	Czech A/L	OK
OM	Monarch A/L	G
OO	Sobelair	OO
OS	Austrian A/L	OE
OV	National A/L	N
OY	Conair	OY
PA	Pan Am	N
PJ	Peregrine A/S	G
PK	Pakistan Intl	AP
PM	Brymon A/W	G
PR	Philippine A/L	RP
PW	Pacific Western	C
QC	Air Zaire	9Q
QF	Qantas	VH
QK	Aeromaritime	F
QU	Uganda A/L	5X
QZ	Zambia A/W	9J
RB	Syrian Arab	YK
RD	Airlift Intl	N
RG	Varig	PP
RH	Air Zimbabwe	Z
RJ	Alia	JY
RM	McAlpine	G
RO	Tarom	YR
RU	CTA	HB
SA	South African A/W	ZS
SD	Sudan A/W	ST
SF	Air Charter	F
SJ	Southern A/T	N
SK	S.A.S.	SE OY LN
SN	Sabena	OO
SQ	Singapore A/L	9Q
SR	Swissair	HB
ST	S.A.T. Flug	D
SU	Aeroflot	CCCP
SV	Saudia	HZ
TE	Air New Zealand	ZK
TG	Thai Intl	HS
TK	Turkish A/L	TC
TL	Trans Mediterranean	OD
TP	Air Portugal	CS
TQ	Tyrolean A/W	OE
TU	Tunis Air	TS
TV	Transamerica	N
TW	TWA	N
TZ	American T.Air	N
UH	Bristow	G
UJ	Air Lanka	4R
UK	Air UK	G
UP	Air Foyle	G
UQ	LEA	G
UT	UTA	F
UW	Air Rwanda	9XR
UY	Cameroon A/L	TJ
VA	Viasa	YV
VF	British Air Ferries	G
VG	RFG	D
VL	Eagle Air	TF
VO	Tyrolean	OE
VS	Intercontinental	5N
VY	Air Belgium	OO
WD	Wardair	C
WE	WDL Flugdienst	D
WG	Air Ecosse	G
WN	Norfly	LN
WO	World A/W	N
WT	Nigeria A/W	5N
WU	Nether Lines	PH
5W	Nationair	C

British Aircraft Preservation Council Register

The British Aircraft Preservation Council was formed in 1967 to co-ordinate the works of all bodies involved in the preservation, restoration and display of historical aircraft. Membership covers the whole spectrum of national, Service, commercial and voluntary groups, and meetings are held regularly at the bases of member organisations. The Council is able to provide a means of communication, helping to resolve any misunderstandings or duplication of effort. Every effort is taken to encourage the raising of standards of both organisation and technical capacity amongst the member groups to the benfit of everyone interested in aviation. To assist historians, the B.A.P.C. register has been set up and provides an identity for those aircraft which do not qualify for a Service serial or inclusion in the UK Civil Register.

Aircraft on the current B.A.P.C. Register are as follows:

Notes	*Reg.*	*Type*	*Owner or Operator*
	1	Roe Triplane Type IV (replica)	Now G-ARSG
	2	Bristol Boxkite (replica)	Now G-ASPP
	3	Blériot XI	Now G-AANG
	4	Deperdussin monoplane	Now G-AANH
	5	Blackburn monoplane	Now G-AANI
	6	Roe Triplane Type IV (replica)	Manchester Air & Space Museum
	7	Southampton University MPA	The Shuttleworth Trust
	8	Dixon ornithopter	The Shuttleworth Trust
	9	Humber Monoplane (replica)	Airport Terminal/Birmingham
	10	Hafner R.11 Revoplane	Museum of Army Flying
	11	English Electric Wren	Now G-EBNV
	12	Mignet HM.14 Pou-du-Ciel	Museum of Flight/E. Fortune
	13	Mignet HM.14 Pou-du-Ciel	The Aeroplane Collection Ltd
	14	Addyman standard training glider	N. H. Ponsford
	15	Addyman standard training glider	The Aeroplane Collection Ltd
	16	Addyman ultra-light aircraft	N. H. Ponsford
	17	Woodhams Sprite	The Aeroplane Collection Ltd
	18	Killick MP Gyroplane	N. H. Ponsford
	19	Bristol F.2b	Anne Lindsay
	20	Lee-Richards annular biplane (replica)	Newark Air Musem
	22	Mignet HM.14 Pou-du-Ciel (G-AEOF)	Aviodome/Schiphol, Holland
	25	Nyborg TGN-111 glider	Midland Air Museum
	26	Auster AOP.9	S. Wales Aircraft Preservation Soc
	27	Mignet HM.14 Pou-du-Ciel	M. J. Abbey
	28	Wright Flyer (replica)	RAF Museum/Cardington
	29	Mignet HM.14 Pou-du-Ciel (G-ADRY)	J. J. Penney/Aberdare
	31	Slingsby T.7 Tutor	S. Wales Aircraft Preservation Soc
	32	Crossley Tom Thumb	Midland Air Museum
	33	DFS.108-49 Grunau Baby 116	Russavia Collection/Duxford
	34	DFS.108-49 Grunau Baby 116	D. Elsdon
	35	EoN primary glider	Russavia Collection
	36	FZG-76 (V.I) (replica)	The Shuttleworth Trust
	37	Blake Bluetit	The Shuttleworth Trust
	38	Bristol Scout replica (A1742)	RAF St Athan
	40	Bristol Boxkite (replica)	Bristol City Museum
	41	B.E.2C (replica) (6232)	RAF St Athan
	42	Avro 504 (replica) (H1968)	RAF St Athan
	43	Mignet HM.14 Pou-du-Ciel	Lincolnshire Aviation Museum
	44	Miles Magister (L6906)	G. H. R. Johnson (G-AKKY)
	45	Pilcher Hawk (replica)	Stanford Hall Museum
	46	Mignet HM.14 Pou-du-Ciel	Alan McKechnie Racing Ltd
	47	Watkins monoplane	RAF St Athan
	48	Pilcher Hawk (replica)	Glasgow Museum of Transport
	49	Pilcher Hawk	Royal Scottish Museum/Edinburgh
	50	Roe Triplane Type 1	Science Museum
	51	Vickers Vimy IV	Science Museum
	52	Lilienthal glider	Science Museum
	53	Wright Flyer (replica)	Science Museum
	54	JAP-Harding monoplane	Science Museum
	55	Levavasseur Antoinette VII	Science Museum

Reg.	Type	Owner or Operator	Notes
56	Fokker E.III	Science Museum	
57	Pilcher Hawk (replica)	Science Museum	
58	Yokosuka MXY-7 Ohka II	Science Museum	
59	Sopwith Camel (replica) (D3419)	RAF St Athan	
60	Murray M.I helicopter	The Aeroplane Collection Ltd	
61	Stewart man-powered ornithopter	Lincolnshire Aviation Museum	
62	Cody Biplane (304)	Science Museum	
63	Hurricane (replica) (L1592)	Torbay Aircraft Museum	
64	Hurricane (replica)	—	
65	Spitfire (replica) (QV-K)	—	
66	Bf 109 (replica)	—	
67	Bf 109 (replica) (14)	Midland Air Museum	
68	Hurricane (replica)	Midland Air Museum	
69	Spitfire (replica) (QV-K)	Torbay Aircraft Museum	
70	Auster AOP.5 (TJ472)	Aircraft Preservation Soc of Scotland	
71	Spitfire (replica) (P9390)	Norfolk & Suffolk Aviation Museum	
72	Hurricane (replica) (V7767)	N. Weald Aircraft Restoration Flight	
73	Hurricane (replica)	Queens Head/Bishops Stortford	
74	Bf 109 (replica) (6)	Torbay Aircraft Museum	
75	Mignet HM.14 Pou-du-Ciel	Nigel Ponsford	
76	Mignet HM.14 Pou-du-Ciel (G-AFFI)	Bomber County Museum/Cleethorpes	
77	Mignet HM.14 Pou-du-Ciel	P. Kirby/Innsworth	
78	Hawker Hind (Afghan)	Now G-AENP	
79	Fiat G.46-4 (MM53211)	British Air Reserve/Lympne	
80	Airspeed Horsa (TL769)	Museum of Army Flying	
81	Hawkridge Dagling	Russavia Collection/Duxford	
82	Hawker Hind (Afghan)	RAF Museum	
83	Kawasaki Ki-100IB	Aerospace Museum/Cosford	
84	Nakajima Ki-46 (Dinah III)	RAF St Athan	
85	Weir W-2 autogyro	Museum of Flight/E. Fortune	
86	de Havilland Tiger Moth (replica)	Yorkshire Aircraft Preservation Soc	
87	Bristol Babe (replica) (G-EASQ)	Bomber County Museum	
88	Fokker Dr 1 (replica) (102/18)	Fleet Air Arm Museum	
89	Cayley glider (replica)	Manchester Air & Space Museum	
90	Colditz Cock (replica)	Torbay Aircraft Museum	
91	Fieseler Fi 103/FZG.76 (V.I)	Lashenden Air Warfare Museum	
92	Fieseler Fi 103/FZG.76 (V.I)	RAF Museum/Henlow	
93	Fieseler Fi 103/FZG.76 (V.I)	RAF St Athan	
94	Fieseler Fi 103/FZG.76 (V.I)	Aerospace Museum/Cosford	
95	Gizmer autogyro	N.E. Aircraft Museum	
96	Brown helicopter	N.E. Aircraft Museum	
97	Luton L.A.4a Minor	Nene Valley Aviation Soc/Sibson	
98	Yokosuka MXY-7 Ohka II	RAF Museum/Henlow	
99	Yokosuka MXY-7 Ohka II	Aerospace Museum/Cosford	
100	Clarke glider	RAF Museum/Cardington	
101	Mignet HM.14 Pou-du-Ciel	Lincolnshire Aviation Museum	
103	Pilcher glider (replica)	Personal Plane Services Ltd	
104	Blériot XI (replica)	Now G-AVXV/St Athan	
105	Blériot XI (replica)	Aviodome/Schiphol, Holland	
106	Blériot XI (164)	RAF Museum	
107	Blériot XXVII (433)	RAF Museum	
108	Fairey Swordfish (HS503)	RAF Museum/Henlow	
109	Slingsby Kirby Cadet	RAF Museum/Henlow	
110	Fokker D.VII replica (static) (5125)	Leisure Sport Ltd	
111	Sopwith Triplane replica (static) (N5492)	Leisure Sport Ltd	
112	D.H.2 replica (static) (5964)	Leisure Sport Ltd	
113	S.E.5A replica (static) (B4863)	Leisure Sport Ltd	
114	Vickers Type 60 Viking (static)	Leisure Sport Ltd	
115	Mignet HM.14 Pou-du-Ciel	Essex Aviation Group/Andrewsfield	
116	Santos-Dumont Demoiselle (replica)	Cornwall Aero Park, Helston	
117	B.E.2C (replica)	N. Weald Aircraft Restoration Flight	
118	Albatross D.V. (replica)	N. Weald Aircraft Restoration Flight	
119	Bensen B.7	N.E. Aircraft Museum	
120	Mignet HM.14 Pou-du-Ciel (G-AEJZ)	Bomber County Museum/Cleethorpes	
121	Mignet HM.14 Pou-du-Ciel (G-AEKR)	S. Yorks Aviation Soc	
122	Avro 504 (replica)	British Broadcasting Corp	
123	Vickers FB.5 Gunbus (replica)	British Broadcasting Corp	

Notes	Reg.	Type	Owner or Operator
	124	Lilienthal Glider Type XI (replica)	Science Museum
	125	Clay Cherub	Midland Air Museum
	126	D.31 Turbulent (static)	Midland Air Museum
	127	Halton Jupiter	Shuttleworth Trust
	128	Watkinson Cyclogyroplane Mk IV	British Rotorcraft Museum
	129	Blackburn 1911 Monoplane (replica)	Cornwall Aero Park/Helston
	130	Blackburn 1912 Monoplane (replica)	Cornwall Aero Park/Helston
	131	Pilcher Hawk (replica)	C. Paton
	132	Blériot XI (G-BLXI)	Aerospace Museum/Cosford
	133	Fokker Dr 1 (replica) (425/17)	Torbay Aircraft Museum
	134	Pitts S-2A static (G-RKSF)	Torbay Aircraft Museum
	135	Bristol M.IC (replica) (C4912)	Leisure Sport Ltd
	136	Deperdussin Seaplane (replica)	Leisure Sport Ltd
	137	Sopwith Baby Floatplane (replica) (8151)	Leisure Sport Ltd
	138	Hansa Brandenburg W.29 Floatplane (replica) (22912)	Leisure Sport Ltd
	139	Fokker Dr 1 (replica) 150/17	Leisure Sport Ltd
	140	Curtiss R3C-2 Floatplane (replica)	Leisure Sport Ltd
	141	Macchi M.39 Floatplane (replica)	Leisure Sport Ltd
	142	SE-5A (replica) (F5459)	Cornwall Aero Park/Helston
	143	Paxton MPA	R. A. Paxton/Staverton
	144	Weybridge Mercury	Cranwell Gliding Club
	145	Oliver MPA	D. Oliver
	146	Pedal Aeronauts Toucan MPA	Shuttleworth Trust
	147	Bensen B.7	Norfolk & Suffolk Aviation Museum
	148	Hawker Fury II (replica) (K7271)	Aerospace Museum/Cosford
	149	Short S.27 (replica)	Fleet Air Arm Museum
	150	SEPECAT Jaguar GR.1 (replica) (XX718)	RAF Exhibition Flight
	151	SEPECAT Jagaur GR.1 (replica) (XX824)	RAF Exhibition Flight
	152	BAe Hawk T.1 (replica) (XX162)	RAF Exhibition Flight
	153	Westland WG.33	British Rotorcraft Museum
	154	D.31 Turbulent	Lincolnshire Aviation Museum
	155	Panavia Tornado GR.1 (replica) (ZA322)	RAF Exhibition Flight
	156	Supermarine S-6B (replica) (S1595)	Leisure Sport Ltd
	157	Waco CG-4A	Pennine Aviation Museum
	158	Fieseler Fi 103/FZG.76 (V.I)	Joint Bomb Disposal School
	159	Fuji Oka	Joint Bomb Disposal School
	160	Chargus 108 hang glider	Museum of Flight/E. Fortune
	161	Stewart Ornithopter Coppelia	Bomber County Museum
	162	Goodhart Newbury Manflier MPA	Science Museum/Wroughton
	163	AFEE 10/42 Rotabuggy (replica)	Wessex Aviation Soc Wimborne
	164	Wight Quadraplane Type 1 (replica)	Wessex Aviation Soc Wimborne
	165	Bristol F.2b	RAF Museum/Cardington
	166	Bristol F.2b	Shuttleworth Trust
	167	S.E.5A replica	Torbay Aircraft Museum
	168	D.H.60G Moth static replica (G-AAAH)	Hilton Hotel/Gatwick
	169	SEPECAT Jaguar GR.1 (static replica) (XX110)	No 1 S. of T.T. RAF Halton
	170	Pilcher Hawk (replica)	A. Gourlay
	171	BAe Hawk T.1 (replica) (XX262)	RAF Exhibition Flight/Abingdon
	172	Chargus Midas Super 8 hang glider	Science Museum/Wroughton
	173	Birdman Promotions Grasshopper	Science Museum/Wroughton
	174	Bensen B.7	Science Museum/Wroughton
	175	Volmer VJ-23 Swingwing	Manchester Air & Space Museum
	176	SE-5A (replica) (A4850)	S. Yorkshire Aircraft Preservation Soc
	177	Avro 504K (replica) (C1381)	(*Stored*)/Henlow
	178	Avro 504K (replica) (E373)	(*Stored*)/Henlow
	179	Sopwith Pup (replica)	N. Weald Aircraft Restoration Flight

Note: Registrations/Serials carried are mostly false identities. MPA = Man Powered Aircraft.

Future Allocations Log (In-Sequence)

The grid provides the facility to record future in-sequence registrations as they are issued or seen. To trace a particular code, refer to the left hand column which contains the three letters following the G prefix. The final letter can be found by reading across the columns headed A to Z. For example, the box for G-BMPD is located five rows down (BMP) and then four across to the D column.

G-	A	B	C	D	E	F	G	H	I	J	K	L	M	N	O	P	R	S	T	U	V	W	X	Y	Z
BML																									
BMM																									
BMN																									
BMO																									
BMP																									
BMR																									
BMS																									
BMT																									
BMU																									
BMV																									
BMW																									
BMX																									
BMY																									
BMZ																									
BNA																									
BNB																									
BNC																									
BND																									
BNE																									
BNF																									
BNG																									
BNH																									
BNI																									
BNJ																									
BNK																									
BNL																									
BNM																									
BNN																									
BNO																									
BNP																									
BNR																									
BNS																									
BNT																									
BNU	A	B	C	D	E	F	G	H	I	J	K	L	M	N	O	P	R	S	T	U	V	W	X	Y	Z

Credit: *Wal Gandy*

Future Allocations Log (Out-of-Sequence)

This grid can be used to record out-of-sequence registrations as they are issued or seen. The first column is provided for the ranges prefixed with G-B, ie from G-BNxx to G-BZxx. The remaining columns cover the sequences from G-Cxxx to G-Zxxx and in this case it is necessary to insert the last three letters in the appropriate section.

G-B	G-C	G-E	G-G	G-I	G-K	G-M	G-O	G-P	G-S	G-U
										G-V
										G-W
	G-D	G-F	G-H	G-J	G-L	G-N		G-R		
									G-T	
										G-X
										G-Y
										G-Z

Overseas Airliner Registration Log

This grid may be used to record airliner registrations not included in the main section.

Reg	*Type*	*Operator*

ADDENDA

New overseas registrations

Notes	Reg.	Type	Owner or Operator
	C-FTIV	Douglas DC-8-73AF (878)	Air Canada
	C-GQBG	Douglas DC-8-54	Quebecair
	D-AERY	L.1011-385 TriStar I	LTU
	D-AM..	Boeing 757-2G5	Luftransport Sud (LTS)
	EI-ABI	D.H.84 Dragon	Aer Lingus
	F-GEBU	SA.226TC Metro II	Air Languedoc
	F-GMFM	Douglas DC-8-71	Pointair *Region Reunion*
	N603FF	Boeing 747-130	Tower Air
	N723PA	Boeing 747-212B	Pan Am *China Clipper II*
	OE-ILE	Boeing 737-2TS	Lauda-Air
	OO-ILF	Boeing 737-3QB	Air Belguim
	OY-MB	Fokker 50	Maersk Air
	OY-MB	Fokker 50	Maersk Air
	OY-MB	Fokker 50	Maersk Air
	OY-MB	Fokker 50	Maersk Air
	TF-FLB	Douglas DC-8-55F	Icelandair

V8 (Brunei)

	V8-RBA	Boeing 757-2M8ER	Royal Brunei Airlines
	V8-RBB	Boeing 757-2M8ER	Royal Brunei Airlines
	V8-RBC	Boeing 757-2M8ER	Royal Brunei Airlines

British Aircraft Preservation Council Register

	185	Waco CG-4A	Museum of Army Flying, Middle Wallop
	186	D.H.82B Queen Bee (K3584)	M. Eastman

We've made choosing an aviation enthusiast magazine a lot easier.

- Of all the magazines produced in the U.K. only about one in four give monthly updates on a wide range of topics, covering military aviation, production news, airline fleet and route news, register reviews and extensive movement reports.
- Only a handful of these are professionally printed and illustrated with topical photographs.
- Of these, just four can boast a readership well into four figures and a coverage of the whole U.K.
- And of the 'big four' only one has proved popular and progressive enough to double its readership in under two years, and comprehensive enough to list over 50,000 Heathrow movements during 1985 (more than double its nearest rival). In the same year it ran to over 500 pages, with more than 100 packed with airline, airport and licencing news.

You might be surprised to learn which one it is .....

Stansted Aviation News
Growing bigger by being better!

For a sample magazine write to the Secretary, 1 Kings Head Court, Sawbridgeworth, Herts.
Or to receive your copy every month for a year send £7.50 if you live in the U.K., or £9.50 overseas.

MACH III PLUS
AVIATION BOOKS

FREE! POST & PACKING on BOOKS in the UK & Eire
overseas rates on application

DO YOU want up to the minute information on new book releases and a free comprehensive booklist crammed with titles of interest to both the civil and military enthusiast?

DO YOU want to know what books are available for the aviation historian or the aero-modeller?

THEN you must join the MACH THREE PLUS computerised mailing service by simply sending us your name and address!

OR Visit the MACH THREE PLUS sales stand at the major airshows in 1986 to see our full range of aviation goods including . . .

Books, kits, posters, postcards, patches, stickers, key fobs, squadron prints, aircraft soundtrack tapes, printed tea towels, tee-shirts, pens, pencils, tie pins, airband radios, etc, etc . . .

MACH THREE PLUS (Mail Order), 87 Pole Hill Road, Hillingdon, Middlesex UB10 0QB.

Are you in the Club?

LAAS INTERNATIONAL

One of the World's leading Aviation Societies

Receive LAAS Aviation News & Review every month.

- Packed with information including:
- Airline and Operators – Worldwide survey.
- United Kingdom Register Review – New Registration and all the changes.
- Ports of Call – Outstanding coverage of UK airfields.
- Touchdown – Aircraft movements at the major airports.
- Industrial and Production news.
- International Register news – All the changes from far afield.
- Biz Jet Update.
- Display Report.
- Special Features – Aircraft Deliveries – Articles – Travel News – Photographs – Sales and Wants and much, much more.

Annual Subscription £7.00* *U.K. only

AVIATION News & Review

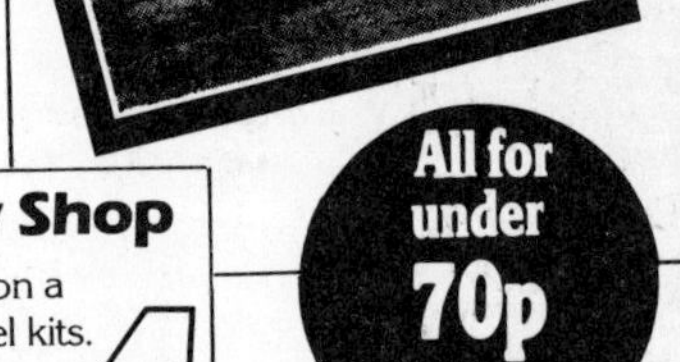

All for under 70p a month!

Visit The Aviation Hobby Shop

- LAAS members receive special discounts on a selected range of aviation books and model kits.

The Aviation Hobby Shop, Horton Parade, Horton Road, West Drayton, Middlesex.

tahs

Send today for FREE sample copy of Aviation News & Review and full details of how to become a member of LAAS – *Your* Aviation Society.

LAAS INTERNATIONAL

Send to:
LAAS International Dept CAM6, 26 Blomfield Court, Westbridge Road, Battersea, London, SW11 3TL.